PHYSICAL FOUNDATIONS OF SOLID STATE AND ELECTRON DEVICES

McGraw-Hill Series in Electrical Engineering

Consulting Editor

Stephen W. Director, *Carnegie-Mellon University*

Circuits and Systems
Communications and Signal Processing
Control Theory
Electronics and Electronic Circuits
Power and Energy
Electromagnetics
Computer Engineering
Introductory
Radar and Antennas
VLSI

Previous Consulting Editors

Ronald N. Bracewell, Colin Cherry, James F. Gibbons, Willis W. Harman, Hubert Heffner, Edward W. Herold, John G. Linvill, Simon Ramo, Ronald A. Rohrer, Anthony E. Siegman, Charles Susskind, Frederick E. Terman, John G. Truxal, Ernst Weber, and John R. Whinnery

Electronics and Electronic Circuits

Consulting Editor

Stephen W. Director, *Carnegie-Mellon University*

Colclaser and Diehl-Nagle: *Materials and Devices for Electrical Engineers and Physicians*
Fabricius: *Introduction to VLSI Design*
Ferendeci: *Physical Foundations of Solid State and Electron Devices*
Franco: *Design with Operational Amplifiers and Analog Integrated Circuits*
Grinich and Jackson: *Introduction to Integrated Circuits*
Hamilton and Howard: *Basic Integrated Circuits Engineering*
Hodges and Jackson: *Analysis and Design of Digital Integrated Circuits*
Long and Butner: *Gallium Arsenide Digital Integrated Circuit Design*
Millman and Grabel: *Microelectronics*
Millman and Halkias: *Integrated Electronics: Analog, Digital Circuits, and Systems*
Millman and Taub: *Pulse, Digital, and Switching Waveforms*
Paul: *Analysis of Linear Circuits*
Roulston: *Bipolar Semiconductor Devices*
Schilling and Belove: *Electronic Circuits: Discrete and Integrated*
Smith: *Modern Communication Circuits*
Sze: *VLSI Technology*
Taub: *Digital Circuits and Microprocessors*
Taub and Schilling: *Digital Integrated Electronics*
Wait, Huelsman, and Korn: *Introduction to Operational and Amplifier Theory Applications*
Yang: *Microelectronic Devices*
Zambuto: *Semiconductor Devices*

PHYSICAL FOUNDATIONS OF SOLID STATE AND ELECTRON DEVICES

Altan M. Ferendeci

University of Cincinnati

McGraw-Hill Publishing Company

New York St. Louis San Francisco Auckland Bogotá Caracas Hamburg Lisbon London Madrid Mexico Milan Montreal New Delhi Oklahoma City Paris San Juan São Paulo Singapore Sydney Tokyo Toronto

This book was set in Times Roman by Publication Services.
The editors were Roger Howell and John M. Morriss;
the production supervisor was Kathryn Porzio.
The cover was designed by John Hite.
Project supervision was done by Publication Services.
Arcata Graphics/Halliday was printer and binder.

PHYSICAL FOUNDATIONS OF SOLID STATE AND ELECTRON DEVICES

2 3 4 5 6 7 8 9 0 HAL HAL 9 0 9 8 7 6 5 4 3 2 1

ISBN 0-07-020478-0

Library of Congress Cataloging-in-Publication Data

Ferendeci, A. M.
Physical foundations of solid state and electron devices / Altan M. Ferendeci.
p. cm.—(McGraw-Hill series in electrical engineering. Electronics and electronic circuits)
Includes index.
ISBN 0-07-020478-0
1. Solid state physics. 2. Electronics. I. Title. II. Series.
QC176.F44 1991
530.4′1—dc20 90-38851

ABOUT THE AUTHOR

Altan M. Ferendeci received his Ph.D. degree from Case Institute of Technology. His previous academic and administrative duties included the positions of acting head of the Physics Division at the Cekmece Nuclear Research and Training Center, Chairman of the Physics Department at Bogazici University (both in Istanbul) and the Electrical Engineering and Applied Physics Department at Case Western Reserve University. In 1988, he joined the Microelectronics Group of the Electrical and Computer Engineering Department at the University of Cincinnati. He is involved in research work on microwave and millimeter wave solid state devices, optical devices, applied superconductivity and electron beam devices. He is the coeditor of the book *Atomic and Molecular Processes in Controlled Thermonuclear Fusion*. He is also the author of many publications and conference presentations. He received the 1989 NASA Inventions and Contributions Award on the work done in "GaAlAs Traveling-Wave Electro-Optical Modulators." He is a member of IEEE, APS, Sigma Xi, and Eta Kappa Nu.

This book is dedicated
To the people
who effected the flow of my life—
my wife and
my mother—
and who would have been so proud if they were alive—
my father and
Amca Bey.

CONTENTS

PREFACE

The main purpose of this book is to present the basic physical concepts about the operational principles of Solid State and Electron Devices to Electrical Engineering, Physics, and Applied Physics majors. Many of the device characteristics are introduced in various undergraduate circuit courses in which devices are viewed as multiport networks, with emphasis on their electrical terminal characteristics. With the exception of introductory physics courses, there are not many unified textbooks that introduce students to the physical concepts on which the operational principles are based. It is my belief that if the students have a thorough understanding of the operation and the physical limitations of a given device, they can also design better electronic systems.

This book is written for the undergraduate senior- and advanced junior-year students. The main emphasis of the book is to teach the students a sufficient but unified set of underlying physical principles without going into too many detailed theoretical derivations. What students need is a solid background through which they can understand the basic operational characterization of any given electronic device. The intent is not to make this purely a physics oriented book, but to lay the physical foundations of those principles using simple models so that the students will be able to comprehend the detailed operational mechanisms of even the more complicated devices. As the technology advances, many new devices or variations of an old device are introduced. With a solid physical background, the student should be able to understand easily how the new device operates. A student who wants to specialize further in the solid- state electronics area can take advanced-level related courses. With the knowledge of the fundamental principles learned in this course, the student will be able to follow the theoretical details of the advanced-level courses more easily.

The properties and the devices associated with the traditional monocrystalline semiconductors—especially silicon, which is still the most widely used material for device fabrication—are covered in detail in this book. But because of the recent technological breakthroughs in epitaxial layer growth (even down to perfect atomic layers and, in certain applications, electrical properties equal or superior to that of Si), the properties of compound-semiconductor and associated new devices are also included throughout this book.

Although most of the book is devoted to modern solid-state devices, some important concepts related to electron devices are also included. At present, it seems that many of these devices have been replaced by their solid-state counterparts, but recently introduced semiconductor devices, especially the heterojunction devices, are based on concepts, associated with electron beam devices, that are novel but nevertheless similar to the old.

This book can be divided into two separate but integral parts: physical principles and devices. The initial chapters of the book introduce the fundamental physical concepts, with emphasis on topics that are necessary for understanding the operation of the electronic devices that are introduced at the latter chapters. After a short introduction of the quantum mechanical treatment of the free and bound particles, statistical concepts are introduced, followed by an introduction to interacting particle concepts. The idea behind these chapters is to introduce the student to the microscopic behavior of the charged particles and then transform these concepts through averaging processes into macroscopic behavior of an ensemble of many particles.

Bound particle concepts are extended into understanding the general properties of solids, with special emphasis on semiconductors. The nonideal behavior of semiconductors with defects and impurities are then discussed, followed by a chapter on various electron emission mechanisms.

Device concepts start with Chapter 8, which is completely devoted to various junction phenomena. Here, the many fundamental junction-related concepts that form the basis for the operational principles of many semiconductor devices are introduced. Operational principles of bipolar (BJT), junction (JFET), and MOS (MOSFET) field effect transistors are then discussed in the following chapters. Modern transistor developments are introduced in Chapter 12 separately rather than as a part of the specific chapters on bipolar, FET, and MOS transistors. It is intended to show the common high-frequency limitations of these devices and present new developments aimed at extending the operation of these devices to higher frequencies. Chapter 13 is devoted to electro-optic devices, with emphasis on recent developments that are bringing closer to reality the realization of purely optical communication systems. This is followed by an overview chapter on device and integrated circuit processing. Chapter 15 is devoted to the basic principles of gas discharges and vacuum and electron beam devices. Appendix A discusses noise, and Appendix B lists useful physical constants.

In many instances, explanations of the various device characteristics are based on the fundamental physical concepts given in the earlier chapters, so the book has a unified approach to the overall device physics and characterization. Because of the limited time for a course of this nature, it is not possible to include all the devices in existence today; some important devices are omitted from the book. But it is hoped that the student will be able to understand the operation of any device by recalling the physical principals learned from this book. Detailed illustrations are included to help clarify the various topics that are discussed throughout the book.

The complete contents of the book cannot be covered in one quarter or in one semester. It is hoped, however, that the book will satisfy the needs of various departments with their course offerings under different names. These courses are generally referred to as either "Physical Electronics" or "Solid State Electronic Devices." Along with Chapters 1–7 and Chapter 15, selected topics from other chapters can easily satisfy the requirements for a Physical Electronics–type course.

By excluding starred (*) sections in Chapter 2, the first seven chapters can be used as a first course in "Solid State Electronics" where fundamental physical concepts are given, followed by a second semester or quarter with emphasis on devices, especially Chapters 8–13. If the students are previously introduced to modern physics in other courses, the first two chapters can also be omitted from such a course and Chapter 8 could be included in the first solid-state course. Chapter 14 is included as an overview of semiconductor processing techniques and could be used in either of the courses discussed above.

Whenever necessary, detailed mathematical derivations are included. On the other hand, in many cases, such as in the discussion of statistical concepts, physical discussions with simple examples are preferred over lengthy derivations. It was felt that, whenever possible, understanding of a device operation becomes more clear if energy band diagrams associated with that device are used. Because of its extreme importance, the band theory of solids is introduced from two different points of view—with physical arguments as well as quantum mechanical derivations.

Many illustrative and complementary examples are included in each chapter. In addition to the topics introduced in a given chapter, the problems given at the end of each chapter also introduce additional concepts related to the topics discussed in that chapter. Some of the problems at the end of the chapters may require extensive numerical calculations. No attempt is made to seperately mark these problems. It is left to the discretion of the instructor to pick and assign this type of problem as computer-oriented problems.

The SI (MKS) system of units is used throughout the book, with one exception. In order not to deviate too much from the well-established common usage, many parameters that contain the *distance unit* are given in centimeters (such as cm^{-3}) rather than meters. In simple calculations, where conversion to meters (m) is not necessary, cm units are carried through the given examples. But whenever dielectric permittivity, magnetic permeability, or Planck, Boltzmann, or any other constants with derived units are involved, all units in a calculation are converted to the MKS system of units (cm $\rightarrow$ m) to be consistent with the units and to avoid any calculational errors.

To reach the final form of this book, I have received continuous encouragement, constructive criticism, and helpful suggestions from various sources. First of all, without my wife Sükran's constant support, understanding, and encouragement in the preparation and writing of the book, this endeavor would not have been possible. Fortunately, during this time my two sons' Derya and Deniz colorful daily activities entertained and helped her endure the long manuscript-writing days and nights. I would also like to thank all my students for their insight, the following reviewers for their candid criticism and suggestions: Bob Darling, University of Washington; M. A. Littlejohn, North Carolina State University; Andrew Robinson, University of Michigan; Patrick Roblin, Ohio State University; T. E. Schlesinger, Carnegie-Mellon University; and Cary Yang, University of Santa Clara, my colleagues, J. T. Boyd and T. Mantei at the University of Cincinnati, Vik J. Kapoor and C. Papadakis for their understanding and encouragement, Ann Patrella for the partial typing of the manuscript, and Shin-Lin Lu for his careful reading of the final manuscript.

Altan M. Ferendeci

INTRODUCTION

A typical communication channel is shown in Fig. A. An image of an object is to be transmitted to a distant location by means of a transmission system. The object is illuminated by a spotlight, and the image is focused on a video camera by a regular optical system. The light intensity, reflected by the object, is converted to an electrical signal, processed by the camera electronics, and sent to a transmitter. The antenna of the transmitter directs the signal toward a synchronous satellite. The satellite has both receiving and transmitting capabilities and is powered by solar cells. The received signal is down- or up-converted in frequency and retransmitted back to a receiver on earth. The received signal is detected, processed, and finally displayed on a video screen.

The signal could have also been sent through a fiber optic link. At the transmitting end, the signal modulates a semiconductor laser, which is then coupled to a fiber optic cable. The transmitting and the receiving ends are connected by an underground or trans-oceanic fiber optic cable. At the receiving end, the signal is detected, demodulated, and processed in the usual way before reaching the final display unit.

A closer look at the system components of Fig. A shows a variety of components whose operational characteristics are based on different physical principles, although they all eventually employ the control of charge flow. When the object is illuminated at the studio, the light that is reflected by the object falls on the photo cathode of an imaging device. The intensity of light is converted into photoelectrons, which are amplified by various stages of bipolar or field-effect transistor amplifiers. The signal is fed into a transmitter, where it modulates a microwave tube in which high-power microwaves are generated. The composite signal is then fed to the transmitting antenna. The signal propagates through the atmosphere and ionosphere of the earth and is picked up by the antenna of the satellite.

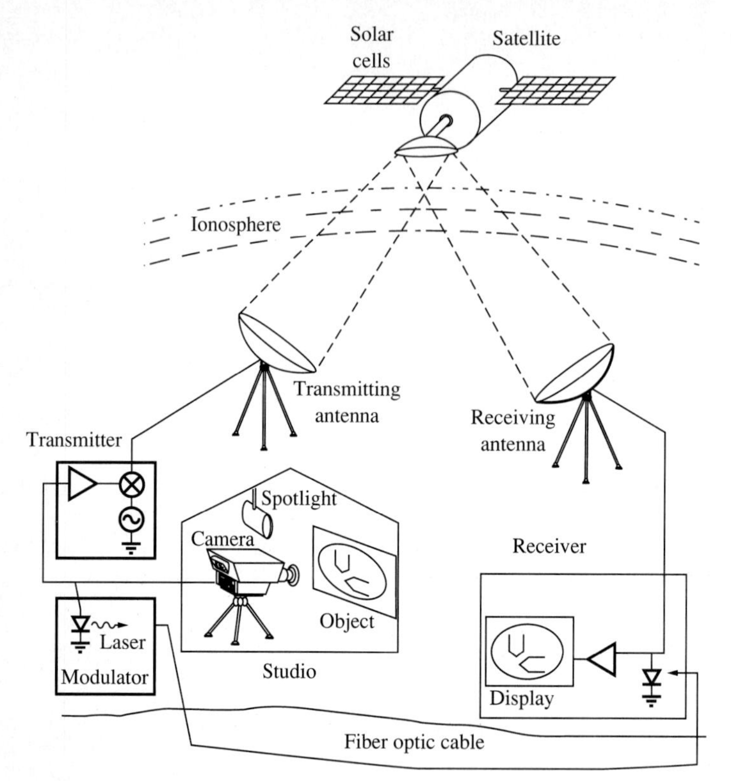

FIGURE A
Basic components of a communication system.

The signal is retransmitted back to earth after it is up- or down-converted and amplified by a traveling wave tube in the satellite. The satellite is usually powered by a photocell array illuminated by the sun. The signal is finally received by the earth receiving station, preamplified, detected, amplified, and processed by various electronic circuits and finally fed into a display device, which is presently a cathode ray tube.

The signal path may follow an alternate path. Once it is amplified at the transmitter end, it can be used to modulate a laser light, which is then coupled to a fiber optic transmission line. The cable connects the transmitter to the receiver at the other end. The signal is then detected, demodulated, and processed in the usual way at the receiver end.

Overall, the transmission of a signal from its starting point to its final destination is easily identified with just a few basic units: TV camera, transmitter, satellite and earth station receiver (or the fiber optic link), and the display unit. But a closer look at these units shows that they are made up of subunits and each subunit is made up of discrete components, each of which performs a separate task. The operational characteristics of these are different, but all employ the control and manipulation of charged particle flow from one end to the other. The simple task of charge control appears in many different forms and thus results in a different electrical characteristic for each particular device.

A system designer is usually confronted with the task of designing a system that will respond to a given input and process the signal in the form best suited for that task. The designer is usually given the operational terminal characteristics (linear, nonlinear, thermal, etc.) of these devices and then tries to put together a system that will satisfy the requirements of the task. If the designer is aware of the advantages and disadvantages of a group of similar devices, system design can be accomplished in the best possible way.

In some cases the component designer must come up with the best possible component that will fulfill the requirements of better temperature performance, highest gain, wide bandwidth response, and lower noise figure. Designers who are aware of the detailed operational principles of a device can improve on the performance of the device. Designers with a strong background in the physical principles and a clear general understanding of the operational characteristics of the devices can implement novel applications that result in new devices.

From its initial generation to its final destination, the signal, which contains the desired information, travels through various electronic devices that process or modify the form of the signal through conversion, modulation, amplification, propagation, or detection. Each device performs a different function, and their operational characteristics are based on different physical phenomena. All these devices, when initially invented and implemented, were bulky, inefficient, and unreliable, but now they have been improved so much in every respect that complex and sophisticated electronic systems can be put together within a volume of a few cubic meters and sent to orbits around the earth as satellites that function very well for many years without maintenance. These developments are due, of course, to the patient and persistent work of many scientists and engineers who pursued the task of understanding the physical basis for the operational characteristics of these electronic devices. Present versions of typical electronic devices have been realized through proper modeling, identifying of important parameters that determine the electrical characteristics of a device, and developing new and improved materials and processing technology. This does not mean that the developments related to a particular device have stopped. On the contrary, efforts continue to better the present technology, and as a result new and better devices are developed as offspring of the old technology.

Physical electronics is the study of motion of charged particles in a vacuum or in matter under the application of electric and magnetic fields. Classically, an electron is considered to be a particle with definite mass and size. When free electrons are subjected to electromagnetic fields, the forces on the electrons can be found by equating the electromagnetic forces to Newton's law of motion. If one places an electron inside a crystal made up of a three-dimensional array of atoms, the motion of the electron becomes very complicated. If one further considers not one but an ensemble of electrons, it becomes impossible to trace the motion of each individual electron even if the interaction of the electrons with each other is neglected. For a more realistic model, the collective motion of all these electrons should be considered. This leads to the concepts of ensemble averages and statistical mechanics.

In many cases, even with the aid of statistical mechanics, the classical equations of motion and Maxwell's equations are not sufficient to explain the experimentally observed behavior of charged particles. Fortunately, these discrepancies are removed by resorting to quantum mechanics. Because of the nature of quantum mechanics, changes and replacement of classical concepts at the atomic scale are required. For example, such ideas as wave-particle duality, uncertainty principle, and wave function do not have any classical counterparts. Quantum mechanics becomes a simpler tool in explaining the operational principles of many electronic devices. In some cases, use of a quantum mechanical concept such as tunneling through a potential barrier becomes the only means of explaining the resultant characteristic of an electronic device such as a tunnel diode. Thus, understanding of the basic quantum mechanical concepts is necessary to carry out the overall task of studying the physics of electron devices.

We are now ready to start an exciting journey into the fundamentally important physical principles of electron devices.

CHAPTER 1

INTRODUCTORY PHYSICAL CONCEPTS

There are fundamental physical concepts that we should be familiar with before attempting to understand the basic operational characteristics of electronic devices. We will assume that in every interaction process universal conservation laws such as the conservation of energy and momentum are satisfied. We will also assume that Maxwell's equations are applicable at the atomic level. This has been verified with all atomic level experiments. But even as we introduce some of the simplest electronic devices, it will often be necessary to use quantum mechanical concepts to explain the operational principles of these devices, and in some cases, only quantum theory will satisfactorily explain the physical basis of the electrical characteristic of a particular device.

A highly sophisticated field of study, quantum theory has been severely tested by many challenging experiments and passed these tests with flying colors. Although it is beyond the scope of this book to cover all general aspects of quantum theory, the reader should be familiar with some of the fundamental concepts in order to understand the operational principles of many of the devices introduced in the later chapters. In this chapter, we will introduce the fundamental principles of quantum theory from a conceptual point of view rather than give detailed mathematical theory. Emphasis will be placed on some quantum mechanical concepts, such as the tunneling effect, that have no counterparts in classical mechanics.

1.1 CONCEPTS OF QUANTUM MECHANICS

Electromagnetic radiation such as light, x-rays, and so on, is known to behave as waves in nature. This has been classically established through experimental verification of interference and diffraction, but certain physically observed experimental phenomena are not explainable by the wavelike nature of radiation. Some of these are

black body radiation
photoelectric effect
Compton effect

Attempts to explain the results of these experiments using classical mechanics and classical thermodynamics in conjunction with electromagnetic theory predicted results that conflicted with the experimental observations at the atomic level. When it came down to the interaction of radiation with matter or matter with matter at the atomic level, it was necessary to change the classical way of thinking. For example, in order to explain the observed black body radiation, Planck hypothesized that, contrary to the classical concepts, the energy of exchange between the electromagnetic waves of frequency ν and material bodies can only take place in discrete amounts in multiples of $E = h\nu$. Here, $h = 6.63 \times 10^{-34}$ joule-seconds (J·s) is known as Planck's constant. (h is now considered one of the universal constants of nature).

To explain the photoelectric phenomenon, Einstein hypothesized that light acted like a particle with an equivalent energy given by

$$E = h\nu \tag{1.1}$$

and he called these particles *photons*. As will be shown in Chapter 5, Eq. 1.1 will be the basis of explaining the photoelectric effect and the corresponding devices.

From these two hypotheses, two new concepts have evolved. One was the startling concept that electromagnetic waves, which were known to be waves, behaved in many interactions with the physical medium as if they were particles. It was also discovered that interactions in nature occurred in discrete quantities contrary to the classically accepted continuous way. Classical mechanics has no restrictions on the interactions, provided that they satisfy the conservation relations.

A second dilemma in nature appeared when interactions of particles with matter were studied. In addition to its usual electrical charge, an electron is classically known to be a particle with a definite mass and a finite size. When subjected to electromagnetic fields, the behavior of a free electron can be described by classical equations of motion. On the other hand, in some experiments performed with electrons, as well as with other classically known atomic particles such as neutrons and protons, these particles interacted with matter in such a way that the results were not explainable by classical equations of particle motion. To explain these observed experimental results, DeBroglie hypothesized that particles behaved like waves with an equivalent wavelength given by the relation

$$p\lambda = h \qquad \text{(DeBroglie relation)} \tag{1.2}$$

where $p = mv$ is the linear momentum of the particle and λ is its associated wavelength. For macroscopic bodies (e.g., a baseball) λ is so small that it cannot be directly or indirectly measured by existing instruments, but for microscopic particles such as an electron, a proton, or a neutron, the associated wavelength can possess a value that can be indirectly measurable. Electron and neutron diffraction experiments are examples of this wavelike nature of the particles.

One asks the question, "Are these particles so smart that they have a built-in memory and, depending on the interaction process, know when to act as a particle and when to act as a wave?" Of course, this is not even a plausible question. Therefore, we should have a consistent and a sound means of studying the behavior of particles on the atomic scale, irrespective of the type of experiment we perform. This is provided by the quantum theory. In the following section, without giving the details, we will first introduce the wave function, then give some simple solutions of the Schrodinger equation.

Example 1.1. A photon has an energy of 10.0 eV.* Find the wavelength of the radiation.

Solution

$$E = h\nu$$

$$\nu = \frac{E}{h} = \frac{10\text{eV} \times 1.6 \times 10^{-19}\ \text{J/eV}}{6.63 \times 10^{-34}\ \text{J} \cdot \text{s}}$$

$$= 2.413 \times 10^{15}\ \text{Hz}$$

$$\lambda = \frac{c}{\nu} = \frac{3 \times 10^{8}\ \text{m/s}}{2.413 \times 10^{15}\text{s}^{-1}} = 1.243 \times 10^{-7}\ \text{m}$$

$$= 1243\text{Å}$$

In spectroscopy, it is common to use the unit cm^{-1} for wavelength; λ' represents the number of wavelengths in one centimeter.

$$\lambda' = \frac{1}{1.243 \times 10^{-5}\text{cm}} = 80{,}442\ \text{cm}^{-1}$$

---

*In SI, mks, system of units, the unit of energy is joules. Atomic particles have very small masses, so the resulting energy in joules becomes a very small number. A common unit of energy for atomic particles is the electron volt (eV), which is the equivalent energy that an electron acquires when it moves through a potential difference of 1 volt. The conversion between joules and electron volts is 1 eV = 1.6×10^{-19}J.

Example 1.2. Find the DeBroglie wavelength of a neutron that has an energy of 10.0 eV.

Solution. The momentum of the neutron is (from classical mechanics)

$$p = \sqrt{2mE}$$

$$= \sqrt{2 \times (1.673 \times 10^{-27}\text{kg}) \times (10\text{eV} \times 1.6 \times 10^{-19}\text{J/eV})}$$

$$= 7.317 \times 10^{-23}\ \text{kg} \cdot \text{m/s}$$

Using the DeBroglie relation,

$$\lambda = \frac{h}{p}$$

$$= \frac{6.63 \times 10^{-34} J \cdot s}{7.317 \times 10^{-23}\text{kg} \cdot \text{m/s}} = 9.061 \times 10^{-12}\text{m} = 0.09061\ \text{Å}$$

1.2 WAVE FUNCTION AND WAVE PACKETS

We know that, in optics, plane waves are used to study diffraction and interference phenomena. Let us, therefore, try to see whether we can represent a particle as a plane wave. Let us take a one-dimensional system and assume that the particle propagates in the $(+x)$ direction. We can assume a plane-wave representation for the particle in the form

$$\Psi(x, t) = A \exp[j(kx - \omega t)] \tag{1.3}$$

where $k = 2\pi/\lambda$ is the propagation constant and $\omega = 2\pi\nu$ is the corresponding radian frequency of the wave. Note that from the DeBroglie relation $p\lambda = h$, we can write k in terms of p as

$$k = \frac{2\pi}{\lambda} = \frac{2\pi}{h/p} = \frac{p}{h/2\pi} = \frac{p}{\hbar} \tag{1.4}$$

where we define $\hbar = h/2\pi$ (since $h/2\pi$ comes up in many problems, a shorthand notation is used). Also from Planck's relation

$$E = h\nu = \frac{h\omega}{2\pi} = \hbar\omega \tag{1.5}$$

Replacing ω and k in Eq. 1.3 by their equivalent values as given in Eq. 1.4 and 1.5, we write the wave function as

$$\Psi(x, t) = A \exp\left[j\left(\frac{p}{\hbar}x - \frac{E}{\hbar}t\right)\right] \tag{1.6}$$

We can assume that since Eq. 1.6 contains p and E explicitly, a plane wave can be used to simply represent a particle possessing momentum p and energy E. But we also know that a particle has a finite size. Although a plane wave seems to represent a particle with momentum $p = mv$ and energy $E = h\nu$, the plane wave has infinite extent and cannot at the same time represent the finite size of the particle.

We next try a superposition of an infinite number of plane waves whose wavelengths are centered around ω_o. This implies that the waves have propagation constants centered around k_o. Assuming that these k's have a very small range Δk centered around k_o (Fig. 1.1), we can superimpose and write the sum of the plane waves as

$$\Psi(x, t) = \sum_{k_o-\Delta k/2}^{k_o+\Delta k/2} A(k)\, e^{j(kx-\omega t)} = \int_{k_o-\Delta k/2}^{k_o+\Delta k/2} C e^{j(kx-\omega t)} dk \tag{1.7}$$

Here for simplicity, we have assumed that $A(k) = C =$ constant.*

* Note that in general, wave amplitude $A(k)$ is a function of k; that is, amplitude of each plane wave may be different.

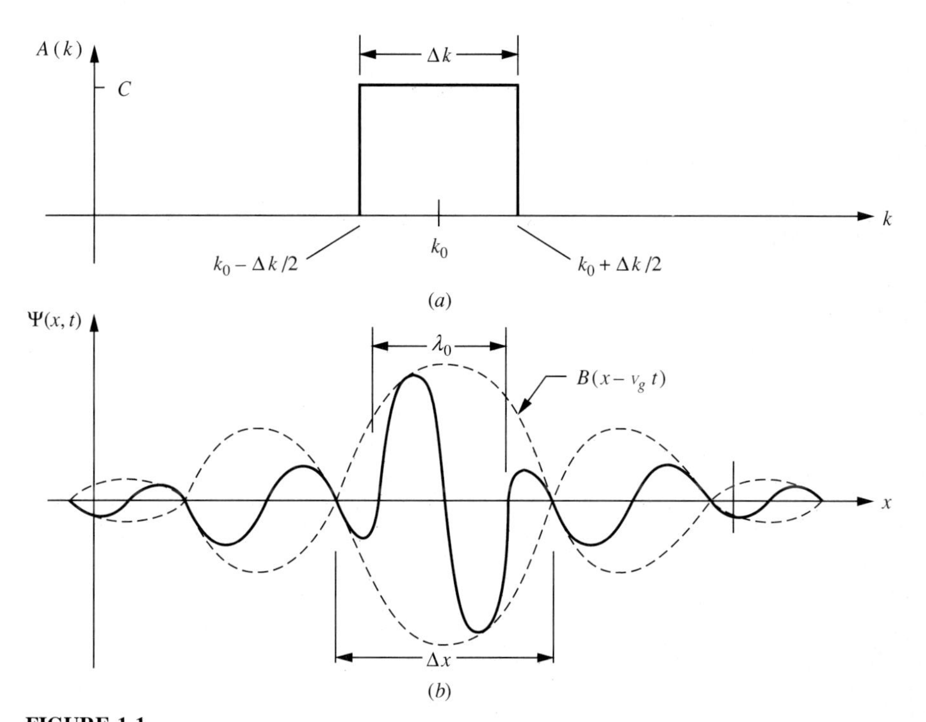

FIGURE 1.1
(a) Plane wave amplitudes and (b) the resulting wave packet at an instant of time t. $B(x - v_g t)$ is the envelope and Δx is the extent (size) of the particle.

We know that ω is a function of the propagation constant through the DeBroglie relation. We can, therefore, expand ω in the vicinity of ω_o as a function of k. Using Taylor's series expansion, ω can be written as

$$\omega = \omega_0 + (k - k_0)\frac{\partial \omega}{\partial k}\bigg|_{k=k_0} + \cdots \tag{1.8}$$

$$\omega \approx \omega_0 + (k - k_0)v_g$$

Substituting Eq. 1.8 into Eq. 1.7, adding $j(k_o x - k_o x)$ into the exponent, rearranging terms, and taking the constant terms out of the integral, Eq. 1.7 becomes

$$\Psi(x, t) = C e^{j(k_o x - \omega t)} \int_{k_o - \Delta k/2}^{k_o + \Delta k/2} e^{j(x - V_g t)(k - k_o)} dk \tag{1.9}$$

Substituting the variable $u = k - k_o$, performing the integration, and manipulating the terms, we finally obtain

$$\Psi(x, t) = B(x - v_g t) \exp\left[j(k_o x - \omega_o t)\right] \tag{1.10}$$

where

$$B(x - v_g t) = C\Delta k \left[\frac{\sin[(x - v_g t)\Delta k/2]}{(x - v_g t)\Delta k/2}\right] \tag{1.11}$$

One sees that the term $B(x - v_g t)$ behaves as the envelope of a plane wave. The envelope also propagates in the $(+x)$ direction with a velocity

$$v = v_g = \frac{\partial \omega}{\partial k}\bigg|_{k=k_o} \tag{1.12}$$

This velocity is called the *group velocity* of the particle. The superposition of waves given by Eq. 1.8 and the resulting wave function given in Eq. 1.10 is known as a *wave packet*.

From these simple considerations, if we identify Δx as the approximate size of the particle, we see that superposition of plane waves can be used to represent a free particle possessing an average energy $E_o = \hbar\omega_o$ and average momentum $p_o = \hbar k_o$. However, in general, particles are under the influence of electromagnetic (and gravitational) forces, and their behavior should be analyzed in a more consistent and general manner. In order for a theoretical model to be successful, it should predict or agree with all the experimentally observed results. Fortunately, there is now a very well-established general theory that is used to study the behavior of atomic particles, whether they are freely moving in space or moving under the influence of other forces or interacting with each other. This is the area of quantum mechanics, based on the solution of the Schrodinger equation.

In the following sections, we will restrict our study of quantum mechanics to the simple solutions of the Schrodinger equation. The Schrodinger equation may be thought of as the counterpart of Newton's equation of motion for studying the spatial and time behavior of atomic particles.

Before introducing the Schrodinger equation, we can gain additional insight into the atomic problems from the simple representation of a particle by a wave packet. Simple wave packet representation of a free particle given by Eq. 1.10 shows one of the essential differences between classical and quantum mechanics. If we hypothetically identify the width of the function $\Psi(x, t)$ as the extent (size) of the wave packet, from Fig. 1.1 this width can be approximated by

$$\Delta x \approx 2\pi/\Delta k$$

On the other hand, the width of the function $A(k)$ which represents the magnitude of the plane waves in k space is Δk. If the product of $\Delta x \Delta k$ is taken

$$\Delta x \Delta k = \frac{2\pi}{\Delta k}\Delta k = 2\pi$$

Using the DeBroglie relation $\Delta k = \Delta p/\hbar$, this product can finally be written as

$$\Delta x \Delta k = \hbar 2\pi = h \tag{1.13}$$

This is called *Heisenberg's uncertainty principle.** It is interpreted in the following way: Small Δx means that the extent of the wave packet is very small—in other words, the corresponding location of the particle in space can be determined with minimal uncertainty. If, on the other hand, Δx is large, it implies that the whereabouts of the particle in space is not certain because of the large extension of the wave packet in space. Similarly, the smaller the Δk is, the smaller the Δp; therefore, the particle momentum can be determined more accurately. If we thus define Δx and Δp as the uncertainties in the measurement of the position and momentum of the particle rather than the extent (size) of the particle and the range of momentum, Heisenberg's uncertainty principle can be stated thus: The momentum and the position of a particle cannot be simultaneously measured with the same precision. If we want to measure the momentum of a particle precisely, we cannot measure its location with the same precision at all and vice versa.

An alternate form of the uncertainty principle is given by

$$\Delta E \Delta t \geq \hbar/2 \tag{1.14}$$

This implies that if we want to measure the energy of a system precisely—that is, if we want the uncertainty in the measurement of the energy to be as small as possible ($\Delta E \Rightarrow 0$)—then the observation time for this measurement should be very long.

---

* Actually the correct expression for the uncertainty principle is

$$\Delta q_i \Delta p_i \geq \hbar/2$$

where q_i and p_i are the i' components of canonically conjugate momentum and coordinate. For position and momentum

$$\Delta x \Delta p \geq \hbar/2$$

Example 1.3. Show that the group velocity defined by Eq. 1.12 is consistent with the kinetic energy of the particle.

Solution. From the definition of group velocity

$$v_g = \frac{\partial \omega}{\partial k}$$

Using the DeBroglie wavelength (dropping the subscript g), we can write

$$\hbar \frac{\partial \omega}{\partial p} = v$$

$$\frac{\hbar}{m} \frac{\partial \omega}{\partial v} = v$$

$$\hbar \partial \omega = m v \, d v$$

Integrating both sides of the last equation and using Plank's relation, we obtain

$$\hbar \omega = m \frac{v^2}{2} + \text{constant}$$

$$E_t = \text{K.}E. + \text{P.}E.$$

The integration constant is set equal to the potential energy. Thus the total energy of the particle is equal to the sum of the kinetic and potential energies.

1.3 THE SCHRODINGER EQUATION

The Schrodinger equation can be regarded as the counterpart of Newton's equation of motion applied to the study of the behavior of microscopic bodies. In three dimensions, the Schrodinger equation is given by

$$-\frac{\hbar^2}{2m^*} \nabla^2 \Psi(\mathbf{r}, t) + \Phi(\mathbf{r}, t)\Psi(\mathbf{r}, t) = j\hbar \frac{\partial \Psi(\mathbf{r}, t)}{\partial t} \tag{1.15}$$

where m^* is the effective mass of the particles in an isolated system under the influence of the potential energy $\Phi(\mathbf{r}, t)$, ∇^2 is the Laplacian operator, and $\mathbf{r}$ represents a shorthand notation for a three-dimensional orthonormal set of coordinates. Note that for conservative systems

$$\mathbf{F} = -\nabla \Phi(\mathbf{r}) \tag{1.16}$$

and thus, the Schrodinger equation can be considered to be an equation of motion for the atomic system of particles. Also, the Schrodinger equation can be shown to be equivalent to an energy eigenvalue equation with the left-hand side being the sum of the kinetic and the potential energies and the right-hand side being the total energy of the system.

In Eq. 1.15, $\Psi(\mathbf{r}, t)$ is the complete wave function representing the isolated system of particles satisfying Eq. 1.15. If $\Psi(\mathbf{r},t)$ is known for the system, it is then assumed that everything about the system under consideration is known. One notes that the Schrodinger equation is different from the known wave equations. Its solution results in a wave function $\Psi(\mathbf{r}, t)$ which is, in general, a complex quantity. Therefore, $\Psi(\mathbf{r}, t)$ cannot be by itself directly related to the physically real quantities. To add physical interpretation to the wave function, a probability function is defined by

$$P(\mathbf{r}, t)d^3r = \Psi(\mathbf{r}, t)\Psi^*(\mathbf{r}, t)d^3r \tag{1.17}$$

where $\Psi^*(\mathbf{r}, t)$ is the complex conjugate of $\Psi(\mathbf{r}, t)$. $P(\mathbf{r},t)d^3r$ is interpreted as the probability of finding the particle under consideration in a volume element d^3r near $\mathbf{r}$ at time t. $P(\mathbf{r}, t)$ is known as the probability density function. If this probability function is integrated over all space,

$$\int_{\text{over all space}} \Psi(\mathbf{r}, t)^*\Psi(\mathbf{r}, t)d^3r = \text{finite} \tag{1.18}$$

that is, the total probability of finding the particles in the entire space should be finite. We now talk in terms of probabilities rather than precise and definite parameters.

In many atomic problems, the potential energy function acting on an isolated system of particles is independent of time, that is, $\Phi(\mathbf{r}, t) = \Phi(\mathbf{r})$ only. Under these conditions, the Schrodinger equation can be considerably simplified by writing the wave function as a product of two functions that are functions of position and time only. The resulting equation is then separable into two differential equations that are functions of the respective variables t and $\mathbf{r}$. The time dependence simply gives an exponential solution, and the total wave function can be written as

$$\begin{aligned}\Psi(\mathbf{r}, t) &= u(\mathbf{r})f(t)\\ &= u(\mathbf{r})e^{-j\omega t} = u(\mathbf{r})e^{-j(E/\hbar)t}\end{aligned} \tag{1.19}$$

The Schrodinger equation then reduces to an equation that depends on $u(\mathbf{r})$ only and is given by

$$-\frac{\hbar^2}{2m^*}\nabla^2 u(\mathbf{r}) + \Phi(\mathbf{r})u(\mathbf{r}) = Eu(\mathbf{r}) \tag{1.20}$$

Equation 1.20 is known as *the time-independent Schrodinger equation*. Here E is the total energy of the system. The goal of quantum mechanics is then to find the time-independent wave function $u(\mathbf{r})$ for a system of particles subjected to $\Phi(\mathbf{r})$ consistent with the total energy E of the system. When Eq. 1.19 is used in Eq. 1.17, the probability function becomes

$$P(\mathbf{r}, t)d^3r = u(\mathbf{r})u^*(\mathbf{r})d^3r \tag{1.21}$$

and is independent of time. The resulting spatial part of the wave function is known as the stationary state of the system.

In many problems, as the particle moves in space it may encounter changes or discontinuities in potential energy, and the Schrodinger equation will have different solutions in different regions of space. In order for these problems to be solvable, it is necessary that the wave function have certain properties: it should be a single-valued function of coordinates, and the wave function as well as its first derivative should be continuous at the point of potential discontinuity. These are then the boundary conditions on the wave functions, that is,

$$\left.\begin{aligned} u(x_i) &= \text{continuous} \\ \frac{du(x_i)}{dx} &= \text{continuous} \end{aligned}\right\} x_i \text{ is where the potential discontinuity occurs.} \tag{1.22}$$

Solution of the Schrodinger equation (which is a boundary value problem) for a free particle results in simple plane wave functions instead of the wave packet discussed for the general free particle. This does not mean that the solution of the Schrodinger equation is inconsistent. As was shown before, the wave packet is composed of plane waves whose propagation constants do not differ too much from a central value k_o. Therefore, even if each plane wave that makes up the wave packet is considered separately, the resulting reflections and transmissions from potential discontinuities will not differ much from a single plane wave whose average propagation constant is k_o.

Nor are the plane wave solutions inconsistent with Heisenberg's uncertainty relation. We only specify the particle's energy and its momentum. The time interval during which the energy E of the particle is observed and the spatial location of the particle are unimportant.

1.4 REFLECTION FROM A POTENTIAL BARRIER

As an illustrative example of the solution of the Schrodinger equation and some of the resulting concepts, consider a free particle of mass m that is incident on a one-dimensional potential discontinuity. Assume that the free particle, moving toward the right, experiences a change in potential energy at $x = 0$. The potential changes from zero to a constant value Φ_0 as shown in Fig. 1.2.

Let the total energy of the incoming particle be E and assume that $E > \Phi_0$. The Schrodinger equation should be solved in two distinct regions: Region I, where $\Phi(x) = 0$, and Region II, where $\Phi(x) = \Phi_0$ and constant. The Schrodinger equations in one dimension and resulting solutions in the two respective regions can be written as

Region I

$$-\frac{\hbar^2}{2m}\frac{d^2u_1}{dx^2} = Eu_1 \tag{1.23a}$$

$$u_1(x) = Ae^{jk_1x} + Be^{-jk_1x} \tag{1.23b}$$

where

$$k_1^2 = \frac{2m}{\hbar^2}E > 0 \tag{1.23c}$$

Region II

$$-\frac{\hbar^2}{2m}\frac{d^2u_2}{dx^2} + \Phi_0 u_2 = Eu_2 \tag{1.24a}$$

$$u_2(x) = Ce^{jk_2x} + De^{-jk_2x} \tag{1.24b}$$

where

$$k_2^2 = \frac{2m}{\hbar^2}(E - \Phi_0) > 0 \tag{1.24c}$$

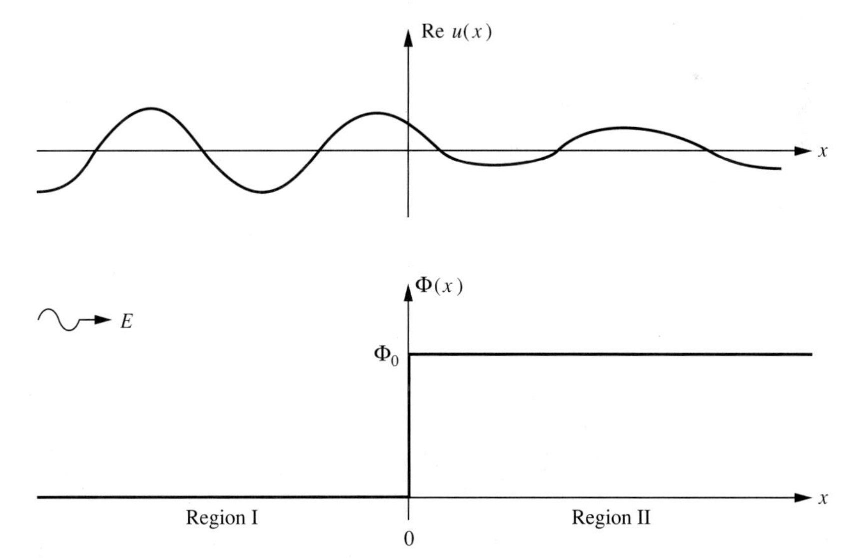

FIGURE 1.2
A particle incident from $-\infty$ with energy E is subjected to a potential discontinuity Φ_0 at $x = 0$. The energy of the incoming particle $E > \Phi_0$. The real part of the resulting wave function is also plotted in the two regions. Note the difference in wavelength in the two regions.

The total time-dependent wave functions in the two regions are then obtained from Eq. 1.19 by multiplying $u(x)$'s by $e^{-j\omega t}$. This leads to the complete wave functions in the two regions

$$\Psi_1(x, t) = Ae^{j(k_1 x - \omega t)} + Be^{-j(k_1 x + \omega t)} \qquad -\infty < x \le 0 \tag{1.25}$$

and

$$\Psi_2(x, t) = Ce^{j(k_2 x - \omega t)} + Be^{-j(k_2 x + \omega t)} \qquad 0 \le x < \infty \tag{1.26}$$

The first terms in these equations represent waves propagating in the positive x direction; the second terms represent waves propagating in the negative x direction. The particles initially originate from $-\infty$ with zero initial potential energy and experience the potential discontinuity at $x = 0$. Therefore, reflections should arise because of the presence of the potential discontinuity. For classical particles, change in potential energy results in a change in the velocity (or momentum) of the particle only and no reflection occurs. Therefore, reflections at a potential discontinuity is a quantum mechanical phenomenon.

To solve for the constants of integration appearing in Eqs. 1.23 and 1.24, in addition to the usual boundary conditions, some physically plausible arguments should be used. It was concluded that the reflected waves occurred due to changes in potential energy. Since Φ_0 extends all the way into $x = +\infty$, there is no potential discontinuity in Region II and therefore, there should be no reflected wave in this region; that is, D should be set equal to zero. Also, the term with coefficient A in Region I should represent the incoming particles; thus, A can be assumed to be known. To find B and C at $x = 0$, we use the conditions

$$u_1(0) = u_2(0) \qquad \text{(continuity of the wave function)}$$

and

$$\left.\frac{du_1}{dx}\right|_{x=0} = \left.\frac{du_2}{dx}\right|_{x=0} \qquad \text{(continuity of the first derivative of the wave function)}$$

The resulting equations are

$$A + B = C$$

$$k_1(A - B) = k_2 C$$

which yield

$$\frac{B}{A} = \frac{k_1 - k_2}{k_1 + k_2} \tag{1.27}$$

$$\frac{C}{A} = \frac{2k_1}{k_1 + k_2} \tag{1.28}$$

Here $|B/A|$ is interpreted as the probability of reflection, $|B/A|^2$ as the reflection coefficient, $|C/A|$ as the probability of transmission, and $|C/A|^2$ as the transmis-

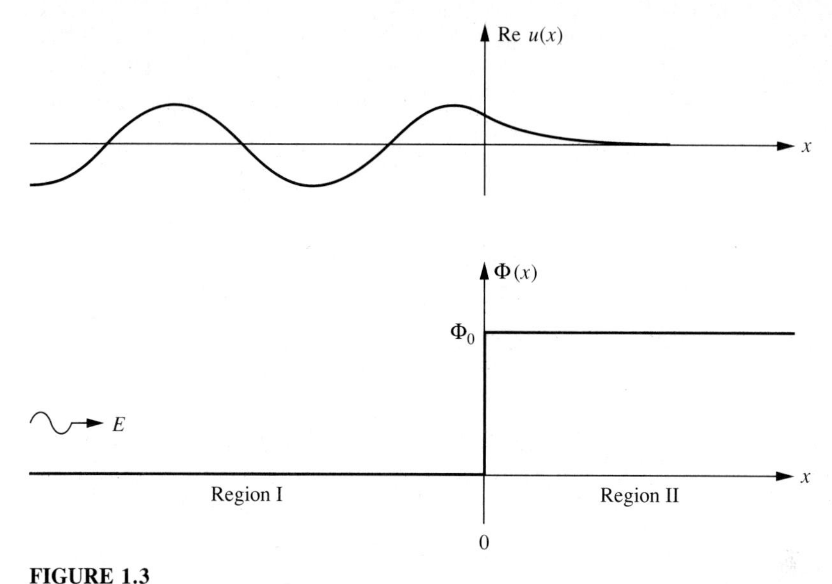

FIGURE 1.3
A particle incident from $-\infty$ with energy E is subjected to a potential discontinuity Φ_0 at $x = 0$. The energy of the incoming particle $E < \Phi_0$. The real part of the resulting wave function is also plotted in the two regions. Note that Region II is classically forbidden.

sion coefficient. These probabilities should be interpreted as the probability of reflection and transmission of N incoming particles. They give the probabilities for fractions of N particles that are likely to be reflected and transmitted. The real part of the resulting wave functions are drawn in Fig. 1.2. Note that λ in the two regions is different due to different k's.

If initially the incoming particle has an energy E which is smaller than Φ_0, that is, $E < \Phi_0$, then classically we expect particles to be completely reflected back from the discontinuity. Region II is a classically forbidden region, since the kinetic energy of the particle in this region is negative. For this case, the procedure of solving the Schrodinger equation in the two regions is the same as before but, for the solution in Region II, becomes

$$u_2(x) = Ce^{-\kappa x} + De^{+\kappa x} \tag{1.29}$$

where $\kappa^2 = -k_2^2 = 2m(\Phi_0 - E)/\hbar^2 > 0$. For large values of x, we do not expect the particles to have very large probabilities of being at this classically forbidden region. Therefore, D should be set to zero to keep this probability from being very large at $x = +\infty$. The solution in Region II is then an exponentially decaying solution. Note that if the wave function $\Psi_2(x, t)$ is considered, the wave function is not a traveling wave but an oscillation whose amplitude decays exponentially with increasing distance. The potential energy and resulting real part of the wave function is plotted in Fig. 1.3. The reflection coefficient due to the potential barrier when $E < \Phi_0$ is given by

$$R = \left|\frac{A}{B}\right|^2 = \left|\frac{k_1 - j\kappa}{k_1 + j\kappa}\right|^2 = 1 \tag{1.30}$$

The particle is completely reflected. On the other hand, the probability of finding the particle per unit length in Region II is given by

$$|u_2|^2 = |C|^2 e^{-2\kappa x} \tag{1.31}$$

That is, from the point of view of quantum mechanics, there is a finite probability of finding the particle in a classically forbidden region. Although the particle appears in Region II, it is eventually reflected back completely.

1.5 THE TUNNELING PHENOMENON

In some physical problems, a particle may be under the influence of a potential energy variation that can be approximated as shown in Fig. 1.4. Assume that the particle has an energy $E < \Phi_0$, and it is incident from the left. There are three separate solutions of the Schrodinger equation in the three regions. Assuming that there are no reflections from Region III, these are

$$\begin{aligned}
u_1(x) &= A e^{-jkx} + B e^{+jkx} \qquad & k^2 &= \frac{2m}{\hbar^2} E > 0 \\
u_2(x) &= C e^{-\kappa x} + D e^{+\kappa x} \qquad & \kappa^2 &= \frac{2m}{\hbar^2}(\Phi_0 - E) > 0 \\
u_3(x) &= F e^{-jkx} \qquad & k^2 &= \frac{2m}{\hbar^2} E > 0
\end{aligned} \tag{1.32}$$

Equating the respective wave functions and their first derivatives at $x = 0$ and $x = d$, we find the transmission coefficient into Region III as

$$T = \left|\frac{F}{A}\right|^2 = \frac{4}{4\cosh^2(\kappa d) + [(\kappa/k) - (k/\kappa)]^2 \sinh^2(\kappa d)} \tag{1.33}$$

This is the probability that a particle will appear in Region III. Classically, for a particle to appear in this region, the particle should, one way or another, gain an energy in excess of $(\Phi_0 - E)$ to overcome the barrier height. Since no energy exchange is involved in the problem, the appearance of the particle in Region III is again a purely quantum mechanical phenomenon. This is referred to as *tunneling* of the particle through a classically forbidden region. There are many devices such as the tunnel diode, high electric field emission, α-particle decay, Josephson junctions, and so on, whose operational characteristics are explainable only in terms of this phenomenon. Note: as can be seen from Eq. 1.33, the probability of tunneling into Region III increases with decreasing potential barrier width.

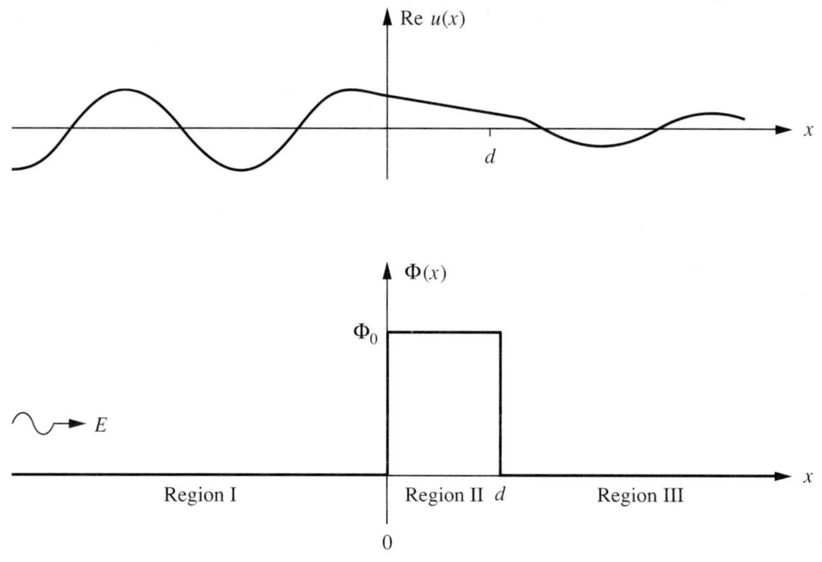

FIGURE 1.4
Potential energy barrier. The particle with energy $E < \Phi_0$ tunnels through the barrier and appears in Region III.

FURTHER READING

Leighton, R. B.: *Principles of Modern Physics*, McGraw-Hill, New York, 1959.
Powell, J. L., and B. Crasemann: *Quantum Mechanics*, Addison-Wesley, Reading, MA, 1961.
Schiff, L. I.: *Quantum Mechanics*, 2d ed., McGraw-Hill, New York, 1955.
Shankland, R. S.: *Atomic and Nuclear Physics*, 2d ed., The Macmillan Corp., New York, 1960.

PROBLEMS

1.1. Derive Eq. 1.20.

1.2. Derive Eq. 1.33.

1.3. Compute the equivalent wavelength of
(*a*) a 1 MeV proton
(*b*) a "thermal" proton at 300 K (mean kinetic energy 1/40 eV)
(*c*) a 10 eV electron
(*d*) a 70 kg man walking at 3 km/hr

1.4. Consider two waves of equal amplitude but slightly different frequencies; f and $f + \Delta f$, where $\Delta f \ll f$.
(*a*) Show that the resultant disturbance is an amplitude-modulated oscillation at a frequency nearly the original frequency, and that the crest of the primary wave travels with a velocity $v_p = \omega/k$, whereas the crest of the modulation (that is, the group of the crests) travel with a velocity $v_g = \Delta\omega/\Delta k$.
(*b*) If $v_p = v_g$, show that the velocity of the waves must be independent of frequency.

1.5. A free particle with an energy $E < \Phi_1$ travels under the influence of a potential energy variation as shown in Fig. P.1.5. If the particle is initially coming from $-\infty$, find the reflection coefficient for the particle in Region I.

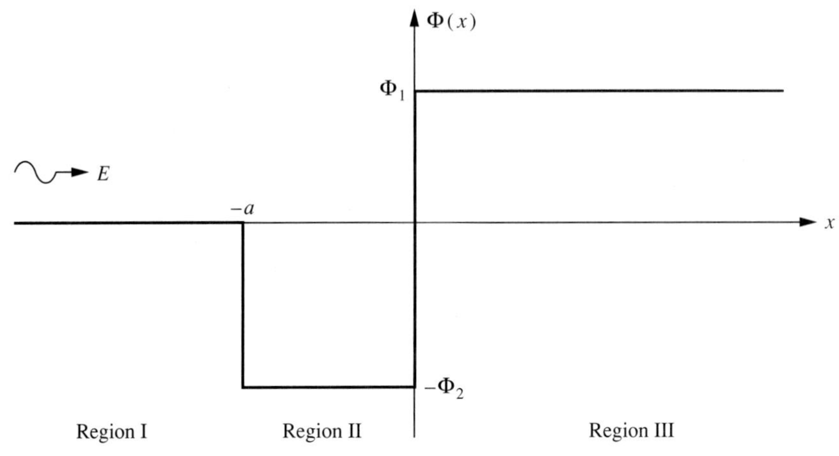

FIGURE P.1.5

1.6. Electrons are emitted by a cathode, move toward an anode 1 cm away, pass through a hole in the anode, and impinge on a crystal, as shown in Fig. P.1.6. The voltage difference V_A between the cathode and anode is 50 V. If the electrons are emitted from the cathode with zero initial velocity, when they reach the anode their velocity is given by $v = (2eV_A/m)^{1/2}$.

(*a*) Calculate the DeBroglie wavelength of electrons before the electrons hit the crystal.

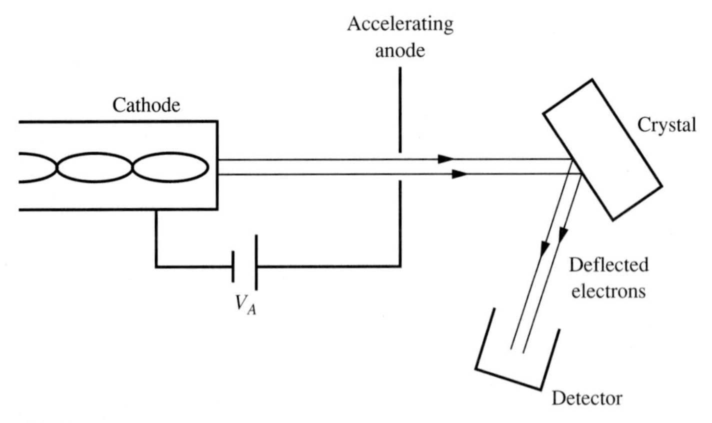

FIGURE P.1.6
Davisson–Germer Experiment. When deflection of electrons from the crystal is analyzed, the scattered electrons do not follow classical laws of particle collisions.

(*b*) Do you have to apply classical or quantum mechanical theories when
(i) the electron is between the cathode-anode space, and
(ii) it interacts with the crystalline atoms (the crystal atom spacing is 1.2 Å)?

1.7. An electron is incident on a triangular potential barrier as shown in Fig. P.1.7 with an energy $E < \Phi_0$. The tunneling probability is given

$$\left|\frac{F}{A}\right| = \exp\left\{\frac{4(2m)^{1/2}}{3e\,\mathscr{E}_x h}\Phi_0^{3/2}\right\}$$

here $\mathscr{E}_x$ is the slope of the potential—that is, the electric field—at $x = 0$. Show that as t gets smaller, the probability of tunneling gets larger.

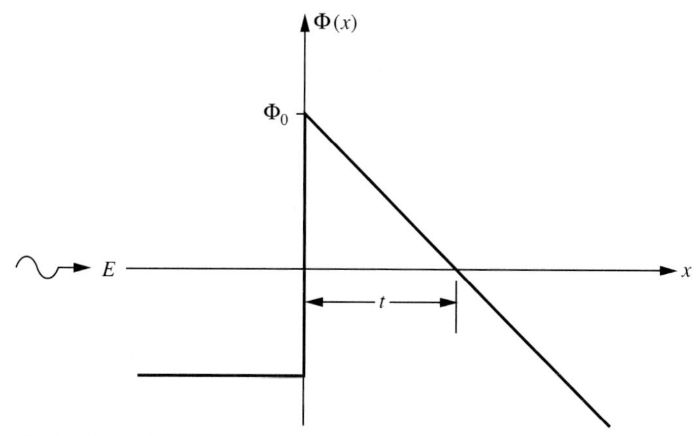

FIGURE P.1.7

1.8. If a flux of electrons with a current of 10 mA is incident on a potential barrier, as shown in Fig. 1.4., find the fraction of current that will appear on the other side of the barrier if the barrier height is 1.0 eV, the energy of the electrons is $E = 0.5$ eV and $d = 20$ Å. If E is almost zero, what is the new current? Does the initial electron energy have a noticeable influence on the tunneling probability as long as $E < \Phi_0$?

1.9. Show that as Φ_0 goes to infinity in Fig. 1.3, the wave function inside the forbidden region goes to zero, and the standing wave with nulls at $x = -n\pi/k$ forms in Region I (n = integer).

CHAPTER 2

BOUND PARTICLES

We know that particles become free if they overcome the binding forces with which they are bound to a nucleus, a solid, or a crystal. Most of the time particles are not free, and their motion is usually governed by internal binding forces, which are expressed in terms of an equivalent potential energy. This potential function can be due to the Coulomb forces between an electron and an ion, between more than one electron and an ion, an electron and the atoms of a solid, and so on. It can also include the interaction potentials among various electrons. We assume that however complicated the interaction is, an equivalent potential function can be written for that system of interacting particles. The wave function that is obtained from the solution of the Schrodinger equation can then be used, in principle, to obtain all the necessary information about the system under consideration. Even for the hydrogen atom, which contains only a single ion and a single electron, the solution of the Schrodinger equation is very complicated indeed. If a second electron is added to the system, as in the helium atom, the complications multiply, and one then looks for the approximate solutions of the Schrodinger equation. Fortunately, for many complicated systems there are highly sophisticated mathematical approximation techniques to solve the Schrodinger equation.

In this chapter, we will give additional basic quantum mechanical concepts that arise when we consider bound particles. As before, these results are based on the fundamental properties of the wave function rather than detailed mathematical solutions.

2.1 BOUND PARTICLES

We now consider bound particles which are restricted to move only in certain regions of space. An example of this is a particle trapped in a potential well, as shown in Fig. 2.1. Although a rigorous mathematical solution of the Schrodinger equation for this problem exists, we will look into the solution of this problem in terms of the previously established results related to the wave function.

Two possible assumed solutions are shown in Fig. 2.1*a*. Assume that the particle inside the well has an arbitrary energy E'' as shown in Fig. 2.1*b*. There will be three solutions to the Schrodinger equation corresponding to this given energy and the associated potential energy in the three respective regions. We know from the results introduced in Chapter 1 that the solutions in Regions I and III will be exponentially decaying solutions. The solution in Region II, which is expected to be sinusoidally varying is also shown in the same figure. We see that these solutions satisfy the condition of continuity of the wave functions at $x = -d/2$ and $x = d/2$ but do not satisfy the requirement of continuity of the first derivatives at the connecting points. Therefore, these assumed solutions of the Schrodinger equation are not acceptable solutions for the particle possessing energy E''. If we continue to look for acceptable solutions of the wave functions, we see that a different trial wave function solution, corresponding to the energy E' as shown in Fig. 2.1*a*, will satisfy both requirements of the boundary conditions at $x = -d/2$ and $x = d/2$. From these considerations, we conclude that a particle trapped in a potential well cannot have any arbitrary energy. It can exist inside the well provided the particle possesses only *discrete* allowed energies which are, of course, determined from the solution of the Schrodinger equation satisfying the proper boundary conditions.

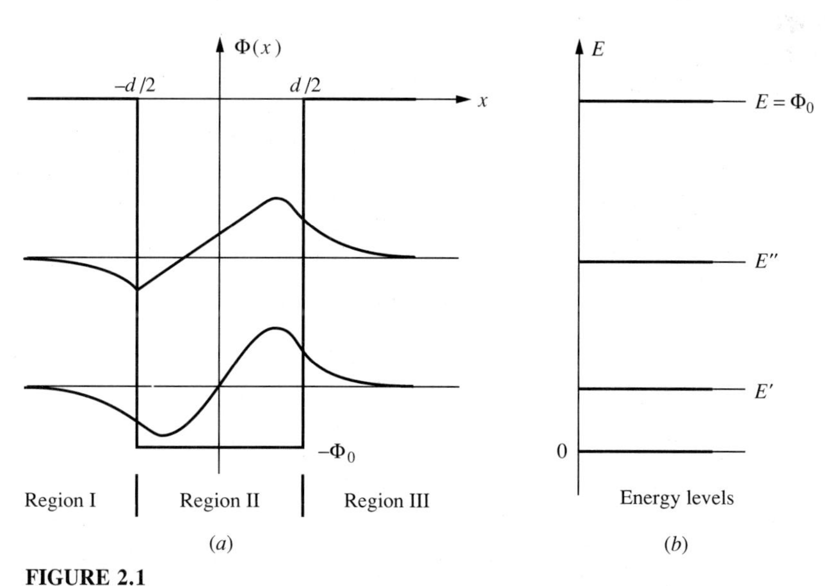

FIGURE 2.1
Particle in a potential well. Wave function corresponding to energy level E'' is not acceptable.

2.2 ONE-DIMENSIONAL INFINITE-POTENTIAL WELL

Suppose a particle is trapped in a potential well where the potential is infinite at both $x = 0$ and $x = d$, as shown in Fig. 2.2. Before attempting a solution for the energy of the particles, we can again use the previously obtained solutions of the Schrodinger equation to estimate the wave function solutions for the two regions at $x < 0$ and $x > d$. If the potentials were finite at these points, there would be exponentially decaying wave function solutions, similar to Eq. 1.29, in the two regions. But we see that penetration into the classically forbidden regions approaches zero as Φ_0 becomes very large ($\kappa \to \infty$). At the same time, the amplitude C also approaches zero from Eq. 1.28. (See Prob. 1.9) We therefore conclude that we only need the solutions of the Schrodinger equation inside the well and determine the allowable energies corresponding to those wave functions that become zero at $x = 0$ and $x = d$.

The solution of the Schrodinger equation inside the well gives

$$u(x) = A\sin(kx) + B\cos(kx) \tag{2.1}$$

where $k^2 = 2mE/\hbar^2$. Applying the boundary conditions $u(0) = u(d) = 0$, we find $B = 0$ and $kd = n\pi$ where n is an integer. Using this result in the definition of k, we find that the particles can exist inside the well only when they possess discrete energies given by

$$E_n = \frac{h^2}{8md^2}n^2 \qquad n = 1, 2, 3, \ldots \tag{2.2}$$

The constant A in Eq. 2.1 is calculated from the *normalization condition*, that is, if the particle is in an energy state E, the probability of finding the particle in that state should be one. A is obtained from

$$\int_0^d u_n^*(x)u_n(x)\,dx = 1 \tag{2.3}$$

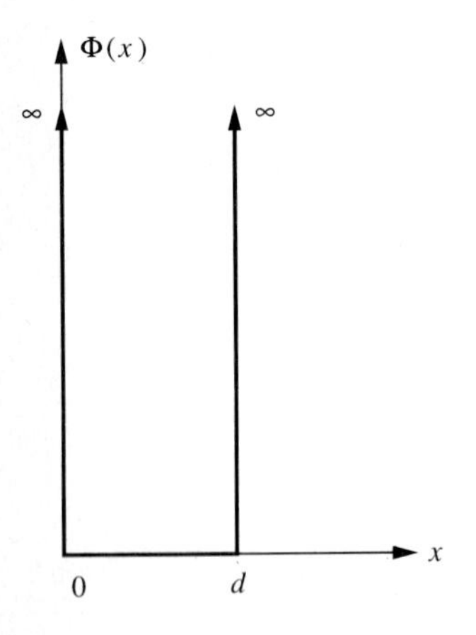

FIGURE 2.2
One-dimensional infinite well.

Using $A = \sqrt{2/d}$, the corresponding wave functions become*

$$u_n(x) = \sqrt{\frac{2}{d}} \sin\left(\frac{n\pi}{d}x\right) \tag{2.4}$$

We have, therefore, mathematically verified the results that we have speculated from the previous section: that is, the bound particles can only exist inside a potential well if they possess discrete allowed energy values; for an infinite well, these energies are given by Eq. 2.2.

2.3 HYDROGEN ATOM

In a hydrogen-like atom (an atom where there is only a single orbiting electron), the potential energy of a single electron due to the presence of the nucleus of charge Ze is given by

$$\Phi(r) = -\frac{Ze^2}{4\pi\epsilon_0 r} \tag{2.5}$$

where Z is the atomic number of the atom. For the simplest case, $Z = 1$ for the hydrogen atom.

Since $\Phi(r)$ is independent of time and is spherically symmetric (no θ and ϕ dependence), we can use the time-independent Schrodinger equation in spherical coordinates which is given by

$$\frac{\hbar^2}{2m*}\nabla^2 u(r,\theta,\phi) + \phi(r)u(r,\theta,\phi) = Eu(r,\theta,\phi) \tag{2.6}$$

where the Laplacian operator is given by

$$\nabla^2 = \frac{1}{r^2}\frac{\partial}{\partial r}\left(r^2\frac{\partial}{\partial r}\right) + \frac{1}{r^2}\frac{1}{\sin\theta}\frac{\partial}{\partial\theta}\left(\sin\theta\frac{\partial}{\partial\theta}\right) + \frac{1}{r^2}\frac{1}{\sin^2\theta}\frac{\partial^2}{\partial\phi^2} \tag{2.7}$$

We now make the assumption that the wave function can be written as a product of three functions which are only functions of the respective spatial variables, that is,

$$u(r,\theta,\phi) = R(r)\Xi(\theta)\Phi(\phi) \tag{2.8}$$

---

* Rather than using the sinusoidal functions, for the solution of $u(x)$ we could have written, as we have done before,

$$u(x) = A'e^{jkx} + B'e^{-jkx}$$

Applying the boundary condition $u(0) = 0$ yields $B' = -A'$. Applying the condition $u(d) = 0$ yields

$$u(x) = A\sin\left(\frac{n\pi}{d}x\right)$$

where $A = 2jA'$ is a new constant. This is the same as Eq. 2.4.

The Schrodinger equation can then be separated into three separate differential equations, one for each of the variables. These equations are coupled through constants known as separation constants. The resulting equations are

$$\frac{d^2\Phi}{d\phi^2} + m^2\Phi = 0 \tag{2.9}$$

$$\frac{1}{\sin\theta}\frac{d}{d\theta}\left(\sin\theta\frac{d\Xi}{d\theta}\right) + \left[l(l+1) - \frac{m^2}{\sin^2\theta}\right]\Xi = 0 \tag{2.10}$$

$$\frac{1}{r^2}\frac{d}{dr}\left(r^2\frac{dR}{dr}\right) + \frac{2m^*}{\hbar^2}\left[E + \frac{Ze^2}{4\pi\epsilon_0 r} - \frac{\hbar^2 l(l+1)}{2m^* r^2}\right]R = 0 \tag{2.11}$$

Eq. 2.11 has a solution only for discrete values of the energy of the electron. Labeling this energy E_n, it is given by

$$E_n = -\frac{me^4Z^2}{32\pi^2\epsilon_0^2\hbar^2}\frac{1}{n^2} \quad \text{(J)} \tag{2.12}$$

$$= -\frac{13.6}{n^2}\text{(eV)} \quad \text{(for } Z = 1\text{)}$$

provided n is an integer ≥ 1.

Solutions of Eqs. 2.9 and 2.10 are also possible for only restricted values of separation constants m and l. The following relations hold among the separation constants

$$n \geq l + 1 \qquad l \geq |m| \tag{2.13}$$

$$m = 0, \pm 1, \pm 2, \pm 3, \ldots$$

The corresponding wave function for the hydrogen atom can be written as

$$u_{nlm}(r, \theta, \phi) = R_{nl}(r)\Xi_{lm}(\theta)\Phi_m(\phi) \tag{2.14}$$

where each function is dependent on the separation constants n, l and m, and their dependences are represented by the corresponding integer subscripts.

The constants n, l and m, known as quantum numbers, are defined as

n = principal quantum number
l = angular momentum quantum number
m = z component of angular momentum quantum number

For any combination of these quantum numbers, the wave functions given by Eq. 2.14 are known to represent the *atomic state* of the system. (A solution for such a combination of quantum numbers is an independent solution of the system). From the relations among the various quantum numbers given by Eq. 2.13, we can show that there are n^2 independent states for the energy state E_n. Those independent states possessing the same energy E_n are called *degenerate states*. The degeneracy of a given hydrogen atom energy level is, therefore, n^2. In writing Eq. 2.5, it was

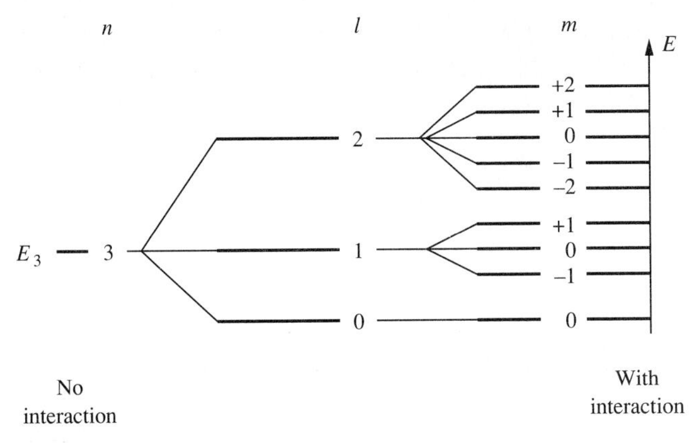

FIGURE 2.3
Splitting of the electronic energy level E_3 in a hydrogen atom ($n = 3$). Note: the quantum numbers n, l, m satisfy conditions given by Eq. 2.13.

assumed that the proton and the electron were the only particles that led to the potential energy given by Eq. 2.5. If, on the other hand, the influence of external forces or the influence of neighboring atoms is taken into account, the degeneracy in energy level E_n will be lifted and this will result in n^2 slightly different energy levels available for the electrons in a hydrogen atom instead of a single energy level E_n. This is referred to as splitting of energy level E_n into n^2 discrete levels. Fig. 2.3 shows this splitting for energy level E_3 ($n^2 = 9$).

When $n = 1$ ($l = 0$ and $m = 0$), the state of the system is known as the ground state of the hydrogen atom. The corresponding wave function for u_{100}, which is spherically symmetric, is shown in Fig. 2.4 as a function of radius.

For each physical variable, an equivalent quantum mechanical operator can be defined. The *expectation value* of a physical parameter $q(\mathbf{r})$ is defined by

$$< q(\mathbf{r}) > = \int_{\text{over all space}} \Psi^*(\mathbf{r}) Q(\mathbf{r}) \Psi(\mathbf{r})\, d^3r \qquad (2.15)$$

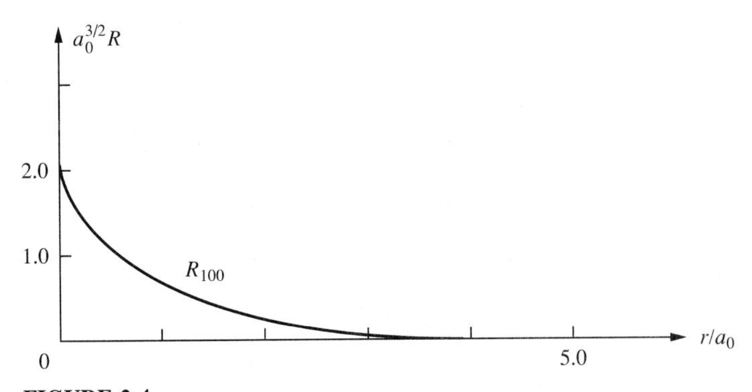

FIGURE 2.4
Radial dependence of the wave function u_{100}. a_0 is the first radius in the Bohr model of the hydrogen atom.

where $Q(\mathbf{r})$ is the corresponding operator for the physical quantity $q(\mathbf{r})$.* If the expectation value for the position of the electron in the hydrogen atom is calculated for the ground state of the hydrogen atom, we find (see Problem 2.1)

$$< \mathbf{r} > = \frac{3a_0}{2} \tag{2.16}$$

where

$$a_0 = \frac{4\pi\epsilon_0 h^2}{m^* e^2} = 0.53 \times 10^{-10} \text{ m} \tag{2.17}$$

and is the classical electron first Bohr orbit radius of the hydrogen atom.

When electrons are excited to the upper energy levels by absorbing energy from external sources, they make transitions back to the lower energy levels emitting photons with frequencies given by

$$\nu_{if} = \frac{(E_i - E_f)}{h} \tag{2.18}$$

where i and f represent the quantum numbers for the initial upper and final lower energy states, respectively.

Quantum mechanical calculations show that the rate of transition $\dot{P}_f$ for the spontaneous emission of an electron making a transition from an initial energy state i to a final energy state f is given by

$$\dot{P}_f = \frac{2\pi}{\hbar} | < f|H'|i > |^2 \rho(E_i) \tag{2.19}$$

where $\rho(E_i)$ is the density of the initial state, $< f|H'|i >$ is called the interaction matrix between the final and the initial states, and H' is the perturbing Hamiltonian operator for the interaction.†

---

* Eq. 2.15 is generally written in short hand notation as

$$< q(\mathbf{r}) > \equiv < \Psi|Q(\mathbf{r})|\Psi > \tag{2.15.b}$$

Where $< \Psi|$ is known as the bra vector and $|\Psi >$ as the ket vector. The notation implies the proper variables and the integration related to the states of the system. If the initial and final states of the system are different, bra and ket vectors represent the final and the initial states of the system, that is, as $< \mathbf{f}|$ and $|\mathbf{i} >$. Instead of Ψ, any of the quantum numbers of the corresponding states can be used inside the bra and ket vectors to represent the state of the system.

† Schrodinger equation given by Eq. 1.15 can be written as (see Problem 2.3)

$$H\Psi = E\Psi \tag{2.19a}$$

where H is the Hamiltonian (total energy) operator, and E's are the corresponding eigenvalues (energy of the system). For an isolated atom, Eq.2.19a can be written

$$H_0 \Psi_{0n} = E_{0n}\Psi_{0n} \tag{2.19b}$$

0 subscript represents the unperturbed system. If the isolated atom is perturbed by an external means, in many cases the total Hamiltonian operator can be written as

$$H = H_0 + H'$$

Here, the perturbation Hamiltonian H' arises from the presence of the perturbing source.

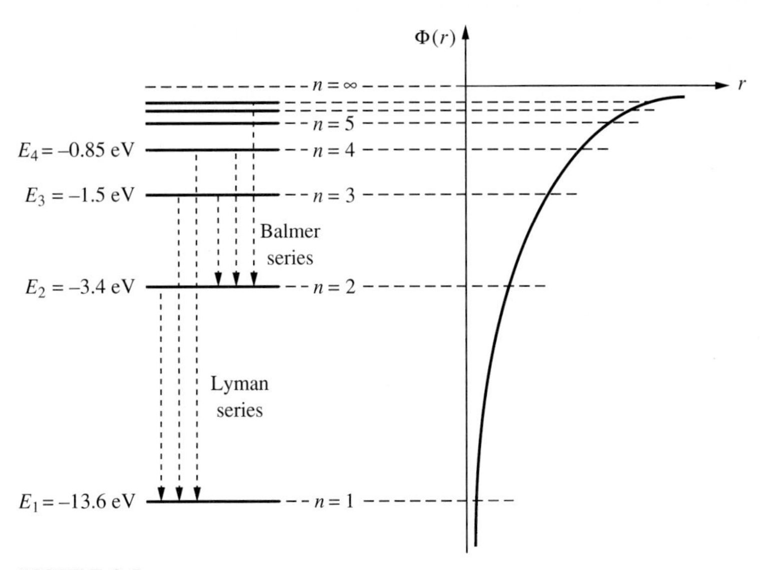

FIGURE 2.5
Energy levels of hydrogen atom. Transitions from higher levels to $n = 1$ level are known as Lyman series, to $n = 2$ as Balmer series, and to $n = 3$ as Pashen series. Electrons occupying the higher energy states are said to be in excited states.

Emitted light due to transitions among certain energy levels is classified into series of spectra. For the hydrogen atom, transitions from higher levels to the $n = 1$ level are known as Lyman series; to the $n = 2$, as Balmer series; and so on. When the interaction matrix given by Eq. 2.19 is calculated for a given atomic system, the integral becomes zero between transitions among certain energy levels. These are said to be "forbidden" transitions, dependent on the quantum numbers of the initial and final states. The restrictions on the resulting quantum numbers make up the *selection rules* for first-order transitions among energy levels of a given specific atomic system.

Figure 2.5*a* shows the relative energy levels for an isolated hydrogen atom. The potential energy given by Eq. 2.5 is plotted in Fig. 2.5*b*. The ground state ($n = 1$) is -13.6 eV below zero energy level. If the electron absorbs an energy equal to or greater than 13.6 eV, it overcomes the binding forces of the positive nucleus and becomes free. The electron and ion then behave as independent free charges with opposite signs and the atom is now said to be in the *ionized state*. Energy level $E = 0$ is known as the *vacuum level.*

2.4 PAULI EXCLUSION PRINCIPLE

The angular quantum number l, found from the solution of Eq. 2.10, arises because of the angular motion of the electron around the nuclear charge. It was also found experimentally that an electron possesses an intrinsic angular momentum (as if it was rotating around its own axis) and a corresponding magnetic moment. This intrinsic momentum for the electron is called the *electron spin*. The component of the spin angular momentum along the z axis has two possible eigenvalues

$$\sigma_z = s\hbar \tag{2.20}$$

where $s = \pm 1/2$. These are referred to as *spin up* $(+\ 1/2)$ and *spin down* $(-\ 1/2)$.

If classical definitions of angular momentum and associated magnetic moment were used to calculate the spin angular momentum, the electron charge would have to spin around its axis with a velocity greater than the speed of light in order to possess a spin magnetic moment that would agree with the experimentally measured electron spin. This would violate the principles of the special theory of relativity; thus, the spin is a purely quantum mechanical phenomenon.

In an atom with more than one electron, the solution of the Schrodinger equation gives rise to stationary states with discrete quantum numbers similar to n, l and m. If we also include the spin quantum number s along with these, the *Pauli exclusion principle* states that no two electrons can occupy the same energy level having the identical set of quantum numbers n, l, m, and s. In other words, no two electrons can have the same spatial distribution and spin orientation, or, when spin is excluded, no more than two electrons can have the same spatial wave functions. The Pauli exclusion principle predicted the classification of the atoms in the periodic table and even predicted presence of atoms with certain electronic properties which were not known at that time.

2.5 PARTICLES IN A THREE-DIMENSIONAL INFINITE-POTENTIAL WELL

The electrons move almost freely inside a metal in the background of positive charges of the lattice atoms. When an electron approaches the boundary of the metal, it can move a short distance out of the solid, but a positive image charge is immediately created to maintain charge neutrality within the solid. The resulting Coulomb force created between the electron and the image charge attracts the electron back into the solid. Unless the electron can gain energy from an external source, it can not overcome this binding force and thus is compelled to move back into the solid. The electron motion can be approximated by an equivalent model of free particles trapped by infinite potential barriers located at the physical boundaries of the metal. Otherwise, they are free to move between these barriers. This model is used in calculating the energy levels and wave functions of electrons in a block of metal.

In a one-dimensional infinite potential well, the wave functions are given by

$$u(x) = \sqrt{\frac{2}{L}}\sin(k_x x) \qquad k_x = \frac{n\pi}{L}; \qquad n = 1,2,3,\ldots \tag{2.21}$$

Using a product solution for each of the coordinates, this can be generalized to a three-dimensional infinite-potential well. The corresponding wave functions are

$$u(x,y,z) = \sqrt{\frac{8}{L_x L_y L_z}} \sin(k_x x) \sin(k_y y) \sin(k_z z) \tag{2.22}$$

where L_x, L_y, and L_z are the dimensions of the box in x, y, and z directions and

$$k_x = \frac{n_x \pi}{L_x} \qquad k_y = \frac{n_y \pi}{L_y} \qquad k_z = \frac{n_z \pi}{L_z} \tag{2.23}$$

$$n_i = 1,2,3,4,\ldots,\infty \qquad \text{positive integer } (i = x,y,z)$$

The total energy of the particle can be written as

$$\begin{aligned} E &= \frac{\hbar^2}{2m^*}(k_x^2 + k_y^2 + k_z^2) \\ &= \frac{\hbar^2 \pi^2}{2m^*}\left[\left(\frac{n_x}{L_x}\right)^2 + \left(\frac{n_y}{L_y}\right)^2 + \left(\frac{n_z}{L_z}\right)^2\right] \\ &= \frac{h^2}{8m^*}\left[\left(\frac{n_x}{L_x}\right)^2 + \left(\frac{n_y}{L_y}\right)^2 + \left(\frac{n_z}{L_z}\right)^2\right] \end{aligned} \tag{2.24}$$

To simplify calculations further, we assume that the solid is a square box of the dimensions $L_x = L_y = L_z = d$, thus

$$E = \frac{h^2}{8m^* d^2}(n_x^2 + n_y^2 + n_z^2) \tag{2.25}$$

where, the integers n_x, n_y, and n_z are referred to as quantum numbers.

2.6 DENSITY OF AVAILABLE STATES

The number of available energy states for electrons in a metal can now be calculated from our simple three-dimensional well model of the metal. From Eq. 2.25, the energy of an electron can be written as

$$E = \frac{h^2}{8m^* d^2} n^2 \tag{2.26}$$

with

$$n^2 = n_x^2 + n_y^2 + n_z^2$$

$$n_i = 1,2,3,4,\ldots,\infty \qquad \text{positive integer } (i = x,y,z)$$

For arbitrary integer values of n_x, n_y and n_z, the corresponding energy states can be plotted as lattice points in a coordinate system made up of n_x, n_y, and n_z coordinates, as shown in Fig. 2.6. Every combination of integers n_x, n_y, and n_z may be plotted as a point in this coordinate system. Each point will thus represent an energy state of the system. On the other hand, an energy level E depends on n through Eq. 2.25, and various combinations of integer n_i' s$(i = x,y,z)$ will give the same energy. As n becomes very large, the combinations of n_i' s that will give the same energy E will also be very large. Also, as n becomes larger, the discrete points in the n_x,n_y,n_z coordinate system get closer, and the changes in n can be considered to be a continuous, rather than discrete, function of n. This is equivalent to saying that the energy E varies continuously for large values of E because of the relation given by Eq. 2.26. Using the coordinate system n_x, n_y, and n_z, and assuming that there are a large number of electrons in the conductor, the number of available energy states for electrons in the interval E and $E + dE$ can now be calculated. This will be done first by calculating the number of available states between n and $n + dn$, since n space is related to E space through Eq. 2.26.

From Fig. 2.6, the number of available states $dS(n)$ between n and $n + dn$ is the same as the number of states that lie in the spherical shell of radius n and thickness dn. This is only restricted to the first quadrant of the coordinate system since the negative integers in Eq. 2.26 do not give any additional information about the system. Therefore, the number of available states between n and $n + dn$ is

$$dS(n) = \frac{1}{8}(4\pi n^2\, dn) \tag{2.27}$$

To convert this to the number of available states between E and $E + dE$, Eq. 2.26 is solved for n, and n and dn are substituted into Eq. 2.27. This yields

$$dS(E) = \frac{4\sqrt{2}\pi(m^*)^{3/2}}{h^3} d^3 E^{1/2} dE \tag{2.28}$$

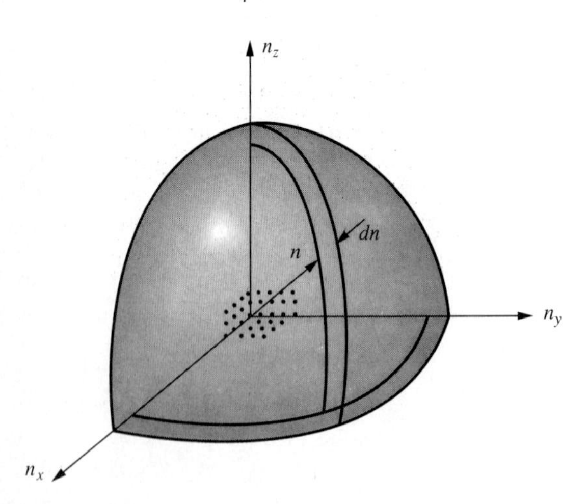

FIGURE 2.6
Coordinate system to calculate the available energy states.

If an energy interval dE near E is considered, the available number of energy levels will be larger for a box of large dimensions. This is consistent with the energy levels given by Eq. 2.26. The larger the d of a box, the smaller will be the energy level difference between consecutive energy levels. Therefore, more levels will be available within an energy level difference dE. This is shown in Fig. 2.7.

To obtain the density of available states independent of physical size of the solid, density of available states per unit energy interval dE between E and $E + dE$ can be found by dividing both sides of Eq. 2.28 by the physical volume $V = d^3$. This gives

$$\frac{ds(E)}{dE} = \frac{d}{dE}\left(\frac{S(E)}{V}\right) = \frac{4\sqrt{2}\pi(m^*)^{3/2}}{h^3}E^{1/2} \tag{2.29}$$

Note also that the density of available energy states is proportional to the 3/2 power of the mass m^* of the particles. In solids, especially in semiconductors, it will be shown in Chapter 5 that the mass of the electrons is not equal to the free electron mass m_0. They are usually smaller than m_0. Therefore, the number of available energy states within an energy interval dE m_0 will be smaller for electrons with the effective mass $m^* < m_0$.

Let us assume that there are N electrons within the metal, and that these electrons are not given additional energy from external sources. Assume also that the electrons are at 0 K, that is, there is no thermal motion. In accordance with the Pauli exclusion principle, only two electrons can be placed at a given energy level. The electrons will occupy the lowest energy levels first. We will also assume that the N electrons can only fill energy levels up to a maximum energy value, which will be labeled as E_F. Although there are available energy states above this energy E_F, there are no more electrons to fill these levels.

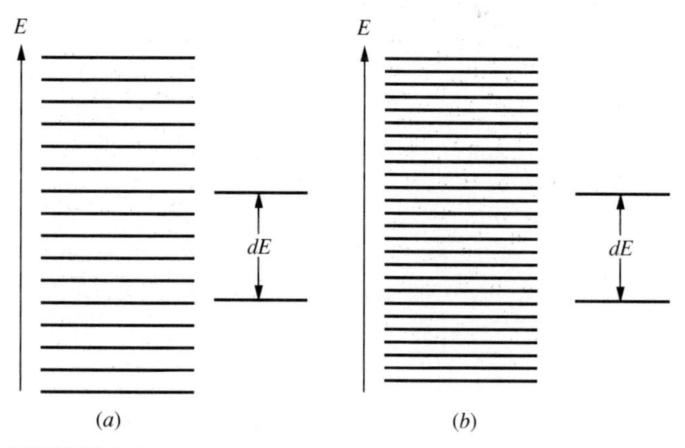

FIGURE 2.7
Energy levels within the same energy level dE for a box of (*a*) small (*b*) large dimensions.

Let $G(E)/dE$ be the number of available electronic states per unit energy interval dE. To calculate the maximum energy at $T = 0\ K$, Eq. 2.28 is multiplied by 2 (to put two electrons with spin up and spin down in that state) and integrated from 0 to E_F to give the total number of available electronic states.

$$G(E_F) = \int_0^{E_F} \frac{dG(E)}{dE} dE = \int_0^{E_F} 2\left(\frac{dS(E)}{dE}\right) dE$$

or

$$G(E_F) = \int_0^{E_F} \frac{8\sqrt{2}\pi(m^*)^{3/2}}{h^3} VE^{1/2} dE \tag{2.30}$$

This equation is integrated and set equal to the total number of free electrons N in the metal. Defining the density of electrons as $n = N/V$, and solving for E_F, we find

$$E_F = \left\{\frac{3h^3 n}{16\sqrt{2}\pi(m^*)^{3/2}}\right\}^{2/3} \tag{2.31}$$

E_F is called the *Fermi energy* of the metal. This is the maximum energy level occupied by the electrons in a solid at $T = 0$ K. Energy levels above E_F are empty at absolute zero.

*2.7 IONIZATION AND RECOMBINATION IN GASES

In Section 2.3, it was shown that 13.6 eV of energy is required to move a bound electron in a hydrogen atom from its ground state to a vacuum state where the electron becomes free and is no longer bound to the hydrogen atom. The process, known as ionization, produces an *electron-ion* pair for each of the freed electrons. If the atom contains more than one electron, it is possible to free one, two, or all of the bound electrons from the influence of the nucleus. The energy required to move a single electron from the ground state of a given atom to the vacuum level is called the ionization energy E_i of that atom. Ionization energy depends on the shell structure of a particular atom and is lowest for alkali atoms that contain only a single electron in their outermost orbits, and highest for gases that have closed shell orbits like helium, argon, neon, and so on. Agents that produce ionization in gases are photons, thermal vibrations, and collisions.

When the incident photon energy $h\nu$ is greater than the ionization energy of the atom, electrons gain energy in the interaction process and become free. Photon spectra that produce ionization in gaseous media cover a broad range, from that of the visible light all the way up to gamma rays.

Particles possess internal thermal energy which is characterized by a thermodynamic temperature T, and the average thermal energy of a particle is in the order of kT (k is the Boltzmann constant). As will be shown in Chapter 3, the particle energy distribution at equilibrium extends from zero to very large values.

At room temperature, $kT = 0.026$ eV, and very few particles have energy greater than E_i in a given atomic system to produce any noticeable ionization. As the temperature is increased, the number of particles at the high-energy tail of the particle distribution increases well above E_i, and appreciable ionization can be produced. There is no threshold for thermal ionization, but the number of ionized atoms increases with rising temperature. As the temperature is increased to greater than 10^5 K, practically all the atoms are ionized, and the gaseous medium becomes fully ionized.

Ionization can also be produced by collisions of atoms with ions or electrons. It was found experimentally that approximately 30 eV of energy is required to knock out an electron from its bound orbit by a collisional process to produce an electron-ion pair in the gaseous medium. This process is responsible for the slowing down of high-energy particles as they traverse a gaseous medium. In addition to being used as a detector, the number of electron-ion pairs produced in a medium can be used to estimate the energy of the incoming charged particle.

Whatever the source of ionization, there is also a competing process known as *recombination* which removes the electron-ion pairs from the medium by recombining an electron with an ion and forming a neutral atom. Initially, this newly formed neutral atom may be in an excited state, but eventually decays down to its ground state.

The number of ions created per unit of time is equal to the rate of ionization q minus the ions removed from the medium through ionization. Thus, the rate of change of ion density can be written as

$$\frac{dn_i}{dt} = q - \alpha_i n n_i \tag{2.32}$$

where α_i is called the recombination coefficient, and the recombination rate is dependent on the density of electrons n and the density of ions n_i. Note that if the ionizing source is removed, $q = 0$ and if, to start with, there are equal numbers of electrons and ions (electrically neutral, i.e., $n = n_i$), the recombination equation reduces to

$$\frac{dn}{dt} = -\alpha_i n^2 \tag{2.33}$$

which gives, after integration,

$$n = \frac{n_0}{1 + \alpha_i n_0 t} \tag{2.34}$$

where n_0 is the initial ion density at the instant the ionizing source is turned off at $t = 0$.

In any ionized medium, the recombination process competes with the ionization process. If the ionizing source is kept constant, the system reaches an equilibrium state, and the number of electrons, ions, and neutral atoms becomes constant. For each of these species, a thermodynamic temperature T can be assigned. For low ionization and laboratory gas discharges, the electron temperature

T_e is usually greater than the corresponding ion and neutral temperatures T_i and T_n. Increasing the kinetic energy of the system or using preferential excitation, the ion temperature can be made equal to or greater than the electron temperature. At very high temperatures, the presence of neutral atoms can be neglected.

*2.8 PLASMA SPACE CHARGE FIELDS

Plasmas, sometimes referred to as the fourth state of matter, include such naturally occurring objects as the ionosphere, all regions of the sun, the lightning channel, the atmospheres, and the interiors of the majority of stars, planetary nebula, and both hot and cold interstellar gas regions. In addition, they include laboratory plasmas such as positive columns of glow discharges, R.F. and electrodeless discharges, flames, and various plasmas involved in thermonuclear fusion research. In solids, the behavior of free electrons under a background of positive lattice ions may also be treated as a plasma, and many phenomena that take place in ionized gases can be shown to take place in solid state plasmas as well. In this section, only ionized gases will be considered.

Plasma contains approximately equal numbers of positive and negative charges. Any disturbance giving rise to a departure from neutrality generates large electrostatic or *space-charge* forces that counteract the disturbance. The individual charged particles, and, through them, the plasma as a whole, can also interact with electric and magnetic fields. The electromagnetic fields may be generated externally or may rise from the motion of the charged particles themselves within the plasma.

The plasma may have a number of characteristic lengths and frequencies which play an important role in its dynamics. High-frequency electrical properties of a gaseous discharge arise mostly from the motion of the electrons. Ions and neutral atoms, because of their heavier masses, contribute mainly to the mechanical properties of the plasma. Ions in many cases may also contribute to the electrical properties of the ionized medium.

In this section, we give a simplifying plasma model containing only electrons, neutral particles, and one species of singly charged positive ions with mass m_i. We further assume that all the particles are characterized by kinetic temperatures T_e, T_i, T_n. In many laboratory plasmas, we can approximate the temperatures of electrons, ions and neutrals as either $T_e = T_i$, $T_n = 0$ or $T_e >> T_i$ and T_n. The latter case assumes that the ions are immobile and just form a positive neutralizing background for the electrons. This may also be a model for the electrons in a solid. As a first approximation, neutral particles may be neglected in studying the high-frequency behavior of plasmas.

The charged particles in a plasma interact with each other by means of long-range Coulomb forces so that each particle may be interacting simultaneously with many other particles. Properties of the plasma which arise from this kind of interaction are called collective phenomena. Two of the parameters that play a very important role in the collective behavior of the plasma are (*a*) a characteristic length called the *Debye length*, and (*b*) a characteristic frequency called the *plasma frequency*. In discussing these parameters, collisional effects will be neglected although they are implicitly included in our assumption of the existence of the temperatures T_e and T_i.

*2.8.1 Debye Length

Consider an isolated positively charged test particle Q. The potential (relative to a reference point at infinity) in the vicinity of this charged particle at a distance r is given by the usual Coulomb potential

$$\Phi(r) = \frac{1}{4\pi\epsilon_0}\left(\frac{Q}{r}\right) \tag{2.35}$$

We now consider a neutral plasma containing an equal number of negative charges (electrons) and positive charges (ions) and insert the above test charge Q into this plasma. To maintain the charge neutrality, the plasma electrons and ions align themselves to shield the influence of the external charge from the rest of the plasma. The potential distribution of the charge Q is modified and is given by

$$\Phi(r) = \frac{Q}{4\pi\epsilon_0}\frac{e^{-(r/\lambda_d)}}{r} \tag{2.36}$$

where λ_d

$$\lambda_d^{-1} = \left[\frac{e^2 n_0}{\epsilon_0 k}\left(\frac{1}{T_e} + \frac{1}{T_i}\right)\right]^{1/2} \tag{2.37}$$

is called the *Debye length*. Compared with the isolated potential of a point charge, we see that the Coulomb potential near Q is reduced by the rearrangement of plasma charges in the neighborhood of Q, and at distances $r \gg \lambda_d$, the potential drops to zero. In other words, the plasma shields itself from the disturbance effect of Q at distances $r > \lambda_d$. In deriving Eq. 2.36 it was assumed that the average interparticle distance between the plasma particles is a lot smaller than λ_d, that is, many plasma particles are needed within the spherical radius λ_d to effectively shield the test particle from the rest of the plasma. Note that if any ion or electron of the plasma itself is isolated from the plasma, neutrality will be violated and space-charge-restoring forces will again act within the plasma to maintain the neutrality in the vicinity of this isolated charge.

The Debye length is an important parameter of the plasma, and may be physically interpreted in several different ways. It is

the screening distance for charges in the plasma;

the effective distance of interaction between charged plasma particles;

the distance over which substantial deviations from neutrality take place; and

the order of magnitude of the thickness of the space-charge sheaths which form near electrodes or walls in contact with the plasma.

The Debye length was originally used by Longmuir and Tonks to define *plasma*. They called an ionized medium a plasma if the Debye length for that medium was small compared to the size of the medium.

*2.8.2 Plasma Frequency

Any plasma perturbation violating charge neutrality will be eliminated and neutrality restored, but, because of their inertia, the displaced charged particles will tend to overshoot their equilibrium positions, and the perturbation will oscillate about these equilibrium particle positions. Because of their small masses, if electrons are displaced from their equilibrium positions they will oscillate about their equilibrium positions with such a high frequency that to a first-order approximation the plasma ions can be considered stationary.

We now consider a simple case where we will neglect the thermal motion of both the electrons and the ions and consider the plasma to be infinite in extent. Referring to Fig. 2.8, equal numbers of electrons and ions having densities $n_e = n_i = n_0$ are uniformly distributed throughout the space. We assume that an electron layer is moved a distance ζ to the right from its equilibrium position x_0. We know that in the region between x_0 and $x_0 + \zeta$ the charge neutrality is destroyed, and an internal electric field $\mathscr{E}$ is generated as a result of the displacement of the charges. This produces a restoring force on the displaced electrons directed toward x_0. The electrons are forced to move back toward x_0, but since there are no damping forces on the assumed model, they overshoot their equilibrium positions and continue to move in the negative direction. This time, the restoring force is reversed, and they are again forced to move back toward x_0.

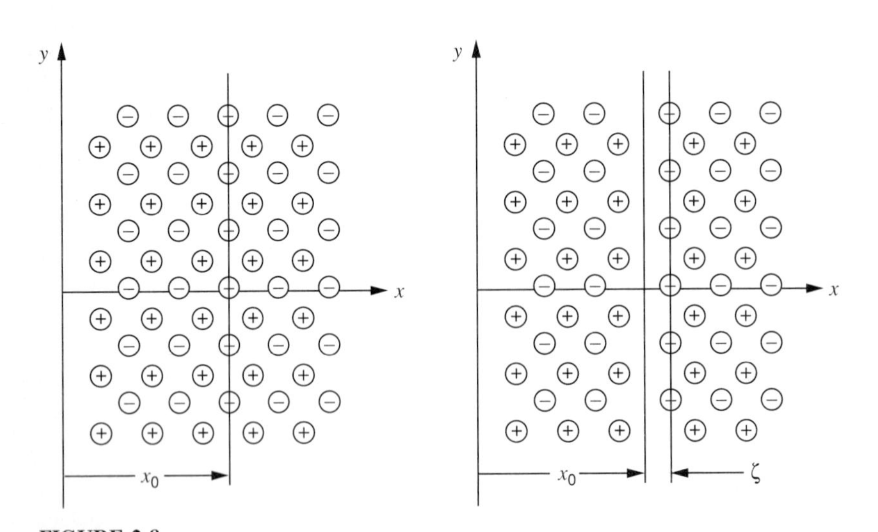

FIGURE 2.8
(*a*) Cold, infinite plasma at equilibrium. (*b*) A layer of electrons is displaced a distance ζ from their equilibrium position x_0.

The equation of motion for these electrons can be shown to be a simple harmonic motion whose angular frequency is given by

$$\omega_{\text{pe}} = \left(\frac{e^2 n_0}{m_e \epsilon_0}\right)^{1/2} \tag{2.38}$$

ω_{pe} is called the plasma frequency of the electrons. This natural frequency of oscillation is modified for a thermal plasma, as well as for a plasma of finite size.

In various ways, plasma frequency is another very important parameter that enters into the description of collective charged-particle behavior in a plasma:

1. It is the natural resonant frequency of the plasma.
2. $(1/\omega_{\text{pe}})$ is a measure of the shortest time in which the plasma can respond to shielding itself against externally applied fields. We note from Eqs. 2.37 and 2.38, (assuming ion thermal motion is negligible), that

$$\omega_{\text{pe}} \lambda_d = \left(\frac{kT_e}{m_e}\right)^{1/2} \approx v_{eT} \tag{2.39}$$

   where v_{eT} is the mean electron thermal speed, so that the electrons with mean thermal energy traverse the Debye distance in a time interval $1/\omega_{\text{pe}}$, that is, they can move through distances to achieve screening within this time.
3. It gives the plasma a dielectric constant

$$\kappa_e = \left[1 - \left(\frac{\omega_{\text{pe}}}{\omega}\right)^2\right]$$

   so that the medium has a cutoff frequency for transmission of electromagnetic waves at $\omega = \omega_{\text{pe}}$. ($\omega$ is the radial frequency of the electromagnetic wave). Alternately, we may think of the plasma as behaving somewhat like a dielectric at frequencies greater than ω_{pe}, and like a conductor at frequencies below ω_{pe}.

Two of the various scientific and engineering applications of gaseous discharges will play an important role in our future technology. These are the achievement of controlled thermonuclear fusion reactions in the laboratory, and the production of flat-screen display devices to replace bulky cathode ray tubes. Plasmas are most frequently mentioned in relation to research in controlled thermonuclear fusion. In its present form, the plasma display screen comes very close to being a real competitor to the cathode ray tube.

FURTHER READING

Cobine, J. D., *Gaseous Conductors, Theory and Engineering Applications,* Dover, Chaps VII and VIII, 1958.

Leighton, R. E., *Principles of Modern Physics,* McGraw-Hill, New York, 1959.

McDowell, M. R. C., and Altan M. Ferendeci, eds. *Atomic Processes in Thermonuclear Controlled Fusion,* Plenum Press, NATO Advanced Study Institute Series B, vol. 53B, 1980.
Sherr, Sol, *Electronic Displays,* Wiley, New York, 1979.

PROBLEMS

2.1. The ground state wave function for the hydrogen atom ($l = 0, m = 0$) is given by

$$u_{100} = A_{100} e^{-Zr/a_0} \tag{P.2.1}$$

where a_0 is the classical Bohr electron orbit radius, and Z is the atomic number ($Z = 1$ for hydrogen).
(*a*) Show that Eq. P.2.1 satisfies the Schrodinger equation.
(*b*) Find A_{100} from the normalization relation.

$$\int_0^\infty u_{100}^* u_{100}\, d^3r = 1.0 \tag{P.2.2}$$

(*c*) Using Eq. 2.15, show that $<\mathbf{r}>= 3a_0/2$

2.2. When an electron in a hydrogen atom is excited to $n = 4$ state, find the frequency and wavelength of the emitted photon when the electron makes a transition to $n = 1$ (Lyman δ line) or to $n = 2$ (Balmer β line) or $n = 3$ (Pachen α line).

2.3. An eigenvalue equation is given by

$$\boldsymbol{H}\Psi = E\Psi$$

where $\boldsymbol{H}$ is the Hamiltonian operator, and E and Ψ are the corresponding eigenvalue and eigenfunctions. If one-dimensional momentum and potential operators are defined by

$$\boldsymbol{p} \Rightarrow -i\hbar\frac{\partial}{\partial x} \qquad \Phi(x) \Rightarrow \Phi(x)$$

show that one can obtain a Hamiltonian operator given by

$$\boldsymbol{H} = \left(\frac{1}{2m^*}\boldsymbol{p}^2\right) + \Phi(x)$$

$$\boldsymbol{H} = -\frac{\hbar^2}{2m^*}\frac{\partial^2}{\partial x^2} + \Phi(x)$$

and the one-dimensional Schrodinger equation becomes an energy eigenvalue equation.

2.4. The iron atom has an atomic number $Z = 26$. When all but one of the electrons of the iron atom are removed, the energy required to remove the last electron can be calculated using Eq. 2.12. Find the necessary energy to remove this last electron from the binding forces of the positive nucleus of the iron.

2.5. Copper and gold metal atoms have one valance electron per atom to contribute to the electrical conduction. The atomic information for each metal is given by

Element	Atomic number (Z)	Atomic weight	Mass density (gm/cm^3)
Copper	29	63.54	8.9
Gold	79	197.0	19.3

For each element, find

(*a*) the number density of conduction electrons,

(*b*) the number of available energy states, and

(*c*) the Fermi level for each metal.

(Assume the mass of each electron is equal to m_0)

Note: One gram-mole of any substance contains the same number of atoms or molecules, that is, 2 gm of H_2, 32 gm of O_2 and 4 gm of He contain 6.02486×10^{23} molecules (Avogadro's number N).

2.6. If the recombination coefficient for air is 2×10^{-7} cm^3/sec at 100 torr of pressure, find the time t it takes for the ion density to drop to 10 percent of its original value of $n_0 = 10^{10}$ cm^{-3} after the ionizing source is turned off. (1 Torr = 1mm of Hg = 1/760 atm.).

2.7. Compute the Debye length for a partially ionized gas with electron and positive ion concentrations of 10^{12} cm^{-3}

(a) at room temperature

(b) at 10,000 K

In each case, how small can the container be and still have the gas retain the plasma properties?

2.8. A certain vacuum tube with a volume of 50 cm^3 contains a gas at a pressure of 1 torr and a temperature of 5000 K. If 0.1 percent of the gas is ionized, and electron and ion temperatures are the same as the gas temperature, determine whether the contents behave as a plasma or not.

2.9. Show that a plasma medium acts like a conductor if $\omega < \omega_p$, and as a dielectric if $\omega > \omega_p$ for a plane wave of frequency of ω. Hint: the propagation constant of a plane wave is given by $\beta = \omega/v$ where $v = 1/(\mu\epsilon)^{1/2}$

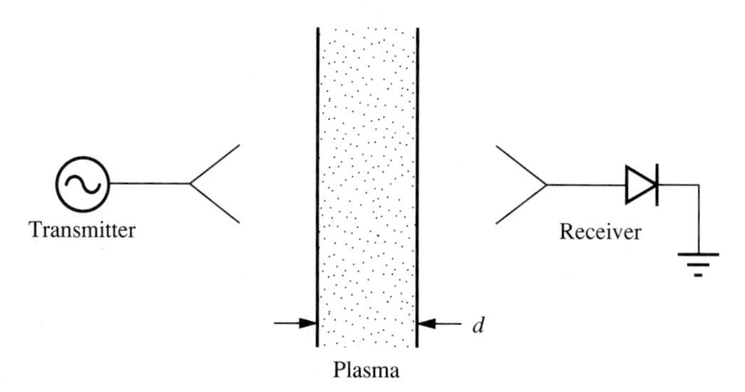

FIGURE P.2.9

2.10. A variable frequency microwave signal is incident on a plasma slab of thickness $d = 10$ cm. A second horn placed on the other side of the plasma slab starts detecting the signal at a frequency $f = 3 \times 10^9$ Hz,

(*a*) Find the electron density of the plasma.

(*b*) What is the wavelength of the microwave signal in the plasma medium at $f = 5 \times 10^9$ Hz?

2.11. If, in a thermonuclear machine called a Tokomak, the container is a donut-shaped chamber with a total volume of 500 cm^3 and contains a gas concentration of 10^{14}

atoms/cm^3, find the average energy per particle and the total energy required to raise the temperature of all the atoms from room temperature to 10^7 K. Note: assume that the medium will be fully ionized at that temperature.

2.12. Find the electron plasma frequency and Debye length in a silicon semiconductor at room temperature if the electron density (*a*) 10^{15}cm^{-3} and (*b*) 10^{17}cm^{-3}. (For silicon, the dielectric constant is 11.8).

CHAPTER 3

EQUILIBRIUM STATISTICAL MECHANICS

When we considered the energy levels of electrons in a solid, we assumed that the electrons had no internal energy and occupied all the lowest available energy levels up to the Fermi Energy E_F. In reality, the solid is at a temperature $T > 0$ K, and therefore the electrons possess internal kinetic energy. Since the absolute temperature T is a measure of this internal kinetic energy of the electrons, we expect the energy distribution of the electrons inside the solid to be modified. Because there are a large number of electrons that contribute to the electrical properties of an electronic device, it is impossible to follow the individual motion of each of these electrons. Even if it were possible, the inclusion of the interaction of electrons with each other and with lattice atoms of the solid would make this treatment a complicated and formidable task. The motion of a few hundred particles may be followed in time by using a numerical computer, but this computation will take a very long time, even with today's highly advanced computers.

If we are interested in certain properties of a system—for example, the average energy of all the particles—it is nearly impossible to determine this average from the individual energies of the particles. We, therefore, should find ways to effectively take into account the collective behavior of the particles through proper averaging processes. These concepts lead to the ideas of statistical mechanics and probability distributions. In this chapter, fundamental concepts of quantum statistics will be presented by physical arguments rather than detailed mathematical analysis.

3.1 FUNDAMENTALS OF STATISTICAL MECHANICS

Let us introduce statistical concepts by considering a coin toss. A coin toss in a single football game, to determine who gets the football first, is a highly non-statistical phenomenon. This result is either a heads or a tail, but people usually accept this result as a consequence of a probabilitistic chance of getting 50 percent heads and 50 percent tails. This would be roughly true if one counted the number of times a team got tails through a single season or, even better, through a span of a few seasons. As the number of coin tosses increases, the chance of getting tails becomes a 50 percent probability. If one tosses a coin a million times, one can speak of probability of getting heads or tails as 0.5. If one tosses the coin one million one times, again it is a 50 percent chance that it can come up either heads or tails. Furthermore, if one tosses one million coins at one time and counts the heads and tails, the result may be very close to half of the coins being heads and the other half being tails. This shows that in order to talk about statistical averages, we must talk about the average behavior of a large number of particles. Because of constraints imposed on a physical system—such as the kind of particles, the number of particles, the total energy of a system—the particles may be distributed in a restricted fashion throughout a parameter space. If we can determine this distribution of the particles under consideration, we can, with proper averaging techniques, calculate any other average quantity associated with the same system of particles.

To gain additional insight into the probability concepts, consider a slot machine that has three wheels, each having zero to nine divisions. When the machine arm is pulled down, the wheels start rotating. After a while, they stop, and three digits are displayed on the machine. When all the wheels show the same number, the player wins the jackpot. Let us raise some questions as to the chances of hitting the jackpot. Say that the machine does not differentiate between the numbers: as long as three digits are the same, you get the jackpot. There are ten ways that the three digits can be the same: from 000 to 999. But consider the chance of getting two digits the same but the third different. This chance increases if we don't restrict which two of the rows will come up the same. Finally, if we consider the chances of getting all different numbers every time, the probability increases considerably because any combination of numbers that are displayed is more likely to occur than the others. The probability of obtaining three identical digits is 0.01, of obtaining two the same is 0.1, and all different is 0.89. The probability that any one of the above combinations will occur is 1.00—that is, it will definitely occur.

In the example of the slot machine, suppose that in two successive tries the resulting numbers are 015 and 627. If these two numbers are considered separately, although the numbers are completely different, both results satisfy the same condition from the gambling point of view: that is, the two tries resulted in all different numbers. The two combinations 015 and 627 are called *microstates* of the system, but from the point of view of having all different numbers (it doesn't matter what combination of numbers we have as long as they are all different), these states are the same and are called *macrostates* of the system. Identification of these probable states in a parameter space results in a probability function for that particular parameter space.

Fundamental concepts related to the probability of occurrences lead to the distribution functions in a given parameter space. These parameters may be energy, momentum, speed, and so on. Before introducing the possible distributions of particles in a parameter space, let us elaborate the advantages of using the distribution functions in estimating certain properties in a given parameter space.

Consider the world human population. There are more than four billion people now living on the surface of the earth. The population of different countries is different. Their birth and death rates are all different, depending on many factors, such as the availability of health care. This is also true for regions or cities within a given country. As an example, take country X and assume that the census bureau of that country has a reliable recordkeeping system that will supply us with all the necessary information about the population of that country. If we are interested in knowing the number of people between the ages of 20 and 21, we can deduce the result from the information supplied by the census bureau records of country X by counting the number of people within the ages of 20 and 21. If we are interested in knowing the number of people in a different age bracket, we will again refer to the census bureau and perform another counting process. On the other hand, to start with, we can put the complete information about the population of that country in the form of a distribution function—that is, as a graph—representing the number of people per age interval around a given age. The smaller this interval, the more accurate will be our distribution curve, but such a detailed representation may provide more information than we need. The best solution is to represent the age distribution in a mathematical functional form. Suppose that such a population distribution function is known for that country. Here the parameter for the distribution function is years or age of people. If we are asked to find the number of people between the ages of 20 and 21 years old we simply multiply the average number of people per year between 20 and 21 years old by one year and, as shown in Fig. 3.1, get the average number of people that are in this age bracket. This result will not be much different from that obtained by a straightforward counting of the people in the census bureau's records. Other information about the general behavior of the population can be deduced from the established population distribution function for that country. Country Y will have a different population distribution from country X because of the many differences between the two countries.

3.2 GENERAL PROPERTIES OF DISTRIBUTION FUNCTIONS

Let $f(\mathbf{q})$ be the distribution function for a system of particles within the parameter space $\mathbf{q}$ which may be a vector quantity such as momentum, or may be a scalar

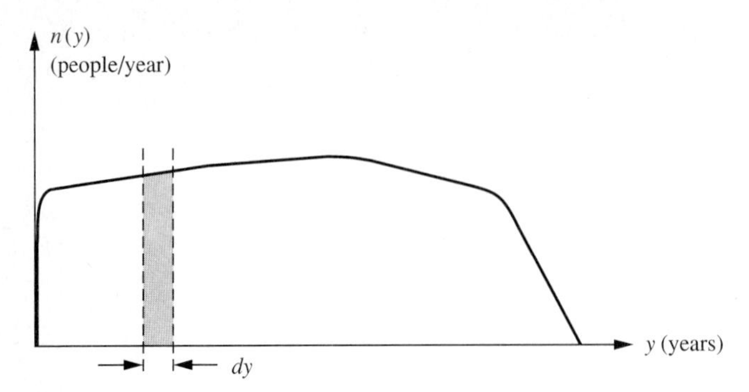

FIGURE 3.1
Population distribution per age of country X. The number of people within the range of y and $y + dy$ years is given by the area under the dashed region.

quantity such as energy. In more detailed and general treatments, the distribution function may be a function of more than one parameter. Only a single parameter dependence will be assumed here. This is more than adequate for the treatment of statistical concepts in this book.

The distribution function $f(\mathbf{q})$ is defined as the number density of particles having the parameter range between q and $q + dq$. The total number of particles within the parameter space is then equal to

$$N = \int_{\substack{\text{over all}\\ \text{parameter}\\ \text{space}}} f(\mathbf{q})\, d\mathbf{q} \tag{3.1}$$

where $d\mathbf{q}$ is a three-dimensional volume element that is written in its most general form, that is, $d\mathbf{q} = d^3q = dq_i\, dq_j\, dq_k$. ($i,j,k$ is an orthonormal coordinate system).

Using the distribution function, we can calculate the average quantities associated with a system of particles. Suppose we want to calculate the average value of a variable Q. As an example, let the variable Q represent the momentum $\mathbf{p}$ of particles. We use the usual method of averaging quantities to find the average momentum of the total of N particles in the system. The total momentum of the particles having momenta between $\mathbf{p}$ and $\mathbf{p} + d\mathbf{p}$ is the product of momentum $\mathbf{p}$ multiplied by the number of particles $f(\mathbf{p})\, d\mathbf{p}$ that have momentum between $\mathbf{p}$ and $\mathbf{p} + d\mathbf{p}$. This product is $\mathbf{p}f(\mathbf{p})\, d\mathbf{p}$. Adding all these contributions of the momentum is equivalent to finding the total momentum of the system of N particles. If we have N particles and divide the total momentum of N particles by N, we will get the average value of the momentum of a system of N particles. This is written as

$$< \mathbf{p} > = \frac{\int \mathbf{p} f(\mathbf{p})\, dp}{\int f(\mathbf{p})\, d\mathbf{p}} \tag{3.2}$$

where the integrals are taken over all the parameter space.

In general, when a parameter $Q(\mathbf{q})$ is a function of the variable $\mathbf{q}$, its average value can be written as

$$< Q > = \frac{\int Q(\mathbf{q}) f(\mathbf{q})\, d\mathbf{q}}{\int f(\mathbf{q})\, d\mathbf{q}} \tag{3.3}$$

The quantity

$$dp(\mathbf{q}) = \frac{f(\mathbf{q})\, d\mathbf{q}}{\int f(\mathbf{q})\, d\mathbf{q}} \tag{3.4}$$

can be thought of as the probability of finding N particles between the parameter $\mathbf{q}$ and $\mathbf{q} + d\mathbf{q}$. The average value of a quantity $Q(\mathbf{q})$ can then be calculated by multiplying $Q(\mathbf{q})$ with the probability that the particles will have that parameter value at $\mathbf{q}$ and integrating all these probabilities over the parameter space. That is

$$< Q > = \int Q(\mathbf{q})\, dp(\mathbf{q}) \tag{3.5}$$

3.3 PHYSICAL BASIS OF DISTRIBUTION FUNCTIONS

Before considering the consequences of statistical arguments, a conceptual understanding of the distribution functions will be given to show how different distribution functions occur depending on the type of particles used in calculating these distribution functions.

In quantum statistics, the distribution function in a parameter space depends on the type of particles considered. From a quantum mechanical point of view, particles can be classified as

1. identical but distinguishable particles
2. identical but indistinguishable particles of integral spin
3. identical but indistinguishable particles of half-integral spin

The distinguishable and indistinguishable classification of particles arises purely from quantum mechanical considerations, such as the associated-wave functions, degeneracy of energy states, Pauli Exclusion principle, and so on. The resulting distribution functions are respectively called

Maxwell–Boltzmann distribution
Bose–Einstein distribution
Fermi–Dirac distribution

We take energy E as the parameter in question for a system of particles. To see physically how the different types of the particles lead to different distribution functions, let us restrict the total number of particles to three and distribute these particles in every possible way among equally spaced energy (parameter) intervals of one unit of energy with a total energy of five units. Irrespective of the type of particles that are considered, the following physical constraints should always be satisfied for any distribution process: For an isolated system, the total number of particles and total energy should be conserved. If n_s is the number of particles in the energy interval s, these constraints can be written as

$$N = \sum_s n_s = \text{constant} \tag{3.6}$$

$$E = \sum_s n_s E_s = \text{constant} \tag{3.7}$$

where E_s is the energy value of the s interval.

Particles are now to be distributed among the given energy states, keeping the total energy of the system and the total number of particles constant throughout the distribution process. As will be shown next, this distribution of particles will largely depend on the type of particles considered.

Table 3.1 shows all possible distributions of three identical but distinguishable particles with a total energy of five units and an energy interval of one unit. Zero energy is also a part of the system. Each horizontal line represents a different distribution and occurs 1/21 of the time. It will be assumed that all these distributions are equally probable. We can now make an averaging process similar to the usual time averaging to find the probability that a given energy state will be occupied. The time average for the number of particles occupying a state of energy E_s is denoted by P_s and is defined like any average as

$$P_s = \frac{t_1}{T} + 2\frac{t_2}{T} + \cdots + N\frac{t_N}{T} \tag{3.8}$$

where t_1 = the time one particle occupies sth state
t_2 = the time two particles occupy sth state
. . .
t_N = the time N particles occupy sth state and
T = the total time

This type of averaging process, when interpreted as the probability per unit energy interval that the three particles will occupy an energy state E_s with a total energy of five units, results in a distribution of the three particles as shown in Fig. 3.2*a*. The area under the distribution curve should be equal to three (the number of particles).

Table 3.2 and Fig. 3.2*b* show the possible distribution of three identical but indistinguishable particles. Similarly, Table 3.3 and Fig. 3.2*c* show the possible distribution of three identical but indistinguishable particles obeying the Pauli exclusion principle.

From these results, it can be seen that the character of the particle makes a big difference in the way particles can be distributed in a parameter space. For distinguishable particles, usually applied to an ideal gas, there are 21 probable

TABLE 3.1
Possible arrangement of three distinguishable particles with a total energy of five units.

	Energy					
	0	1	2	3	4	5
	○ ● ○ ◐ ● ◐					◐ ● ○
	● ● ○ ○ ◐ ◐	○ ◐ ● ◐ ● ○	○ ○ ○		◐ ◐ ○ ● ○ ●	
	● ● ○ ○ ◐ ◐		○ ◐ ● ◐ ● ○	● ● ○ ○ ◐ ◐		
		◐ ● ○	○ ● ○ ◐ ● ◐			
		○ ● ○ ◐ ● ◐		◐ ● ○		
$P_s(E)$	0.857	0.714	0.571	0.429	0.289	0.143

distributions, for indistinguishable particles with integral spins, called *bosons*, there are 6; and for indistinguishable particles with half-integral spins, called *fermions*, there are only two probable distributions. The distribution curves given in Fig. 3.2 do not seem to be much different from each other for three particles, but as the number of particles is increased, marked differences in the three distribution functions occur.

3.4 QUANTUM DISTRIBUTION FUNCTIONS

As the number of particles as well as the number of available states gets large, an ensemble average can be made of the population number n_s of the various

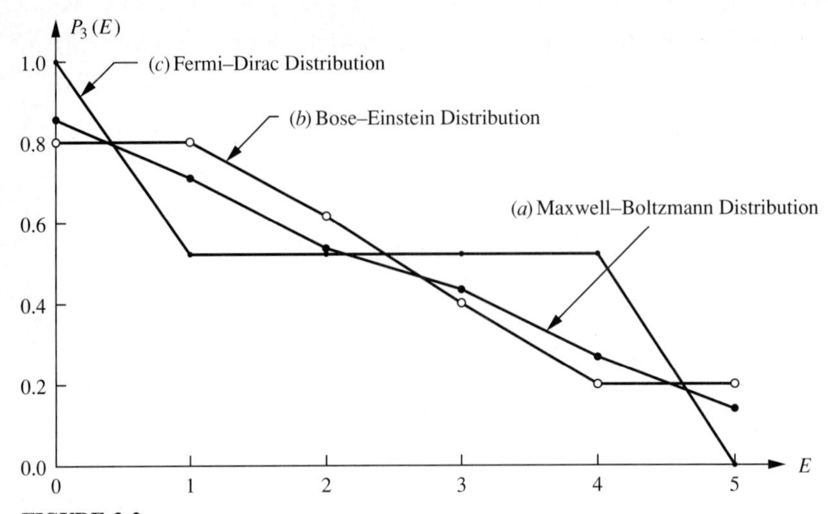

FIGURE 3.2
Energy distribution for three particles distributed to five energy units. (*a*) distinguishable, (Maxwell-Boltzmann) (*b*) indistinguishable, (Bose-Einstein) and (*c*) indistinguishable (Fermi-Dirac) obeying Pauli exclusion principle.

energy states of a free particle in a box obeying the three statistical distributions. Possible distribution of particles over possible energy states are calculated. Since the energy states involved cover the range from zero to infinity, the resulting distribution functions are calculated by making the general assumption that in nature isolated systems occur that minimize the energy of the system. Each of these distribution functions can be represented with an analytical expression.

The usual derivation involves finding $dn(E)/dE$, the number density of particles that occupy the energy interval dE, which lies between E and $E + dE$.

TABLE 3.2
Possible arrangement of three indistinguishable particles with a total energy of five units.

	Energy					
	0	**1**	**2**	**3**	**4**	**5**
	○ ○					○
	○	○			○	
		○ ○		○		
	○		○	○		
		○	○ ○			
$P_s(E)$	0.80	0.80	0.60	0.40	0.20	0.20

TABLE 3.3
Possible arrangement of three indistiguishable particles obeying Pauli Exclusion Principle with a total energy of five units.

	Energy					
	0	1	2	3	4	5
	○		○	○		
	○	○			○	
$P_s(E)$	0.10	0.50	0.50	0.50	0.50	0.00

In general, all three distribution functions have the form

$$\frac{dn(E)}{dE} = \frac{dg_D(E)/dE}{\{\exp[(E/kT) + \nu] + \beta\}} \tag{3.9}$$

Depending on the value of β, the distribution functions are identified as

$\beta = 0$ Maxwell–Boltzmann distribution
$\beta = -1$ Bose–Einstein distribution
$\beta = 1$ Fermi–Dirac distribution

Here $dg_D(E)/dE$ and ν are parameters that are to be determined to give the correct particle density for the system. T is the thermodynamic temperature in degrees K, and k is the Boltzmann constant. Without $dg_D(E)/dE$, the expression given by Eq. 3.9 can be interpreted as the probability that the given system of particles will occupy the energy state E.

Population number density of the various energy states of free particles in a box obeying the three statistics is shown in Fig. 3.3. In Fig. 3.3*b*, the density of the gas represented is many times that represented in Fig. 3.3*a*. Extreme degeneration effects are seen in both the Fermi and the Bose systems. From Eq. 3.9, we see that as the energy under consideration increases, all distribution functions behave similarly to the Maxwell–Boltzmann distribution.

Two of these distribution functions, Fermi–Dirac and Maxwell–Boltzmann, will be considered in some detail.

3.5 THE FERMI–DIRAC DISTRIBUTION FUNCTION

The Fermi–Dirac distribution is applicable to particles obeying the Pauli exclusion principle, such as electrons in a solid. In Eq. 3.9, for Fermi–Dirac distribution $\nu = \exp(-E_F/kT)$ where E_F is a material-dependent parameter,

$$\frac{dn(E)}{dE} = \frac{dg_{FD}(E)}{dE} P_{FD}(E) = \frac{dg_{FD}(E)}{dE} \frac{1}{[1 + \exp\{(E - E_F)/kT\}]} \tag{3.10}$$

$P_{FD}(E)$ is interpreted as the probability that an electron will occupy an energy

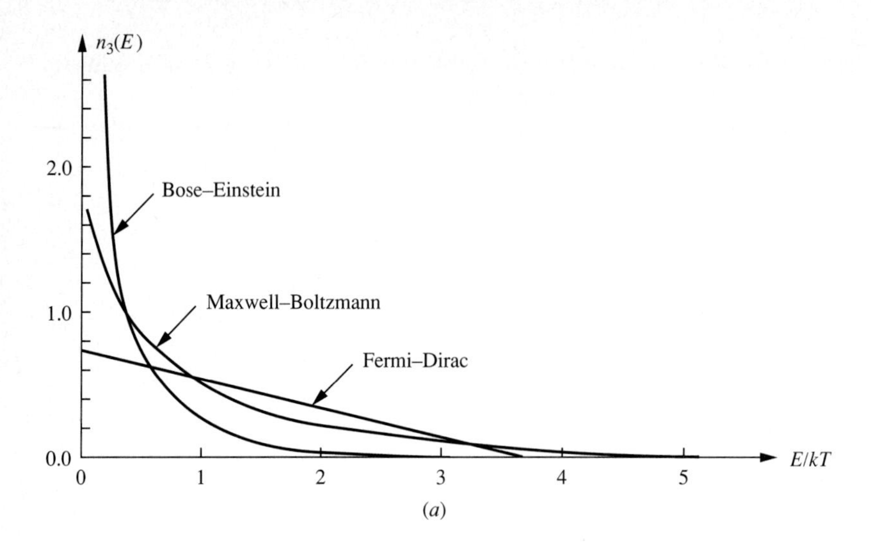

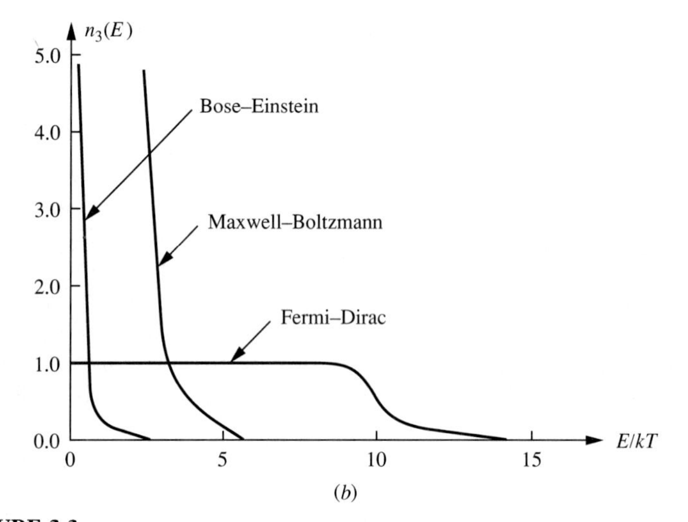

FIGURE 3.3
Distribution functions for various energy states obeying the three statistics. In (*b*) the number density of the gas is many times that in (*a*). (Leighton, 1959).

level E at a given temperature T. General properties of $P_{FD}(E)$ are shown in Fig. 3.4 for various temperatures. These are

1. At $T = 0$

$$
\begin{aligned}
P_{FD}(E) &= 1 \qquad && E < E_F \\
P_{FD}(E) &= 1/2 \qquad && E = E_F \\
P_{FD}(E) &= 0 \qquad && E > E_F
\end{aligned}
$$

2. As T increases, electrons start to occupy energy levels above E_F.

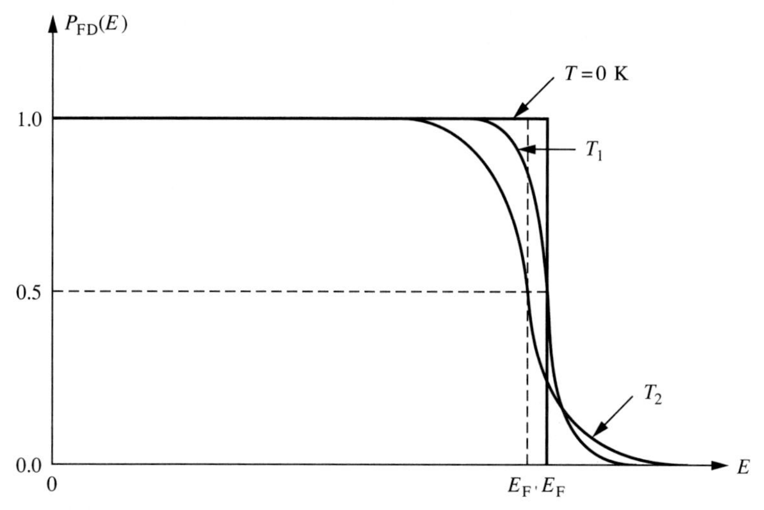

FIGURE 3.4
Fermi–Dirac distribution as a function of energy E. At very low temperatures $P_{FD}(E)$ is equal to unity for $E < E_F$ and equal to zero for $E > E_F$. At higher temperatures $P_{FD}(E)$ drops from unity toward zero less and less sharply and at sufficiently high temperatures and at low densities closely resembles a Maxwell–Boltzmann distribution. ($T < T_1 < T_2$). $E_F^{\prime}$ is the Fermi energy at T_2.

3. As T increases to larger values, E_F decreases, although not much different than its 0 K value. For practical purposes, and for the temperatures that will be treated in this book, E_F will be assumed to be constant.
4. Whatever the temperature is, the area under each curve multiplied by $dg_{FD}(E)/dE$ should be constant for a given particle density.

Using the results of Chapter 2, the density of available energy states between E and $E + dE$ per unit energy interval dE of the electrons in a metal can be written as

$$\frac{dg_{FD}(E)}{dE} = 2\frac{ds(E)}{dE} = \frac{8\sqrt{2}\pi m^{3/2}E^{1/2}}{h^3} \tag{3.11}$$

where the factor 2 is introduced to take care of the Pauli exclusion principle.

Eq. 3.10 can be interpreted in the following way. At a given temperature T, $dn(E)/dE$—the number density of electrons per unit energy interval dE that will occupy an energy level E—will be equal to the availability of electronic states within that interval, multiplied by the probability that an electron can occupy that energy state, that is,

$$\frac{dn(E)}{dE} = \frac{dg_{FD}(E)}{dE}P_{FD}(E)$$

Thus, using Eq. 3.11, Eq. 3.10 reduces to

$$\frac{dn(E)}{dE} = \frac{8\sqrt{2}\pi m^{3/2}}{h^3}E^{1/2}\frac{1}{\exp[(E - E_F)/kT + 1} \tag{3.12}$$

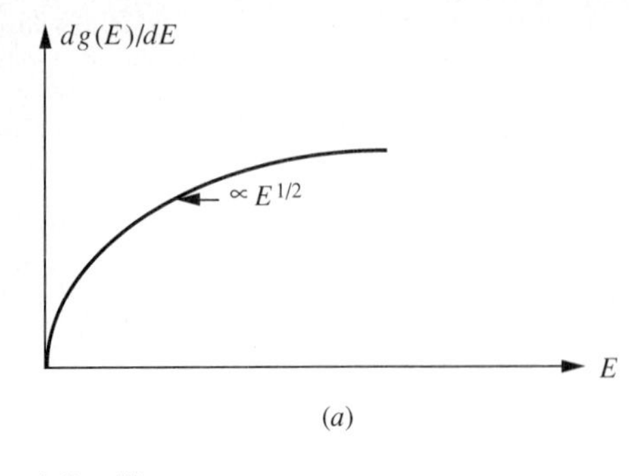

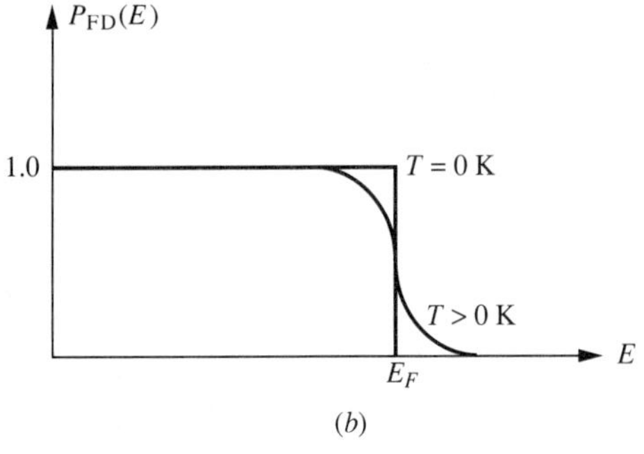

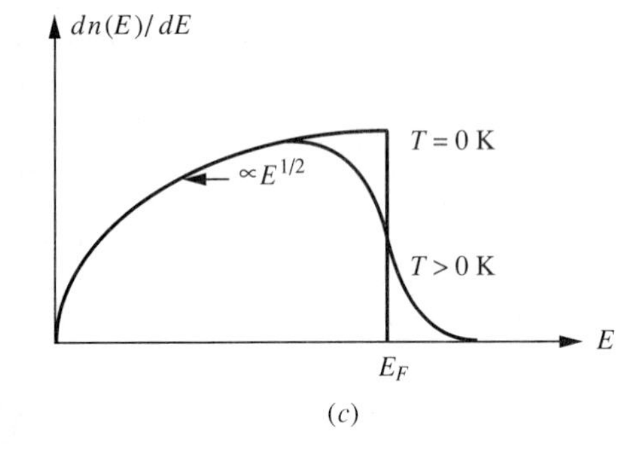

FIGURE 3.5
(*a*) Available density of states per unit energy interval. (*b*) Fermi–Dirac distribution function, the probability that energy level E is occupied. (*c*) Volume density of electrons per unit energy interval between E and $E + dE$.

This is the density of electrons per unit energy interval between E and $E + dE$, shown in graphical form in Fig. 3.5.

In a solid, the density of electrons per unit volume will be equal to the integral of Eq. 3.12 over all energies. Thus

$$n = \int_0^\infty \frac{dn(E)}{dE}\, dE = \int_0^\infty \frac{8\sqrt{2}\pi m^{3/2}}{h^3} \frac{E^{1/2}\, dE}{\{[\exp(E - E_F)/kT] + 1\}} \tag{3.13}$$

Evaluation of this integral is complicated but can be simplified for the case of $T = 0$ K where $P_{FD}(E) = 1$ for $E \leq E_F$ and zero elsewhere. With this approximation, Eq. 3.13 can be written as

$$n = \int_0^{E_F} \frac{8\sqrt{2}\pi m^{3/2}}{h^3} E^{1/2}\, dE$$

$$n = \frac{16}{3}\frac{\sqrt{2}\pi m^{3/2}}{h^3} E_F^{3/2} \tag{3.14}$$

which can be solved for the Fermi energy level E_F

$$E_F = \frac{h^2}{8m}\left(\frac{3n}{\pi}\right)^{2/3} \tag{3.15}$$

where $n = (N/V)$. We see that in a solid which has n electrons per unit volume, the Fermi energy level E_F will be the maximum energy that electrons in that solid can occupy at $T = 0$ K. This result is the same as Eq. 2.31 in Chapter 2.

3.6 SPEED AND VELOCITY DISTRIBUTION FUNCTIONS

The electrons in a solid are in random motion in all directions. In the derivation of some physical phenomena, it may be necessary to use the speed or the velocity distribution of the electrons rather than the energy distribution. We start with the energy distribution of the electrons in a metal already derived in Sec. 3.5. (Eq. 3.12). The number of electrons per cubic meter with energy between E and $E + dE$ is

$$\left(\frac{dn(E)}{dE}\right) dE = \frac{8\sqrt{2}\pi m^{3/2} E^{1/2}}{h^3} \frac{dE}{\{1 + \exp[(E - E_F)/kT]\}} \tag{3.16}$$

It is convenient to convert $(dn(E)/dE)dE$ to a speed distribution first. This is done by expressing the electron energy in terms of its speed s using

$$E = \frac{1}{2}ms^2 \tag{3.17}$$

Substituting Eq. 3.17 into Eq. 3.16, $(dn(E)/dE)dE$ becomes $(dn(s)/ds)ds$

$$\left(\frac{dn(s)}{ds}\right) ds = \frac{8\pi m^3}{h^3} \frac{s^2\, ds}{\{\exp[(ms^2/2) - E_F]/kT + 1\}} \tag{3.18}$$

where the meaning of $dn(s)/ds$ is clearly the number of electrons per cubic meter having speeds between s and $s + ds$. Eq. 3.18 can now be converted to a velocity distribution. The resulting distribution will be the number of electrons per cubic meter with velocity components of the velocity vector between v_x and $v_x + dv_x$, v_y and $v_y + dv_y$, and v_z and $v_z + dv_z$, where

$$s^2 = v_x^2 + v_y^2 + v_z^2 \tag{3.19}$$

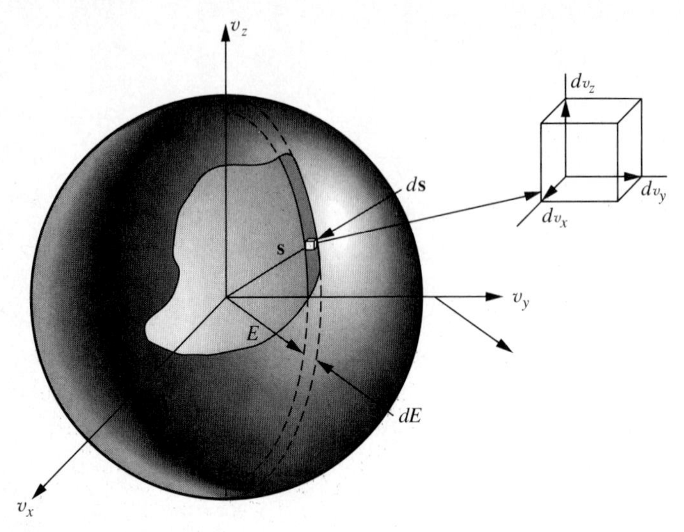

FIGURE 3.6
Relation between the velocity distribution and the speed distribution.

and v_x, v_y, and v_z are the respective components of the velocity vector. The velocity distribution takes into account the direction of each electron motion in addition to its energy or speed. Referring to Fig. 3.6, the speed distribution in Eq. 3.18 refers to all the electrons in the spherical shell bounded by s and $s + ds$. The velocity distribution, on the other hand, is concerned with the electrons contained in the volume element $dv_x\,dv_y\,dv_z$. Therefore, the velocity distribution is simply the ratio of these volumes times the speed distribution, or

$$\left(\frac{dn(\mathrm{v})}{d\mathrm{v}}\right) d\mathrm{v} = \left(\frac{dn(s)}{ds}\right) ds \frac{dv_x\,dv_y\,dv_z}{4\pi s^2\,ds}$$

$$\left(\frac{dn(\mathrm{v})}{d\mathrm{v}}\right) d\mathrm{v} = \frac{2m^3\,dv_x\,dv_y\,dv_z}{h^3\{\exp[(E - E_F)/kT] + 1\}} \tag{3.20}$$

where E in Eq. 3.20 is given by

$$E = \frac{1}{2}m(v_x^2 + v_y^2 + v_z^2) \tag{3.21}$$

3.7 QUANTUM MECHANICAL MAXWELL–BOLTZMANN DISTRIBUTION FUNCTION

From Eq. 3.9, the Maxwell–Boltzmann distribution function for an ideal gas can be written as

$$\frac{dn(E)}{dE} = \frac{dg_{MB}(E)}{dE}P_{MB}(E) = \frac{dg_{MB}(E)}{dE}e^{-(E/kT)} \tag{3.22}$$

Here, the constant $\exp(-\nu)$ is included in $dg_{MB}(E)/dE$. If Eq. 3.22 is interpreted as the number density of particles per unit energy interval that fall between E and $E + dE$, then $dg_{MB}(E)/dE$ becomes the density of available states per energy interval dE at the energy E and $\exp(-E/kT)$ is the probability that an energy state E will be occupied by the particles. Using Eq. 2.27 of Chapter 2 for the density of available states, Eq. 3.22 can be written as

$$\frac{dn(E)}{dE} = \frac{4\sqrt{2}\pi m^{*3/2}}{h^3}E^{1/2}e^{-(E/kT)} \tag{3.23}$$

With the exception of the multiplying constant, this has the same form as the Maxwell–Boltzmann energy distribution for an ideal gas in classical thermodynamics, which is written as $f_{MB}(E)$ and is given by

$$f_{MB}(E) = \frac{2n}{\pi^{1/2}(kT)^{3/2}}E^{1/2}e^{-E/kT} \tag{3.24}$$

3.8 CLASSICAL MAXWELL–BOLTZMANN DISTRIBUTION FUNCTION

The classical Maxwell–Boltzmann distribution is calculated by using the momentum of the particles as the coordinates of the parameter space. This leads to a momentum distribution function given by

$$f_{MB}(\mathbf{p}) = Ke^{-(p_x^2+p_y^2+p_z^2)/2mkT} \tag{3.25}$$

This is the number of particles per unit momentum interval between p_x and $p_x + dp_x$, p_y and $p_y + dp_y$, and p_z and $p_z + dp_z$. Multiplying Eq. 3.25 with the unit volume $d^3p = dp_x\,dp_y\,dp_z$ in momentum space and integrating over all possible particle momenta should give the number density n of the particles

$$n = \int_{-\infty}^{\infty}\int_{-\infty}^{\infty}\int_{-\infty}^{\infty} Ke^{-(p_x^2+p_y^2+p_z^2)/2mkT}\,dp_x\,dp_y\,dp_z \tag{3.26}$$

Integrating Eq. 3.26 and solving for the constant K gives

$$f_{MB}(\mathbf{p}) = \frac{n}{(2\pi mkT)^{3/2}}e^{-p^2/2mkT} \tag{3.27}$$

where $p^2 = p_x^2 + p_y^2 + p_z^2$. A similar expression can also be written for a velocity space with $p^2/2m$ replaced by $mv^2/2$ in Eq. 3.25.

To convert Eq. 3.27 into an energy distribution for classical particles, we note that energy is a scalar quantity and momentum is a vector quantity. Momentum space is a three-dimensional space, and any particles, irrespective of direction, which have a momentum of magnitude $|\mathbf{p}|$ will have the same energy. We

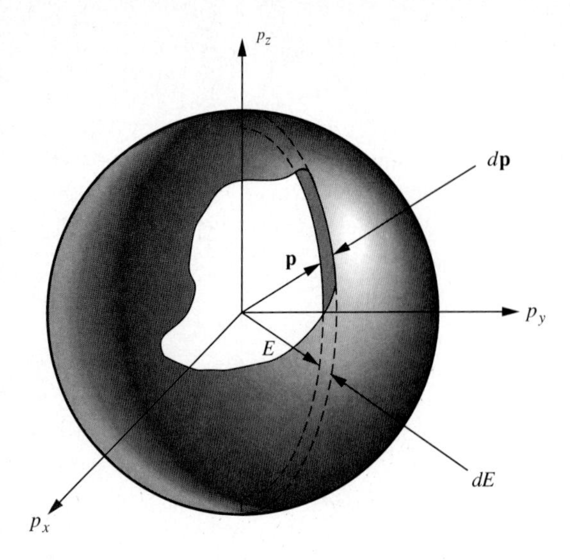

FIGURE 3.7
Momentum space for classical particles and the energy interval dE corresponding to the momentum interval $d\mathbf{p}$.

now consider a spherical shell of thickness $d\mathbf{p}$ between $\mathbf{p}$ and $\mathbf{p} + d\mathbf{p}$, as shown in Fig. 3.7. The particles in this shell will be equivalent to particles having energies between E and $E + dE$. The volume of the spherical shell can be written as

$$dp_x\,dp_y\,dp_z = 4\pi p^2\,dp \tag{3.28}$$

Using $p^2/2m = E$ and $dp = (m/\sqrt{2mE})\,dE$, Eq. 3.25 can now be written in the equivalent form

$$f_{MB}(E) = K'e^{-E/kT}E^{1/2} \tag{3.29}$$

where other multiplying constants are included in K'.

Integrating this equation from 0 to infinity, we can now write down the equivalent Maxwell–Boltzmann energy distribution function as

$$f_{\text{MB}}(E) = \frac{2n}{\pi^{1/2}(kT)^{3/2}}E^{1/2}e^{-E/kT} \tag{3.30}$$

Example 3.1 Calculate the average energy $< E >$ of a system of classical particles.

Solution. From Eq. 3.3, the average energy is

$$< E >= \frac{\int_0^\infty E f_{\text{MB}}(E)\,dE}{\int_0^\infty f_{\text{MB}}(E)\,dE} \tag{3.31}$$

Substituting $f_{\text{MB}}(E)$ into Eq. 3.31

$$< E >= \frac{\int_0^\infty \frac{2n}{\left[\sqrt{\pi}(kT)^{3/2}\right]}E^{3/2}e^{-E/kT}\,dE}{\int_0^\infty \frac{2n}{\left[\sqrt{\pi}(kT)^{3/2}\right]}E^{1/2}e^{-E/kT}\,dE} \tag{3.32}$$

The denominator is simply equal to n; Evaluation of the integral in the numerator gives

$$< E > = \frac{3}{2}kT \tag{3.33}$$

This is the average energy of a particle. Note that the particle has three degrees of freedom. Therefore, the average energy of the particle per degree of freedom would be $kT/2$. If the particle possesses potential energy as well as kinetic energy, Eq. 3.33 should be multiplied by two.

FURTHER READING

Hemenway, C. L., R. W. Henry, and M. Coulton, *Physical Electronics*, 2d ed., Wiley, New York, 1967. The introductory material presented in Section 3.3 follows the excellent treatment of the distribution functions given in Chapter 3 of this reference.

Leighton, R. E. *Principles of Modern Physics*, McGraw-Hill, New York, 1959.

Uman, M. F., *Introduction to Physics of Electronics*, Prentice Hall, Englewood Cliffs, N. J., 1974.

van der Ziel, Aldert, *Solid State Physical Electronics*, 3d ed., Prentice Hall, Englewood Cliffs, N. J., 1976.

PROBLEMS

3.1 Using four particles and a total energy of 10 units, distribute these particles among possible energy units if the particles are

(a) distinguishable,

(b) indistinguishable

(c) indistinguishable and obeying the Pauli Exclusion principle

Plot the corresponding particle energy distributions.

3.2 Show that all the distribution functions have similar exponential energy distributions at very high energies (i.e., $E >> kT$)

3.3 Using $ms^2/2 = E$, show that the Maxwell–Boltzmann particle speed distribution can be written as

$$f_{\mathrm{MB}}(s)\, ds = n\left(\frac{m}{2\pi kT}\right)^{3/2} e^{-ms^2/2kT} s^2\, ds \tag{P.3.1}$$

3.4 Find the average velocity $< v >$ of particles obeying Maxwell–Boltzmann distribution.

3.5 For particles obeying Maxwell–Boltzmann distribution, find

(a) the average speed,

(b) root mean square speed, $< s_{\mathrm{rms}} > = < s^2 >^{1/2}$

(c) the most probable speed

3.6 Find the fraction of particles of a Maxwell–Boltzmann distribution that have x components of velocity greater than $(2kT/m)^{1/2}$.

3.7 According to Planck's hypothesis, the energy exchange between oscillators occurs in discrete multiples of $e = h\nu$, that is, as $e, 2e, 3e, 4e, \ldots$. Show that the average energy of the harmonic oscillators can be written as

$$< E > = \frac{e}{e^{e/kT} - 1}$$

Hint: In Eq. 3.31 use the relations $E = ne$ and the following summations

$$1 + 2x + 3x^2 + \cdots = \frac{1}{(1 - x)^2}$$

$$1 + x + x^2 + x^3 + \cdots = \frac{1}{1 - x}$$

CHAPTER 4

INTERACTING PARTICLES CONCEPTS

The previous chapters dealt with topics mostly related to the behavior of free electrons or electrons bound inside a solid. Statistical concepts were introduced to make a transition from the microscopic to the macroscopic representation of large numbers of particles. So far, interactions among the particles that constitute the solid have not been considered. In this chapter, we will discuss the basic concepts related to the interaction among various particles.

Under the application of external fields, charged particles are forced to move from one electrode to the other in any form of matter, and they usually perform this motion in the presence of atoms that make up the matter and/or other charged particles. Therefore, the particle paths are not only governed by the applied electric and magnetic fields but altered by the presence of other particles in that state of matter. There are four states of matter: solids, liquids, gases, and plasmas (fully ionized gases). Some of these states may appear simultaneously at a given temperature T, but the studies of these individual states are topics in thermodynamic analysis.

The forces exerted by the applied fields effectively move the charged particles in the direction of the resultant force, but the actual paths of these particles are far from the trajectories of the particles that would result when the particles were singly moving in vacuum. If a charged particle moves in a medium where the particles that it interacts with are small in number, it is possible that the charged particle may reach its destination without any alteration of its single-particle path. Under these conditions, the particles are treated as single particles, and their overall effects are obtained simply by multiplying the single-particle effects by the total number of the particles. Actually in almost all devices, interactions among various particles always occur but these may be small enough to be neglected for practical purposes.

4.1 COLLISION CROSS SECTION

Assume that a particle with radius R_2 is at rest, and a second particle with radius R_1 is moving toward the first particle (see Fig. 4.1). Actually, both particles can move in arbitrary directions, but we can always find a frame of reference where only one of the particles moves relative to the other one. The choice of this frame of reference is not necessary in the actual analysis but simplifies the introduction of the concepts related to the collision cross section.

If particle 1 moves with a velocity v_1 toward particle 2, a collision will take place, provided that the two particles make contact with each other. This may be a head-on collision when the velocity direction of the incoming particle coincides with the line joining the centers of the particles or it may be a grazing collision when the incoming particle just touches the outer surface of the second particle. As long as the center of the first particle falls within the area of a circle of radius $(R_1 + R_2)$, there will be a definite collision between the two particles. As shown in Fig. 4.1*a*, the area of this circle is $\pi(R_1 + R_2)^2$, and we define this equivalent

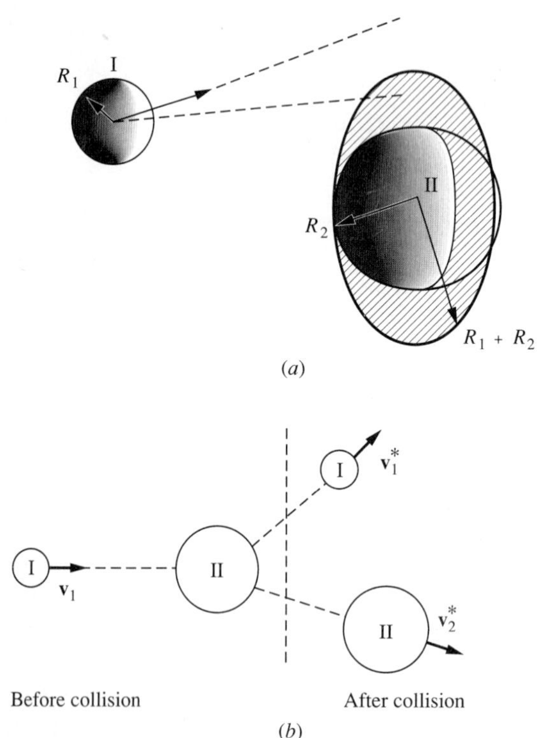

FIGURE 4.1
(*a*) Collision of two particles of radii R_1 and R_2. (b) Scattering of the two particles after the collision.

area as the *collision cross section* σ for this two-particle interaction. If the center of particle 1 does not pass through this area, there will be no collision between the particles, and particle 1 will continue to move without any change in its direction of motion. This type of collision is identical to the collision of two billiard balls. If a collision takes place, the particles will be scattered, as shown in Fig. 4.1*b*.

Actually, particles that determine the electrical characteristics of electronic devices are charged particles, and we know that the interaction potentials among these particles vary from the long-range Coulomb potentials to the shorter-range van der Waals potentials. In many interaction processes, the particles do not have to come into physical contact with each other in order for scattering to take place. Figure 4.2 shows the trajectory of an incoming electron in the presence of a stationary positive ion. Because the mass of the ion is very large compared to that of the electron so that it can be considered stationary, and due to the long range of the Coulomb force, only the electron path will be altered and the electron will be scattered Ω degrees from its original path of travel.

In quantum mechanics, the scattering cross section σ is given by

$$\sigma = \int_\Omega \left(\frac{d\sigma}{d\Omega} \right) d\Omega$$

where $d\sigma/d\Omega$ is the differential scattering cross section, which is defined as the scattering cross section per unit solid angle $d\Omega$ and is related to the scattering amplitude $f(\theta)$ by

$$\frac{d\sigma}{d\Omega} = |f(\theta)|^2$$

where $f(\theta)$ is proportional to

$$f(\theta) \propto \langle f|\mathbf{H}'|i \rangle$$

We can see from this equation that the scattering cross section depends on

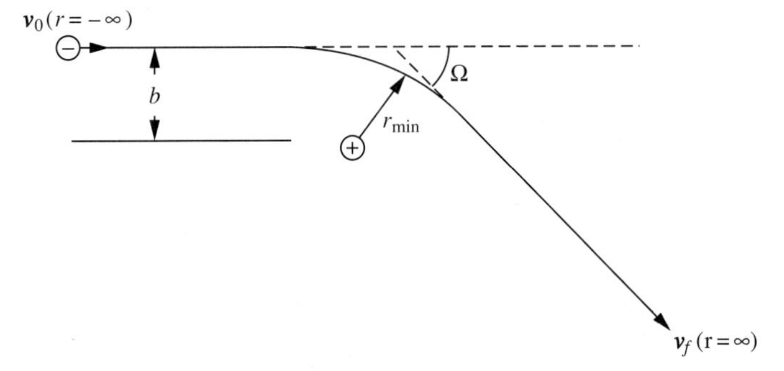

FIGURE 4.2
Trajectory of an electron in the presence of an ion. Ω is the scattered angle of the electron, b is the impact parameter, and r_{min} is the minimum distance of approach.

the final $< f|$ and initial $|i >$ states of the system before and after the interaction, and to the interaction Hamiltonian $\mathbf{H}'$ between the interacting particles. For the Coulomb scattering shown in Fig. 2.2, the initial and final states will be the electronic wave functions at $-\infty$ and $+\infty$ and the interaction Hamiltonian will be related to the Coulomb potential between the electron and the ion.

Eventually, for every interaction process, an equivalent interaction cross section can be defined. These are usually derived from detailed quantum mechanical calculations of the interaction process. The magnitude of this cross section, which may depend on many parameters, such as energy, momentum, temperature, and so on, of the particles, gives an indication to the probability that such an interaction will take place. Examples of some of these interactions are elastic collisions of different particles, such as the electron-molecule, electron-ion, and so on, and inelastic collisions, such as electron-capture, particle-fusion, and so on.

4.2 MEAN FREE PATH AND COLLISION FREQUENCY

When a charged particle (primary particle) moves in a material medium, it makes collisions with other particles in that medium. The other particles can be the same charged particles, or different particles with opposite charge and/or with different masses, or molecules, or bound electrons of atoms, and so on. The collision process will be different for each pair of these particles.

To simplify the explanation of the resulting parameters in a collisional process, an isolated charged particle will be assumed to move in a medium filled with neutral atoms and with no external fields present, as shown in Fig. 4.3. Let us follow the path of the charged particle initially located at A. After the collision with particle 1, it travels a distance d_1 before it collides with particle 2. It is then scattered in an arbitrary direction and travels a distance d_2 before it collides with particle 3, and so on. The direction of the particle after a collision is always arbitrary. If one measures the distances the charged particle travels between each collision, these distances will vary over a wide range due to the randomness of the scattering process. But as shown in Chapter 3, an averaging process can be carried

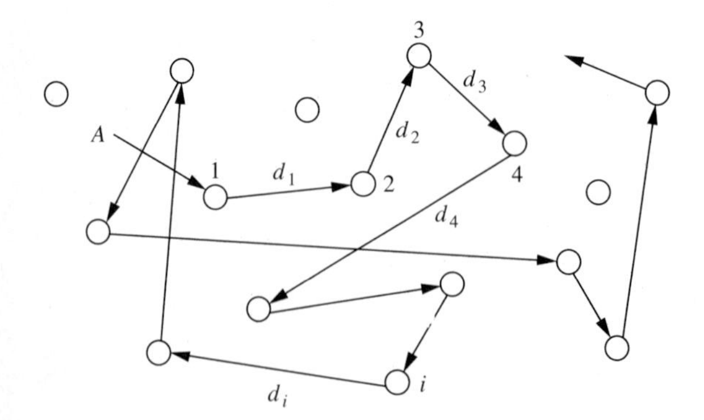

FIGURE 4.3
Collision of a particle A with other particles in a given material. The distance traversed after the i'th collision is d_i.

out for these distances provided the appropriate particle distribution functions are known. We define this average distance as the *mean free path* λ_m for that particle and identify it as the average distance that a charged particle will travel before it makes a collision with a neutral atom. When the magnitude of the mean free path λ_m is compared with the average size L of its container, the relative magnitudes of λ_m and L give an indication of whether, on the average, many or few or even no collisions are experienced by the particle in moving from one boundary of the container to the other. If the mean free path is large relative to L, the electron motion can be treated on an individual basis, and the collisional effects can be ignored. The resulting motion is referred to as the *ballistic motion* of the electron.

A finite time passes between each collision of the particles. In the example of Fig. 4.3, it takes t_1 seconds to travel d_1, t_2 seconds to travel d_2, and so on. Depending on these distances and the velocity after each collision, these time intervals will also vary over a wide range. As with the averaging process done for the definition of the mean free path, we can define an average collision time for the interaction process. The *mean collision time* τ_c is therefore defined as the average time the particle travels before it makes a collision. *Collision frequency* ν_c is defined as the reciprocal of the mean collision time

$$\nu_c = \frac{1}{\tau_c} \tag{4.1}$$

and represents the average number of collisions that take place per second in that collisional process.

4.3 DRIFT VELOCITY AND MOBILITY

When an external field is applied to a medium containing charges, the paths of these particles are altered in accordance with the force relation between the charged particles and the externally applied field. Figure 4.4 shows the motion of an electron with and without an externally applied electric field. When there is no field, the particle motions are always random and the average vector displacement of the particle from its initial position after many collisions averages to zero. On the other hand, although the initial direction of the particle after each collision is still arbitrary in the presence of an external electric field, the particle is under the influence of the electric field and will acquire an additional velocity in the direction (opposite if the particle is an electron) of the electric field due to the constant presence of this field. The force effectively displaces the particle from its initial position a distance $\Delta\mathbf{d}$ within the time interval Δt.

We now concentrate on the effective displacement of the particle from its initial position within the time interval Δt due to the external force. We can define an equivalent velocity, known as the *drift velocity*, for the particle, as

$$\mathbf{v}_d = \lim_{\Delta t \to 0} \frac{\Delta\mathbf{d}}{\Delta t} \tag{4.2}$$

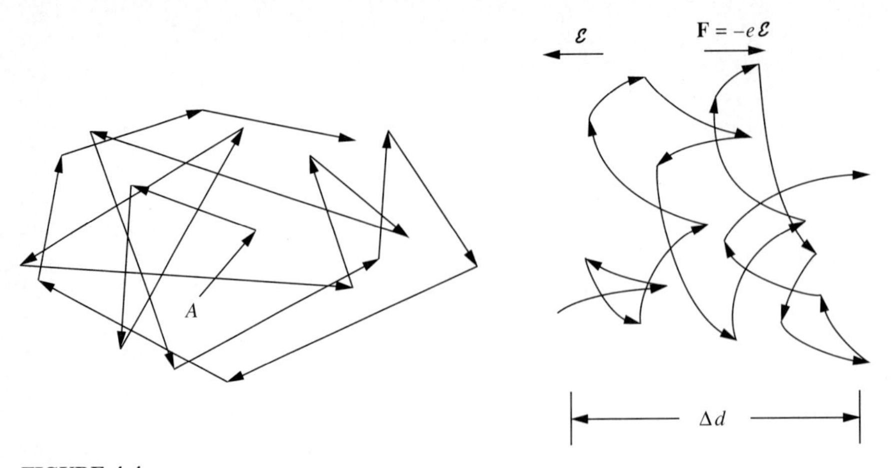

FIGURE 4.4
(*a*) Charged-particle motion with no applied field. (*b*) Charged-particle (electron) motion under an externally applied electric field produces a displacement Δd within the time interval Δt.

We also expect that the magnitude and direction of the drift velocity will be directly proportional to the externally applied electric field $\mathcal{E}$. Within the same time interval, the effective distance $\mathbf{\Delta d}$ traveled by the particle will be large for larger electric fields. We can, therefore, consider a linear relation between the drift velocity and the electric field and write

$$v_d = \mu \mathcal{E} \tag{4.3}$$

where μ is the proportionality constant and is called the *mobility* for the charged carriers. Once we calculate or measure the mobility of a charged carrier inside a medium, we can ignore the details of the collisional process and simply study the overall motion of the charged particle in terms of its drift velocity.

We can arrive at a relation between the mobility of the charged particle and the related collisional process that controls the magnitude of this mobility. Consider a charged particle q that is acted upon by an electric field $\mathcal{E}$. Assume that between each collision, the particle acquires a velocity

$$v_f = v_0 + \frac{q \mathcal{E}}{m^*}\tau_c \tag{4.4}$$

where m^* is the effective mass of the charged particle and $\mathbf{v}_0$ is the initial random velocity of the particle after the collision. When we average out the final velocity of the particle after many collisions, we get

$$< v_f > = \frac{q \mathcal{E}}{m^*}\tau_c \tag{4.5}$$

since $< v_0 > = 0$ due to randomness of the initial velocity. The drift velocity is then one half of this final velocity* or

* $\mathbf{v}_d = (< \mathbf{v}_f > + < \mathbf{v}_0 >)/2$ as the average velocity of the particle under constant acceleration.

$$v_f = \left(\frac{q\tau_c}{2m^*}\right)\mathcal{E} \tag{4.6}$$

Comparing Eq. 4.3 with Eq. 4.6, we find that the mobility is

$$\mu = \frac{q\tau_c}{2m^*} \tag{4.7}$$

Although this is an idealization of the collisional process, dependence of mobility on τ_c has been experimentally verified. If it takes a longer time between each collision, the motion of the particles will not be interrupted as often, and the particles will move faster in the direction of the field due to the continuous presence of the applied field. If τ_c is short, the collisions will alter the paths of the electrons at a much faster rate, and the electrons will not have enough time to acquire velocity in the direction of the field. Therefore, as expected, the mobility will be directly proportional to τ_c.

4.4 PARTICLE FLUX AND CURRENT DENSITY

We now consider an assembly of particles made up of individual particles obeying the relation given by Eq. 4.3—that is, all the particles have the same drift velocity v.*

Assume that the particles of the same species with a volume density n move with a velocity v as shown in Fig. 4.5. We would like to calculate the number of

*From now on, we will use exclusively the drift velocity in our calculations of the transport properties of the charged particles. We drop the subscript d in $\mathbf{v}_d$ in subsequent calculations.

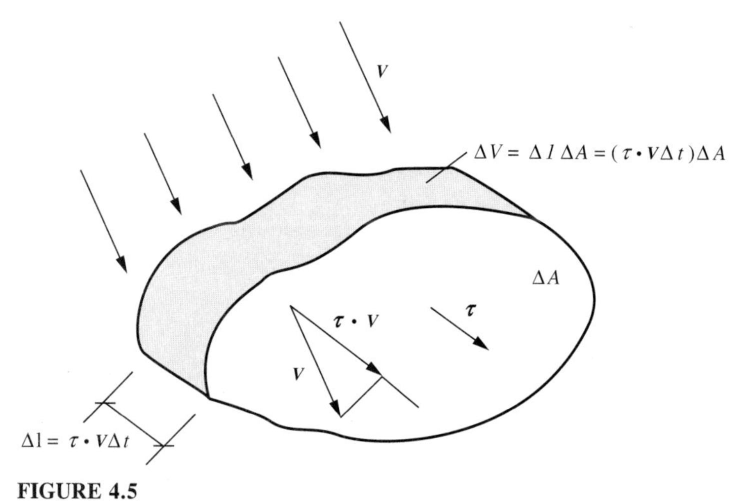

FIGURE 4.5
Flux of particles crossing ΔA in a time terminal Δt. $\boldsymbol{\tau}$ is the unit vector normal to the surface element ΔA.

particles that cross the area ΔA within the time interval Δt. Those particles that fall within the volume element

$$\Delta V = \Delta l \Delta A = (\boldsymbol{\tau} \cdot \mathbf{v} \Delta t) \Delta A$$

will cross the surface ΔA in a time interval Δt. Here $\boldsymbol{\tau}$ is the unit vector perpendicular to the surface element ΔA. Denoting this flow rate by $\Delta N / \Delta t$, we can write

$$\frac{\Delta N}{\Delta t} = \frac{n(\boldsymbol{\tau} \cdot \mathbf{v} \Delta t) \Delta A}{\Delta t} \tag{4.8}$$

where ΔN is the total number of particles that cross ΔA within Δt. In the limit of small time interval Δt, Eq. 4.8 becomes

$$\begin{aligned} \frac{dN}{dt} &= n\mathbf{v} \cdot d\mathbf{A} \\ &= \boldsymbol{\Gamma} \cdot d\mathbf{A} \qquad (\text{where } \boldsymbol{\Gamma} = n\mathbf{v}) \end{aligned} \tag{4.9}$$

where $d\mathbf{A} = \boldsymbol{\tau}\, dA$ and $\boldsymbol{\Gamma}$ is the *particle flux* which is equal to the number of particles per unit area per second moving in the direction of particle motion.

If the particles are charged, we can define an equivalent current density for the moving charges by multiplying $\boldsymbol{\Gamma}$ by the charge q of the particles, that is,

$$\begin{aligned} \mathbf{J} &= q\boldsymbol{\Gamma} = qn\mathbf{v} \\ \mathbf{J} &= \rho \mathbf{v} \end{aligned} \tag{4.10}$$

where ρ is the charge density and has the units of C/m^3. Depending on the sign of q, charge density can be negative or positive (i.e., $q = -e$ for electrons and $q = e$ for holes). The current density $\mathbf{J}$ is equal to the current flow per unit area (A/m^2).

4.5 CONTINUITY EQUATION

Consider a volume V surrounded by a closed surface A, as shown in Fig. 4.6. It is possible to relate the flux of particles entering and leaving the volume to the time rate of change (increase or decrease) of the particle density inside the volume.

The change in the total particles per unit time inside the volume V is

$$\frac{\Delta N}{\Delta t} = \int_V \frac{n(t + \Delta t) - n(t)}{\Delta t}\, dV \tag{4.11}$$

A similar expression can be obtained from the flux of particles entering and leaving the volume element. Using Eq. 4.9, the change in the total particles per unit time inside the volume can be written in terms of the net particles flowing in and out of surface A. This gives

$$\frac{\Delta N}{\Delta t} = -\oint \boldsymbol{\Gamma} \cdot d\mathbf{A} \tag{4.12}$$

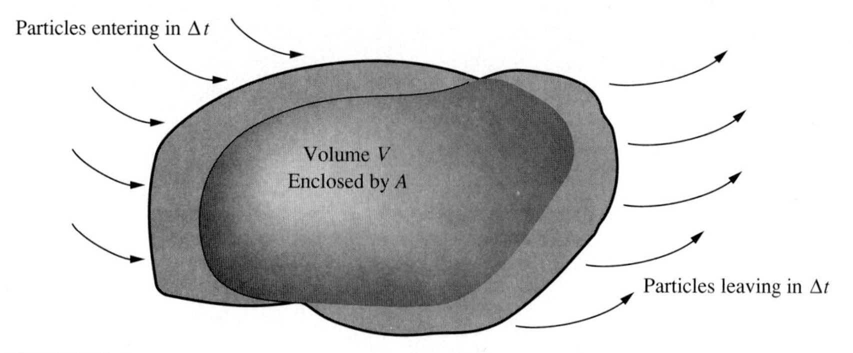

FIGURE 4.6
Particle flow in and out of the volume V enclosed by the surface A.

The negative sign is introduced here to take care of the decrease in the total particle flux if the charge density within the volume V increases.

Equating Eqs. 4.11 and 4.12, and taking the limit $\Delta t \to 0$, we obtain

$$\int_V \left(\frac{\partial n}{\partial t}\right) dV = -\oint_A \mathbf{\Gamma} \cdot d\mathbf{A}$$

Using the divergence theorem, the right-hand side of this equation is changed to a volume integral, and the two terms are collected on the same side.

$$\int_V \left(\frac{\partial n}{\partial t} + \nabla \cdot \mathbf{\Gamma}\right) dV = 0$$

Since V is arbitrary, the integrand should be zero, which gives

$$\frac{\partial n}{\partial t} + \nabla \cdot \mathbf{\Gamma} = 0 \tag{4.13}$$

Equation 4.13 is known as the *continuity equation*. If the particles are charged, we can multiply each term of the equation by q and write the continuity equation for charged particles as

$$\frac{\partial \rho}{\partial t} + \nabla \cdot \mathbf{J} = 0 \tag{4.14}$$

In deriving Eq. 4.14, it was assumed that the increase in charged-particle density is due only to the change in the flux of particles entering and leaving the closed surface A. In many problems, the charged-particle density inside the volume can increase or decrease due to nonlinear processes known as charge creation or charge recombination. These two processes can occur simultaneously for a given system. Denoting the charge density generation and the recombination rates by g and r, respectively, the general continuity equation can now be written as

$$\frac{\partial \rho}{\partial t} + \nabla \cdot \mathbf{J} = g - r \tag{4.15}$$

The corresponding charge creation or charge removal is equivalent to a source or a sink for the change of charge carrier density at a point in space. Charges are usually generated in the volume V by external means. If the source of excitation is maintained, then due to the competing recombination process a steady-state condition may be reached. In all instances, charge creation and charge recombination compete with each other and depending on the magnitudes of the respective rate mechanism, one may dominate over the other. These concepts are very useful in analyzing the transport behavior of charged carriers and will be dealt with in greater detail in the chapters on semiconductor devices.

If a constant external source, such as optical illumination, is responsible for generating electron-ion pairs in a gaseous system or electron-hole pairs in a semiconductor, the generation rate may also be assumed to be constant under this excitation process. On the other hand, due to the nature of the processes involved, the charged particles will recombine with each other to reach the lowest energy states available to them. The recombination rate of these charged particles will depend on the number density of the particles available for recombination. The recombination rate can therefore be written as

$$r = Rn_-n_+ \tag{4.15a}$$

where R is the recombination constant and depends on the detailed nature of the recombination process between the charged particles. In Eq. 4.15a, n_- and n_+ are the respective concentrations of negative and positive charged particles. (R is related to α_i (Eq. 2.32) by $R = q\alpha_i$.)

4.6 ELECTRICAL CONDUCTIVITY

Whenever transportable charge carriers are present in a conductive medium, the application of an external electric field $\mathcal{E}$ induces a drift velocity for the charged carriers, which in turn produce an electrical current in the medium. The relation between the current density $\mathbf{J}$ and the electric field $\mathcal{E}$ is given by *Ohm's Law*

$$\mathbf{J} = \sigma \mathcal{E} \tag{4.16}$$

where σ is defined as the *conductivity* of the medium and is only a function of the physical parameters of the medium. In some instances, for example, for very high fields, σ can be a nonlinear function of the applied electric field. But for almost all the electronic devices that we will consider in this book, a linear dependence between $\mathbf{J}$ and $\mathcal{E}$ as given by Eq. 4.16 will be valid. The form of Ohm's Law given by Eq. 4.16 is more general than the well-known circuit equation $v = iR$ since R depends not only on the physical parameters, but also the geometrical shape of the conducting medium. Current density given by Eq. 4.16 is known as the *conduction current density*.

Using Eqs. 4.10 and 4.3, we can relate the conductivity to the mobility of the charged carriers by

$$\mathbf{J} = qnv = qn\mu \mathcal{E} \tag{4.17}$$

Therefore

$$\sigma = qn\mu \tag{4.18}$$

If there is more than one species of charged particles, the conductivity is the sum of the contributions of all the species, and it can be written as

$$\sigma = \sum_i q_i n_i \mu_i \tag{4.19}$$

Note that the current defined by Eq. 4.17 gives the proper direction of the conventional current. It is opposite to the electron flow.

Example 4.1 Find the proper directions, signs for the currents, drift velocities, and mobilities produced by the motion of both negatively and positively charged particles moving in the same medium under the application of a constant external electric field $\mathcal{E}$.

Solution. Consider a region of a medium where due to the application of a potential difference v, an electric field $\mathcal{E}$ is produced, as shown in Fig. E.4.1. The conventional electrical current flows from the positive toward the negative terminal.

From Fig. E.4.1, $\mathcal{E} = \mathcal{E}_0(-\mathbf{a}_x)$, where $\mathcal{E}_0 = v/L$ and the corresponding velocities are given by

$$\begin{aligned} \mathrm{V}_- &= \mathrm{V}_-\mathbf{a}_x && \text{(electron)} \\ \mathrm{V}_+ &= \mathrm{V}_+(-\mathbf{a}_x) && \text{(holes)} \end{aligned} \tag{E.4.1}$$

where V_- and V_+ are the magnitudes of the respective velocity vectors. By definition, the current density is given by Eq. 4.10. Using $q = -e$ for an electron and $q = +e$ for a singly charged positive charge, respective current densities are

$$\begin{aligned} \mathbf{J}_- &= -en_-\mathrm{V}_- = -en_-(\mathrm{V}_-\mathbf{a}_x) && \text{(E.4.2}a\text{)} \\ &= en_-\mathrm{V}_-(-\mathbf{a}_x) && \text{(electrons)} \\ \mathbf{J}_+ &= en_+\mathrm{V}_+ = en_+(-\mathrm{V}_+\mathbf{a}_x) && \text{(E.4.2}b\text{)} \\ &= en_+\mathrm{V}_+(-\mathbf{a}_x) && \text{(positive charge)} \end{aligned}$$

We see that both current densities are in the same direction, and the total current density is

$$\begin{aligned} \mathbf{J} &= \mathbf{J}_- + \mathbf{J}_+ \\ \mathbf{J} &= (en_-\mathrm{V}_- + en_+\mathrm{V}_+)(-\mathbf{a}_x) \end{aligned} \tag{E.4.3}$$

$\mathbf{J}$, given by Eq. E.4.3, has the right sign of the conventional electrical current. The drift velocity is related to the electric field by

$$\mathrm{V} = \mu\,\mathcal{E} \tag{E.4.4}$$

From Fig. E.4.1, $\mathcal{E} = \mathcal{E}_0(-\mathbf{a}_x)$ and from the corresponding velocities, the sign of mobilities is

$$\begin{aligned} \mathrm{V}_-\mathbf{a}_x &= \mu_-E_0(-\mathbf{a}_x) \qquad \mu_- = -\mu_{\mathrm{n}} \qquad \text{(electron)} && \text{(E.4.5a)} \\ \mathrm{V}_+(-\mathbf{a}_x) &= \mu_+E_0(-\mathbf{a}_x) \qquad \mu_+ = +\mu_{\mathrm{p}} \qquad \text{(positive charge)} && \text{(E.4.5b)} \end{aligned}$$

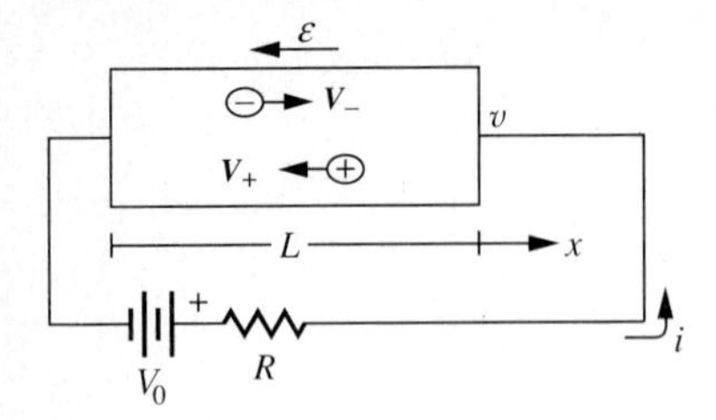

FIGURE E.4.1
Motion of negative and positive charges under the application of an ℰ field.

Similarly, the conductivity due to both charges is

$$\begin{aligned}\sigma &= q_{-}n_{-}\mu_{-} + q_{+}n_{+}\mu_{+} \\ &= -en_n(-\mu_n) + en_p(+\mu_p) \\ &= en_n\mu_n + en_p\mu_p\end{aligned} \tag{E.4.6}$$

Here n_n and n_p stand for the negative- and positive-particle densities. The current density and the conductivity of the sample arise from the contributions of the two charged species, although the particles move in opposite directions.

Example 4.2

(*a*) Calculate the average drift velocity of electrons in copper wire of diameter 1.0 mm carrying maximum available current of 2.32 A (see the Radio Amateur Handbook). Assume $n = 8.5 \times 10^{22}$ electron/cm^3 in copper (see Prob. 2.4).

(*b*) Find the transit time of the electrons if the length of the copper wire is 10 cm.

Solution.

(*a*)

$$J = \frac{i}{A} = \frac{2.32\text{A}}{\pi\left[\frac{(1.0)}{(2)} \times 10^{-3}\text{m}\right]^2} = 2.95 \times 10^6 \text{A/m}^2$$

Therefore, from $\mathbf{J} = nq\mathbf{v}$ we find

$$v = \frac{2.95 \times 10^6 \text{A/m}^2}{1.6 \times 10^{-19}\text{ C} \times 8.5 \times 10^{28}\text{m}^{-3}} = 2.17 \times 10^{-4}\text{m/s}$$

(*b*) The transit time is

$$t = \frac{L}{v} = \frac{10^{-1}\text{m}}{2.17 \times 10^{-4}\text{m/s}} = 460.8 \text{ s}$$

Note that light travels the same distance in 33 picoseconds.

The thermal velocity of electrons (rms velocity) is given by $v_{\text{rms}} = (3\,kT/m)^{1/2}$ (see Problem 3.5). Suppose that all the electrons in copper moved simultaneously in the same direction at the thermal velocity. What current will result in the same copper conductor? ($T = 300$ K; room temperature).

$$\begin{aligned}v_{\text{rms}} &= \left[\frac{3(1.38 \times 10^{-23} J\text{K}^{-1})(300\text{ K})}{9.1 \times 10^{-31}\text{kg}}\right]^{1/2} \\ &= 1.17 \times 10^5 \text{ m/s}\end{aligned}$$

Solution. The resulting current density will be

$$J_{\text{th}} = qnv_{\text{rms}}$$
$$= 1.59 \times 10^{15}\text{A/m}^2$$
$$i = JA = 1.25 \times 10^9\text{A}$$

which is exceedingly high: no conductor will dissipate such a high current.

4.7 DIFFUSION

In the development of conduction current, we showed that charge transport occurs when an electric field acts on the charges. It is also possible to produce charge transport without an electric field, if there are differences in charge concentrations in different parts of the conductive medium. This leads to the concept of *particle diffusion*, which plays a very important role in solid-state device performance, as well as in the actual device fabrication.

Consider a one-dimensional system (Fig. 4.7), where particle density $n(x)$ at x is different (greater) than $n(x + \Delta x)$ at $x + \Delta x$. The particles within Δx see more particles to the left than to the right. Because of the thermal motion, these particles are likely to make more collisions with the particles on the left than on the right. Their effective mean free paths will be larger toward the right. When the random motions of the particles within Δx are averaged out, they will tend to move toward the less populated region of space. The resulting particle flow will depend on the difference in concentrations between x and $x + \Delta x$ and can be written as

$$\Gamma_{xd} = \lim_{\Delta x \to 0} D\left(\frac{n(x) - n(x + \Delta x)}{\Delta x}\right) = -D\frac{\partial n}{\partial x} \tag{4.20}$$

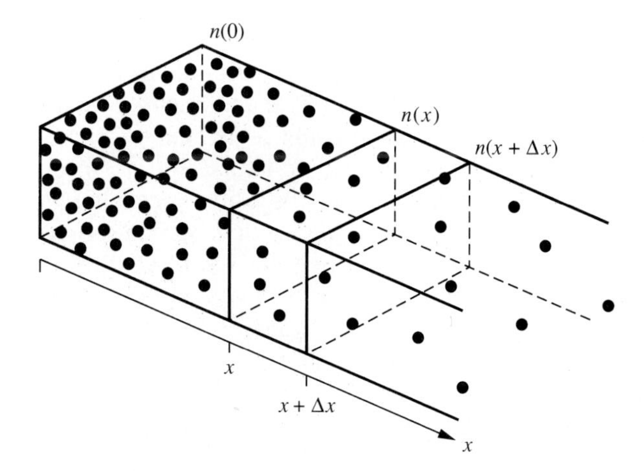

FIGURE 4.7
Transport of charge due to diffusion. Particles diffuse toward the less populated region.

where D, the proportionality constant, is known as the *diffusion constant* and has the units of m^2/s. Note that the negative sign arises from the flow of particles opposite to the particle gradient. For a three-dimensional charged-particle transport, the diffusion current density can be generalized as

$$\mathbf{J}_d = -qD\nabla n \tag{4.21}$$

where $\nabla = (\partial/\partial x)\mathbf{a}_x + (\partial/\partial y)\mathbf{a}_y + (\partial/\partial z)\mathbf{a}_z$ is the gradient operator.

If an electric field $\mathcal{E}$ is also present, the total current density will be the sum of the drift and diffusion currents

$$\mathbf{J} = qn\mu\,\mathcal{E} - qD\nabla n \tag{4.22a}$$

or

$$\mathbf{J} = \rho\mu\,\mathcal{E} - D\nabla\rho \tag{4.22b}$$

where q is the charge for the particle under consideration, $q = +Ze$ for an ion, and $q = -e$ for an electron. Z is the state of ionization of the ion (for a hole, $q = +e$, see Chapter 5), and $\rho = qn$ is the charge density.

If there are no electric fields, and only particle gradients are present in a medium, it is a basic property of the system to eventually come to an equilibrium—that is, to reach a uniform concentration. The spatial and temporal development of the system can be studied by combining Eqs. 4.15 and 4.22. This leads to

$$\frac{\partial n}{\partial t} - D\nabla^2 n = g' - r' \tag{4.23}$$

where ∇^2 is the Laplacian operator, and $g' = g/q$ and $r' = r/q$ are the generation and recombination rates of particles n. If g' and r' are zero for the system under consideration, the resulting equations, Eqs. 4.21 and 4.23, are known as *Fick's* first and second laws. Note that if there are also electric fields present in the system, $\nabla \cdot (n\mu\,\mathcal{E})$ should be added to the left side of Eq. 4.23. Multiplying Eq. 4.23 by q leads to a similar expression for the charge density.

In addition to studying carrier transport in semiconductors, the diffusion process is also used in the manufacture of semiconductor junctions. This process, known as the *diffusion process*, relies on the diffusion of impurity atoms through a host semiconductor material. The diffusion coefficients of the impurity atoms are functions of temperature. Figure 4.8 shows the diffusion coefficient for substitutional diffusers (diffusing atoms that replace a crystalline lattice site) in silicon as a function of $1/T$. By using different impurity materials, separate concentration regions can be obtained within a host crystal. This allows fabrication of various junctions within a crystal to form semiconductor diodes, transistors, and so on, which are discussed in the later chapters of this book.

The *gaseous diffusion process* involves placing a semiconductor wafer in a chamber and flowing a gas containing the impurity atoms through this chamber for a time of t seconds at a constant temperature T. Since this is a *constant replenishment* of the diffusing atoms at the surface of the semiconductor, the

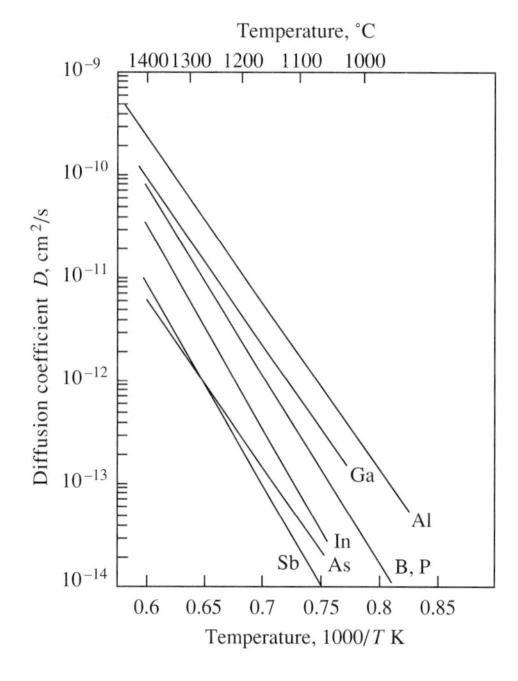

FIGURE 4.8
The diffusion coefficient for substitutional diffusers (diffusing atoms that replace a crystalline lattice site) in silicon as a function of $1/T$ (Tan and Gösele, 1985).

impurity concentration at a given distance x away from the surface and time t is given by (see Problem 4.4)

$$n(x, t) = n_0 \operatorname{erfc}\left(\frac{x}{2\sqrt{Dt}}\right) \tag{4.24}$$

where n_0 is the impurity concentration at the surface— at x = 0—of the semiconductor, and D is the diffusion constant for the impurity atoms in that semiconductor. Erfc(u) is the complimentary error function of argument $u = (x/2\sqrt{Dt})$ and is given by

$$\operatorname{erfc}(u) = 1 - \operatorname{erf}(u) = 1 - \frac{2}{\sqrt{\pi}} \int_0^u e^{-g^2}\, dg$$

If initially a finite quantity of impurity atoms is placed on the surface of the semiconductor at $x = 0$, the diffused impurity concentration at a depth x in a time interval t resulting from this *fixed source* is given by

$$n(x, t) = \frac{Q}{\sqrt{\pi D t}} \exp\left[-\left(\frac{x}{2\sqrt{Dt}}\right)^2\right] \tag{4.25}$$

where Q (atoms per unit area) is the amount of matter placed at $x = 0$ prior to the diffusion process. The resulting diffusion has a gaussian profile.

The following examples will be used to show how various diffusion processes are used in the fabrication of solid-state devices.

Example 4.3 First phosphorus and then gallium atoms are to be diffused as impurity atoms into silicon. (This process is known as *doping*. The idea is to alter the electrical characteristics of a semiconductor, by diffusion or any other process, by introducing controlled amounts of impurity atoms into a semiconductor. See Chapter 6.) The following parameters are used:

$$\mathrm{P}: n_0 = 10^{14}\mathrm{cm}^{-3}, \qquad T = 1500\ \mathrm{K} \qquad \text{and} \qquad t = 5\ \mathrm{h}$$
$$\mathrm{Ga}: n_0 = 10^{16}\mathrm{cm}^{-3}, \qquad T = 1330\ \mathrm{K} \qquad \text{and} \qquad t = 2\ \mathrm{h}$$

Discuss the resulting distribution of these impurities in the semiconductor away from the surface.

Solution. From Fig. 4.8, the diffusion constants for P and Ga are found to be

$$D = 4.8 \times 10^{-12}\mathrm{cm}^2/\mathrm{s} \quad \text{for P}$$
$$D = 2.8 \times 10^{-13}\mathrm{cm}^2/s \quad \text{for Ga}$$

The parameters in Eq. 4.24 are

$$2\sqrt{Dt} = \begin{cases} 5.88 \times 10^{-6}\mathrm{m} & \text{for P} \\ 0.898 \times 10^{-6}\mathrm{m} & \text{for Ga} \end{cases}$$

Therefore, we can write for the spatial dependence of the impurity concentrations in silicon for each impurity atom as

$$n(x) = 10^{20}\mathrm{erfc}(x/5.88 \times 10^{-6})\mathrm{m}^{-3} \qquad \text{for P}$$
$$n(x) = 10^{22}\mathrm{erfc}(x/0.898 \times 10^{-6})\mathrm{m}^{-3} \qquad \text{for Ga}$$

The variations of these impurity profiles as a function of distance into the semiconductor are shown in Fig. E.4.3.* Because of the higher diffusion constant and the longer time, P diffuses further into the semiconductor. If the diffusion time for Ga is increased to 5 hours (same as for P), the resulting impurity profile is shown by the dotted line on the same figure. Although the diffusion times are the same, because of larger D for P, it diffuses further into the semiconductor.

Example 4.4 10^{14} boron atoms are deposited on the surface of a silicon wafer of surface area 10 mm^2. The sample is then placed in a furnace at $T = 1060°$ C.

(a) Find the boron concentration at the surface after 2 hours.

(b) Find the distance from the surface where the boron concentration is 10^{17} atoms/cm^3 after 2 hours.

---

*Error function values are taken from E. Jahnke and F. Emdc, *Table of Functions,* 4th ed., Dover, New York, 1945.

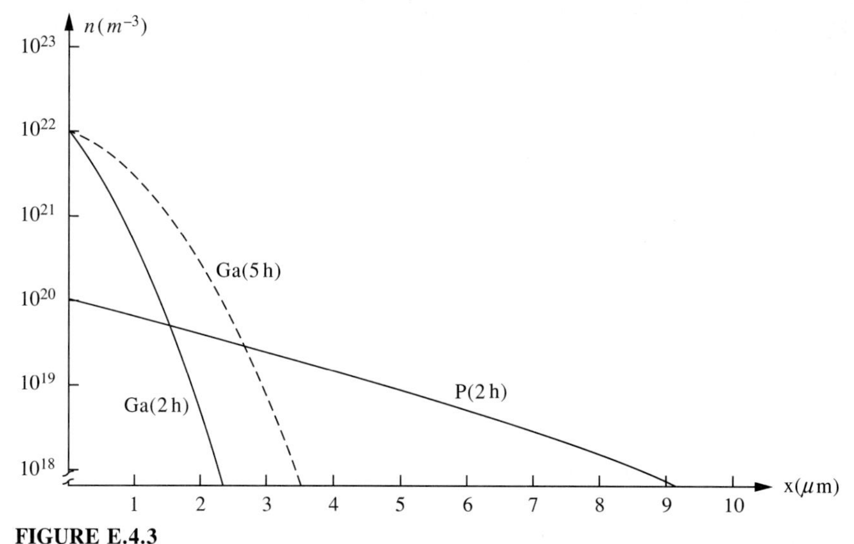

FIGURE E.4.3
Diffusion profiles of P and Ga in silicon by constant replenishment of the impurity atoms after 2 hours. The dotted line shows the Ga profile after 5 hours.

Solution. The diffusion constant for boron at 1333 K is $D = 1.25 \times 10^{-12}\text{cm}^2/\text{s}$ (from Fig. 4.5). Substituting these into Eq. 4.25, we get

$$n(x, 2\text{ h}) = \frac{10^{14}\text{ atoms}/0.1\text{ cm}^2}{\sqrt{\pi(1.25 \times 10^{-12}\text{ cm}^2/\text{s})(2 \times 3600\text{ s})}}$$

$$\times \exp\left[-\left(\frac{x}{2\sqrt{(1.25 \times 10^{-16}\text{ m}^2/\text{s})(2 \times 3600\text{ s})}}\right)^2\right]$$

$$n(x, 2\text{ h}) = 5.95 \times 10^{18}\exp\left[-\left(\frac{x(\mu\text{m})}{1.897}\right)^2\right]\text{ atoms/cm}^3$$

At the surface, this reduces to

$$n(0, 2\text{ h}) = 5.95 \times 10^{18}\text{ atoms/cm}^3$$

The distance x where the concentration is 10^{17} atoms/cm^3 is found from

$$10^{17}\text{cm}^{-3} = 5.95 \times 10^{18}\exp\left[-\left(\frac{x}{1.897}\right)^2\right]$$

Solving for x, we find

$$x = 3.83\mu\text{m}$$

The boron concentration in Si after 2 hours is plotted in Fig. E.4.4. If the diffusion time is increased 3 additional hours, the boron concentration profile changes to that shown by the dashed line in the same figure. Note that the surface concentration is reduced as a function of time since the total amount of matter present at all times is constant.

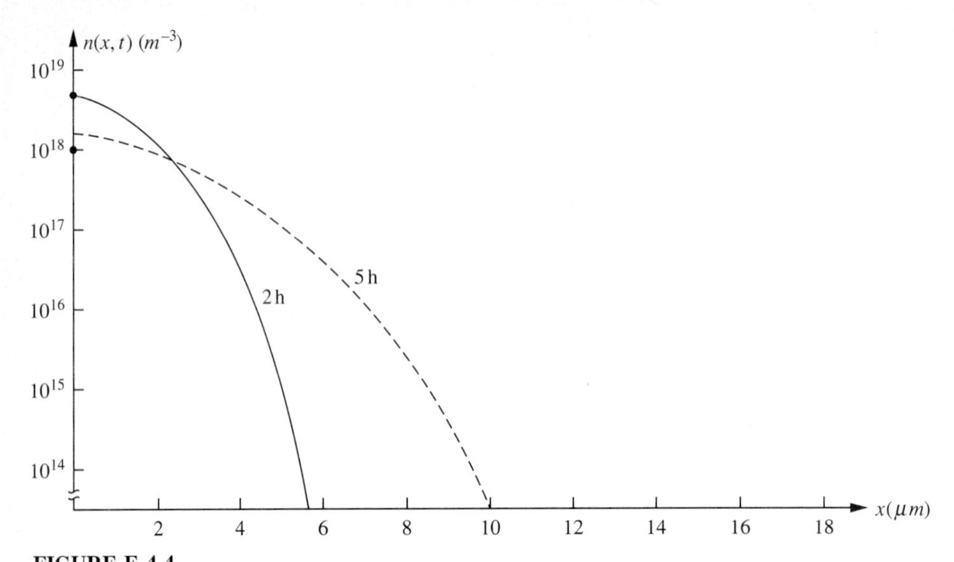

FIGURE E.4.4
Concentration profile of boron atoms in Si after 2 hours (solid line) and 5 hours (dotted line) with a fixed source. Note that the area under each curve is constant.

4.8 EINSTEIN RELATION

The diffusion and drift of charged carriers in a material are related to one another. This results from the randomness of the movement of charge carriers within the material even though there is an additional directed motion for the charged particles due to the presence of the electric field.

The total particle current for charge carriers with charge q is

$$\mathbf{J} = qn\mu\,\mathcal{E} - qD\nabla n \tag{4.26}$$

At equilibrium, $\mathbf{J} = 0$–the drift and diffusion currents cancel each other. Equation 4.26 therefore reduces to (in one dimension)

$$\frac{dn}{dx} - \frac{\mu\,\mathcal{E}}{D}n = 0 \tag{4.27}$$

Integrating this equation gives

$$n(x) = n(0)\exp\left(\frac{\mu}{D}\int_0^x \mathcal{E}\,dx'\right) \tag{4.28}$$

The integral in Eq. 4.28 is by definition related to the potential energy of the particles through the potential difference $v(x)$, that is,

$$\Phi(x) = qv(x) = -q\int_0^x \mathcal{E}\cdot dl = -q\int_0^x \mathcal{E}\,dx' \tag{4.29}$$

Using this in Eq. 4.28

$$n(x) = n(0)\exp\left(-\frac{\mu}{qD}\Phi(x)\right) \tag{4.30}$$

The electron density is also related to the potential energy that the electrons possess through the Boltzmann distribution function as

$$n(x) = n(0)\exp\left(-\frac{\Phi(x)}{kT}\right) \tag{4.31}$$

Comparing Eqs. 4.30 and 4.31, we find that

$$D = \frac{kT}{q}\mu \tag{4.32}$$

Equation 4.32 is known as the *Einstein relation*. It is a very useful relation in studying charged-particle transport in semiconductors, provided that the charge concentrations to which Eq. 4.32 is applied are not very high.

4.9 TRANSPORT OF CHARGED-PARTICLE BEAMS THROUGH MATTER

When an energetic beam of charged particles is incident on a material medium, the charged particles enter the material and, while moving through the material, interact with the various constituents of the medium and lose energy in this process. The major contributions to the energy loss are

direct collision between the incoming particles and the screened nucleus;

excitation of bound electrons in the solid; and

charge exchange processes between the incoming particles and the atoms of the solid.

All these processes are energy dependent and contribute to the energy loss along the path of the particles. The total energy loss per unit length can be written as

$$\frac{dE}{dx} = \left.\frac{dE}{dx}\right|_{\text{nuclear}} + \left.\frac{dE}{dx}\right|_{\text{electronic}} + \left.\frac{dE}{dx}\right|_{\text{exchange}} \tag{4.33}$$

Incoming particles may be electrons, ions, or molecules and, depending on which incoming particle is considered, one or more or all of the above energy-loss mechanisms may become important.

The total loss per unit length (Eq. 4.33) is also equal to the product of the number of scattering centers N and the cross section for energy transfer S, that is,

$$\frac{dE}{dx} = NS(E) \tag{4.34}$$

We can write Eq. 4.34 in the following form

$$\begin{aligned} R(E) &= \int_0^{R(E)} dx = \int_0^E \frac{1}{N} \frac{dE'}{S(E')} \\ &= \int_0^E N^{-1}[S_{\text{nucl}}(E') + S_{\text{el}}(E') + S_{\text{ex}}(E')]^{-1} \, dE' \end{aligned} \tag{4.35}$$

where $R(E)$ is the range for the beam particles having an initial energy E. The range is dependent on the total cross sections for various energy-transfer mechanisms and is the maximum distance that the incoming particles will travel in the material before losing all their energy E to the material itself in the interaction process.

Two important cases of particle beam-material interactions will be dealt with here because of their applicability to electron devices. These are (a) charged particle motion through gaseous media and (b) ion motion through solid materials.

4.9.1 Stopping Power in Gases

The dominant energy-loss mechanism for charged particles going through a gaseous medium is the energy loss that the incoming particle experiences in ionizing the bound electrons of the gas atoms. The theory of this collisional ionization was first derived by Bohr, based on the loss of kinetic energy for the incoming particle in a collision with a bound electron using the momentum transfer from the particle to the electron. When the bound electron acquires enough energy to overcome its binding energy, the atom is said to be ionized and an electron-ion pair is formed. It takes on the average 30 eV of energy to produce one electron-ion pair. The electron-ion pair production per unit length can be written as

$$\frac{dN}{dx} \approx K \frac{nZ^2}{v^2} \tag{4.36}$$

where Z is the charge, n is the number density, $v = (2E/m)^{1/2}$ is the velocity of the incoming particle, and K is a constant. If the incoming particle has a high energy, the particle spends less time in the vicinity of the electron bound to the atom, and it loses less momentum in the collision process. If the energy is low, the particle loses more momentum because of the higher interaction time. This is apparent from l/v^{-2} dependence of the energy-loss rate as given by Equation 4.36.

Nuclear counter tubes, such as Geiger–Mueller tubes, proportional counters, and so on, are practical applications that are based on the ability of the charged particles to produce ionizing collisions in passing through a gaseous medium.

4.9.2 Ion Implantation in Semiconductors

In the energy range 5–500 keV, when a positively charged ion beam is incident on a solid material, the dominant energy loss is by elastic interactions between the beam ions and the screened nucleus of the atoms in the solid. The knowledge of the interatomic forces between the particles plays a very important role in determining this energy loss. The electronic energy loss, which is related to the inelastic energy loss from the passage of an ion through the electronic cloud of the nucleus, also contributes to the overall energy loss of the charged particle. But charge exchange losses represent a small fraction of the total energy loss and for practical purposes can be neglected.

In amorphous substances, the path of an ion entering a solid is not a straight line due to the randomness of the collisional processes involved (see Fig. 4.9). Although the total range of the particle is R_{total}, the projected range R_p along the incoming ion direction is less than R_{total}. There results a lateral spread $R_\perp$ of the particle in a direction perpendicular to the incoming direction. In amorphous substances, the scatter in projected range about the mean projected range is approximately a gaussian function, as shown in Fig. 4.9*b*. If N_D is the number of

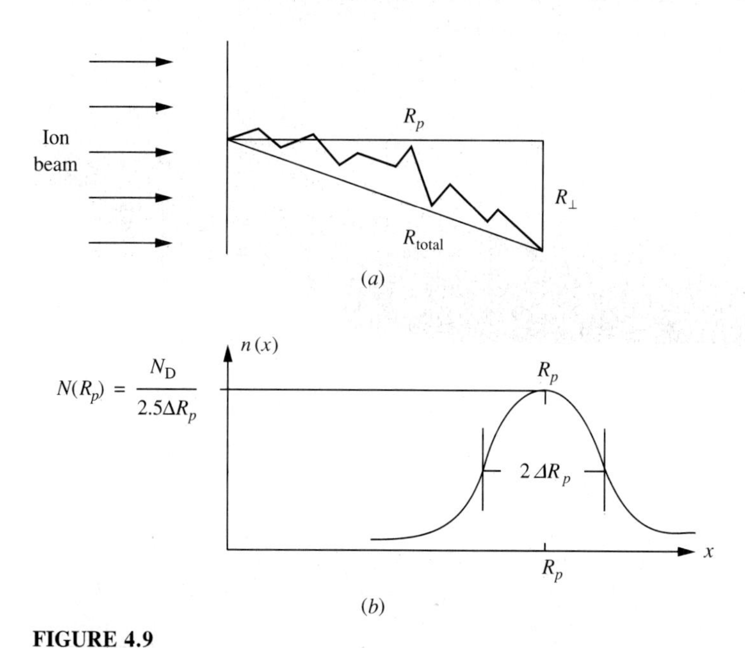

FIGURE 4.9
(*a*) Path of an ion entering a solid. R_{total} is the total range, R_p is the projected range, and $R_\perp$ is the lateral spread. (*b*) For N_D number of implanted ions per cm^2, the resultant implant concentration is a gaussian distribution.

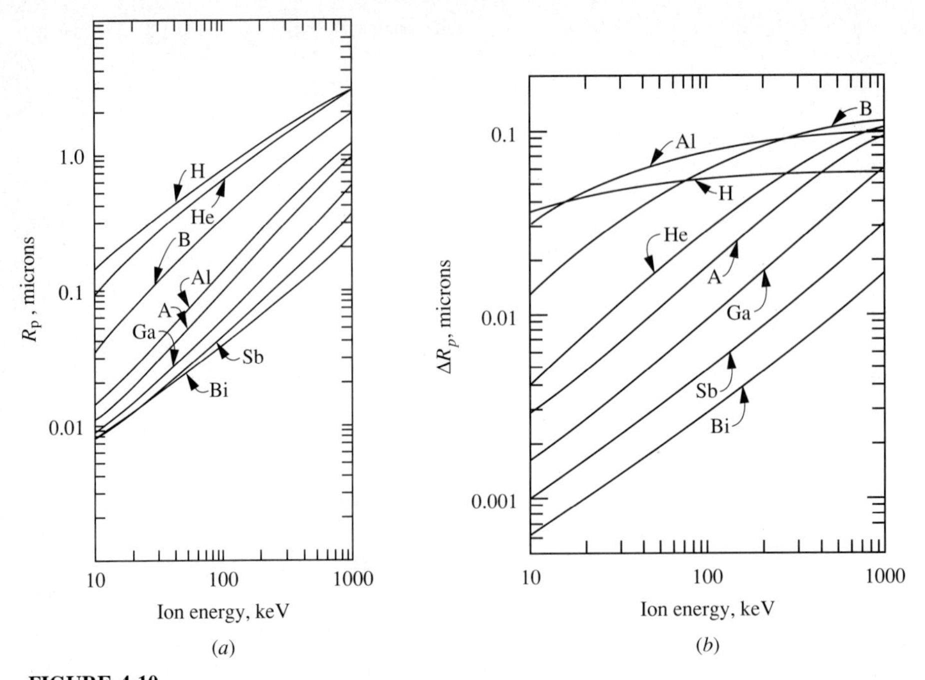

FIGURE 4.10
(*a*) Ion range R_p. (*b*) Mean projected range ΔR_p for various implanted elements in silicon as a function of ion energy (Townsend *et al.*, 1976).

implanted ions per unit area, the concentration N of the implanted ions at x is given by

$$N(x) = \frac{N_D}{2.5\Delta R_p}\exp\left[-\frac{(x - R_p)^2}{2(\Delta R_p)^2}\right] \tag{4.37}$$

The values of R_p and ΔR_p (both in microns) are given in Fig. 4.10*a* and *b* for various implanted elements in silicon.

To achieve a uniform implant concentration with depth, it may be necessary to make several implantations with different energies to approximate a plateau by the sum of gaussian profiles. Figure 4.11 shows the result of three implantations that resulted in a nearly uniform waveguide for light transmission at $\lambda = 633$ nm. The plateau can be smoothed out further by subsequent heat treatment of the host material.

When ions are implanted in crystals, the range of the particles becomes dependent on the orientation of the crystal axis with respect to the incoming ion direction, and the range may be larger due to the phenomenon known as channeling. The details of channeling are beyond the scope of this book; interested readers are referred to the Further Reading listings at the end of this chapter. Ion implantation is becoming a major processing procedure in manufacture of semiconductor devices.

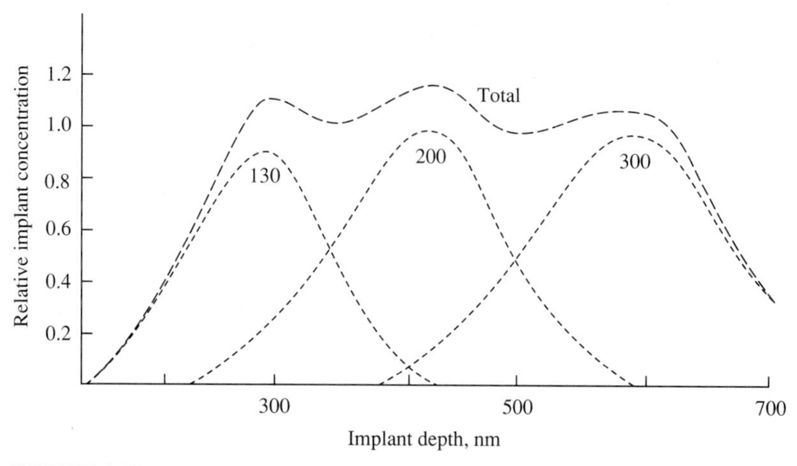

FIGURE 4.11
Three subsequent ion implantations with different energies form nearly uniform waveguide for light transmission at 633 nm (Townsend *et al.*, 1976).

FURTHER READING

Grove, A.S., *Physics and Technology of Semiconductor Devices,* John Wiley and Sons, New York, 1967.

Hemenway, C. L., R. W. Henry, and M. Caulton, *Physical Electronics*, 2d. ed., Wiley, 1967, Chapters 8, 9, and 10.

Shimura, F., *Semiconductor Silicon Crystal Technology*, Academic Press, New York, 1989.

Tan, T.Y., and U. Gösele, "Point Defects, Diffusion Processes, and Swirl Defect Formation in Silicon," Appl. Phys. A, vol. 37, 1–17 (1985).

Townsend, P. D., J. C. Kelly, and N. E. W. Hartley, *Ion Implantation, Sputtering and Their Applications,* Academic Press, 1976.

PROBLEMS

4.1. Suppose a beam of electrons n_b is incident on gaseous atoms of concentration n_s (Fig. P.4.1). The collision cross section for the electron-atom interaction is σ. Neglecting multiple scattering, assume that the fraction dn_b/n_b of the beam particles that are scattered in passing through a distance dx within the volume element Adx is

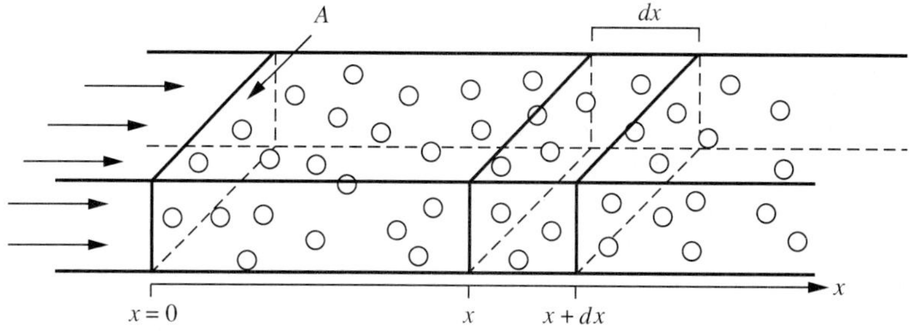

FIGURE P.4.1

equal to the fraction of the area of the element which is obstructed by the scattering atom cross section.

(*a*) If the incident electron concentration at $x = 0$ is n_{b0}, show that

$$n_b(x) = n_{b0}\exp(-n_s\sigma x) \tag{P.4.1}$$

Note: the quantity $(n_s\sigma)$ is known as the *macroscopic cross section.*

(*b*) If one interprets Eq. P.4.1 as the probability that a beam of particles will experience a collision in traversing a distance x, using the definition for an average

$$\lambda_s = \frac{\int_0^\infty x\, dn_b}{\int_0^\infty dn_b}$$

show that the mean free path is related to the collision cross section by

$$\lambda_m = \frac{1}{\sigma n_s} \tag{P.4.2}$$

4.2. The radius of an argon atom is equal to 1.43 Å.

(*a*) Find the macroscopic cross section for electron-argon atom collisions at room temperature if the argon gas is at 1 torr of pressure.

(*b*) What is the mean free path of electrons at (i) 1 torr and (ii) 1×10^{-7} torr of pressure? (1 torr of pressure = 1/760 of an atmosphere)

4.3. Consider a spherical container with a volume of $8 \times 10^3 \text{cm}^3$ which contains nitrogen gas (diameter of N_2 = 3A). Discuss whether a single-particle or a collective-particle approach should be used to study electron motion inside the container if the container pressure is

(*a*) 1 torr

(*b*) 10^{-7} torr

4.4. Show that Eqs. 4.24 and 4.25 satisfy the diffusion equation.

4.5. Silicon is exposed to an impurity source of constant concentration during the diffusion period. Using Fig. 4.8, show that aluminum atoms will diffuse farther than boron atoms in silicon at a given temperature T.

4.6. Aluminum metal is implanted at a thickness of 50 μm on a silicon wafer over an area of 2 mm^2. If the wafer is then placed into a diffusion furnace at a temperature of 1000°C, find

(*a*) the aluminum concentration at the surface after t = 5 hours

(*b*) the distance from the surface where the concentration drops to 10^{16} atoms/cm^3 after 5 hours

4.7. After a diffusion process, the silicon crystal of Problem 4.6 is removed from the furnace and brought to room temperature. Find the change in the surface concentration of the Al atoms after t = 1000 hours if the silicon chip is kept at room temperature.

4.8. Aluminum ions with an initial energy of 200 keV are to be implanted into silicon. The beam flux is 10^{13} ions/(cm^2·s).

(*a*) In order to achieve a total implanted ion dose of $N_D = 10^{16}$ ions/m^2, what should be the implantation time?

(*b*) What is the maximum implanted ion concentration?

(*c*) Find the mean projected range of the Al ions.

(*d*) What is the Al density at a distance R_p = 100 nm?

4.9. The mobilities of electrons and of holes in silicon at room temperature are 0.8 $m^2V^{-1}s^{-1}$ and 0.04 $m^2V^{-1}s^{-1}$ respectively. Find the corresponding diffusion constants for the electrons and the holes.

4.10. If there are 1.96×10^{22} electrons/m^3 and 10^{11} holes/m^3 in a silicon sample, find the conductivity and resistivity of the sample using the constants given in Problem 4.9.

4.11. The mass density of copper is 8.9 g/cm^3, its atomic weight 64. There is one free electron per atom, and the conductivity of copper is $\sigma = 5.8 \times 10^7$ S/m.

(*a*) What is the electron concentration in copper?

(*b*) What is the mobility of electrons?

(*c*) What is the mean collision time between the electrons and the copper atoms?

4.12. A potential difference of 10 V is applied across the two terminals of a copper conductor 1 cm long. Using the results of Problem 4.11,

(*a*) Calculate the drift velocity of electrons in the copper.

(*b*) Calculate the time it takes for an electron to traverse the total length of the metal.

(*c*) The transit time found in (*b*) is very long. We know that if we apply a square pulse across the copper wire, the current flow follows the voltage almost instantaneously. Explain what seems to be a discrepancy between the transit time and the sudden rise in current through the wire.

CHAPTER
5

BASIC PROPERTIES OF SOLIDS

A solid is made up of a large number of atoms and contains equal number of positive and negative charges. Individual atoms may contain any number of electrons, but only the electrons lying in the outer orbits contribute both to the binding of the atoms and to the electrical conduction properties of the solid. Fortunately, when a large number of atoms are brought together, the resulting form of the solid is mostly crystalline in nature—that is, the atoms are spaced in an orderly periodic fashion in three-dimensional-space. This makes it easier to study crystalline solids than noncrystalline solids, where the atoms are randomly distributed. The orderly placement of the atoms within the crystalline solid allows us, from a mathematical point of view, to model the solid with relatively simple approximations and with fewer complications.

In this chapter, we will study the general properties of crystalline solids. The emphasis will be placed on semiconductors and their specific properties. These will be useful in understanding the operational characteristics of the solid-state devices that will be covered in the subsequent chapters.

5.1 CRYSTAL STRUCTURE

Although the concentration of atoms is very high, the atoms of a perfect (ideal) crystal are ordered in a highly uniform manner in the solid. This ordering is so well defined that the overall behavior of the solid can be analyzed in a simpler way by considering only a sample group of neighboring atoms, since this small group of representative atoms repeats itself throughout the solid. Rather than considering all the atoms in the solid simultaneously, the overall crystalline structure can be inferred from a distribution of atoms within simple, three-dimensional geometrical cells. This basic group of atoms constitutes a *unit cell* and is derived from the ordered placement of the atoms in a given solid.

In order to understand the nature of unit cells, we define three coordinate vectors **a**, **b**, and **c** that originate from a common coordinate origin. The various angles associated with these *crystalline coordinates* are labeled as α, β, γ, as shown in Fig. 5.1. The magnitudes and the angles between these vectors are generally dependent on the distances between neighboring atoms and the way they appear naturally in the solid. The symmetry properties of the placement of the atoms simplify the classification of these unit cells.

If a set of hypothetical points is located with equal distances from each other along an axis, these points are referred to as forming a *linear array.* The linear array occurs if the points repeat themselves by the successive translation T_1 along the **a** axis, as shown in Fig. 5.2. If an additional translation occurs along the **b** axis over a distance T_2, and if the linear array formed along the **a** axis repeats itself with respect to T_2 translations, the resulting array is referred to as a *plane lattice*. If every translation T_3 along the **c** axis repeats the plane array formed in the **(a,b)** plane, a three-dimensional *space lattice* forms.

If any of the initial translations T_1, T_2, and T_3 are considered, and if the coordinate origin is shifted accordingly, the rest of the translations will be a repeat of the original translations. Thus, the three-dimensional box that is formed by the shortest translation distances T_1, T_2, and T_3 is referred to as a *unit cell.* The primitive crystalline vectors **a, b,** and **c** are now restricted to the unit cell, and the magnitudes of these vectors are related to the dimensions of the unit cell in the respective directions.

There are four possible symmetry properties. These are

1. Translational (moving along an axis)
2. Rotational (rotating over an angle with respect to an axis)

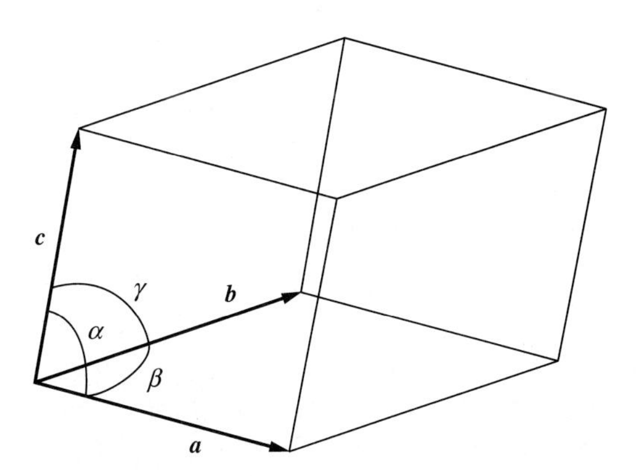

FIGURE 5.1
Crystallographic axes that are used to define the unit cells of a crystalline solid.

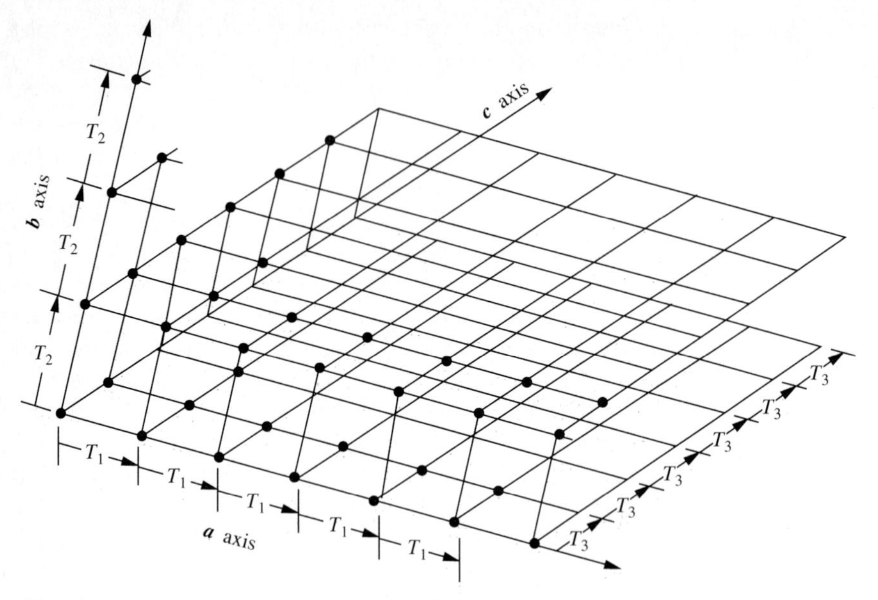

FIGURE 5.2
The translational symmetries lead to different definitions of linear, planar, and space lattice.

3. Reflection (reflection from a mirror)
4. Inversion (both rotation and reflection)

Depending on all the possible combinations of these symmetries, the unit cells are determined and one finds that there are 14 possible unit cells with which all crystalline solids can be identified. These unit cells, which are called Bravais lattices, are shown in Fig. 5.3.

One of the commonly encountered lattices in semiconductors is the simple cubic lattice which has the property that the lengths of the vectors **a, b**, and **c** are all equal, and the corresponding angles α, β, and γ are 90 degrees. Depending on the crystal, all or some of the lattice points (specific points at the corners, surfaces, or within the volume of the unit cell) may be occupied by the atoms of the crystal.

Crystal planes and *crystal directions* are also defined in relation to the crystalline vectors **a, b**, and **c.** The plane shown in Fig. 5.4 can be represented in terms of the multipliers of the vectors **a, b**, and **c** within the unit cell with respect to the origin of the unit cell coordinates. These multipliers are (1, ∞, 1/2) respectively. (Note that the **b** axis is intersected at ∞.) The reciprocals of these numbers are (1, 0 , 2), which may be nonintegers with integer denominators. The reciprocal numbers are then multiplied by the lowest common denominator of the set to remove the fractions. These numbers then represent the given plane. The plane shown in Fig. 5.4 is identified as a (1,0,2) plane. In general, the reciprocal indices are known as the *Miller indices* and are represented by the smallest integers (h,k,l). Some of the sample planes in a cubic lattice are shown in Fig. 5.5.

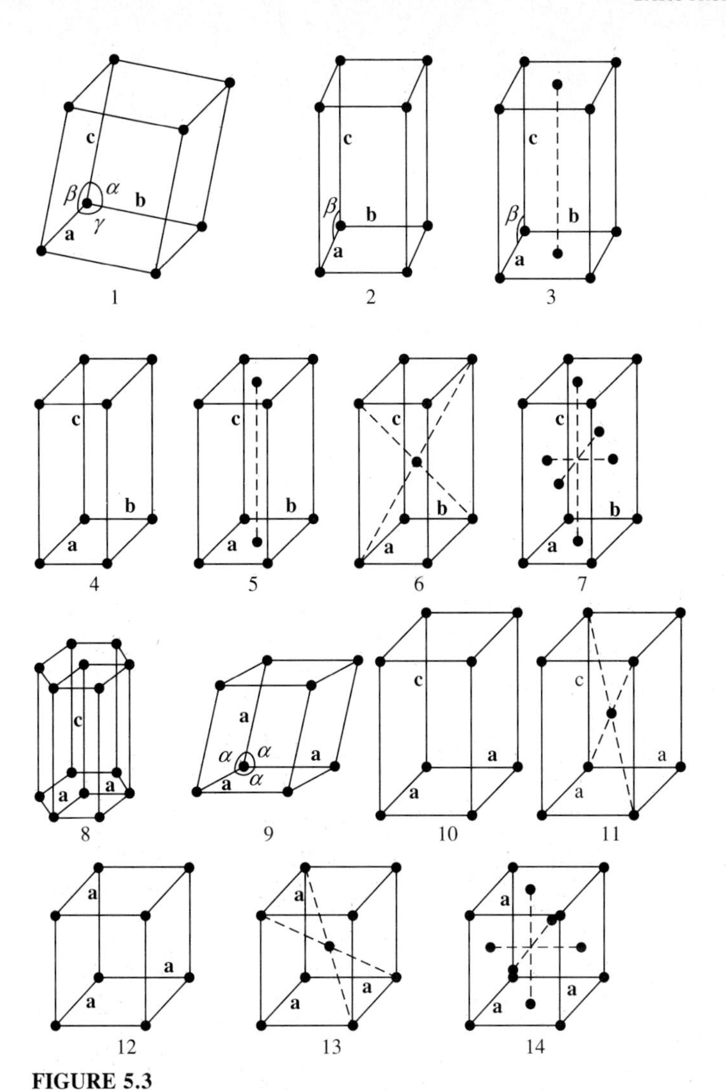

FIGURE 5.3
Basic Bravais crystalline lattices: 1. triclinic, simple; 2. monoclinic, simple; 3. monoclinic, base centered (b.c.); 4. orthorhombic, simple; 5. orthorhombic, b.c.; 6. orthorhombic; b.c.; 7. orthorhombic, f.c.; 8. hexagonal; 9. rhombohedral; 10. tetragonal, simple; 11. tetragonal, b.c.; 12. cubic; simple; 13. base-centered cubic; 14. face-centered cubic (A. J. Dekker, SOLID STATE PHYSICS, © 1965, pp. 7, 244. Reprinted by permission of Prentice-Hall, Inc., Englewood Cliffs, NJ.)

Sometimes for a given plane, one or two intersection points of the primitive lattice vectors fall outside the unit cells and can intersect the lattice vectors in the negative directions. The resulting integer is then labeled with a bar over the Miller index. Such a plane, $(2,\bar{1},2)$, is shown in Fig. 5.5*d*. Crystalline planes repeat themselves over specified distances perpendicular to the crystalline plane.

In a crystal, the directions in which charged particles move also play a very important role in the electrical characteristics of the solid. It is therefore necessary

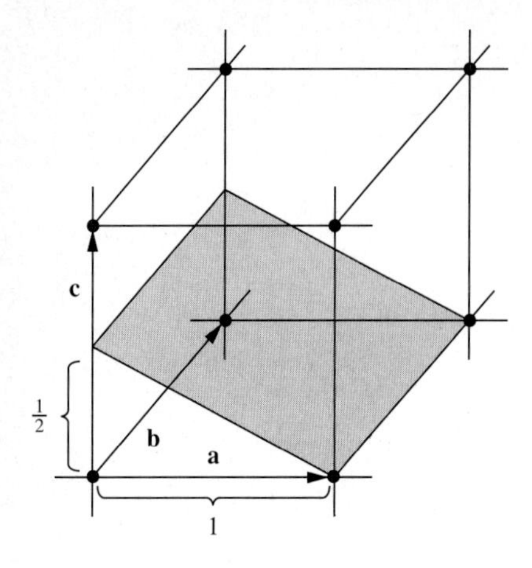

FIGURE 5.4
Determination of (1,0,2) plane in a simple cubic lattice. **a** is intersected at (1), **b** is intersected at (∞), and **c** is intersected at (1/2).

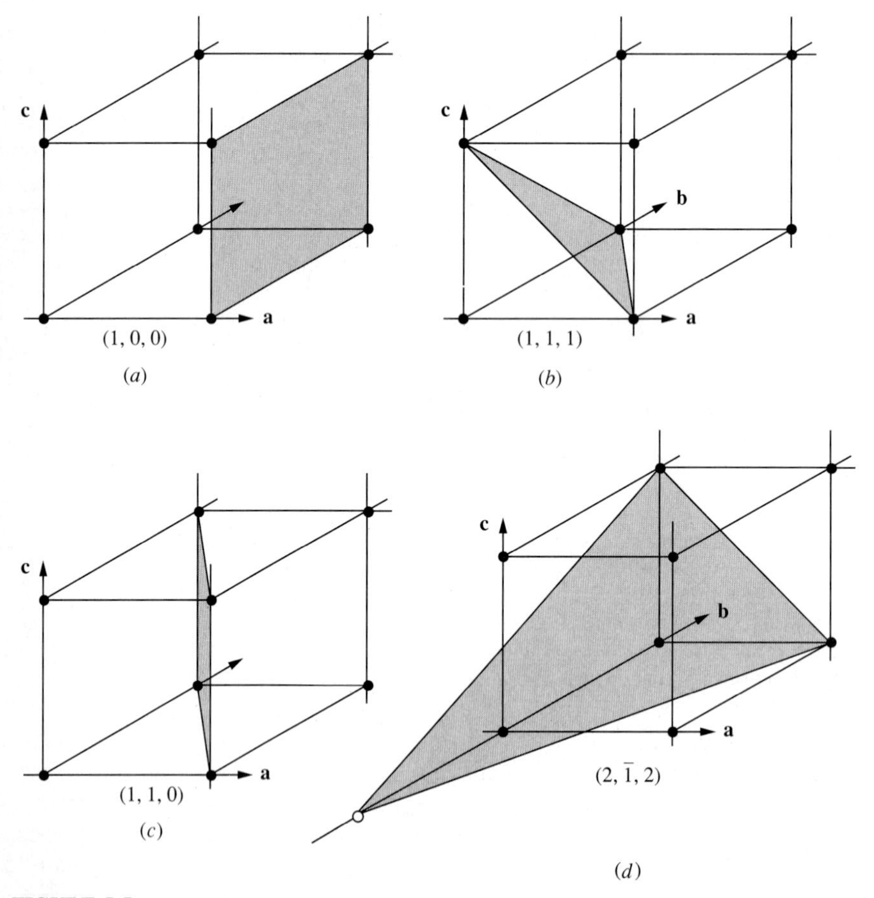

FIGURE 5.5
Shaded areas represent the (*a*) (1,0,0), (*b*) (1,1,1), (*c*) 1,1,0 and (*d*) (2,$\bar{1}$,2) planes in a simple cubic crystal.

to define directions in the crystal. This is accomplished by starting from the unit cell origin and following the displacements along each of the crystalline vectors in the order of the vectors **a, b**, and **c** and representing each displacement with an integer. The brackets are used to label the crystalline directions. Figure 5.6 shows the [100], [011], and [111] directions in a simple cubic lattice. Note that the (1,1,1) plane is perpendicular to the [111] direction.

The binding forces that exist between the atoms of the solid determine the formation of the crystalline structure. Depending on these forces, the crystals can be classified as (*a*) ionic crystals, (*b*) covalent crystals, and (*c*) the metals.

Ionic crystals (i.e., NaCl, KF). One or more electrons of one type of atom are transferred to another, leading to the formation of positive and negative ions. NaCl can be thought to be made up of Na^+ and Cl^- ions. When Na transfers its electron to the Cl atom, they both form stable spherically symmetric noble gas closed-shell orbits. A strong binding also results between the ionic elements. The atoms in the crystal are arranged at the corners and the centers of the faces of an array of the cubes as shown in Fig. 5.7. The resulting crystal structure is known as a *face-centered cubic array.*

Valence crystals. Neighboring atoms share their valence electrons under the formation of homopolar or covalent bonds. Valence crystals are very hard and have poor electrical conductivity (e.g., diamond and carborundum). The diamond structure in which each atom is tetrahedrally coordinated by four like atoms is shown in Fig. 5.8.

Metals. Outer electrons of the constituent atoms have a high degree of mobility leading to high electrical conductivity.

There are also *Van der Waals* crystals, which are related to rare gas atoms that have little or no tendency to give their electrons to the binding forces. At low temperatures, the negative charge and the nucleus form a fluctuating dipole

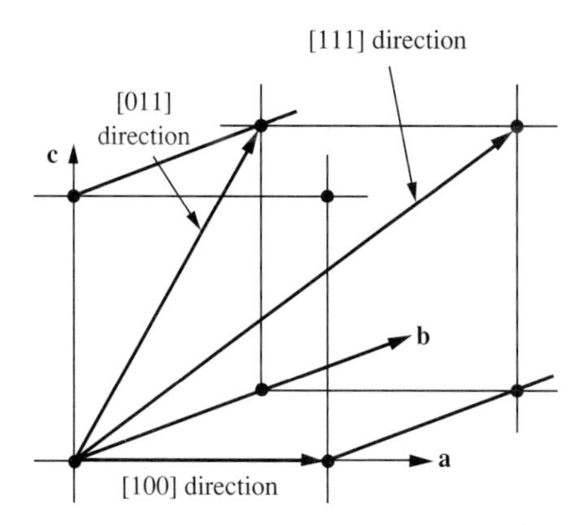

FIGURE 5.6
The [100], [011], and [111] directions in a simple cubic lattice.

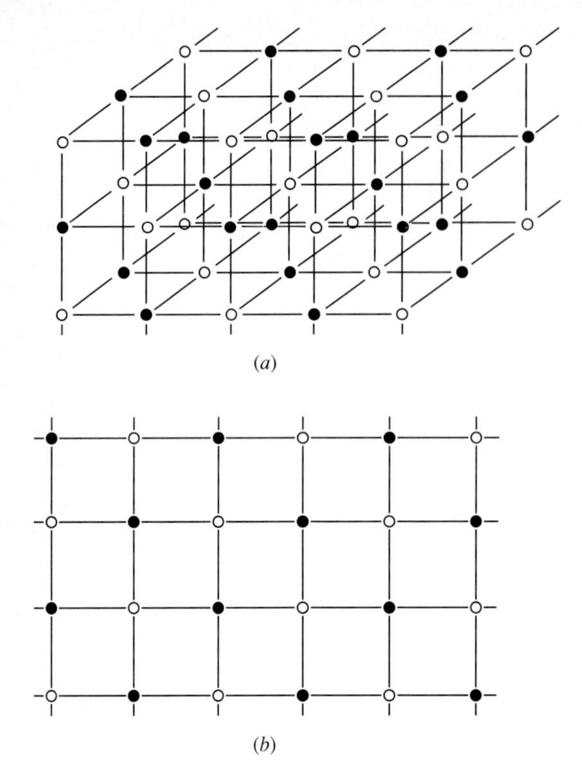

FIGURE 5.7
Ionic bonding in NaCl. (*a*) Three-dimensional and (*b*) two-dimensional representations of a face-centered cubic crystal.

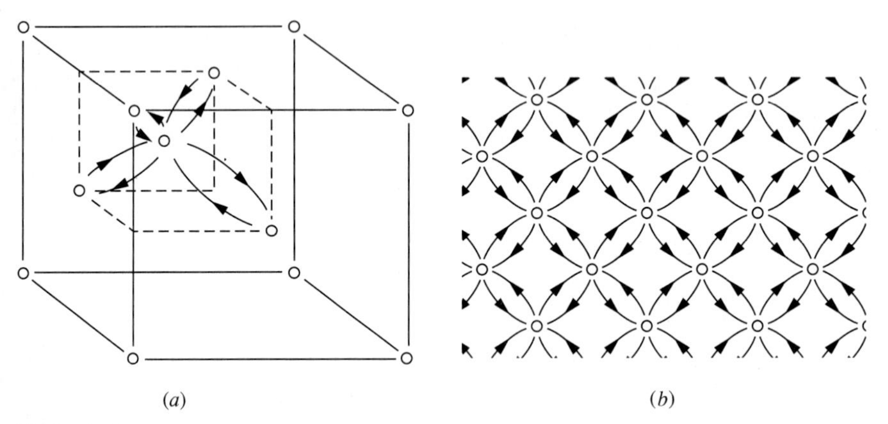

FIGURE 5.8
Diamond structure where valence electrons are shared by the neighboring atoms in the binding process. (*a*) Three-dimensional and (*b*) two-dimensional representations of the diamond structure.

leading to a weak coupling between the dipoles. These forces are called Van der Waals forces.

Semiconductors fall between the valence crystals and the metals. Even though the valence electrons of the semiconductor atoms are shared between each atom, the binding forces are not that strong. At very low temperatures, the semiconductor is an insulator, and at higher temperatures, it conducts current easily. The energy necessary to free an electron from its bound state is called the *band gap energy* (see Section 5.6). The energy gap in these elements is inversely proportional to the interatomic binding length. This explains why diamond is an insulator since it has a large band gap, $E_g = 5.3$ eV and $d = 3.57$ A, compared to germanium which has $E_g = 0.67$ eV and $d = 5.7$ A. At room temperature, the diamond is a very good insulator compared to germanium.

The close interatomic binding length of the diamond makes it more difficult to remove the electron from the bond at the higher temperatures. On the other hand, since the interatomic distance is larger in the germanium crystal, the binding is not as strong as in diamond and, therefore, at reasonable temperatures the covalent binding between the electrons can easily be broken so that these electrons can contribute to the electrical conduction in the crystal.

Metals play a very important part in the implementation of semiconductor devices. Some of the general properties of metals were discussed in Chapter 2. In subsequent chapters, whenever additional properties of metals are needed, these will be explained. Otherwise, our emphasis will be on the general properties of semiconductors.

5.2 BAND THEORY OF SOLIDS

The band theory of solids is very helpful in understanding the operational principles of almost all of the semiconductor devices. It is therefore necessary to understand the physical basis of the band theory. In order to emphasize the importance of this concept, a physical approach followed by a mathematical derivation will be given here.

5.2.1 Physical Approach

The potential energy associated with an electron under the influence of a positively charged nucleus (charge Ze) is given by

$$\Phi(r) = -\frac{Ze^2}{4\pi\epsilon_o r} \tag{5.1}$$

A variation of this potential energy is plotted in Fig. 5.9*a*. We superimpose on the same energy diagram the possible energy levels that the electron can possess in the atom. For an isolated atom, these levels are plotted as discrete energy levels. If we add one more electron into the system, at the same time increasing the nucleus charge by the same amount (neglecting the nuclear forces in the nucleus),

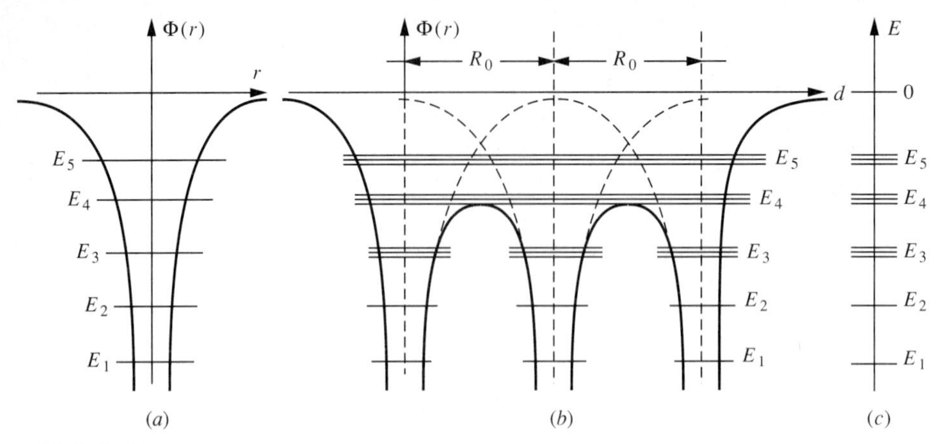

FIGURE 5.9
(*a*) Potential energy variation and discrete energy levels of electrons in an isolated atom. (*b*) The resulting potential energy variation when three atoms are brought close together (R_o is the interatomic distance) and (*c*) corresponding electronic energy levels for the three atom system.

we will now have to consider a three-body problem that includes the interaction between the electrons and the positive nucleus, and between the two electrons. As we add more electrons to the system, we find that the system becomes more complex and the energy-level scheme of the atom deviates considerably from that of the hydrogen-like energy spectrum. We can still come up with an energy-level representation for the complex atom consistent with the Pauli exclusion principle. Because of the interaction between all the particles of the constituent atoms, the equivalent energy level scheme of the complex atom will now have n, l, and m-like states. The complete energy levels and corresponding wave functions will be assumed to be calculated in principle from the solution of the Schrödinger equation. The energy levels for the given atom will also have distinct energy levels. Similar to that of the hydrogen atom, these levels can be schematically represented, as shown in Fig. 5.9*a*.

As these atoms are brought closer together, a further complication arises. This can be explained by referring to Fig. 5.9*b*. Let us assume that in an isolated single atom, there are five discrete atomic energy levels (Fig. 5.9*a*), and four of these levels are occupied by electrons. If we now bring three atoms together so that the separation distance between the atoms is R_0, equivalent to the interatomic distance in a solid, the resultant potential energy of a three-atom system is modified. Since the potential energy between adjacent atoms is added, each electron is subjected to an equivalent potential energy variation as shown by the solid lines in Fig. 5.9*b*. Electrons occupying energy levels E_1, E_2, and E_3 are subjected to equivalent potential barriers formed between the atoms instead of the infinite barriers of the isolated atoms. On the other hand, the electrons occupying the energy level E_4 now have energy greater than the height of the potential barriers and are free to move from one atom to the other. These electrons are not completely "free" in the sense that they move under the influence of a positive background produced by the positively charged nuclei. These electrons contribute to the electronic conduction in the solid.

We know from tunneling phenomena that there is a finite probability that an electron will tunnel through a potential barrier. An electron initially occupying a specific energy level of the isolated atom, say E_3, will now have a finite probability to penetrate the equivalent potential barrier and appear in the adjacent atom. Of course, the electron that was already in this adjacent atom will also have a similar probability of tunneling through the barrier and appearing in the adjacent atom next to itself. There are no rules that say that these tunnelings should occur simultaneously. Therefore, at one instant of time, there may be more than one electron in the level E_3. This contradicts the Pauli exclusion principle. But to allow the phenomena of tunneling, there should be some additional energy levels available in the vicinity of E_3. This arises from the perturbation of the energy levels due to the presence of more than one atom. We now say that the energy level E_3 splits up into three closely spaced energy levels to accommodate all three electrons that were occupying the discrete energy level E_3 in individual atoms. If there are now N atoms instead of the three atoms that we have considered, and if each atom is separated by a distance R_o, there should be N closely spaced energy levels for N electrons in place of that single discrete energy level. Because of the presence of a large number of atoms per unit volume in a solid, the resulting overall energy distribution of the electrons, irrespective of the smallness of the separation of the energy levels, can now be considered to be equivalent to an energy band to accommodate these N electrons. From here on, in analyzing the electrical characteristics of electronic devices, the allowed energy bands, although possessing very closely spaced discrete energy levels, will be assumed to form a continuous band of energies. A schematic illustration of the splitting of the atomic energy levels by the interatomic perturbations as the atomic spacing is reduced is shown in Fig. 5.10*a*. The potential barrier width corresponding to electrons occupying the energy level E_1 is so large that the electrons at the adjacent atoms have very little probability of penetrating through the barriers, and their energy levels stay discrete. Other atomic energy levels split up into bands as shown in Fig. 5.10*b*.

Although electrons corresponding to the energy level E_4 are free to move under the influence of a background of uniformly spaced positive charges, they still form an energy band without violating the Pauli exclusion principle.*At T = 0 K, these electrons occupy only energy levels up to E_F. Although energy levels above E_F are available energy levels for the electrons, they can only occupy these levels if they gain additional energy through external means such as from external forces or from thermal excitation.

Various energy bands in a solid (for the case shown, a metal) resulting from the perturbation of individual atomic levels are shown in Fig. 5.10*c*. The

---

* If the Pauli exclusion principle were not included in the specification of the original energy levels, there would be $2N$ energy levels in the vicinity of the energy level E_3.

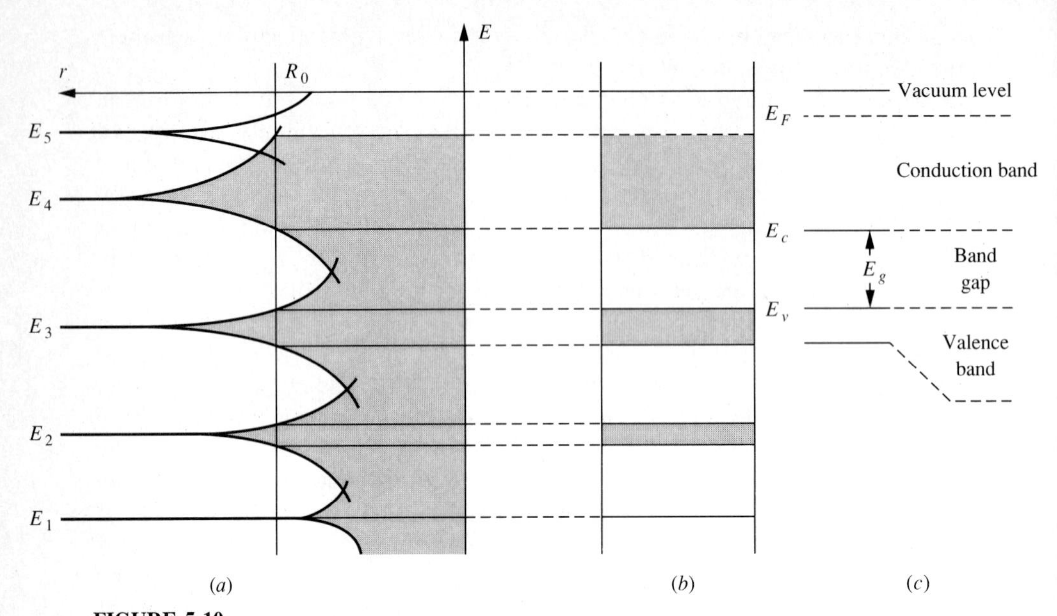

FIGURE 5.10
(*a*) Splitting of energy levels as a function of interatomic distance between the atoms, (*b*) resulting energy bands at the interatomic distance R_0, and (*c*) identification of energy bands.

uppermost band below the vacuum level is called the conduction band. The next lower energy band is called the valence band. Electrons in the conduction band mostly contribute to the charge transport phenomenon in solids. Although energy levels may be available for electrons above E_F, there may not be enough electrons to occupy levels above a certain level labeled as E_F because of the limited number of electrons. This is especially true at $T = 0$ K. E_F is called the Fermi energy. The bottom of the conduction band is labeled as E_c and the top of the valence band is labeled as E_v. In an ideal solid, the electrons are not allowed to occupy the range of energy levels below E_c and above E_v. This forbidden energy interval is called the *energy band gap* and is identified by the band gap energy E_g.

5.2.2 Mathematical Approach

When the three atoms were brought together, it was shown that the equivalent potential energy variation of the three atoms was modified within the adjacent atoms as shown in Fig. 5.9. In an actual crystal, there is a three-dimensional array of a large number of orderly spaced atoms which constitute the solid. The mathematical treatment of a general three-dimensional array is very difficult, but we can simplify the problem by using a one-dimensional model of the solid. Although this is a drastic simplification of the actual solid, it gives many of the essential features related to the band theory of the solids.

In order to represent the crystalline solid more realistically, we can assume that we have an infinite number of atoms in the one-dimensional array or we can assume that a large number of N atoms form an array in such a way that one end joins the other end, forming a closed loop. This type of model is necessary in order to avoid the edge effect which in itself plays an important role in the electrical characteristics of many semiconductor devices. The infinite array or the closed loop guarantees that there are no edge effects.

From a mathematical point of view, we can assume that the wave function associated with the electrons repeats itself at every positional location displaced by the crystal interatomic spacing d. A small portion of this infinite array is shown in Fig. 5.11*a*. Solving the Schrödinger equation for an electron subjected to this potential variation requires mathematical approximations which are themselves complex and lengthy. In order to simplify the problem, the actual potential variation shown in Fig. 5.11*a* is approximated by the equivalent potential barrier as shown in Fig. 5.11*b*. This approximation is known as the *Kronig–Penney* model. If any point in the array is displaced by a distance d, the potential function is repeated, that is

$$\Phi(x) = \Phi(x + d) \tag{5.2}$$

The corresponding time-independent wave function for an electron should also satisfy

$$\psi(x) = \psi(x + d) \tag{5.3}$$

$$\psi'(x) = \psi'(x + d) \tag{5.4}$$

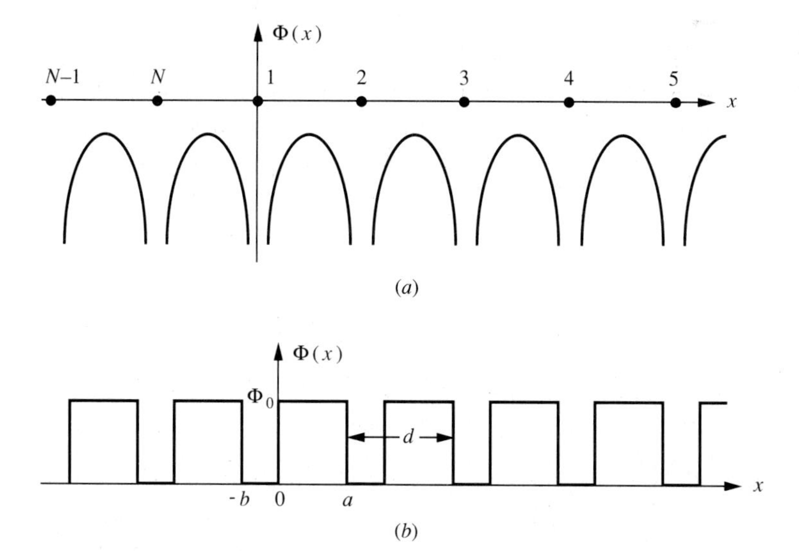

FIGURE 5.11
(*a*) Potential energy variation in an infinite one-dimensional array and (*b*) the equivalent Kronig–Penney model approximation.

It can be shown that for periodic potential variation, the corresponding wave function $\psi(x)$ can be written as

$$\psi(x) = u(x)e^{+jkx} \tag{5.5}$$

where $k = \pi/d$ and $u(x)$ also satisfies the condition

$$u(x) = u(x+d) \tag{5.6}$$

Equations 5.5 and 5.6 are known as the *Bloch functions*. The Schrödinger equations in the two regions of Fig. 5.11b are

$$\left(\frac{\hbar^2}{2m^*}\right)\frac{d^2\psi_1}{dx^2} = E\psi_1 \qquad \text{for} \qquad -b < x < 0 \tag{5.7a}$$

$$\left(\frac{\hbar^2}{2m^*}\right)\frac{d^2\psi_2}{dx^2} + \Phi_0\psi_2 = E\psi_2 \qquad \text{for} \qquad 0 < x < a \tag{5.7b}$$

Substituting Eq. 5.5 into Eq. 5.7, we obtain two equations for u_1 and u_2

$$\frac{d^2u_1}{dx^2} + 2jk\frac{du_1}{dx} - k^2u_1 = \left(\frac{2m^*}{\hbar^2}\right)Eu_1 \qquad \text{for } -b < x < 0 \tag{5.8a}$$

$$\frac{d^2u_2}{dx^2} + 2jk\frac{du_2}{dx} - k^2u_2 = \left(\frac{2m^*}{\hbar^2}\right)(\Phi_0 - E)u_2 \quad \text{for } 0 < x < a \tag{5.8b}$$

If one assumes an exponential solution of the form $Ae^{+\Gamma x} + Be^{-\Gamma x}$ with different A's, B's and Γ's for the two regions and uses the boundary conditions

$$u_1(0) = u_2(0) \qquad \text{and} \qquad u_1'(0) = u_2'(0)$$

$$u_2(a) = u_1(-b) \qquad \text{and} \qquad u_2'(a) = u_1'(-b)$$

one obtains, after a lengthy calculation, the following *dispersion* relation relating α and β to Kronig–Penney potential parameters:

$$\frac{\beta^2 - \alpha^2}{2\beta\alpha}\sinh(\beta b)\sin(\alpha a) + \cosh(\beta b)\cos(\alpha a) = \cos[k(a+b)] \tag{5.9}$$

where

$$\beta^2 = \frac{2m^*(\Phi_o - E)}{\hbar^2} \tag{5.10}$$

and

$$\alpha^2 = \frac{2m^*E}{\hbar^2} \tag{5.11}$$

In order to further simplify this equation, we assume that the potential function $\Phi_0 \to \infty$ and $b \to 0$ but the product $(\Phi_0 b)$ remains constant. This approximation leads to

$$\left(\frac{m^*\Phi_o b}{\hbar^2\alpha}\right)\sin(\alpha a) + \cos(\alpha a) = \cos(ka) \tag{5.12}$$

It can be seen from Eq. 5.12 that the right side of the equation varies between $+1$ and -1. α given by Eq. 5.11 is a function of the energy of the electron. Therefore, equality of the equation can hold for only certain values of the electron energy. Equation 5.12 can be put in the form

$$P\frac{\sin(\alpha a)}{\alpha a} + \cos(\alpha a) = \cos(ka) \tag{5.13}$$

where $P = m\Phi_o ba/\hbar^2$ is a measure of the area of $\Phi_0 b$. Equation 5.13 is plotted in Fig. 5.12 as a function of αa for an arbitrary value of $P = 3\pi/2$. One can see from this plot that there are again allowed and forbidden energy bands for the electrons as expected. As $P \rightarrow \infty$, the probability of tunneling between adjacent atoms becomes remote, and the energy levels become discrete. As $P \rightarrow 0$, the energy bands disappear, and the systems become a continuum, that is, the particles become free. It can also be shown that the total number of possible wave functions in an energy band is equal to the number of unit cells N in the array.

All solids can be classified as metals, insulators, and semiconductors, depending on their energy-band structure. The reasons for this classification are best understood by referring to the individual band structures of the solids. Typical energy levels of these solids are sketched in Fig. 5.13. For a conductor, all the energy levels in the valence band are full, and some electrons also occupy the lower-lying energy levels of the conduction band. On the other hand, in an insulator, all the levels in the valence band are full. Also the band gap energy E_g is so large that at room temperature the probability of an electron in the valence band being excited to the conduction band is very small. Therefore, there are practically no electrons in the conduction band. In a semiconductor, the valence band is also full at $T = 0$ K. The conduction band is empty, but the energy gap E_g is

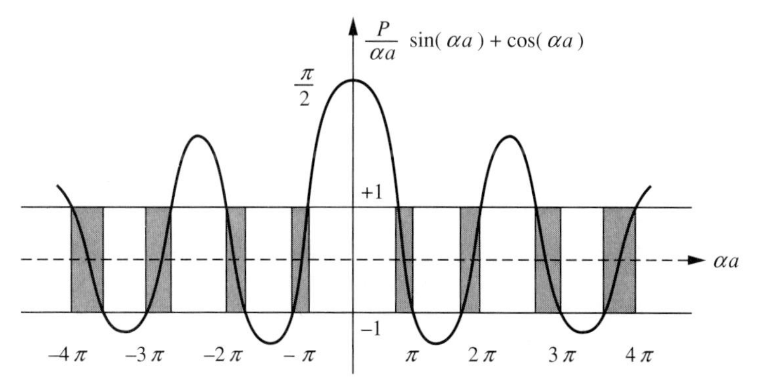

FIGURE 5.12
Plot of Eq. 5.12 for $p = 3\pi/2$ as a function of (αa), which is a function of the electron energy. The energy levels are divided into allowed and forbidden bands (A. J. Dekker, SOLID STATE PHYSICS, © 1965, pp. 7, 244. Reprinted by permission of Prentice-Hall, Inc., Englewood Cliffs, NJ.)

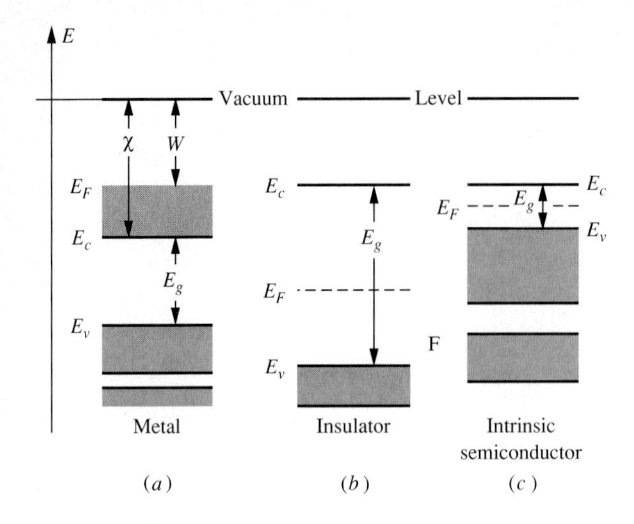

FIGURE 5.13
Typical energy band diagrams for (*a*) a metal, (*b*) an insulator, and (*c*) an intrinsic semiconductor ($T = 0$ K). The dashed areas show levels occupied by electrons. χ is the electron affinity and W is the work function.

small compared to that of an insulator. At room temperature, the electrons lying at the upper levels of the valence band are thermally excited to the conduction band and these electrons, whatever their number is, then contribute to the charge conduction phenomenon in a semiconductor.

Two additional energy values should be identified in studying the response of solids to external excitation and electron emission. These are (*a*) the electron affinity χ, which is the energy difference between the vacuum level and the bottom of the conduction band, and (*b*) the work function W, which is the energy difference between the vacuum level and the Fermi energy. The work function can be thought of as the minimum energy necessary to free an electron from a solid at $T = 0$ K. This is apparent for a metal as shown in Fig. 5.13*a*. The work function and electron affinity are related by

$$W = \chi - E_F \tag{5.14}$$

Not all solids are perfect crystals, and some also contain impurities. When imperfections and impurity effects are included in the band model of the solid, we find that they introduce a few discrete energy levels within the forbidden band gap of the solid. These energy levels are called *traps* and appear in different elements at different energy levels determined largely by the type of imperfections and the impurity atoms. As we will see in Chapter 6, in many cases impurity atoms are purposely introduced into the solid, especially into a semiconductor, in order to change the electrical characteristics of a given sample.

5.3 *E–k* DIAGRAM, REDUCED ZONE

We can infer additional information about the behavior of an electron inside a crystalline solid by considering the wave nature of an electron. The right side of

Eq. 5.12 depends on k, which is inversely proportional to the lattice constant d. Actually, k is related to the propagation constant of an electron in the usual plane wave approximation of an electron. The energy of an electron in free space is related to the propagation constant of the wave by Eq.1.4.

$$E = \frac{\hbar^2 k^2}{2m} \tag{5.15}$$

This energy of the free particle is plotted (dashed lines) as a function of k in Fig. 5.14*a*. When we consider an electron in a crystalline solid, we know that the electron is not absolutely free as we have treated the electron in Chapter 1. As the electron moves within the solid, it is subjected to a periodically changing potential energy produced by the lattice atoms. For a given direction in which the electron is moving, its motion is perturbed by this periodic potential.

To understand the actual behavior of the electron motion in a solid, we will first discuss the Bragg reflection, which describes the reflection of a plane wave from a periodic array of atoms. If the incident plane wave propagation vector makes an angle β with respect to the normal to the plane of a two-dimensional lattice as shown in Fig. 5.15, the condition for the constructive interference of the reflected waves is given by the Bragg condition

$$n\lambda = 2d \cos\beta \qquad (n = \text{integer}) \tag{5.16}$$

When the angle of incidence is zero, the Bragg condition reduces to

$$n\lambda = 2d$$

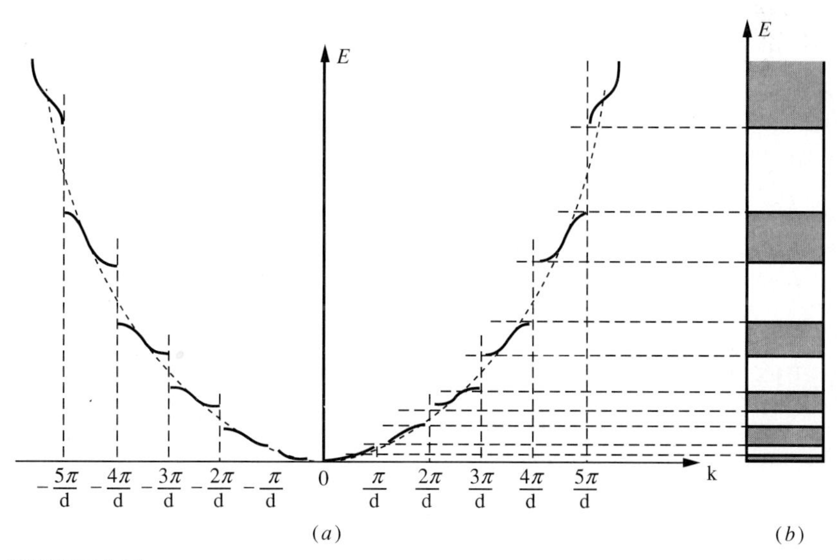

FIGURE 5.14
(*a*) Energy dependence of a free electron as a function of k (dotted line) and under the influence of the perturbing lattice forces (solid line). (*b*) Resulting band structure of the crystal.

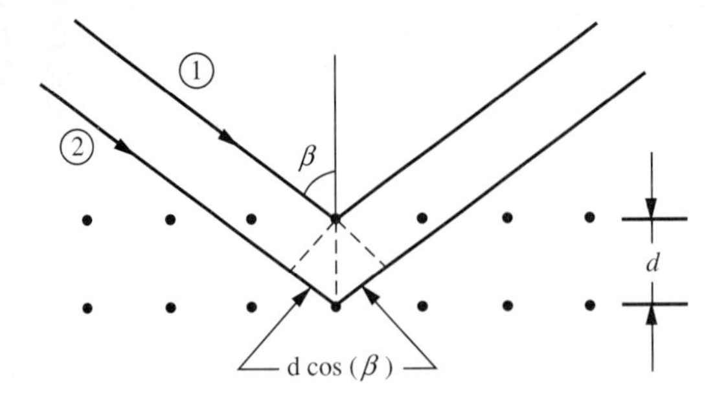

FIGURE 5.15
Bragg reflection of a plane wave from a two-dimensional lattice. Ray (2) travels a distance of 2 $d \cos \beta$ longer than ray (1).

or

$$k \equiv \frac{2\pi}{\lambda} = \frac{n\pi}{d} \tag{5.17}$$

In other words, for values of k equal to the multiples of π/d, a Bragg reflection occurs and the electron is completely reflected back from the lattice.

We can now use the Bragg reflection condition given by Eq. 5.17 for an electron moving inside the periodic lattice. Again, we will assume that there is a one-dimensional array of atoms, and the energy of the electron is given by Eq. 5.15. As the energy of the electron under consideration increases, the corresponding value of k also increases. But as the k of the electron approaches the Bragg condition, the electron suffers a complete Bragg reflection. This introduces discontinuities in energy in the vicinity of $n\pi/k$, where electrons are not allowed. Thus again, there result allowed and forbidden energy bands for the electrons, as shown by solid lines in Fig. 5.14*a*. Figure 5.14*b* shows the projection of the energy levels on a vertical energy scale.

Within a given energy band, the energy of the electron is a periodic function of k. If we replace k by $k \pm \pi n/d$, the energy dependence within that band does not change. It is therefore easier to restrict the analysis of the electron behavior inside the crystal in the range of k values that lie between $-\pi/d$ and $+\pi/d$. Such a model is known as the *reduced zone* and the corresponding k values are known as the *reduced-k* values. The energy dependence of the electron on the reduced-k values is shown in Fig. 5.16.

5.4 EFFECTIVE MASS

When an external force F_{ext} is applied to the electrons, the equation of motion for the electrons can be written as

$$F_{\text{ext}} + F_{\text{int}} = m\frac{dv}{dt} \tag{5.18}$$

Where F_{int} is the internal force exerted on the electrons by the background potential

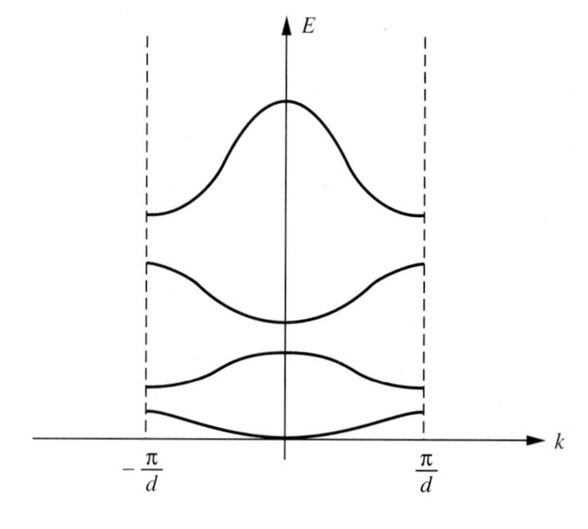

FIGURE 5.16
Reduced zone for the electrons inside the crystal.

of the lattice atoms. We now ask the question of whether we can eliminate the internal forces from Eq. 5.18 by writing it in the form

$$F_{\text{ext}} = m^* \frac{dv}{dt} = m^* a \tag{5.19}$$

where m^* is an effective mass that now includes and takes care of the effects of the internal lattice forces on the electrons.

The work done in moving a particle through a distance of Δs can be written as

$$\Delta E = F_{\text{ext}} \Delta s = F_{\text{ext}} \left(\frac{\Delta s}{\Delta t} \right) \Delta t = F_{\text{ext}} v_g \Delta t \tag{5.20}$$

where the group velocity of the particle is given by

$$v_g = \frac{\Delta \omega}{\Delta k} = \frac{1}{\hbar} \frac{\Delta E}{\Delta k} \tag{5.21}$$

Equation 5.20 then becomes

$$\Delta E = \frac{1}{\hbar} \frac{\Delta E}{\Delta k} \Delta t F_{\text{ext}}$$

Solving for F_{ext}, in the limit $\Delta t \rightarrow 0$, we obtain

$$F_{\text{ext}} = \hbar \frac{dk}{dt} \tag{5.22}$$

Acceleration of a wave packet representing an electron can be written as

$$a = \frac{dv}{dt} = \frac{d}{dt}\left[\left(\frac{1}{\hbar}\right)\frac{dE}{dk}\right]\frac{dk}{dk}$$

$$= \frac{1}{\hbar}\frac{d^2E}{dk^2}\frac{dk}{dt}$$

$$= \frac{1}{\hbar}\frac{d^2E}{dk^2}\frac{F_{\text{ext}}}{\hbar}$$

$$= \frac{1}{\hbar^2}\frac{d^2E}{dk^2}F_{\text{ext}} \tag{5.23}$$

In deriving this equation, we have used Eqs. 5.21 and 5.22. If we define an effective mass m^* by the relation

$$m^* = \frac{\hbar^2}{(d^2E/dk^2)} \tag{5.24}$$

then Eq. 5.23 simply reduces to $F_{\text{ext}} = m^* a$.

If we investigate the variation of the electron velocity (Eq. 5.21) and the effective mass of the electron (Eq. 5.24) within the reduced zone as functions of k, we can deduce additional properties of the electron motion inside the solid. Figure 5.17*a* shows the E versus the reduced wave number k diagram for an allowed energy band of an electron (see also Fig. 5.14). The corresponding group velocity and the effective mass m^* are plotted in Fig. 5.17*b* and *c*, respectively. The velocity of the electron is negative between $-(\pi/d) < k < 0$ and positive between $0 < k < +(\pi/d)$. The effective mass is negative close to the upper band edges and positive close to the lower band edges.

The negative mass concept is not contradictory to the laws of motion. As can be seen from Fig. 5.17, the electron at point A has a negative mass, and its velocity is positive but decreasing in magnitude. Under the application of an external force directed in the $+k$ direction, the resulting deceleration of the electron coupled with the negative mass gives the actual motion in the direction of the applied force. On the other hand, when the electron is on the band edge near $-\pi/d$, the electron has a negative mass, and its velocity is also negative. But the electron is now decelerating in the direction of the applied force. Although the electron is moving in the $-k$ direction, negative mass coupled with deceleration gives the right direction of the force.

5.5 CONCEPT OF A HOLE

Let us assume that the band in question is completely full. If the electrons are acted upon by a force F_{ext}, the energy—and thus k—will increase with time. We now follow the motion of one of these electrons. It can gain enough energy from the external force so that it can reach the upper band edge. The electron can go over into the next higher energy band provided the gain in energy is large enough

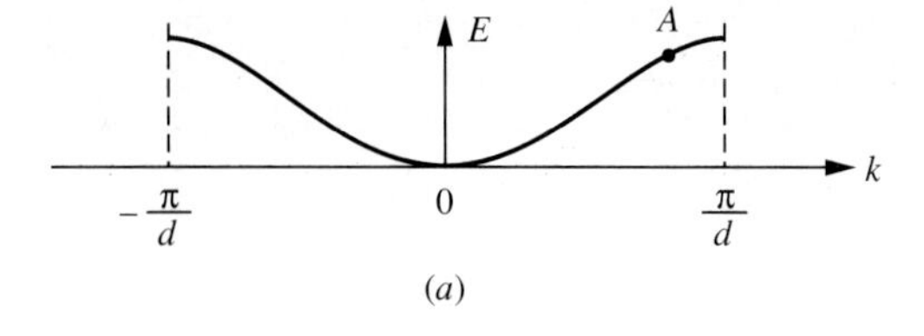

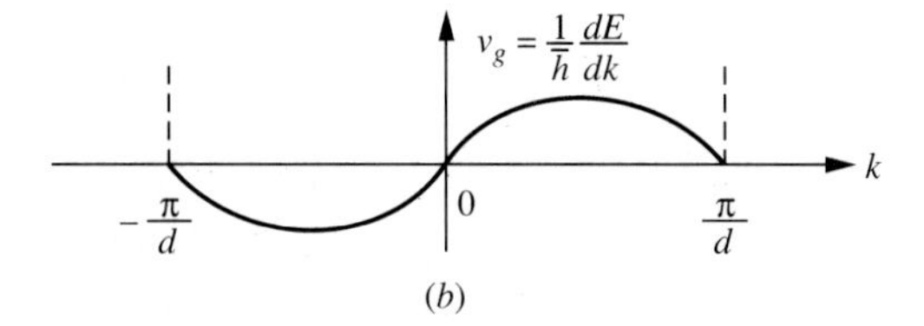

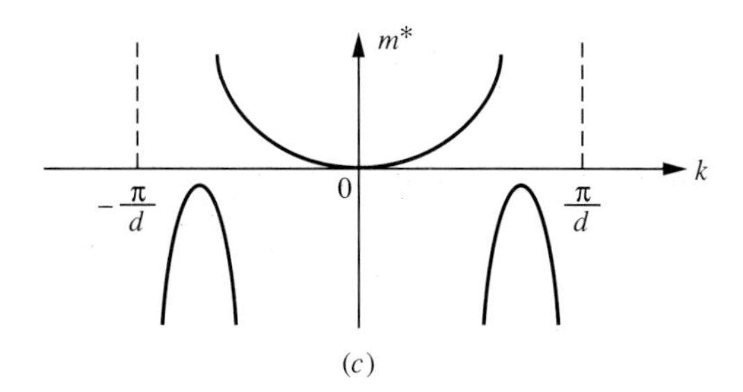

FIGURE 5.17
(*a*) Energy, (*b*) group velocity, and (*c*) effective mass of m^* of electrons as a function of the reduced wave number k.

to jump to the next band. For small forces, this is unlikely to happen. Therefore, the electron encounters a Bragg reflection and appears at the upper left edge of the band with $-k$. Keep in mind that in a completely filled band, there is a pair of electrons at each energy, one with $-k$ and the other with $+k$. Thus, there is no net flow of charge, and the total current is zero.

On the other hand, if there is an electron missing from the band, under the application of an external force all the other electrons in the band will gain energy due to the applied force and move toward the upper band edge. Consider the energy state at the location A in Fig. 5.18 where there is an electron missing. As the electron at a gains energy, it will move into the place of the empty state at A. There is now a state at a where an electron is missing. The electron at b will now move to the empty state at a vacated by the electron that moved into A, and so on. It is easy to see that as the electrons move through the band, the empty state or *hole* moves in the opposite direction.

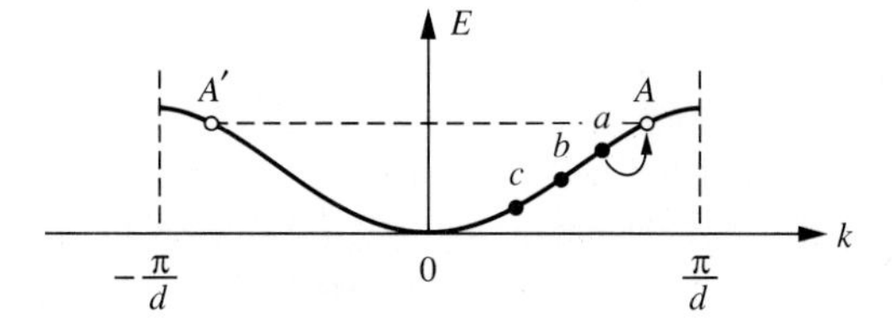

FIGURE 5.18
Electron at a moves into empty state A and consequently the empty state moves to a.

When a band is almost empty, the electrons usually occupy the lower energy levels of the band where the mass of the electrons is positive. In considering the motion of the electrons at these lower energy levels, their displacement through the solid gives the proper sign for the electrical current. The mass of the electrons is positive and the sign of their charges is negative, which is the usual sign for the electrons. But the mass of the electrons is still different from the free mass of the electrons.

When a band is almost full, most of the electrons occupy all the lower energy states and a few of the energy states near the upper band edges are empty. Under the application of an external force, it is better to consider the behavior of the few empty states rather than the motion of the many electrons occupying the lower levels of the band. Assume that only one electron is missing from the band. The total current due to all the electrons in the band except the jth electron can be written as

$$i = \sum_{i=1}^{N-1} -ev_i$$

$$i = \sum_{i=1}^{N} -ev_i + ev_j \tag{5.25}$$

where we have added an electron with $+e$ and $-e$. The first term on the right side of Eq. 5.25 is zero, since a completely filled band does not conduct current. Therefore the current due to the jth electron is simply

$$i = +ev_j \tag{5.26}$$

If we calculate the rate of change of current with time, we can write

$$\frac{di}{dt} = e\frac{dv_j}{dt}$$

$$= e\frac{F_j}{m^*} \tag{5.27}$$

but $F = -e\mathcal{E}$ ($\mathcal{E}$ is the applied field), then

$$\frac{di}{dt} = -e^2\left(\frac{\mathcal{E}}{m^*}\right) \tag{5.28}$$

Since the effective mass of the electron is negative at the upper parts of a given band, the rate of change of current due to the missing electrons is still positive. These empty states are called *holes*.

From here on, instead of dealing with many electrons in an almost-filled band, we will deal with fewer particles by using the concept of a *hole*. A hole will be identified by a positive electrical charge ($+e$) possessing an equivalent positive effective mass. This removes the controversy associated with the negative effective mass for the electrons at the upper energy levels of an almost-filled band. The sign of the current given by Eq. 5.26 is then consistent with the flow of positive charge.

5.6 INTRINSIC SEMICONDUCTORS

An intrinsic semiconductor (perfect and with no externally added impurities) is an insulator at absolute zero temperature. If we consider the binding forces among the atoms in the crystal, this means that all the valence electrons of the atom are very strongly coupled to the next neighboring atom as shown in Fig. 5.19*a*, so that the electrons cannot move within the crystal. If an external source of photons interacts with a tightly bound electron, this electron can overcome the binding forces and become free, leaving behind an empty state. The energy necessary to free an electron is equal to the band gap energy E_g. We say that the electron is excited to the conduction band from the valence band. In this process, an *electron-hole* pair is produced. The recapture of the electrons by the atoms to complete their electron-pair binding is known as a *recombination* of an electron-hole pair. With this process, both charges are removed from circulation.

Under the application of an external electric field, the free electrons will move within the crystal, provided they are not recaptured by the atoms. A bound electron from a neighboring atom can also move into the location of an empty

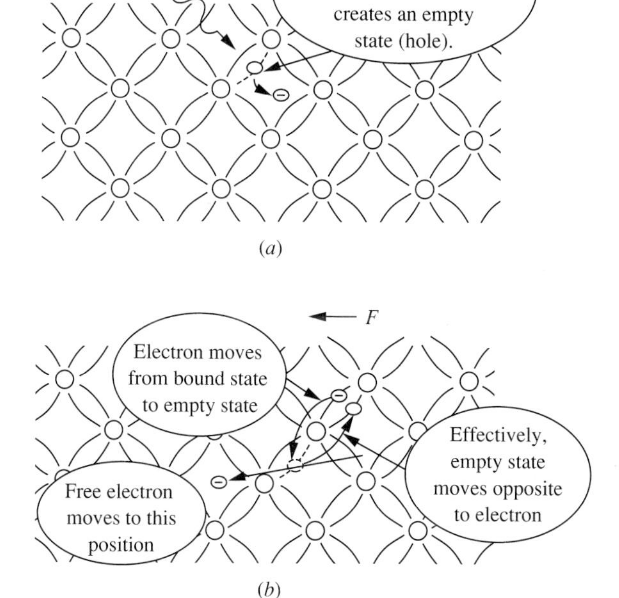

FIGURE 5.19
(*a*) The tightly bound electron gains energy fom the incident photon. It becomes free and a hole (empty state) is created at this location. (*b*) By application of an external field, an electron from a neighboring bond can move into an empty location. This is equivalent to a hole moving within the crystal in addition to the free electron.

state as shown in Fig. 5.19*b*. In this sense, the motion of the empty state is opposite to the motion of the bound electron. The motion of the bound electrons from one empty state to the other is equivalent to holes moving in the opposite direction to the free electrons.

From the perspective of the band theory, the case of Fig. 5.19*a* (without any external excitation) is equivalent to a completely filled valence band and a completely empty conduction band. With photon or thermal excitation, it is equivalent to the excitation of electrons from the valence band into the conduction band generating holes in the valence band, where now the electrical conduction of the charged particles is determined by both the motion of the electrons in the conduction band and the holes in the valence band. Of course, at $T = 0$ K, when the external source of photons is removed, the electrons will recombine with the holes and the semiconductor will not conduct current. When room and higher temperatures are considered, there is a continuous thermal excitation of electrons from the valence band into the conduction band. At the same time, some of these electrons recombine with the holes and are lost from the conduction band. Eventually, an equilibrium steady state is reached where the number of holes in the valence band and the number of electrons in the conduction band are equal.

We can now calculate the temperature dependence of the number density of the electrons and holes in an intrinsic semiconductor at a given temperature. Volume density of available electronic states at an energy E is given by Eq. 3.11 (see Chapter 3). If we use the bottom of the conduction band as zero energy and denote the density of electronic states by $n_c(E)$, we can write Eq. 3.11 as

$$n_c(E) = \frac{8\sqrt{2}\pi}{h^3}(m_e^*)^{3/2}(E - E_c)^{1/2} \tag{5.29}$$

Electrons have less and less available energy states as the top of a band is approached. We can therefore write the available energy-state density near the upper parts of the valence band as

$$n_v(E) = \frac{8\sqrt{2}\pi}{h^3}(m_p^*)^{3/2}(E_v - E)^{1/2} \tag{5.30}$$

where m_p^* is the effective mass of the hole. Note that energy E of an electron is negative below and close to E_v. Fig. 5.20*a* shows the variation of density of available states in the conduction and valence bands in an intrinsic conductor.

The probability that an electron will occupy an energy E is given by the Fermi–Dirac distribution function, that is,

$$f_n(E) = \frac{1}{1 + e^{[(E - E_F)/kT]}} \tag{5.31}$$

The probability that an energy state will not be occupied by an electron, that is, the probability that a hole will occupy the energy state E in the valence band, can be written as

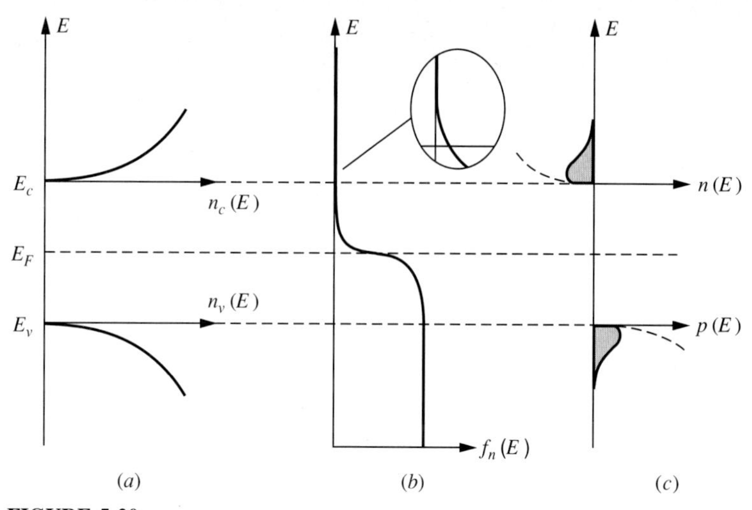

FIGURE 5.20
(*a*) Density of available states, (*b*) Fermi–Dirac distribution, and (*c*) electron and hole densities in an intrinsic semiconductor as a function of energy *E*.

$$f_p(E) = 1 - f_n(E) = \frac{e^{[(E-E_F)/kT]}}{1 + e^{[(E-E_F)/kT]}} \tag{5.32}$$

The variation of the Fermi–Dirac distribution for a semiconductor as a function of energy is shown in Fig. 5.20*b*.

The density of electrons in the conduction band is given by

$$n = \int_{E_c}^{\infty} f_n(E) n_c(E)\, dE \tag{5.33}$$

Similarly, the density of holes in the valence band is

$$p = \int_{-\infty}^{E_v} f_p(E) n_v(E)\, dE \tag{5.34}$$

Variations of the products of $[f_n(E)n_c(E)]$ and $[f_p(E)n_v(E)]$ are plotted in Fig. 5.20*c* as a function of energy E. Note that to distinguish between the polarity of charges, the electron density is plotted to the left and the hole density to the right side of the charge-density axis; also the corresponding charge densities are highly exaggerated.

Using the approximations that for the conduction band $(E - E_F) \gg kT$ and for the valence band $(E_F - E) \gg kT$, and substituting these into Eq. 5.33 and 5.34, we obtain for the density of conduction electrons and valence holes respectively

$$n = \frac{2}{h^3}(2\pi m_e^* k)^{3/2} T^{3/2} e^{[(E_F - E_c)/kT]} \quad 5.35$$

and

$$p = \frac{2}{h^3}(2\pi m_p^* k)^{3/2} T^{3/2} e^{[-(E_F - E_v)/kT]} \quad 5.36$$

If we now calculate the product of n and p, we obtain

$$np = 4\left(\frac{2\pi k}{h^2}\right)^3 (m_e^* m_p^*)^{3/2} T^3 e^{[-(E_c - E_v)/kT]} \quad (5.37)$$

or

$$np = 4\left(\frac{2\pi k}{h^2}\right)^3 (m_e^* m_p^*)^{3/2} T^3 e^{(-E_g/kT)} \quad (5.38)$$

or

$$np = 2.33 \times 10^{43} (m_{er}^* m_{pr}^*)^{3/2} T^3 e^{(-E_g/kT)} \quad (5.39)$$

Where we have used $(E_c - E_v) = E_g$, the band gap energy and m_{er}^* and m_{pr}^* are the ratios of the effective masses of the electrons and holes relative to the rest mass m_0 of the electron. It is apparent that the (np) product for a semiconductor depends on the band gap energy E_g and the absolute temperature T of the crystal.

In an intrinsic semiconductor

$$n = p = n_i \quad (5.40)$$

Thus, the intrinsic electron or hole concentrations are given by

$$n_i = 2\left(\frac{2\pi k}{h^2}\right)^{3/2} (m_e^* m_p^*)^{3/2} T^{3/2} e^{(-E_g/2kT)} \quad (5.41)$$

The (np) product given by Eq. 5.38 holds for all semiconductors, even if they contain purposely introduced impurities. It should be also kept in mind that the larger the band gap of a given semiconductor, the smaller will be the intrinsic concentration n_i.

Example 5.1. Calculate the intrinsic hole concentration of a germanium crystal at room temperature.

Solution. For simplicity, we take $m_e^* = m_p^* = m_e$ (the free electron mass). The band gap energy of germanium is 0.67 eV.

$$p = n_i = \left(\frac{2\pi\ 1.38 \times 10^{-23}\ \text{J/K}}{(6.62 \times 10^{-34}\ \text{J}\cdot\text{s})^2}\right)^{3/2} [(9.1 \times 10^{-31}\ \text{kg})^2]^{3/4}$$

$$\times\ (300\ \text{K})^{3/2} \exp\left[-\frac{(0.67\text{eV})(1.6 \times 10^{-19}\ \text{J/eV})}{2(1.38 \times 10^{-23}\ \text{J/K})(300\ \text{K})}\right]$$

$$p = 5.98 \times 10^{22}\ \text{holes/m}^3 = 5.98 \times 10^{16}\ \text{holes/cm}^3$$

In Fig. 5.20, when the Fermi level E_F in an intrinsic semiconductor was plotted as a function of E, the Fermi energy was placed at the center of the forbidden energy gap, between the conduction and valence bands. We can now mathematically show that this is the actual location of the Fermi energy in an intrinsic semiconductor. When we discussed the Fermi energy in Chapter 2, we concluded that at absolute zero temperature the energy levels below E_F are full and above E_F are empty. Since for an intrinsic semiconductor the valence band is completely full and the conduction band is completely empty, we expect the Fermi level to fall between the conduction and valence bands. The location of the Fermi level inside the forbidden energy band is not in violation of our previously derived results on the occupancy of the electronic states nor the band theory of solids. It merely states that the probability that an electron can occupy a certain state is determined relative to the Fermi energy, even if this energy is not an allowable energy state for an electron.

We can use Eqs. 5.35 and 5.36 to determine the location of the Fermi energy level in an intrinsic semiconductor. At equilibrium, the electron density and the hole density should be equal. If we let $E_I = E_F$ for an intrinsic semiconductor, equating Eq. 5.35 to Eq. 5.36, we find for the intrinsic Fermi level E_I as

$$E_I = \frac{E_c + E_v}{2} + \frac{3kT}{4} \ln\left(\frac{m_e^*}{m_p^*}\right)$$

if we take the effective masses equal,

$$E_I = \frac{E_c + E_v}{2} \tag{5.42}$$

which shows that the intrinsic Fermi level E_I lies midway between the conduction and valence bands.

5.7 PROPERTIES OF COMMON SEMICONDUCTORS

There are various types of semiconductors that are used in implementing solid-state devices. The choice of one type of semiconductor over another depends on the particular application. These selections are determined mostly by the specific required properties of the semiconductors. Some of these semiconductors, such as Ge and Si, by nature behave as semiconductors alone, but if other elements—from the IIIrd and Vth columns in the periodic table—are combined together, they form what are known as compound semiconductors with features that are sometimes superior to single elements like Ge or Si. Note that the elements that fall in the IIIrd column of the periodic table have three valence electrons and the elements that fall in the Vth column have five valence electrons.

When elements are combined to form a new compound, the bonds between various atoms form in such a proportion of these elements that the electron-pair bindings become uniform and complete throughout the compound, and the element acts as an intrinsic semiconductor. That is, all the individual valence electrons of the elements in the compound are shared in the electron-pair formation. If the bond completeness is not satisfied, depending on which element has a higher concentration relative to the other, these compounds may show properties of extrinsic semiconductors—semiconductors whose electrical properties are controlled either by excess electrons or by excess holes. This was one of the reasons why the development of compound semiconductors was delayed so long. But the technology has matured so much in the last few decades that highly sophisticated electronic devices can now be realized using compound semiconductors as a basis.

Properties of some common semiconductors are listed in Table 5.1. In addition to the band gap energy of these elements, the mobilities of electrons and holes are also listed.

5.7.1 Germanium

When semiconductor technology was evolving, germanium was one of the early semiconductors that was easily produced in large quantities and with sufficient purity. Germanium also has higher electron mobility than silicon. As we will show in the later chapters of this book, one of the essential requirements for a semiconductor device to operate at high frequencies is that it have a very high charge mobility. We will also show that in order for a semiconductor to operate at higher temperatures, it should also possess a large band gap energy. With the recent technological improvements in the manufacture of silicon and gallium arsenide, germanium is now generally used in devices which incorporate the bulk properties of the semiconductor for their electrical characteristics. The band structure of germanium is shown in Fig. 5.21*a*. Other important physical parameters of germanium are listed in Table 5.2.

5.7.2 Silicon

Silicon is at present one of the most widely used semiconductors in the semiconductor industry. In addition to possessing a higher band gap energy, the natural

TABLE 5.1
Properties of Semiconductors

Material	Type*	E_g(eV)	μ_e(cm^2/V · s)	μ_p(cm^2/V · s)
GaP	I	2.24	300	100
AlSb	I	1.52	200	550
GaAs	D	1.43	8800	400
InP	D	1.27	4600	150
InAs	D	0.35	33000	460
InSb	D	0.17	78000	750
GaSb	D	0.70–0.74	9000	1400
Ge	I	0.67	3800	700
Si	I	1.12	1400	480

* Direct (D) or Indirect (I) energy gap.

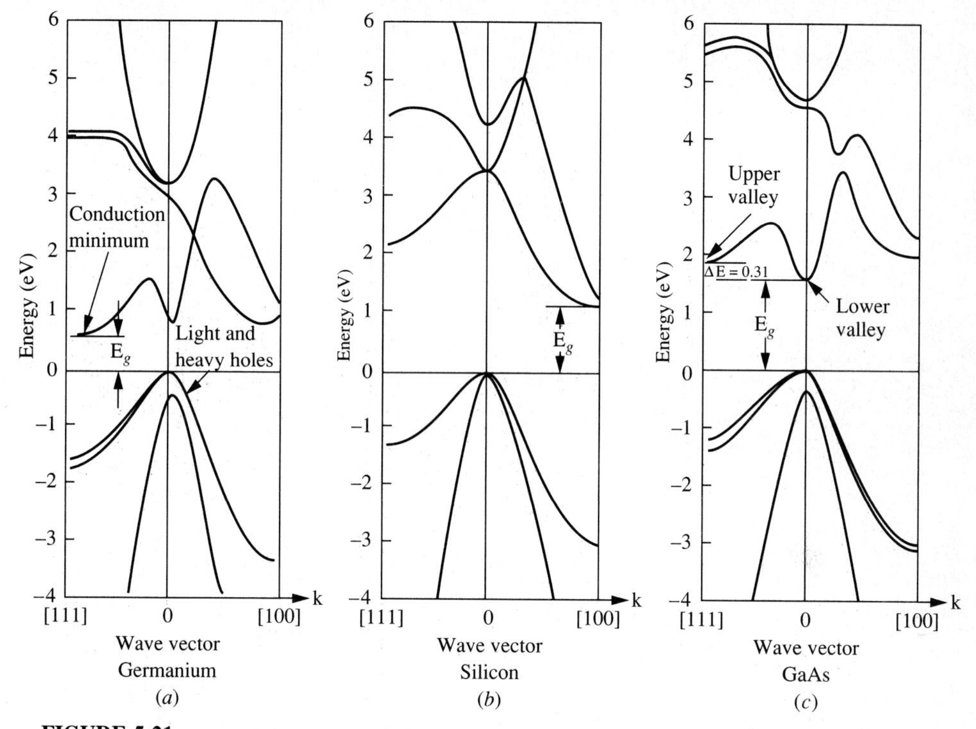

FIGURE 5.21
Energy band diagrams of (*a*) germanium, (*b*) silicon, and (*c*) GaAs as a function of reduced wave vector for [111] and [100] directions (Chelikowski and Cohen, 1976).

oxide of silicon makes the manufacture of semiconductor devices comparatively easy. Silicon also possesses superior thermal characteristics. The energy band diagram of silicon is shown in Fig. 5.21*b*, and other properties are given in Table 5.2.

5.7.3 Compound Semiconductors

GaAs is gradually winning acceptance in the semiconductor device area. Development of GaAs technology was slow because of the difficulty in producing pure and uniform crystal samples. Since GaAs is a compound semiconductor, obtaining proper intrinsic GaAs wafers is difficult and requires considerably more advanced technology. Because of the high electron mobilities and higher band gap energy, GaAs is gradually replacing silicon in many high-frequency devices. GaAs also has other unusual properties, such as possessing negative differential mobility at high electric fields leading to highly interesting applications. These will be discussed in later chapters. The band structure of GaAs is shown in Fig. 5.21*c*. Various other parameters of GaAs are given in Table 5.2.

TABLE 5.2
Properties of Ge, Si, and GaAs

	Ge	Si	GaAs
Atomic number	32	14	(Ga)31,(As)33
Energy gap, E_g (eV)	0.67	1.12	1.43
Dielectric constant	16	11.8	13.1
Effective mass of electrons (m_e^*/m)	0.082	0.19	0.072*
Effective mass of holes (m_p^*/m)	0.28	0.49	0.5
Electron mobility $\mu_e (m^2/V \cdot s)$	0.39	0.15	0.88*
Hole mobility $\mu_p (m^2/V \cdot s)$	0.15	0.05	0.04
Electron diffusion constant D_e (m^2/s)	9.9×10^{-3}	3.80×10^{-3}	2.29×10^{-2}
Hole diffusion constant D_p	4.9×10^{-3}	1.3×10^{-3}	1.04×10^{-4}
Effective intrinsic $n_i^{\dagger}$ (electron/cm^3)	2.5×10^{13}	1.4×10^{10}	1.8×10^{6}
Scatter limit velocity, v_{sL}			
electrons (cm/s)	6×10^{6}	10^{7}	9×10^{6}
holes (cm/s)	8×10^{6}	10^{7}	9×10^{6}
$\mathcal{E}_B$ (field before breakdown) (V/cm)	$2\text{–}3 \times 10^{5}$	$3\text{–}6 \times 10^{5}$	$3.5\text{–}6.5 \times 10^{5}$

Note: all temperature dependent parameters are at $T = 300$ K.
* at the main valley
† measured

As can be seen from Fig. 5.21, as in almost all semiconductors the compound semiconductors have two types of holes traveling in their valence bands: *heavy holes* and *light holes*. The light holes move freely, similar to the electrons in the conduction band, but the heavy holes are very sluggish. Unfortunately, the potential energy of the two holes coincides at $k = 0$; thus, the overall effect of the holes is governed by the sluggish heavy holes, which have effective masses nearly eight times larger than the light holes. By introducing strain into the crystal, the two bands can be separated and light holes can be preferentially used to increase the hole mobility.

New compound semiconductors can be produced using three or four elements. If there are three elements involved in the final compound, it is called a *ternary* compound, and if four elements are involved, it is called a *quaternary* compound.

When the energy band structure of crystals is studied in three dimensions, the electron energy dependence as a function of the reduced vector shows interesting differences. This variation is highly direction dependent, as shown in Fig. 5.21. The minimum energy does not always occur at $\mathbf{k} = 0$, but at other values of $\mathbf{k}$. The crystal is usually cut along certain preferred axes to take advantage of the crystal behavior in these directions. These axes are related to the orientation of the cut planes relative to the unit cell location of the crystals.

When the semiconductor has a minimum of energy at $\mathbf{k} = 0$, it is called a *direct band gap* semiconductor. If the minimum occurs at a different $\mathbf{k}$, then it is called an *indirect band gap* semiconductor. The band gap energy minimum of a semiconductor can be altered by changing the molar concentration of an element in a compound semiconductor. A variation of energy as a function of reduced k for

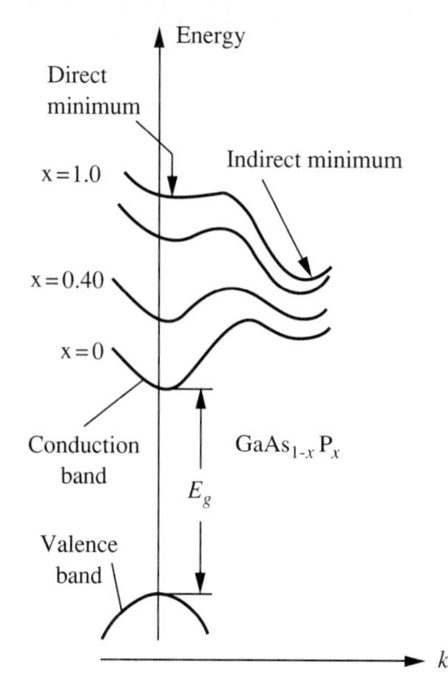

FIGURE 5.22
Variation of E as a function of reduced k for various molar concentrations of P in GaAs/GaP system (*Optoelectronic Applications Manual*).

various compounds of the GaAs/GaP system is shown in Fig. 5.22. As the mole fraction x in $GaAs_{(1-x)}P_{(x)}$ is increased, the compound semiconductor changes from a direct to an indirect band gap semiconductor. Whether a semiconductor possesses a direct or indirect band gap energy makes the semiconductor a prime candidate for certain specific applications, such as optical detectors, light-emitting diodes, lasers, and so on. Figure 5.23 shows the change in the band gap energy as a function of molar concentration in a $GaAs_{(1-x)}P_x$ compound semiconductor. By choosing the molar concentration properly, it is possible to obtain a compound semiconductor with close to the desired band gap energy. This is very important, especially in realizing optical devices.

Figure 5.24 shows the energy gap and lattice constants for several III-V compound semiconductors. As we will show in the later chapters of this book, growing

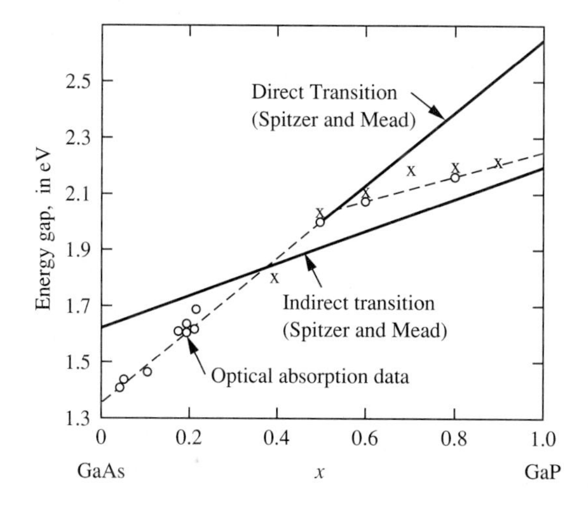

FIGURE 5.23
Energy gap versus composition for $GaAs_{(1-x)}P_{(x)}$. The dashed lines represent room temperature optical measurements. The solid lines represent theoretical calculations (Willardson and Beer, 1966).

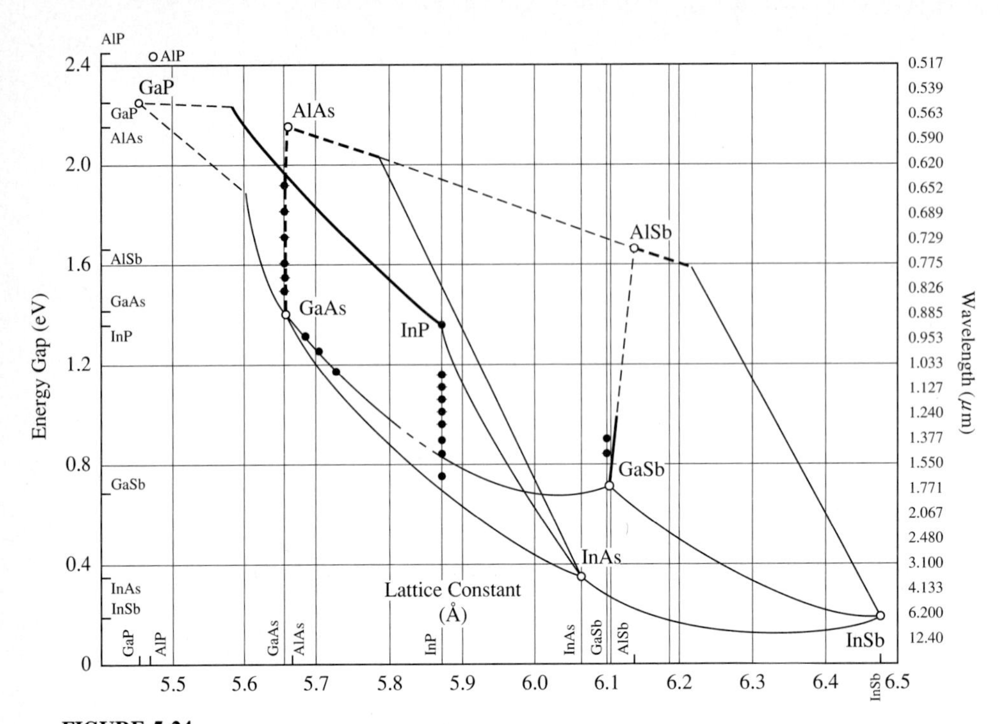

FIGURE 5.24
Energy gap and lattice constants for several III-V compounds. The boundaries joining the binary compounds give ternary energy gap and lattice constants (Tsang, 1985).

alternate layers of compound semiconductors with different compositions—that is, with different band gap energy—over one another produces highly interesting and useful devices. The essential requirements for growing alternate layers with abrupt transitions depend on how well the lattice constants of the two different compound semiconductors match for certain combinations of these semiconductors. As shown in Fig. 5.24, GaAs and AlAs have nearly the same lattice constant. Thus, any combination of $Al_xGa_{1-x}As$ will match GaAs, opening ways for realizing well lattice-matched devices. Other combinations of elements can be found from the figure that will allow growth of lattice-matched compound crystals. For example, $In_xGa_{1-x}As$ can be lattice-matched to InP at a specific molar concentration x.

FURTHER READING

Azaroff, L. V., and J. J. Brophy, *Electronic Processes in Materials*, McGraw-Hill, New York, 1963.

Chelikowski, J. R., and M. L. Cohen, "Nonlocal Pseudopotential Calculations for the Electronic Structure of Eleven Diamond and Zinc-Blend Semiconductors," *Phys. Rev.*, vol. B14, 1976, 556.

Conwell, E. M., *High Field Transport in Semiconductors*, Academic Press, New York, 1967.

Dekker, A. J., *Solid State Physics*, Prentice Hall, New York, 1961.

Long, D., "Energy Band Structures of Mixed Crystals of III-V Compounds," in *Semiconductors and Semimetals*, R. K. Willardson and A. C. Beer, eds., vol. 1, Academic Press, New York, 1966.

Optoelectronic Applications Manual, Hewlett-Packard Optoelectronic Division, McGraw-Hill, New York, 1977.

Tsang, W. T., "MBE for III-V Semiconductors," in *Semiconductors and Semimetals*, R. K. Willardson and A. C. Beer, eds., vol. 22, Academic Press, New York, 1985.

PROBLEMS

5.1. Calculate the intrinsic electron concentration for InP and InSb at room temperature. Compare these results with the electron concentration in copper.

5.2. Find the temperatures at which the intrinsic electron concentrations of Si and GaAs become equal to that of Ge at room temperature.

5.3. Find the intrinsic hole concentration of InSb at $T = 77$ K and compare it with that at room temperature. (See Problem 5.1).

5.4. If the transit time of a charge carrier is defined as the time it takes for an electron to traverse between two points, compare the transit time of electrons and holes in Si and GaAs if the length of the sample is 50μm.

5.5. A typical energy band diagram for an n- and p-type silicon semiconductor is shown in Fig. P.5.5. Find the work function and electron affinity for each semiconductor. What is the band gap energy of the silicon?

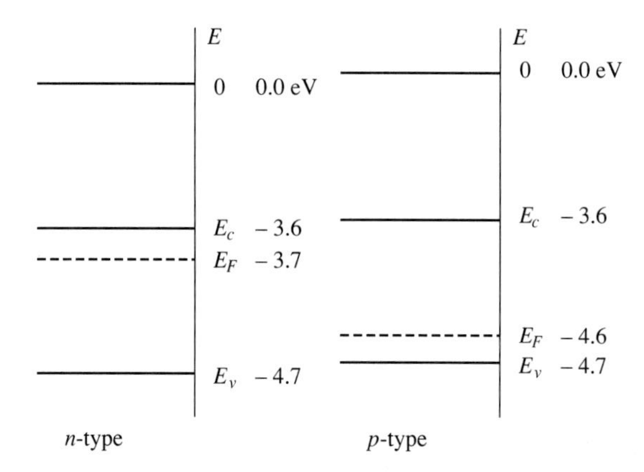

FIGURE P.5.5

5.6. Show that if Eq. 5.13 k is replaced by $k + \pi n/d$, the energy dependence within a given band does not change.

5.7. Show that the distances between successive planes in a cubic crystal are given by

$$d_{hkl} = a/(h^2 + k^2 + l^2)$$

5.8. Show that the simple cubic lattice of NaCl shown in Fig. 5.7 can be thought of as two face-centered cubic lattices.

5.9. The lattice constant of NaCl is 2.81 Å. Find the density of NaCl crystal. The atomic weights of Na and Cl are 22.991 and 35.457, respectively.

5.10. Calculate the theoretical values of the intrinsic electron concentrations for Ge, Si, and GaAs, and compare them with the measured values given in Table 5.2. Explain the reasons for the differences.

CHAPTER 6

EXTRINSIC SEMICONDUCTORS

In Chapter 5, the general physical and electrical properties of intrinsic (pure and ideal) semiconductors were given. In this chapter, we will discuss the changes in these properties when there are impurities or defects in the semiconductors. The defects, which are defined as the local random deviations from the ideal crystalline structure, usually occur during the manufacturing process of even the purest semiconductors. Impurities occur during the crystal-growing phase or may be purposely introduced into the semiconductor to change the electrical characteristics of the whole or part of the semiconductor. In the early days, the production of highly pure crystals of single semiconductor materials was a major problem due to the inefficient removal of unwanted impurity elements during the manufacturing process. Throughout the years, as the technology has matured, the manufacturing process has reached such a sophisticated stage that it is now possible to produce highly pure compound semiconductor crystals with the highest electronic grade purity.

In order to control the electrical characteristics of the semiconductor devices, certain controlled amounts of impurities are purposely introduced into semiconductors during the manufacturing stages of the semiconductor devices. Depending on which impurity elements are introduced into the semiconductor, the electrical characteristics of the semiconductor crystal show drastic changes and consequently either negative (electrons) or positive (holes) charges become the dominant charge carriers within the semiconductor. Any semiconductor that deviates from the ideal and pure crystalline lattice structure is known as an extrinsic semiconductor. In general, dominant extrinsic properties are mostly created by purposely adding controlled amounts of impurity atoms into the semiconductor during the manufacturing phase of a device.

In this chapter, the effects of impurities on the electrical characteristics of semiconductors will be discussed. We will also introduce additional general properties and the parameters of semiconductors necessary to understand many of the device characteristics.

6.1 EXTRINSIC SEMICONDUCTORS

The III–V elements of the periodic table are generally used as impurities in monocrystalline semiconductors such as germanium (Ge) and silicon (Si). Figure 6.1*a* shows the two-dimensional schematic representation of the covalent bonds of the valence electrons of a pure Ge crystal. At absolute zero temperature, all the valence electrons of the Ge atoms are shared by the neighboring Ge atoms. The resulting band diagram of the intrinsic semiconductor is shown in Fig. 6.1*a*. Since there are no free electrons in the medium, the conduction band is completely empty and the valence band is completely full.

Let us now assume that an element from the Vth column of the periodic table such as an arsenic atom, which has five valence electrons, is substituted for one of the Ge atoms within the crystal, as shown in Fig. 6.1*b*. Four out of these five electrons are shared with the four neighboring Ge atoms. The fifth electron of As does not have any other available valence electrons within the crystal to be shared within a covalent or any other binding process. But at absolute zero temperature, this electron is still loosely coupled to its original As atom.

If, instead of an element from the Vth column, an element from the IIIrd column of the periodic table such as a boron atom, which has three valence

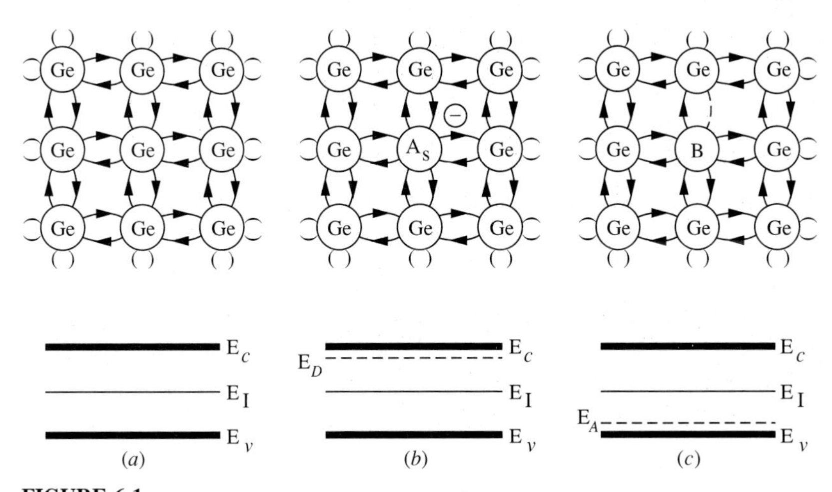

FIGURE 6.1
(*a*) Pure Ge crystal. Substitution of an impurity atom in place of Ge atom from the Vth (*b*) and from the IIIrd (*c*) column of the periodic table. Equivalent band diagrams are given under each diagram.

electrons, is substituted for one of the Ge atoms, the three valence electrons of the boron atom will be shared by the three neighboring Ge atoms. The electron of the fourth neighboring Ge atom will also be used in the binding process with the boron atom, but the boron atom will not have the fourth electron to complete this covalent binding process. The missing electron in this incomplete binding process is shown as a dotted line in Fig. 6.1*c*.

Let us increase the temperature of the impure crystal and look into the behavior of the excess or deficiency of the electrons in the binding process. We first consider the crystal that contains an impurity atom from the Vth column of the periodic table where the fifth valence electron is considered to be loosely coupled to its original atom. The binding energy of this electron can be calculated approximately by comparing this problem to that of a hydrogen-like atom where also only a single electron circulates around a stationary positive nucleus.

Since all of the other four electrons of the impurity atom are compensated by the four electrons of the neighboring Ge atoms, there is a charge imbalance of one positive charge located at the position of the impurity atom as shown in Fig. 6.2*a*. The binding energy of the fifth electron is then calculated by assuming that the loosely coupled electron is circulating around this equivalent positive charge under the influence of the Coulomb attraction.

The ionization energy of the electron in the hydrogen atom is given by

$$E_i = -\left(\frac{me^4}{32\pi^2\epsilon_0^2 h^2}\right)\frac{1}{n^2} \tag{6.1}$$

$$E_i = -\frac{13.6}{n^2}\text{eV} \qquad (n = 1)$$

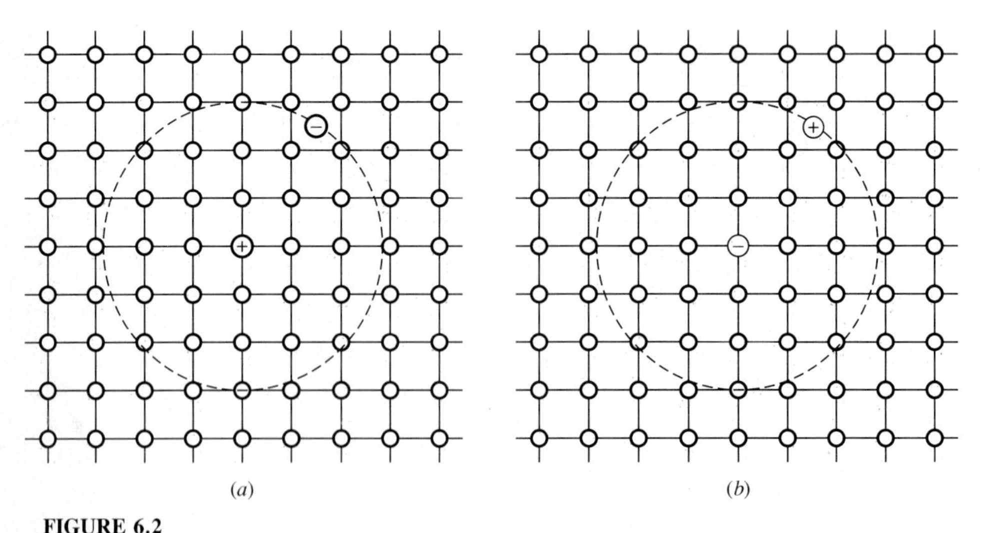

FIGURE 6.2
Orbits of a loosely coupled electron (*a*) and of a hole (*b*) inside an extrinsic semiconductor.

In deriving Eq. 6.1, it was assumed that the surrounding medium in the hydrogen atom problem, where the electron is moving, is a vacuum which has a dielectric permittivity of ϵ_0. The dielectric constant of a semiconductor crystal is usually very high, that is, the dielectric constants of Ge, Si, and GaAs are 16, 11.8, and 12, respectively. If we now assume that the fifth electron of the impurity atom is moving in a medium with a dielectric constant $\kappa_e > 1$, the ionization energy of the electron is proportionately reduced compared to that of the hydrogen atom. The ionization energy now becomes (replacing ϵ_0 by $\epsilon \Rightarrow \kappa_e \epsilon_0$)

$$E_i = -\frac{13.6}{\kappa_e^2} \qquad \text{(for } n = 1) \qquad (6.2a)$$

$$E_i = -0.053 \text{ eV} \qquad \text{(for Ge)} \qquad (6.2b)$$

We see that the equivalent binding energy of the loosely coupled fifth electron is very small. We also note that the ionization energy of the hydrogen atom depends on the mass of the electron. We have shown that due to presence of the lattice forces, the effective mass of an electron is different from the free mass of the electron. Since the effective mass of an electron near the band edges is usually smaller than the free electron mass, the ionization energy given by Eq. 6.2 is further reduced due to the mass dependence in the numerator of Eq. 6.1. We conclude that it requires very little external energy to free the loosely bound electron of the impurity atom, which then freely moves within the crystal.

The electron orbit in a hydrogen atom is also proportional to the dielectric permittivity of the surrounding medium (Eq. 2.16). Consequently, at absolute zero, the first Bohr radius of this electron in Ge also increases from 0.53 Å to 8.5 Å. Therefore, within 17 Å diameter, there are enough Ge atoms to warrant the loosely bound electron approximation.

The equivalent band diagram of a resulting semiconductor is shown in Fig. 6.1*b*. When the electron is free to move inside the crystal under the application of an external field, it means that it moves inside the conduction band. Since it requires approximately 0.05 eV to ionize the loosely bound electron, this also means that it requires 0.05 eV to move the electron from its bound ground-state location to the conduction band. Therefore, the equivalent energy level of the ground state of the bound impurity electron can be placed inside the forbidden energy gap at 0.05 eV below the bottom of the conduction-band energy E_c. Since the impurity atoms of the Vth column can easily give electrons to the conduction band and generate conduction electrons, they are known as *donor* atoms. When a semiconductor is doped by donor atoms the semiconductor is identified as an *n-type* semiconductor, due to the excess negative charges that are introduced into the semiconductor. The energy levels corresponding to the ground-state binding energy of the loosely coupled electrons of the donor atoms are known as *donor energy levels*. An electron which initially occupies a donor-atom energy state at absolute zero degrees Kelvin is said to be "excited" into the conduction band with the addition of thermal energy or any other external energy supplied to the

TABLE 6.1
Properties of impurity elements (Adler, Smith, and Longini, 1964)

			Ionization energy (eV)	
Material	**Atomic number**	**Type**	**Ge**	**Si**
Boron	5	Acceptor	0.0104	0.045
Aluminum	13	Acceptor	0.0102	0.057
Gallium	31	Acceptor	0.0108	0.065
Indium	49	Acceptor	0.0112	0.160
Phosphorous	15	Donor	0.0120	0.044
Arsenic	33	Donor	0.0127	0.049
Antimony	51	Donor	0.0096	0.039

electron. Thermal energy at room temperature is enough to excite almost 100 percent of the electrons from the donor energy levels into the conduction band.

When the impurity is an element from the IIIrd column, the situation is different. Due to the lack of a fourth valence electron to complete the electron-pair binding process, the role of the impurity atom is now reversed. The nucleus of the impurity atom is now treated as a fixed negative charge and the missing electron is replaced by a positive charge circulating around this negative core (Fig. 6.2*b*). When the equivalent hydrogen atom problem is now calculated, the corresponding ionization energy of the positive circulating charge is interpreted as the necessary energy required to move an electron from a neighboring Ge atom into the missing link of the impurity atom. This energy is again in the same order of the ionization energy of the donor atom discussed above. Note also that the effective mass of a hole, which is different from that of an electron, modifies the corresponding ionization energy of acceptor atoms.

In Chapter 5, it was shown that the motion of a hole was equivalent to the motion of a bound electron moving into an empty state. Therefore, with the addition of an impurity atom from the IIIrd column, we have introduced empty energy states above the top of the valence band into which valence band electrons can move and thus generate additional holes in the semiconductor at higher temperatures. The equivalent band diagram for this case is shown in Fig. 6.1*c*. The ionization energy calculated by the hydrogen atom model is now placed in the energy gap just above the valence band. These states are completely empty at absolute zero temperature.

As the temperature is increased, some of the electrons at the top of the valence band are easily excited into these energy states and leave behind in the valence band empty states. These empty states are then equivalent to the generation of holes in the valence band. The resulting semiconductor is known as a *p-type* semiconductor due to the excess positive charge carriers that are introduced into the semiconductor. The impurity atoms of the IIIrd column are known as *acceptor* atoms since they accept electrons from the valence band. The corresponding energy level of the impurity atom is known as the *acceptor energy level*.

Extrinsic semiconductors used for electronic devices contain controlled amounts of either donor or acceptor atoms in the host semiconductor. It is possible to *dope* various regions of a semiconductor with different impurity elements and combine the different electrical characteristics of these regions to implement the desired semiconductor device characteristics.

Properties of some of the commonly used impurity elements in Ge and Si are given in Table 6.1.

6.2 n- AND p-TYPE SEMICONDUCTORS

Band structures of an n-type and a p-type semiconductor are shown in Fig. 6.3. In the same figure are also shown the concentrations of the respective charge carriers

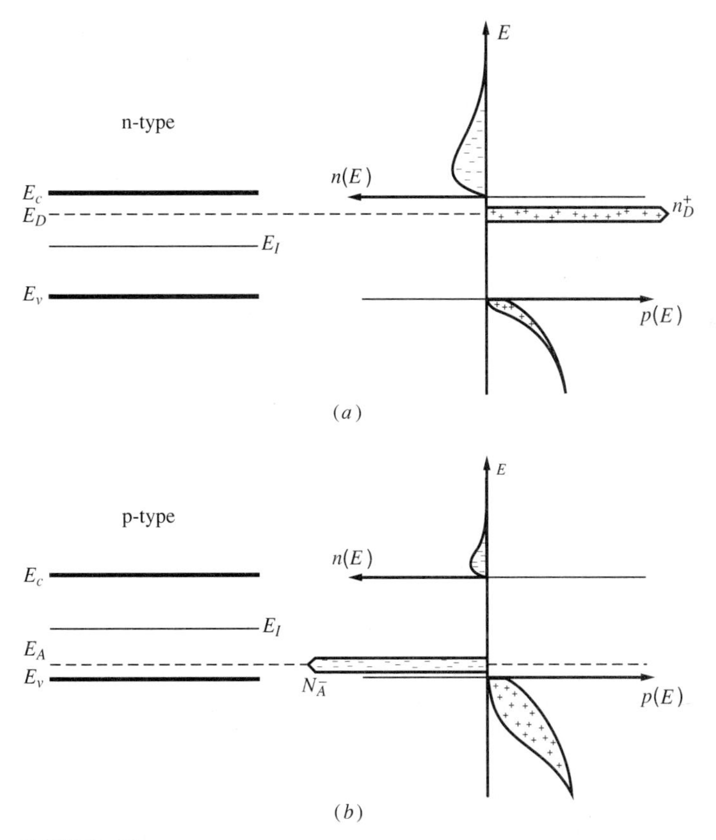

FIGURE 6.3
Energy-band diagram and charge concentrations in (*a*) n-type and (*b*) p-type semiconductors. Note the highly exaggerated energy distribution of the electrons and holes.

as a function of energy. In an n-type semiconductor, most or all of the donor atoms are ionized. The electrons that are excited into the conduction band leave behind immobile positive charges at the donor energy levels, as shown in Fig. 6.3*a*. All of these ionized atoms are placed at the location of the donor energy levels shown as a discrete energy level. But as the concentration of the donor atoms increases, due to the random distribution of the impurity atoms, there results a gaussian-type spread in this energy level. At still higher concentrations, the tail ends of these levels can extend into the conduction band and effectively reduce the band gap energy of the semiconductor. This effect is very important in the operation of semiconductor lasers and will be discussed further in Chapter 13.

At ordinary temperatures, there are also electrons that are thermally excited from the valence band into the conduction band, but their concentration is small compared to those electrons excited from the donor energy levels. The electrons excited from the valence band leave behind empty states in the valence band thus producing holes in the valence band. These are shown in Fig. 6.3*a* as positive charges in the valence band.

The energy distribution of the conduction electrons that are excited from the donor levels and the valence band are shown as negative charges in the conduction band in the same figure. The energy distribution of these electrons is highly exaggerated, especially at the higher energies. Actually the average energy value of the conduction electrons in the conduction band are within a few kT above the bottom of the conduction band.

For a p-type semiconductor, the band diagram and the resulting charge concentrations are shown in Fig. 6.3*b*. When electrons from the valence band are excited into the acceptor energy levels, the acceptor atoms are now negatively charged and like the donor ions are also immobile in the crystal. A large number of holes is created in the valence band due to the excitation of electrons from the valence band into the acceptor energy states. Again, the energy distribution for these holes as shown in Fig. 6.3*b* is highly exaggerated. There are also other holes in the valence band as a result of the thermal excitation of electrons from the valance band into the conduction band. Although at ordinary temperatures, the density of these holes is small compared to the holes generated by transitions into the acceptor levels, their contributions to the electrical characteristics of the semiconductor can become large at high temperatures and completely change the extrinsic properties of the semiconductor.

A selected region of a semiconductor is usually doped with either donor or acceptor impurity atoms. In the following discussions, and in order to make our analysis more complete, we will assume, to start with, that the semiconductor contains both donor and acceptor atoms. We will also assume that the impurity atom concentrations are very low relative to the semiconductor crystalline density. This case is referred to as nondegenerate doping.

We now calculate the equilibrium charge distributions in the semiconductor. For electrical neutrality, the total positive charge should be equal to the total negative charge in the semiconductor. This can be written as

$$n + n_A^- = p + n_D^+ \tag{6.3}$$

where n is the total electron concentration in the conduction band, n_A^- is the negatively charged (ionized) acceptor ion density, p is the total hole concentration in the valence band, and n_D^+ is the ionized donor density.

From Eqs. 5.35 and 5.36, the electron and hole concentrations can be written as

$$n = N_c \exp\left(\frac{(E_F - E_c)}{kT}\right) \tag{6.4a}$$

$$p = N_v \exp\left[-\frac{(E_F - E_v)}{kT}\right] \tag{6.4b}$$

where N_c and N_v are defined by

$$N_c = A_c T^{3/2} = 2\left(\frac{2\pi m_e^* k}{h^2}\right)^{3/2} T^{3/2} \tag{6.5a}$$

$$N_v = A_v T^{3/2} = 2\left(\frac{2\pi m_p^* k}{h^2}\right)^{3/2} T^{3/2} \tag{6.5b}$$

and are constant at a given temperature. The constants A_c and A_v are independent of temperature.

The probability that an energy state E will be occupied by an electron is given by the Fermi–Dirac distribution function. For a lightly doped semiconductor, the ionized acceptor density can be written as the product of the acceptor concentration multiplied by the probability that an electron will occupy an acceptor energy level E_A, that is

$$\begin{aligned} n_A^- &= N_A f(E_A) \\ n_A^- &= N_A\left[\frac{1}{1 + e^{[(E_A - E_F)/kT]}}\right] \end{aligned} \tag{6.6}$$

On the other hand, the ionized donor density can be written as the donor atom concentration multiplied by the probability that the donor atom will not be occupied by an electron at the energy state E_D. This probability can be written as $[1 - f(E_D)]$. Therefore,

$$\begin{aligned} n_D^+ &= N_D[1 - f(E_D)] \\ n_D^+ &= N_D\left[\frac{1}{1 + e^{[(E_F - E_D)/kT]}}\right] \end{aligned} \tag{6.7}$$

In writing down the total electron and hole concentrations in the conduction and valence bands, we have used the electron and ion concentrations that are given for intrinsic semiconductors. These were calculated by assuming that electrons are excited from the valence band into the conduction band through thermal excitation. For the case of an n-type semiconductor, there are additional electrons in the conduction band due to the ionized donor atoms. Electron contributions from the donor atoms are usually many orders larger than that due to the electrons excited from the valence band. There is also an equilibrium between the electron excitation into the conduction band and hole-electron recombination in the valence band. Whatever means we use to excite the electrons into the conduction band, there is always a competing process of recombination between the electrons and the holes, and the recombination requires one electron-hole pairing to take place. Ignoring the physics of the recombination process, we can assume that at equilibrium, the time rate of change of the electron and the hole concentrations are equal and can be written as

$$\frac{dn}{dt} = \frac{dp}{dt} = g' - R'np \tag{6.8}$$

where g' is the generation rate and R' is the recombination coefficient. At steady state, $dn/dt = dp/dt = 0$, that is,

$$np = \frac{g'}{R'} = \text{constant} = n_i^2 \tag{6.9}$$

where we set the constant equal to n_i^2. Because of the two competing processes, the *np* product at equilibrium should always be constant, irrespective of the semiconductor, and thus is set equal to the square of n_i to take into account all types of semiconductors including the intrinsic semiconductor.

Under equilibrium conditions, we can therefore use Eqs. 6.3 through 6.9 to calculate the various charged particle concentrations as well as to determine the Fermi energy E_F for the extrinsic semiconductor.

Substituting these equations into Eq. 6.3, we obtain

$$\begin{aligned} N_c\left[\exp\left(\frac{E_F - E_c}{kT}\right)\right] + N_A\left\{\frac{1}{1 + \exp\left[(E_A - E_F)/kT\right]}\right\} \\ = N_v\left[\exp\left(-\frac{E_F - E_v}{kT}\right)\right] + N_D\left\{\frac{1}{1 + \exp\left[(E_F - E_D)/kT\right]}\right\} \end{aligned} \tag{6.10}$$

When charges in extrinsic semiconductors are considered, depending on which carrier is the dominant charge carrier in the semiconductor, these charges are labeled as

Majority carriers	electrons in an n-type material holes in a p-type material
Minority carriers	electrons in a p-type material holes in an n-type material

6.3 FERMI ENERGY IN EXTRINSIC SEMICONDUCTORS

In an n-type semiconductor, at very low temperatures, close to absolute zero, we know that the conduction band is completely empty, the valence band is completely full, and the donor energy levels are all occupied by the electrons. Therefore, we expect the Fermi energy of the semiconductor to lie between the donor energy level and the conduction band. Similarly, for a p-type semiconductor, we expect the Fermi energy to lie between the valence band and the acceptor energy levels, since for this case all these acceptor energy states are empty at absolute zero. As the temperature is increased, we expect the Fermi levels in the corresponding semiconductors to change.

For an n-type material, there are no acceptor atoms, that is, $N_A = 0$ and Eq. 6.3 reduces to

$$n = p + n_D^+ \tag{6.11}$$

Using Eq. 6.9, we can write Eq. 6.11 as

$$n = \left(\frac{n_i^2}{n}\right) + n_D^+$$

Solving for n, we obtain

$$n = \frac{n_D^+ + \sqrt{(n_D^+)^2 + 4n_i^2}}{2} \tag{6.12}$$

If n_D^+ is large compared to n_i, we can neglect n_i in Eq. 6.12 and write

$$n \approx n_D^+ \tag{6.13}$$

This is called the *extrinsic range* for the semiconductor.

If the temperature of the crystal is very high, it is possible that the concentration of thermally excited electrons from the valence band can exceed the electrons excited from the donor levels. Under these conditions, Eq. 6.11 reduces to

$$n = p = n_i \tag{6.14}$$

This is the *intrinsic range*, and the semiconductor loses its extrinsic properties and acts as if there are no impurities in the crystal. The carriers that are thermally excited from the valence band mask the effects of the impurity electrons.

For a p-type semiconductor, we include only the acceptor atoms and set $N_D = 0$. We then obtain expressions similar to the above equations with n replaced by p and donor concentrations replaced by the acceptor concentrations. Note that for many practical considerations, we can assume that for a given semiconductor the respective donor or the acceptor atoms are completely ionized at room temperature.

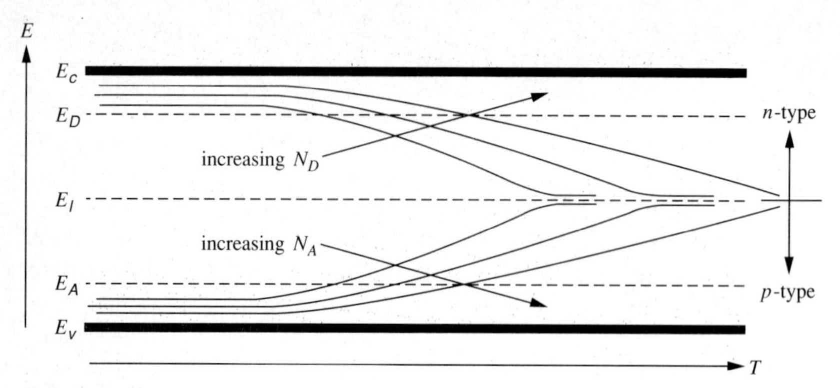

FIGURE 6.4
Fermi energy as a function of temperature (arbitrary units) for various impurity concentrations. The temperature dependence of the band gap is neglected.

The variation of the Fermi energy as a function of temperature is plotted in Fig. 6.4 for various concentrations of the donor and acceptor atoms for the corresponding n- and p-type semiconductors. As the temperature increases, the Fermi level approaches the center of the band gap as expected for an intrinsic semiconductor.

Example 6.1. Calculate the electron density in an n-type silicon semiconductor, if the donor-atom density is 5×10^{16} cm^{-3}. Assume that 90 percent of the donor atoms are ionized. Also find the hole density in this n-type semiconductor.

Solution. For silicon $n_i = 1.4 \times 10^{10}$ cm^{-3}. We can therefore neglect n_i in Eq. 6.12.

$$n = n_D^+ = 0.9N_D = 4.5 \times 10^{16} \text{ cm}^{-3}$$

From Eq.6.8, the hole density is

$$p = \frac{n_i^2}{n} = 4.36 \times 10^3 \text{ holes/cm}^3$$

Note that the hole density in an n-type semiconductor falls way below the intrinsic hole density.

Example 6.2. For Example 6.1, find the location of (*a*) the Fermi level E_F and (*b*) the donor level E_D.

Solution. Using Eq. 6.9, we can write

$$N_c \exp\left(\frac{E_F - E_c}{kT}\right) = n_D^+ = 0.9N_D$$

Solving for E_F, and using $N_c = 2.5 \times 10^{19}$ cm^{-3}, we find

$$E_F - E_c = kT \ln\left(\frac{0.9N_D}{N_c}\right)$$

$$E_F - E_c = (0.026) \ln\left[\frac{(0.9)5 \times 10^{16}}{2.5 \times 10^{19}}\right] \text{ eV}$$

Finally

$$E_F = E_c - 0.164 \text{ eV}$$

The Fermi level is 0.164 eV below the conduction band.

From Eq.6.6

$$n_D^+ = N_D \left\{ \frac{1}{1 + \exp\left[(E_F - E_D)/kT\right]} \right\}$$

Setting

$$n_D^+ = 0.9N_D$$

and solving for E_D and using $E_F = E_c - 0.164$ eV from the above solution, we find

$$E_D = E_c - 0.107 \text{ eV}$$

The donor energy is 0.107 eV below the conduction band.

6.4 MOBILITY AND SATURATION VELOCITY

The mobility of charge carriers was discussed in Chapter 4, and we saw that it depends inversely on the effective mass of the charge carriers (Eq. 4.7). In a semiconductor, the charge carriers are the electrons and holes. Because of the complex lattice forces, it was shown in Chapter 5 that the effective mass of a charge carrier is very much affected by where the charge carrier is located in a given allowed band. In a semiconductor, the electrons usually occupy the lower energy levels of the conduction band and the holes occupy the upper energy levels of the valence band. Table 5.1 lists the effective masses and the mobilities of the electrons and holes in some common semiconductors. As expected, the electron and hole effective masses are different from the free electron mass.

Because of the different lattice forces that arise in different semiconductors due to the physical nature of the solid, we also expect that the mobility of the charge carriers will differ widely from one semiconductor to another. Even the crystal orientation of the semiconductor makes a big difference in the movement of the charged particles.

As we saw in Chapter 4, the drift velocity of the charge carriers is related to the applied electric field by

$$v_d = \mu \mathcal{E} \tag{6.15a}$$

where the mobility μ is related to the average collision time of the charge carriers by

$$\mu = \frac{q}{2m^*}\tau_c \tag{6.15b}$$

where it is assumed that the mobility of the charge carriers is independent of the electric field and the drift velocity increases with increasing electric field.

In a semiconductor, the main scattering sources for charge carriers are the lattice atoms, impurity atoms, and other defects in the semiconductor. Since with the present technology, the presence of defects can be reduced to a minimum level, the lattice atoms are the most important scatterers of charge carriers in intrinsic semiconductors. In extrinsic semiconductors, the impurity atoms, in addition to the host lattice atoms, play a very important role in the scattering of the charged particles.

Although the applied voltages in semiconductor devices are low, because of the very short distances over which these voltages are applied, the resulting electric fields are usually very high. Therefore in many semiconductor devices, the charge carriers are subjected to very high fields. It has been shown theoretically and verified experimentally that both the mobility and the current density of the charge carriers depend very strongly on the applied electric field. Figure 6.5 shows that the current density as a function of the applied electric field in an n-type silicon and a p-type germanium. The current density (or the drift velocity) increases linearly first with $\mathscr{E}$ field, then increases with the square root of $\mathscr{E}$, and then becomes constant irrespective of $\mathscr{E}$. These results suggest that the mobility will have the inverse variation with $\mathscr{E}$ as shown in Fig. 6.6 for Ge.

Although the lattice atoms in a semiconductor are not free to move in the sense of the electrons, the thermal vibrations of these lattice atoms can be shown to be equivalent to harmonic oscillators, known as *phonons*. For a given polarization (longitudinal, or transverse) the frequency of oscillations of the lattice atoms occurs in two distinct frequency ranges at small wave numbers. These are the *acoustic mode,* with a frequency corresponding to the long-wavelength vibrations close to that of sound, and the *optical mode,* with a frequency that falls in the optical range. These phonons can gain or lose energy in an interaction process with a free charge carrier. As the electric field increases, the carriers in the semiconductor interact strongly with the lattice (especially emitting what are known as optical phonons), so that the average velocity of the charge carriers falls much below a linear projection from low field values. When the applied electric field in a semiconductor is increased, the drift velocity of the electrons increases linearly first, then starts to saturate and becomes constant beyond a certain electric field. The value of the electric field where saturation sets in depends largely on the semiconductor that is considered. This saturated velocity is called v_{sL}, *scatter-limited velocity*, and occurs usually at fields greater than 10^5 V/cm.

The drift velocity variation of charge carriers in Ge, Si, and GaAs are shown in Fig. 6.7 as a function of the electric field. The drift velocity of the electrons in the GaAs sample increases up to a maximum value and with further increase in

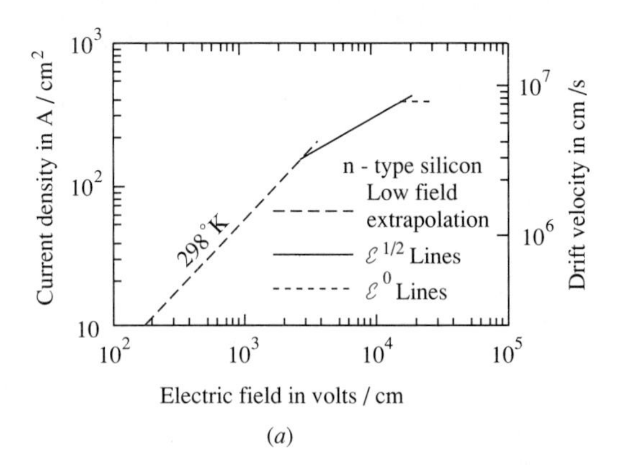

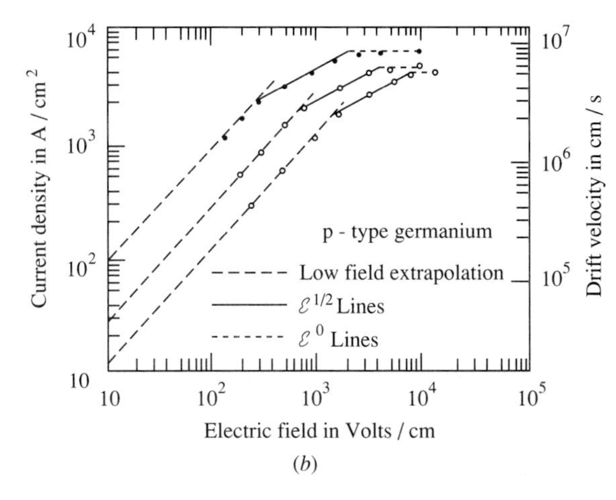

FIGURE 6.5
Current density as a function of electric field in n-type silicon and (*b*) p-type germanium (Ryder, 1953).

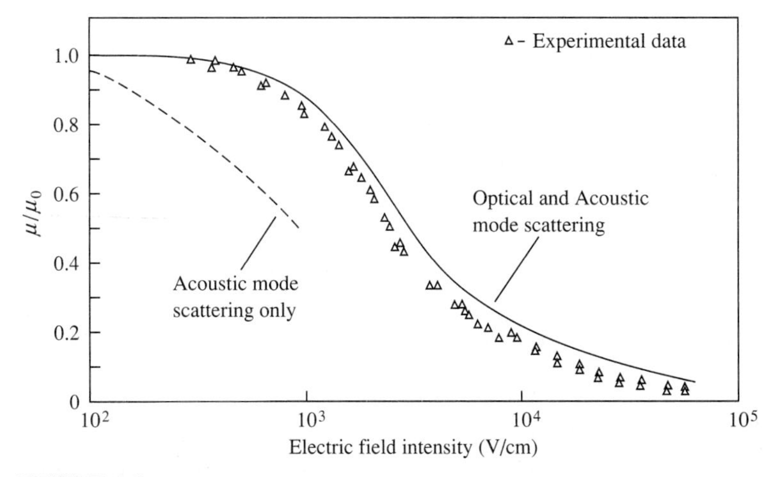

FIGURE 6.6
Ratio of mobility to low field mobility as a function of electric field intensity for n-germanium at 300 K. The solid line represents the theoretical curve obtained including both optical and acoustic phonon scattering, the dashed line that is obtained considering acoustic mode scattering only (Conwell, 1967).

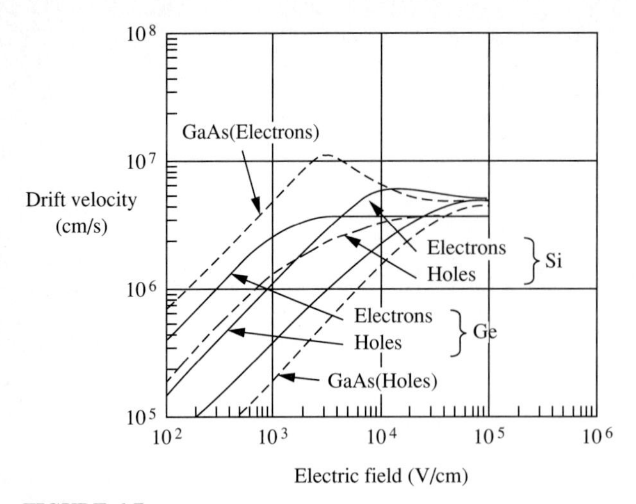

FIGURE 6.7
Drift velocity of carriers as a function of electric field in silicon, germanium, and GaAs. (S. M. Sze and R. M. Ryder, "Microwave Avalanche Diodes," *Proc. IEEE*, vol. 59, 1971, 2240–2254, © 1971 IEEE.)

the electric field, it decreases and finally reaches a saturation value. Between the maximum and the saturation velocity, GaAs shows negative differential mobility. This reduction in drift velocity is explained by the band structure of GaAs and will be given in detail in Chapter 12. It is explained in terms of the transition of the charge carriers from the central valley into the satellite valley of the conduction band of Fig. 5.14. The scatter-limited velocity of the charge carriers is close to or greater than the thermal velocity of the electrons. Thus, the electrons possessing v_{sL} are referred to as "hot" electrons.

6.5 RECOMBINATION OF CARRIERS

At optical wavelengths, when electrons in a p-type semiconductor are photoelectrically excited into the conduction band—for example, by a continuous source of photons—electron-hole pairs are generated. That is, for every electron excited into the conduction band, a hole is generated in the valence band. Since the holes are the majority carriers in the p-type material, the excess number of holes will be small in relation to the hole equilibrium density p_{po} in the bulk semiconductor. On the other hand, because electrons are the minority carriers, the electron density will be highly influenced by the externally generated electrons.

Electrons are constantly excited into the conduction band, and if there were no removal of electrons from the conduction band, the electron density in the conduction band would increase indefinitely for a continuous source of photons. Actually, some of the electrons that are excited into the conduction band recombine with the holes in the valence band by losing energy and disappear from circulation either by emitting an optical photon or by losing energy in a nonradiative interaction process with the lattice atoms. These processes of generation and recombination are shown in Fig. 6.8*a*. Here, only transitions between the conduction and valence bands are assumed to take place.

Let g' be the rate of electron generation due to the incident photons, and let r' be the recombination rate of the electrons. The excess electron density n' can be written as

$$n' \equiv n - n_{\text{po}}$$

where n is the total electron density, and n_{po} is the minority equilibrium electron density in the p-type semiconductor.

If we now let τ_n represent the average recombination time for the electrons, we can think of (dt/τ_n) as the probability that an electron will recombine with a hole within a time interval dt. Within this time interval, the reduction in the excess electron density will be equal to the product of excess electron density multiplied by the probability that the electrons will be lost at that point in space due to the recombination with holes. This can be written as

$$dn' = -n'\left(\frac{dt}{\tau_n}\right)$$

or the change in excess electron density per unit time becomes

$$\frac{dn'}{dt} = -\frac{n'}{\tau_n} \tag{6.16}$$

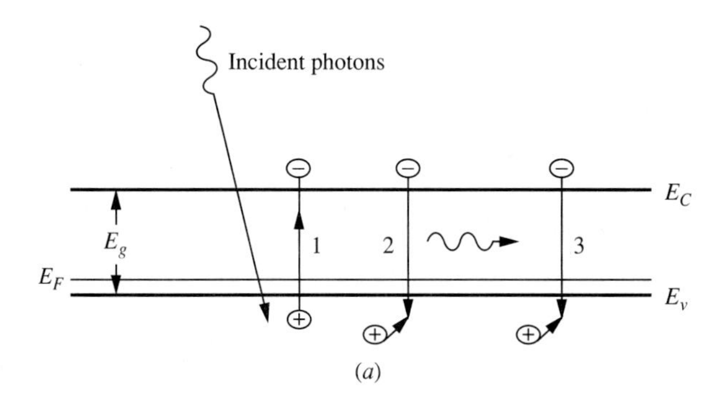

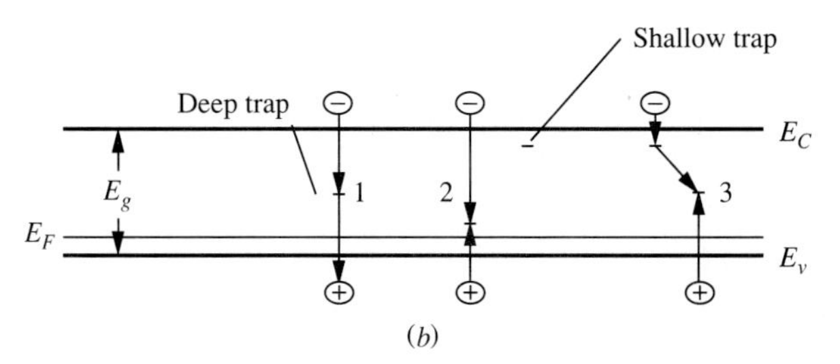

FIGURE 6.8
(*a*) (1) Incident photons generate electron-hole pairs. Electrons recombine with the holes in the valence band (2) by emitting photons and (3) by a nonradiative process. (*b*) Some possible recombinations through traps determine the overall recombination rate.

The negative sign represents the reduction in number density as a function of time. The right side of this equation, n'/τ_n, is interpreted as the recombination rate r' for the electrons in a p-type semiconductor.

For any general system, the resulting rate equation, taking into account the electron generation rate g' of the electrons per unit time, can be written as

$$\frac{dn'}{dt} = g' - \frac{n'}{\tau_n}$$

If we perform an experiment, as shown in Fig. 6.9, where we shine a monochromatic source of light onto a p-type semiconductor sample and later turn it off at $t = 0$, we can measure the recombination time of electrons by observing the decay of the current in the sample as a function of time. At $t = 0$, electron generation stops—that is, $g' = 0$—and the rate equation reduces to

$$\frac{dn'}{dt} = -\frac{n'}{\tau_n}$$

The solution is simply found to be

$$n'(t) = n'(0)\exp\left(-\frac{t}{\tau_n}\right) \tag{6.17}$$

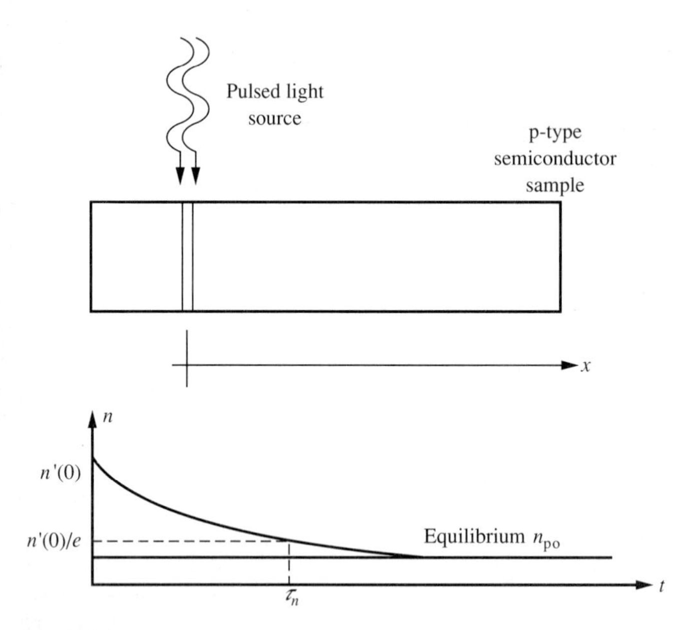

FIGURE 6.9
Experiment to measure the recombination of the electrons under photon illumination. $n'(0)$ is the excess electron density at $t=0$ s at the location of the photon source ($e = 2.7183$, the logarithmic constant).

where $n'(0)$ is the excess electron density at $t = 0$ just before the source is turned off. From the exponential decay of the current, the recombination time can be experimentally measured.

If the assumption that the electrons combine only with the holes in the valence band were true, from theoretical calculations one would expect a shorter recombination time than that measured. Additional photoelectric experiments show that actually there are available electronic discrete energy states in the forbidden gaps of the semiconductors. These may be near the conduction or the valence bands or may lie within any region of the band gap. These energy states are called *traps* and are due to the presence of impurities, as well as imperfections, in the crystal which occur during the manufacture of the semiconductor. There are many forms of these imperfections that act as trapping centers for the electrons and holes. If these trap levels are way below the conduction and far away from the valence band energy levels, they are referred to as *deep trap levels*.

An electron may decay into one of these trap states, spend some time in this state, and drop to the valence band, and the electron-hole pair disappears from the system. There may be other forms of recombinations through these traps. Some are shown in Fig. 6.8*b*. All of these contributions increase the overall recombination time of the electrons.

The band structure of the semiconductor, especially whether it is a direct-band or an indirect-band gap semiconductor, also plays a major role in the recombination process. In a direct band semiconductor, when an electron in the conduction band makes a transition into the valence band, the momentum of the electron before and after the transition is the same, so that the electron either loses its energy by a radiative transition or by a nonradiative process. In this process, the momentum is easily conserved. Therefore, the probability of a band-to-band transition for an electron in a direct-band gap semiconductor is very high. On the other hand, if the electron makes a transition from an indirect-band gap semiconductor, the initial and the final momentums of the electron are different. In order to conserve momentum, the lattice phonons are involved in the recombination process. Therefore the probability of recombination is relatively small in these indirect band gap semiconductors. The dominant recombinations in indirect gap materials are predominantly nonradiative. These factors will be discussed in more detail in Chapter 13.

6.6 DIFFUSION LENGTH

Suppose a continuous source of light illuminates a very small portion of an extrinsic semiconductor sample where excess charge carriers (electron-hole pairs) are generated by these photons (Fig. 6.10). Since the charge concentration away from the source is less than where the light source is located, these excess charge carriers will diffuse away from the source location. If the light source continues to shine on the sample, the concentration of these excess charge carriers will reach a steady-state distribution where the density of excess charge carriers will exponentially decrease as we move away from the source of these carriers.

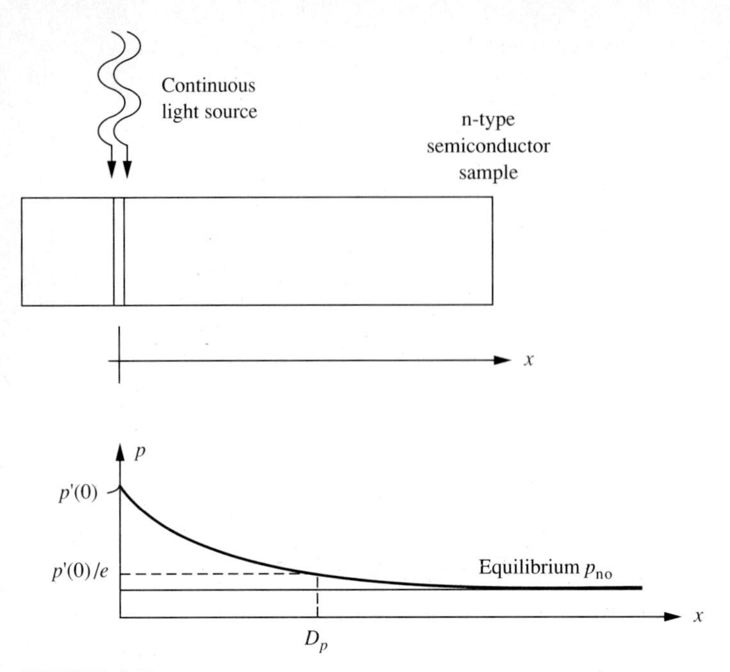

FIGURE 6.10
Diffusion length of holes in an n-type semiconductor. $p'(0)$ is the excess hole density at $x = 0$ ($e = 2.7183$, the logarithmic constant).

If the semiconductor sample is an n-type semiconductor to start with, the sample already has many electrons as majority carriers. The light source generates both electrons and holes, but if we assume that the excess charge density generated by the photons is much smaller than the majority electrons but a lot larger than the minority holes, the excess electrons will not have much effect on the electron concentration at other points in the sample. On the other hand, the photo-excited hole density will far exceed the thermally generated equilibrium hole density. It therefore becomes important that we study the recombination of the excess hole carriers in the sample since as these holes diffuse away from the source, they are most likely to recombine with the abundant electrons.

The change in hole density at any point is governed by the continuity equation

$$\frac{\partial p}{\partial t} + \frac{1}{e}\nabla \cdot \mathbf{J}_p = g' - r' \tag{6.18}$$

where $\mathbf{J}_p$ is the diffusion current density, whose x component can be written as

$$\mathbf{J}_{px,\ \text{diff}} = -eD_p\left(\frac{\partial p}{\partial x}\right) \tag{6.19}$$

If we use $p'/\tau_p = (p - p_{no})/\tau_p$ for the recombination rate, we can write Eq. 6.18 as

$$\frac{\partial p}{\partial t} = g' - \frac{p - p_{\text{no}}}{\tau_p} + D_p\left(\frac{\partial^2 p}{\partial x^2}\right) \tag{6.20}$$

where p_{no} is the equilibrium charge density without excess charges, τ_p is the recombination time of the holes.

For steady state, we set $\partial p/\partial t = 0$. Away from the source location ($x > 0$), $g' = 0$. If we let $p' = p - p_{\text{no}}$, equal to the excess hole density, Eq. 6.20 can be written as

$$\frac{d^2 p'}{dx^2} = \frac{1}{D_p \tau_p} p' \tag{6.21}$$

where for the last term in Eq. 6.20, p is replaced by $p' = p - p_{\text{no}}$. The solution to this equation is easily found to be

$$\begin{aligned} p' &= p'(0) \exp\left(-\frac{x}{(D_p \tau_p)^{1/2}}\right) \\ p' &= p'(0) \exp\left(-\frac{x}{L_p}\right) \end{aligned} \tag{6.22}$$

where $p'(0)$ is the excess hole concentration at $x = 0$. The hole concentration decreases exponentially as a function of distance. The quantity

$$L_p \equiv (D_p \tau_p)^{1/2} \tag{6.23}$$

is called the diffusion length for the holes and is interpreted as the average distance that the holes travel before they recombine with the electrons. A similar diffusion length is defined for electrons, if the electrons are the minority carriers in a semiconductor.

6.7 ELECTRICAL CONDUCTIVITY IN SEMICONDUCTORS

The electrical conductivity of a semiconductor can be written as (Eq. E.4.6)

$$\sigma = en\mu_n + ep\mu_p \tag{6.24}$$

for an intrinsic semiconductor $n = p = n_i$, where n_i is given by Eq. 5.37. We can then write Eq. 6.24 as

$$\begin{aligned} \sigma_i &= e(\mu_n + \mu_p)n_i \\ \sigma_i &= e(\mu_n + \mu_p)\,(A_c A_v)^{1/2} T^{3/2} \exp\left(-\frac{E_g}{2kT}\right) \end{aligned} \tag{6.25}$$

A_c and A_v are defined by Eq. 6.5.

If the $\ln(\sigma_i/T^{3/2})$ is plotted against $(1/kT)$ the slope of the resulting curve is proportional to $-E_g$ (Fig. 6.11). Therefore, measurement of the intrinsic conductivity can be used to determine experimentally the band gap energy of a semiconductor.

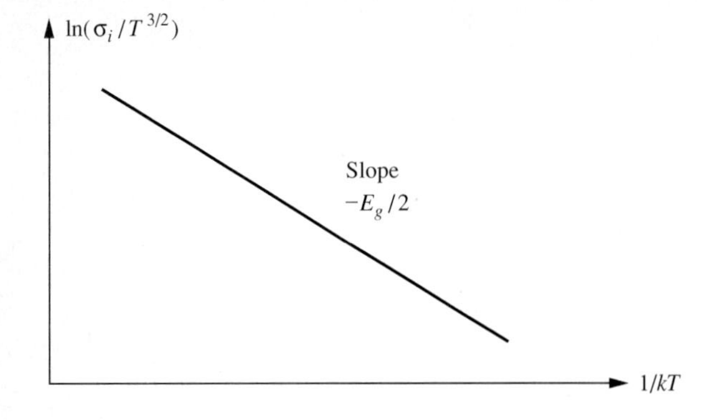

FIGURE 6.11
Logarithmic plot of intrinsic conductivity of a semiconductor versus $1/kT$ (arbitrary units). Slope gives E_g.

The variation of the logarithm of the conductivity of an n-type extrinsic semiconductor is shown in Fig. 6.12 as a function of the inverse temperature. At low temperatures, the concentration of charge carriers is low. As the temperature increases, the donor atoms are ionized and the charge concentration increases. As the temperature is further increased, the semiconductor becomes an intrinsic semiconductor due to the large number of thermally excited electrons from the valence band. Of course, variations of μ_n and μ_p with temperature also effect σ, but this effect is not as great as the change in concentrations that occurs with temperature. The temperatures between the low-, moderate-, and high-conductivity regions of Fig. 6.12 depend on the host semiconductor and usually shift to higher temperatures with increasing band gap energy.

6.8 HALL EFFECT

Some of the evidence for the concept of a hole is provided by the Hall effect experiments. Two pairs of electrodes (primary and secondary) are connected at opposite sides of the sample material, as shown in Fig. 6.13*a*. Due to the voltage

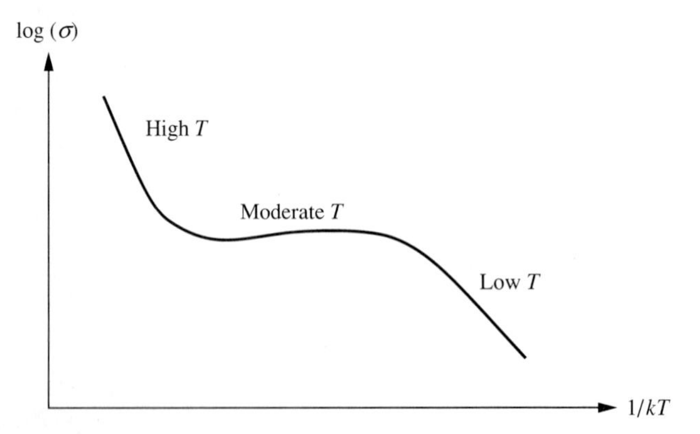

FIGURE 6.12
Variation of conductivity of an extrinsic semiconductor versus $1/kT$ (arbitrary units).

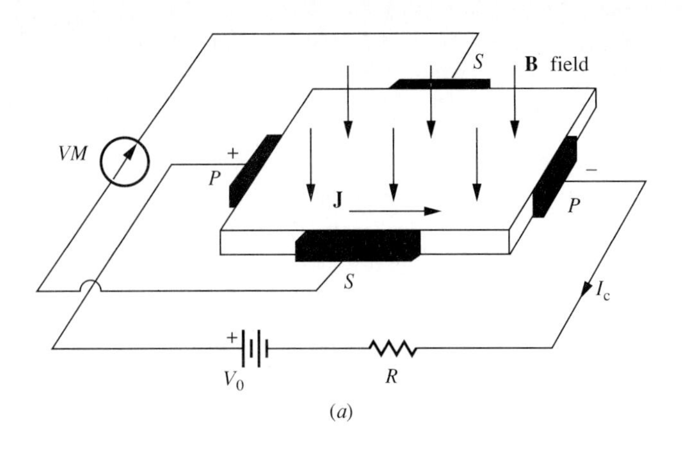

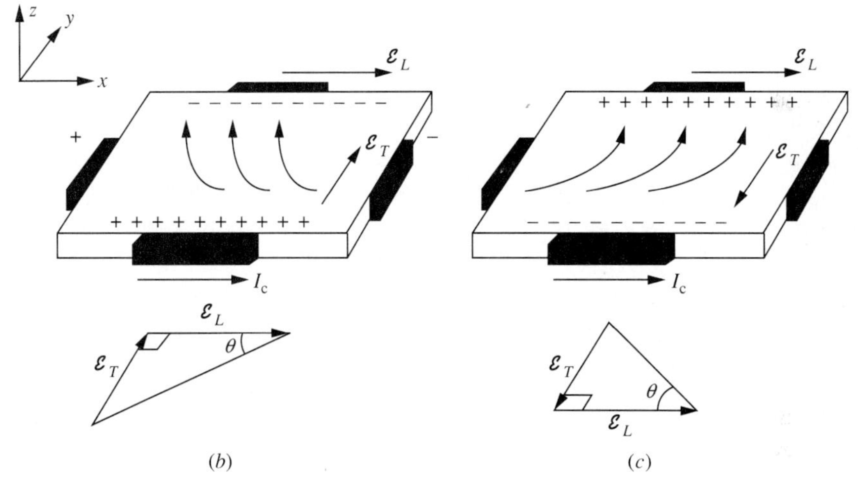

FIGURE 6.13
(*a*) Hall coefficient experiment and resulting fields for (*b*) electrons and (*c*) holes. P = primary and S = secondary electrodes.

source externally applied across the primary electrodes, a dc current I_c flows between these two electrodes. A magnetic field is then applied perpendicular to the sample. As a result of the Lorentz force, the charged particles are forced to move toward the set of oppositely placed secondary electrodes. When charge builds up between these electrodes, an electric field is produced whose magnitude increases until the opposing force produced by this electric field cancels the force due to the magnetic field. An equilibrium is finally reached, and a potential difference between these electrodes results from the transverse electric field $\mathcal{E}_T$ produced across the secondary electrodes. If electrons were the only type of charge carriers in all the materials, we would have observed the same generated polarity of potential across the secondary electrodes, irrespective of the material used. On the contrary, experimental results show that this polarity is negative for some elements and positive for others.

We will now show that for electrons and holes we get different deflections in the sample and therefore obtain different polarities for the potentials. The Hall effect can thus be used as a means of determining the presence of holes in a material.

Under the application of the external potential V_0 across the primary electrodes, a longitudinal electric field $\mathscr{E}_L$ is set up directed toward the $+x$ direction, as shown in Fig. 6.13*a*. As a result of this, a constant current I_c is flowing through the sample. Since the electrons are the charge carriers, to produce the indicated current direction, they should move from right to left under the application of the external electric field. Let the magnetic field point downward. The Lorentz force acting on the electrons

$$\begin{aligned} \mathbf{F} &= q(v \times \mathbf{B}) \\ \mathbf{F} &= -e\,[v(-\mathbf{a}_x) \times B(-\mathbf{a}_z)] = evB\mathbf{a}_y \end{aligned} \tag{6.26}$$

is thus directed toward the upper edge of the sample, that is, in the $+y$ direction. This means that a transverse electric field $\mathscr{E}_T$ is set up across the sample also in the $+y$ direction as shown in Fig. 6.13*b*.

If the charge carriers were holes, for the same primary current direction (holes moving from left to right), the holes would also have been deflected toward the same terminal, as shown in Fig. 6.13*c*. But since the holes charge the upper terminal of the sample to a positive potential, the polarity of the transverse field is opposite to that of the electrons as shown in Fig. 6.13*c*. Thus, the polarities of the Hall voltage for the two cases are different.

Irrespective of the sign of the charge, the drift velocity for the charged particles at equilibrium (when $\mathscr{E}_T$ balances the magnetic force) can be written as (see Problem 6.16)

$$v_d = \frac{|\mathscr{E}_T|}{|\mathbf{B}|} \tag{6.27}$$

Substituting the definition of mobility $\mu = v_d/\mathscr{E}_L$ into this equation, we get

$$\mu = \frac{\mathscr{E}_T}{B\,\mathscr{E}_L} = \frac{\tan\theta}{B} \tag{6.28}$$

where the Hall angle θ is defined by

$$\tan\theta \equiv \frac{\mathscr{E}_T}{\mathscr{E}_L} \tag{6.29}$$

The current density J in the sample can be written as

$$J = \rho v_d = nq\mu\,\mathscr{E}_L \tag{6.30}$$

Using Eq. 6.28, we can solve for $\mathscr{E}_T$ and find

$$\mathscr{E}_T = \left(\frac{J}{nq}\right) B$$

TABLE 6.2
Hall coefficients for materials

	R_H (ohm/mT)
Li	-17.0×10^{-11}
Be	$+24.4 \times 10^{-11}$
Na	-25.0×10^{-11}
Cu	$-\ 5.5 \times 10^{-11}$
Ag	$-\ 8.4 \times 10^{-11}$
Cd	$+\ 6.0 \times 10^{-11}$

or

$$\mathscr{E}_T = R_H J B \tag{6.31}$$

where R_H is the hole coefficient and is given by

$$R_H = \frac{1}{nq} = \frac{\mathscr{E}_T}{JB} \tag{6.32}$$

For a material which contains both holes and electrons, Eq. 6.28 can be written as

$$\tan\theta = +\frac{(n_p \mu_p^2 - n_n \mu_n^2)}{(n_p \mu_p + n_n \mu_n)} B \tag{6.33}$$

Measured Hall coefficients for several metals are given in Table 6.2. Semiconductor materials show higher Hall coefficients, but their value depends on the impurity doping in the host semiconductor.

FURTHER READING

Adler, R. B., Smith, A. C., and Longini, R. L., *Introduction to Semiconductor Physics*, Wiley, New York, 1964.

Conwell, E. M., *High Field Transport in Semiconductors*, Academic Press, New York, 1967.

Haynes, J. R., and W. Shockley, "The Mobility and Life of Injected Holes and Electrons in Germanium," *Phys. Rev.*, vol. 81, 1951, 835.

Pryor, A. C., "The Field Dependence of Carrier Mobilities in Silicon and Germanium," *J. Phys. Chem. Solids*, vol. 12, 1959, 175–180.

Ryder, E. J., "Mobility of Holes and Electrons in High Electric Fields," *Phys. Rev.*, vol. 90, 1953, 766–769.

Sze, S. M., and R. M. Ryder, "Microwave Avalanche Diodes," *Proc. IEEE*, vol. 59, 1971, 2240-2254.

Sze, S. M., *Physics of Semiconductor Devices*, Wiley, 2d ed., New York, 1981.

Van der Ziel, A., *Solid-State Physical Electronics*, Prentice Hall, Englewood Cliffs, N.J., 1976.

PROBLEMS

6.1. If the resistivity of a Ge semiconductor is given as 0.05 ohm-m, find the density of holes and electrons in the sample at 300 K. Calculate the results for both n- and p-types. Are the resulting densities different? Explain.

6.2. Boron atoms are used to dope a Si crystal. If for every 1×10^7 Si atoms, there is one B atom in the semiconductor, find the density of holes and electrons in the sample. Assume that the B atoms are completely ionized. Also find the mass of the total B atoms in 1 cm^3.

6.3. For problem 6.2, find the Fermi energy level, and show that the approximation of 100 percent ionization at room temperature of the B atoms is correct.

6.4. Calculate the diffusion length of holes and electrons in Si and Ge at room temperature. Use $\tau_p = 10\ \mu s$ and $\tau_n = 50\ \mu s$ for both semiconductors.

6.5. If the conductivity of an n-type GaAs sample is 20 S/m, find the electron, hole, and impurity concentrations; then find the Fermi energy of the semiconductor.

6.6. An n-type Si sample with a resistivity of 0.03 ohm-m and length of 1.0 cm is illuminated at one end by a continuous light source. The excess charge density is found to be $1 \times 10^{14}\ cm^{-3}$. Find the diffusion current density of the charge carriers at $x = 0$, $x = 0.5$ cm, and at $x = 0.98$ cm.

6.7. In an n-type Si sample of thickness t, a voltage of 3 V is applied across the opposite ends. Find the corresponding velocity of the electrons if (see Fig. P.6.7.)

(*a*) $t = 500\ \mu m$
(*b*) $t = 10\ \mu m$
(*c*) $t = 0.25\ \mu m$

Find the thermal velocity of the electrons at room temperature and compare it with the velocities calculated above.

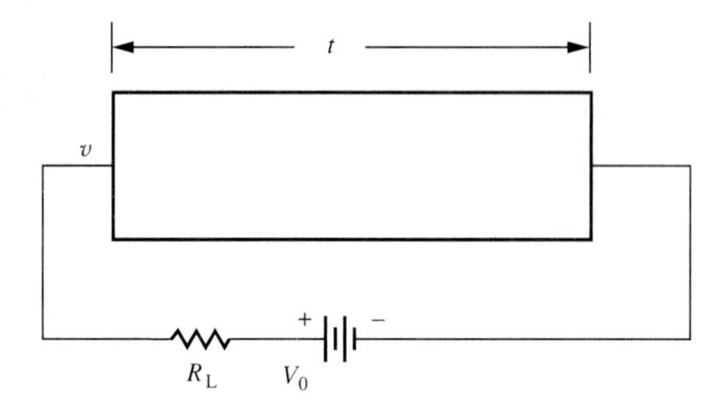

FIGURE P.6.7

6.8. At room temperature, the collision frequency ν_c of the electrons in an n-type Si is 300 GHz. Find the mean free path for the electrons if the donor concentration is $10^{17}\ cm^{-3}$. Assume 100 percent ionization.

6.9. If the temperature of the semiconductor in Problem 6.8 is reduced to 77 K, the electron collision frequency is reduced to 1 GHz. Find the new mean free path for the electrons. Will the change in the electron concentration due to lower temperature have any effect on this collision process?

6.10. If a p-type silicon crystal has $N_D = 10^{22}$ atoms/m^3, find the temperature at which the crystal loses its intrinsic properties.

6.11. Repeat Problem 6.10 for Ge and GaAs if the donor concentration is the same.

6.12. Find the average energy of the electrons in the conduction band. Assume that $|E_c - E_F| \gg kT$.

6.13. Find the intrinsic resistivity of Si at room temperature. If, through a doping process, one As atom replaces every one millionth Si atom, find the change in resistivity and the mass density of the Si.

6.14. Using Cu, design a gauss meter that will measure magnetic fields between 0 to 2 T. What will be the output voltage through this range? Is this a sensitive instrument? If you replace copper by an n-type silicon semiconductor of $N_D = 10^{17}\ \text{cm}^{-3}$, what will be the range of the Hall voltage?

6.15. (*a*) Design an experiment and show the circuit diagram of how you can experimentally determine (i) the recombination time and (ii) the diffusion length of the electrons in a p-type material. What simple modifications are necessary in a single experimental arrangement to differentiate between the recombination time and diffusion length?

Note: the electron density as a function of time t and position x can be written as (Haynes and Shockley, 1951)

$$n_p(x,t) = \frac{N}{\sqrt{4\pi D_n t}} \exp\left(-\frac{(x-\mu_n \mathcal{E} t)^2}{4D_n t} - \frac{t}{\tau_n}\right) + n_{po}$$

where $\mathcal{E}$ is the applied electric field.

(*b*) Plot the given electron density distribution as a function of position at different times.

6.16. Using the Lorentz force equation $\mathbf{F} = q(\mathcal{E} + \mathbf{v} \times \mathbf{B})$, show that charged particles, irrespective of the sign of their charges, drift with a constant velocity given by

$$\mathbf{v} = \frac{\mathcal{E} \times \mathbf{B}}{\mathbf{B}\cdot\mathbf{B}} = \frac{\mathcal{E} \times \mathbf{B}}{B^2}$$

6.17. You are to design a temperature sensor using a semiconductor.

(*a*) What type of semiconductor are you going to use?

(*b*) What will be the temperature sensitivity of this thermometer?

(*c*) What kind of temperatures (low or high) can you measure with this thermometer? Explain.

CHAPTER 7

ELECTRON EMISSION

Electrical characteristics of all electronic devices are based on controlling the flow of charged particles, generally electrons, between two regions in space. This flow is usually controlled by externally applied electric and/or magnetic fields. The electron motion can take place either inside a solid or in a vacuum between two or more electrodes. Devices that operate on the principle of charge motion in a solid material are referred to as solid-state devices. These will be covered in detail in the later chapters of this book. Electronic devices based on the free-charge flow in a vacuum between appropriate electrodes are referred to as vacuum tube devices. These require an understanding of the different ways in which free electrons can be generated. This is usually accomplished by the emission of electrons from solids where originally the electrons are bound to the solid by binding forces. The emission mechanisms are therefore based generally on the concept of imparting additional energy to the electrons by some external means so that the electrons can overcome these binding forces in the solid and can be emitted out of the solid as free electrons. High field emission is an exception to this general rule. Various emission mechanisms will be discussed.

7.1 PHOTOELECTRIC EMISSION

Figure 7.1*a* shows an experimental arrangement for observing the photoelectric effect. Two electrodes are placed in an evacuated glass tube. Monochromatic light of frequency ν is incident on the cathode. Photoelectrically emitted electrons move

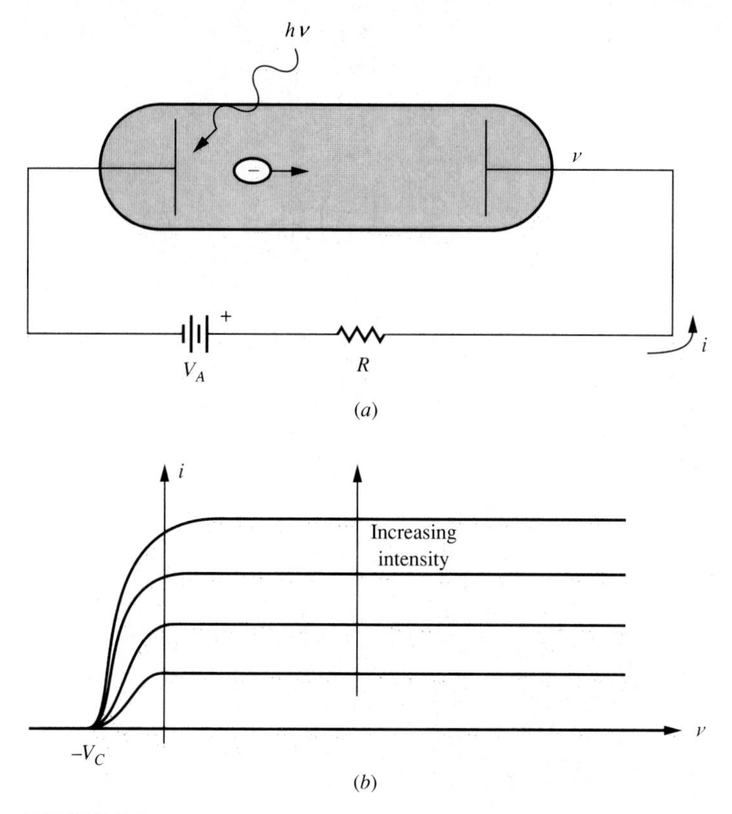

FIGURE 7.1
(*a*) Experimental arrangement for observing photoelectric effects and (*b*) i versus v with light intensity as a parameter. V_C is the negative voltage below which current is zero.

toward the anode and a current i flows in the external circuit. For a given intensity of light I, the current is constant for positive values of the anode potential as shown in Fig. 7.1*b*.

If the applied potential is reversed, and the anode electrode is made negative with respect to the cathode, the field produced between the electrodes repels the emitted electrons and the number of electrons reaching the anode decreases. Further increase in this negative voltage repels more electrons. When the potential becomes equal to $-V_C$, the electron current flow stops completely. This implies that the electrons do not have sufficient kinetic energy to overcome the repelling potential at the anode and the current flow ceases in the external circuit. For a monochromatic (single frequency) light, no current flow is observed in the external circuit below $-V_C$, irrespective of the light intensity.

The characteristic energy associated in freeing an electron from a solid is known as the work function. A typical energy diagram for a metal is shown in Fig. 7.2. The bottom of the conduction band is taken as the reference energy level. The

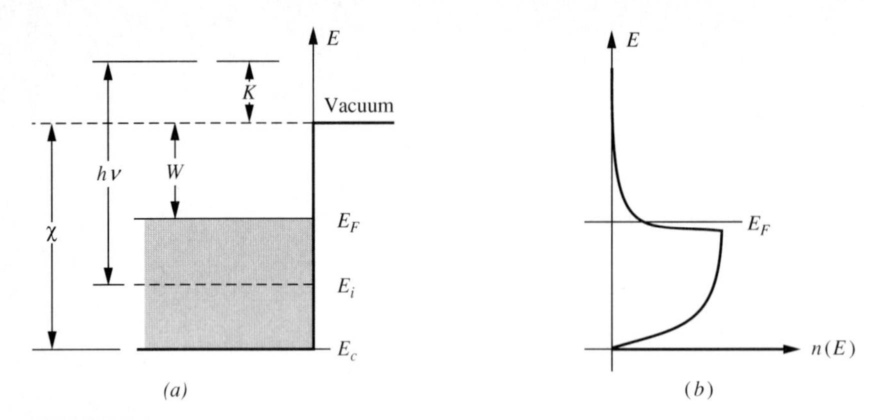

FIGURE 7.2
(*a*) Idealized band structure of a metal and (*b*) the number density of electrons in the conduction band.

conductor surface is assumed to be ideal so that there is no variation in the surface energy levels associated with the solid. Actually, the barrier energy near the surface can be appreciably modified by the space-charge forces, by the presence of absorbed gases, or by bringing the material in contact with other materials possessing different physical parameters. Such modifications may drastically effect the emission characteristics of the solid. In this discussion of the introductory level, the idealized band model as shown in Fig. 7.2 is sufficient in understanding the basics of emission mechanisms. When necessary, the idealized model will be modified to explain the resulting important changes in the corresponding emission process.

Assume that a monochromatic light of frequency ν is incident on a solid. From the wave-particle duality of matter, consider light as photons of energy $h\nu$. From the conservation of energy, the kinetic energy of the ejected electrons can be written as

$$K = (E_i + h\nu) - (E_F + W) \tag{7.1}$$

where E_i is the energy the electron had in the solid before the interaction (see Fig. 7.2*a*). The maximum kinetic energy of the ejected electrons will be due to those electrons which initially have energies near E_F, that is, $E_i \sim E_F$. Thus

$$K_{\text{max}} = \frac{1}{2}mv_{\text{max}}^2 = h\nu - W \tag{7.2}$$

Maximum kinetic energy should then be equal to

$$K_{\text{max}} = eV_C$$

Note that V_C will also depend on frequency ν of the incident light. If ν is decreased, V_C will be smaller. For frequencies below ν_c, which is defined as

$$h\nu_c = W \tag{7.3}$$

there will be no emitted photoelectrons from the solid, irrespective of the intensity of the light falling on the emitting surface.

The fundamental concept of the photoelectric effect described above is a highly simplified approach to a more complicated problem. In addition to metals, the photoelectric effect is also observed in insulators and semiconductors.

The photoelectric effect can be treated as *surface photoemission* and *volume photoemission* phenomena. In the surface photoelectric effect, the electrons are liberated from the surface by the incident photons. In the dominant volume photoelectric effect, the photons interact with the valence electrons of the solid and lose energy to these electrons. The photoelectrons generated move toward the surface in a zigzag manner and lose energy while approaching the surface. If these electrons, when they reach the surface, still have enough energy left, they can overcome the barrier potential and escape from the surface as photoelectrons.

Photoelectron emission efficiency is highly dependent on the physical properties of the material used as a cathode. For high efficiency:

1. The escape probability for the electrons from the surface should be high;
2. *The escape depth* (the average distance the photon-generated electrons have to travel before reaching the surface) should be relatively large; and
3. The material should have a large photon absorption coefficient or a short *absorption length*, so that the electrons are generated close to the surface.

All of these properties are determined mostly by the electron affinity χ and the energy gap E_g of the material. Electron affinity can largely be manipulated by covering the surface of a semiconductor or insulator by a material which effectively lowers the work function of the material. In many cases, even a monolayer of surface atoms can appreciably affect the work function of a material. This results in a change of the work function of a metal or the electron affinity of a semiconductor. For a Cs-O layer on a silicon target or other III–V compounds, the electron affinity becomes negative and the escape probability becomes very large.

Photoelectron emission from various cathode materials are highly frequency (wavelength) dependent. Since the lowest frequency of light that can eject photoelectrons is approximately equal to the work function of the material (Eq. 7.3), various compounds show different spectral responses. The spectral response of some of the common photocathodes used in present day electron devices are shown in Fig. 7.3.

The photoelectric effect is explained by assuming that the monochromatic light acts as a particle-possessing energy $h\nu$. If the wave nature of the light were used to describe the photoelectric effect, there would be discrepancies between the experimentally observed results and the predictions based on the interactions between an electron and an electromagnetic wave. The photon concept explains all the experimentally observed effects.

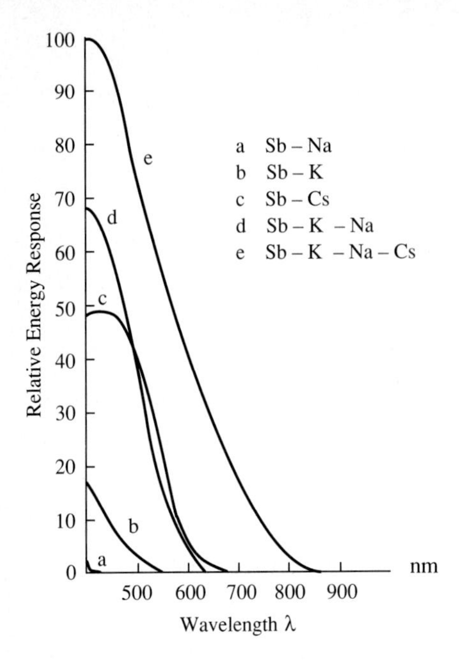

FIGURE 7.3
Spectral distribution for multi-alkali cathodes (Gorlich, 1956).

The intensity of light is then the number of photons per second multiplied by the photon energy $h\nu$. Whatever the number of photons are, even when there are a few photons, the photoelectrons are emitted from the solid in a very short time (the emission time is in the order of 10^{-9} s). For low-intensity light, wave interaction requires a very long time for the electrons to gain energy from the field and be ejected. Also, the classical wave approach has no limitation on the frequency of incident light and electrons should be ejected irrespective of the frequency. The photoelectric phenomenon was explained by Einstein using the photon concept and was one of the major contributions to the development of modern quantum theory.

7.2 THERMIONIC EMISSION

When the temperature of a metal is increased, some of the electrons that are bound to the solid gain enough thermal energy to overcome the binding forces within the solid and appear as free electrons outside the solid. This process is known as thermionic emission. As the temperature is further increased, the number of electrons that are emitted also increases, therefore showing a strong dependence on the temperature of the solid. Thermionic emission also depends on the physical parameters of the solid, especially to the work function W of the solid.

The thermionic emission phenomena can be understood basically by referring to the simplified energy-band diagram and the electron energy-distribution function in the metal. Figure 7.4*b* shows the electron energy-distribution in a solid for two different temperatures with $T_2 > T_1$. The electron distribution function for $E > E_F$ is highly exaggerated to show the increase in the number of electrons that will have thermal energies greater than χ so that these electrons will overcome the binding forces and be emitted as free electrons from the solid. A necessary but not sufficient condition for emission to take place is that the electrons must possess an energy greater than $\chi = E_F + W$. The insufficiency of this condition can be understood by referring to Fig. 7.5.

If the components of the momentum in the direction normal to the surface is less than the critical value $p_{xc} = mv_{xc}$, the electrons will not be emitted from the surface. Figure 7.5*a* shows two electrons with the same energy but with two different directions. Whenever the x-component of the velocity is greater than

$$v_{xc} = \frac{p_{xc}}{m} = \sqrt{\frac{2(E_F + W)}{m}} \tag{7.4}$$

the electrons will overcome the surface forces along the normal to the surface and be emitted. Figure 7.5*b* shows the same conditions in momentum space. Electron A will not be emitted from the solid as long as its x-directed momentum is less than $p_{xc} = mv_{xc}$.

Referring to Fig. 7.6, we can now calculate the current density of the thermally emitted electrons in the following way. Assume that electrons have a velocity v_x. The fraction of electrons contributing to the current flow through the surface area A in a time interval Δt will be the electrons within the volume element $(\Delta l\ A) = v_x \Delta t A$ multiplied by the number of electrons having velocity v_x or $\Delta N = (v_x \Delta t A)n(v)$, where $n(v)$ is the velocity distribution of the electrons. The thermally emitted current density is by definition equal to

$$\Delta J_{th} \equiv \frac{\Delta i_{th}}{A} \equiv \frac{e(\Delta N/\Delta t)}{A}$$

or

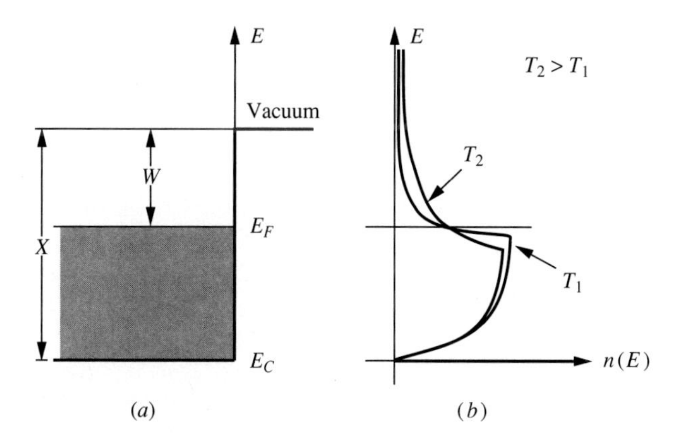

FIGURE 7.4
(*a*) Energy-band diagram and (*b*) electron energy distribution in a typical solid for various temperatures. These are highly exaggerated to show relative electron density at higher temperatures.

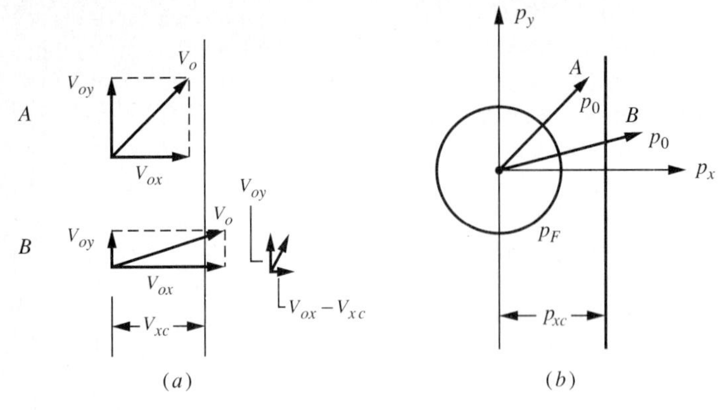

FIGURE 7.5
(*a*) Required directed velocity for emission. (*b*) Same energy in momentum space. p_F is the momentum corresponding to the Fermi energy E_F.

$$dJ_{th} = ev_x n(v) \tag{7.5}$$

where $n(\mathbf{v})$ is given by Eq. 3.2. The total current density will be due to contributions of all the electrons, or

$$J_{th} = \int_{\text{all}} \int \int_{\text{velocities}} dJ_{th} d^3v$$

$$J_{th} = \int_{-\infty}^{\infty} \int_{-\infty}^{\infty} \int_{-\infty}^{\infty} e \frac{2m^3}{h^3} \frac{v_x dv_x dv_y dv_z}{1 + \exp[m(v_x^2 + v_y^2 + v_z^2)/2kT - (E_F/kT)]} \tag{7.6}$$

Effectively, only electrons with $v_x > v_{xc}$ will contribute to the emission current. Since the required energy for the electrons to be emitted is greater than E_F, we can neglect the one in the denominator and write Eq. 7.6 as

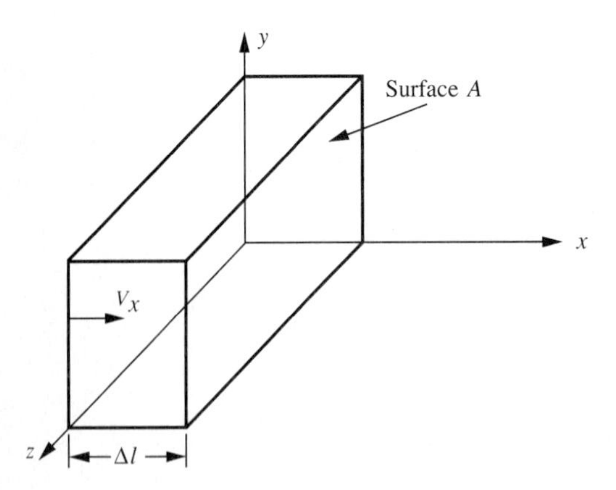

FIGURE 7.6
Electrons within volume $A\Delta l$ contribute to the emission current density within time interval Δt.

$$J_{th} \approx e\frac{2m^3}{h^3}\exp\left(\frac{E_F}{kT}\right)\int_{v_{xc}}^{\infty}\exp\left(-\frac{mv_x^2}{2kT}\right)v_x\,dv_x\int_{-\infty}^{\infty}\exp\left(-\frac{mv_y^2}{2kT}\right)dv_y$$
$$\times\int_{-\infty}^{\infty}\exp\left(-\frac{mv_z^2}{2kT}\right)dv_z \qquad (7.7)$$

The integrals on v_y and v_z are the same:

$$\int_{-\infty}^{\infty}\exp\left(-\frac{mv_y^2}{2kT}\right)dv_y = \sqrt{\frac{2\pi kT}{m}}$$

Substituting $u = mv_x^2/2kT$ into the v_x integral and carrying out the integration, we get

$$\int_{v_{xc}}^{\infty}\exp\left(-\frac{mv_x^2}{2kT}\right)v_x\,dv_x = \frac{kT}{m}\exp\left(-\frac{mv_{xc}^2}{2kT}\right) \qquad (7.8)$$

Finally, Eq. 7.7 becomes

$$J_{th} = \frac{4em\pi k^2T^2}{h^3}\exp\left(\frac{E_F}{kT}\right)\exp\left(-\frac{E_F+W}{kT}\right) = \frac{4em\pi k^2T^2}{h^3}\exp\left(-\frac{W}{kT}\right) \qquad (7.9)$$

where we used $\frac{1}{2}mv_{xc}^2 = E_F + W$. We can write Eq. 7.9 as

$$J_{th} = \mathcal{A}T^2\exp\left(-\frac{W}{kT}\right)\ \text{A/m}^2 \qquad (7.10a)$$

Defining the constant $\mathcal{A}_0$ by

$$\mathcal{A}_0 = \frac{4\pi emk^2}{h^3} = 1.2\times10^6\ \text{A/m}^2\text{K}^2 = 120\ \text{A/cm}^2\text{K}^2$$

We can write J_{th} as

$$J_{th} = 120T^2e^{-W/kT}\ \text{A/cm}^2 \qquad (7.10b)$$

Equation 7.10 is known as the Richardson–Dushmann equation. We see that at a given temperature, a solid with a lower work function will emit more electrons than a solid with a higher work function. Also, we see a marked increase in emission-current density with temperature. The constant $\mathcal{A}_0 = 1.2\times10^2$ A/cm^{2}K calculated theoretically may be different for actual solids because of the idealized model that is used in the above derivation. Surface forces, material processes, and surface roughness may contribute to the actual value of $\mathcal{A}_0$. But the temperature behavior given by Eq. 7.10 has been experimentally verified to hold for all emitting cathodes.

The plot of the logarithm of the experimental thermionic emission-current density as $\ln(J_{th}/T^2)$ against $1/kT$ as shown in Fig. 7.7 can be used to determine

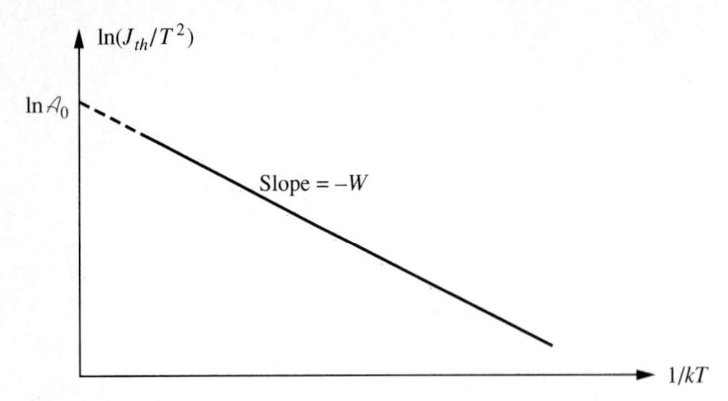

FIGURE 7.7
$\ln(J_{th}/T^2)$ versus $1/kT$. The slope of the curve is $-W$ and zero intercept of $\ln(J_{th}/T^2)$ is $\ln(\mathcal{A}_0)$.

TABLE 7.1
Thermionic work function of various elements (from Kohl, 1967)

Group I	Li	Na	K	Rb		Cs	Cu	Ag	Au
W_0, eV	2.4	2.3	2.2	2.15		1.9	4.42	4.5	4.9
Group II	Be	Mg	Ca	St		Ba	Zn	Cd	Hg
W_0, eV	3.67	3.6	3.2	2.6		2.5	4.3	4.1	4.5
Group III	Al		Ga				La	Ce	Pr
W_0, eV	4.2		3.8				3.3	2.8	2.7
Group IV	Ti	Zr	Hf	Th	C	Si	Ge	Sn	Pb
W_0, eV	3.9	3.57	3.65	3.4	4.4	3.6	4.8	4.4	4.0
$\mathcal{A}_0\exp(-\alpha/k)^a$		120	31.9	70	48	8			
Group V	V	Cb	Ta	As	Sb	Bi			
W_0, eV	4.1	4.0	4.1	5.2	4.0	4.6			
$\mathcal{A}_0\exp(-\alpha/k)^a$		37	37						
Group VI	Cr	Mo	W	U					
W_0, eV	3.9	4.2	4.5	3.3					
$\mathcal{A}_0\exp(-\alpha/k)^a$	48	55	70						
Group VII	Mn	Re							
W_0, eV	4.0	5.1							
$\mathcal{A}_0\exp(-\alpha/k)^a$		51							
Group VIII	Fe	Co	Ni	Rh	Pd	Os		Ir	Pt
W_0, eV	4.5	4.4	4.41	4.8	4.99	4.7		5.4	5.32
$\mathcal{A}_0\exp(-\alpha/k)^a$	26	41	30	33	69			63	32

[a] Temperature dependence of the work function is given by $W = W_0 + \alpha T$ and is included in the expression $\mathcal{A}_0\exp(-\alpha/k)$ (see Problem 7.8). Units of $\mathcal{A}_0 exp(-\alpha/k)$ are in $(A\ cm^{-2}\ K^{-2})$.

the work function and $\mathcal{A}_0$ by for a specific material. The slope of the curve and the zero intercept of the current density gives respectively the work function W and the constant $\mathcal{A}_0$ of that material. Table 7.1 gives the work function and the effective $\mathcal{A}_0$ for various materials.

7.3 SCHOTTKY EFFECT

In the previous section, the derivation of thermionic emission was based only on electrons overcoming the work function of the solid due to the thermal energy associated at a given temperature T. Once the electrons become free, they are generally forced to move from the emitting material, called the cathode, toward a different electrode, called an anode, by applying an external potential between these two electrodes. At low potential differences, the emission characteristic of the electrons is not altered, but at higher potentials—that is, at higher electric fields—thermionic emission-current density of the cathode material is noticeably modified. This phenomenon is known as the Schottky effect.

The idealized surface potential barrier that an electron sees at the boundary of the solid is shown in Fig. 7.8 as a dotted line. An electron at an energy E_i is free to move inside the solid. When it reaches the surface, it tries to move away and out of the surface. The solid then lacks a negative charge and the resulting Coulomb force attracts the electron back into the solid. The magnitude of this attractive force can be calculated by using the image theory used in static field calculations.

When the electron moves a distance x meters away from an initially neutral solid, an equivalent positive charge is generated in the solid which attracts the electron back to the solid. This induced charge can be replaced by an equivalent positive charge $+e$ located $-x$ distance away from the surface in the solid. The resulting Coulomb force on the electron due to these two charges separated by a distance of $2x$ can be written as

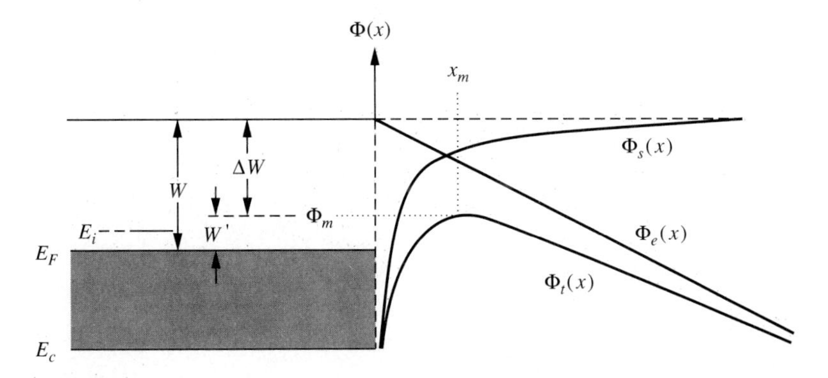

FIGURE 7.8
Potential energy modification of the idealized potential barrier (dashed lines) of a solid by image potential, $\phi_s(x)$, and the additional potential, $\phi_e(x)$, due to applied external electric field.

$$F_x = -\frac{e^2}{4\pi\epsilon_0(2x)^2} \tag{7.11}$$

The associated potential energy is

$$\Phi_s(x) = -\int_\infty^x F_x(x')\,dx' = -\frac{e^2}{16\pi\epsilon_0 x} \tag{7.12}$$

The variation of this potential energy as a function of distance away from the surface is shown in Fig. 7.8.

Consider now an externally applied electric field directed toward the emitting surface. The electron will experience an additional potential energy due to this applied field $\mathcal{E}$. The resulting potential at a distance x away from the surface is

$$\Phi_e(x) = -\int_0^x F_x(x')\,dx' = -\int_0^x e\,\mathcal{E}\,dx' = -e\,\mathcal{E}x \tag{7.13}$$

The electron experiences the total equivalent potential energy $\Phi_t(x)$, which is the sum of $\Phi_s(x)$ and $\Phi_e(x)$. The externally applied field effectively lowers the equivalent surface potential barrier of the electron in the solid as shown in Fig. 7.8. This reduction in the equivalent work function is found by calculating the maximum value of Φ_t which occurs at x_m and which can be calculated from

$$\left.\frac{d\Phi_t(x)}{dx}\right|_{x=x_m} = 0 \tag{7.14a}$$

This gives

$$x_m = \left(\frac{e}{16\pi\epsilon_0\,\mathcal{E}}\right)^{1/2} \tag{7.14b}$$

Substituting this back into $\Phi_t(x)$, the maximum potential energy at x_m becomes

$$\begin{aligned}\Phi_m &= \Phi_t(x_m) = \Phi_s(x_m) + \Phi_e(x_m)\\ &= -\frac{e^2}{16\pi\epsilon_0 x_m} - e\,\mathcal{E}x_m\end{aligned}$$

or

$$\Phi_t(x_m) = -e\left(\frac{e\,\mathcal{E}}{4\pi\epsilon_0}\right)^{1/2} = -\Delta W \tag{7.15}$$

The thermally emitted electrons now experience an equivalent work function W' and resulting current density can be written as

$$J_e = \mathcal{A}_0 T^2 e^{-W'/kT} \tag{7.16}$$

Using $W' + \Delta W = W$

$$J_e = \mathcal{A}_0 T^2 e^{-(W-\Delta W)/kT} = J_{th} e^{\Delta W/kT} \tag{7.17}$$

Finally, using Eq. 7.15, the emission-current density becomes

$$J_e = J_{th} \exp\left[\frac{e(e\,\mathcal{E}/4\pi\epsilon_0)^{1/2}}{kT}\right] \tag{7.18}$$

The externally applied field then enhances the thermal current density of the emitted electrons. To verify the correctness of the Schottky effect, the $\log(I_e)$ is plotted versus $\mathcal{E}^{1/2}$ for fixed values of temperature T as shown in Fig. 7.9. The extrapolation of the Schottky plots to zero field permits an evaluation of the zero field emission-current densities from the cathodes.

7.4 FIELD EMISSION

When a very high electric field is directed toward a solid at room temperature, electrons are emitted from the solid by a completely different mechanism called the high field emission. To understand the physical basis for the high field emission, Fig. 7.8 is redrawn with a larger electric field applied between the cathode and the anode material (see Fig. 7.10). Because of the high field strength, the slope of $\Phi_e(x)$ is very steep. The electrons inside the solid see on the boundary of the solid an equivalent potential barrier that has a very small width. In Chapter 1, we saw that an electron has a finite probability of penetrating a potential barrier and appears on the other side of the barrier provided that the width of the potential barrier is small. The same concept can be applied to the electrons in the solid that see an equivalent potential barrier $\Phi_t(x)$ on the boundary of the solid. An

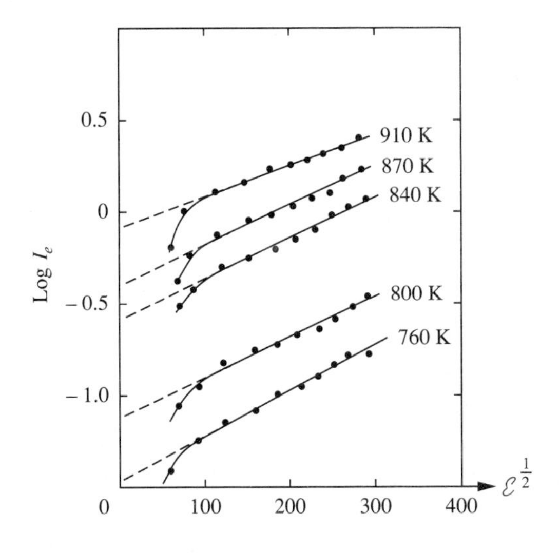

FIGURE 7.9
Variation of electron emission in the Schottky region as a function of the square root of the electric field intensity (Eisenstein, 1948).

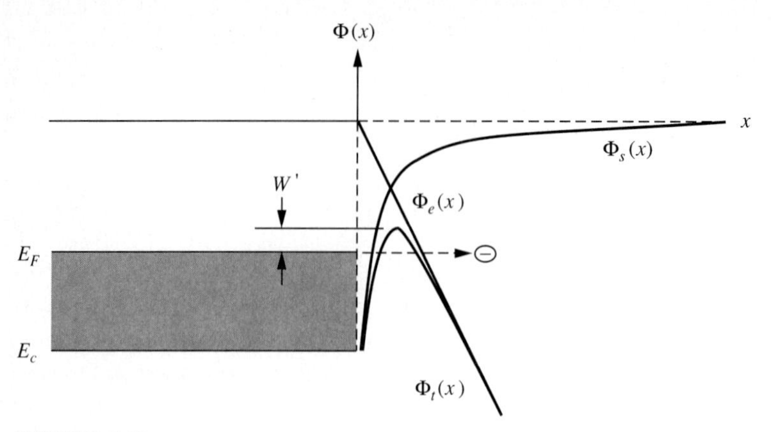

FIGURE 7.10
Electron tunneling due to the narrow barrier produced by the very high electric field.

electron possessing an energy in the vicinity of the Fermi level E_F now has a finite probability of tunneling through the surface potential barrier and escaping from the solid without requiring the excess energy needed to overcome the equivalent potential barrier height W'. The mathematics involved in calculating the current density due to the high electric field is beyond the level of this book, and only the results will be given here.

The high field current density is related to the applied electric field by

$$J_f = C\,\mathcal{E}^2 e^{-D/\mathcal{E}} \tag{7.19}$$

where $\mathcal{E}$ is the applied electric field magnitude and C and D are constants dependent on the physical parameters of the materials that emit the electrons.

For tungsten metal

$$C = 1.26 \times 10^5 \text{ A/V}^2$$

$$D = 2.76 \times 10^{10} \text{ V/m}$$

The constant D is proportional to $W^{3/2}$ and C is proportional to $(E_F/W)^{1/2}$ of the solid. Note that the high field emission is independent of the temperature T of the solid.

Based on field emission, a new cathode concept is being developed. It utilizes planar-integrated circuit technology and does not require a heater filament—which is a source of high power consumption. A thin layer of oxide is sandwiched between two plates as shown in Fig. 7.11. The lower plate has conical islands, which are grown by a special technique. The upper plate comprises circular holes at the location of conical islands. When a voltage is applied between the two plates, very high fields are generated in the vicinity of the cone tips, which emit electrons with reasonable current density. The potential differences used are not too high. The configuration can be thought of as a nonthermal solid-state cath-

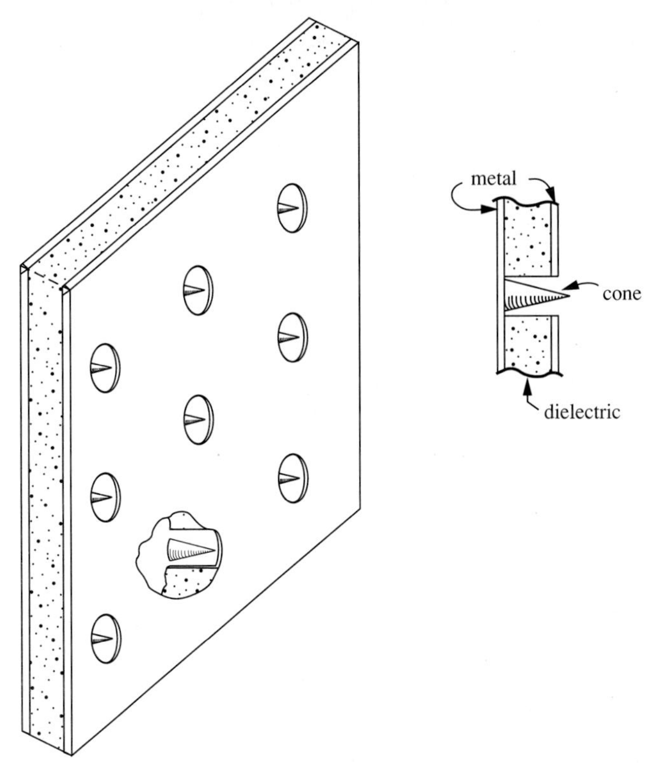

FIGURE 7.11
Solid-state field emitting cathode. High electric field is set between the holes of the upper plate and the cone tips.

ode. If successfully developed, these cathodes could replace the thermally heated cathodes used in all present day vacuum-tube devices.

7.5 SECONDARY ELECTRON EMISSION

When electrons are incident on a solid, it was experimentally observed that the bound electrons in the solid can gain enough energy through collisional processes from the incoming electrons, can overcome the binding surface potentials and emerge as free electrons from the solid surface. The phenomenon is known as secondary electron emission, which is a blessing or a nuisance depending on whether the effect is used to an advantage or threatens the electrical characteristic of an electronic device. In this section, the basic results of secondary electron emission and related important application areas will be given.

Figure 7.12 shows the experimental arrangement commonly used for measurement of the secondary electron emission. Primary electrons produced by an electron gun and having a mono (single) energy $E_p = ev_p$ impinge on the target element D. The true secondary electrons liberated from the target as well as

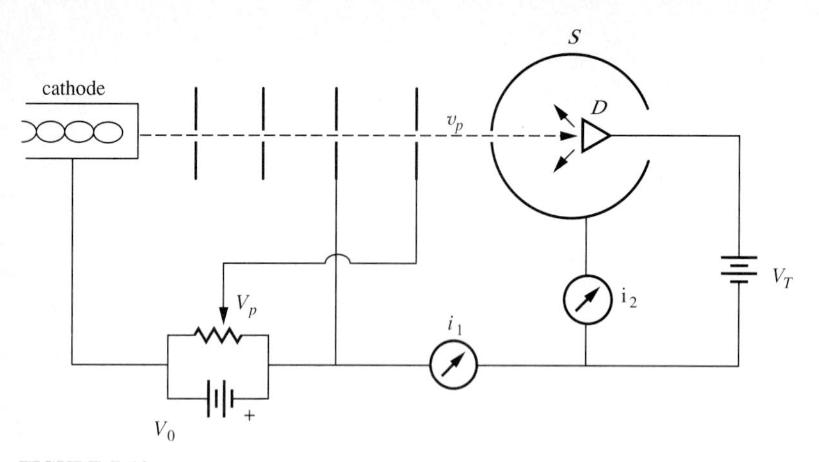

FIGURE 7.12
Experimental setup to measure the secondary electron yield of a solid placed at D. By adjusting the potentials, true secondary electron current is measured.

the elastically or inelastically scattered primary electrons are collected by the spherical electrode S. The manipulation of the potentials V_p and V_T allows the separation of the true secondary electrons from the incident electrons. The general properties of the secondary electrons are investigated by studying the secondary electron yield versus the primary electron beam energy E_p as shown in Fig. 7.13. The secondary electron emission yield is defined as the total number of emitted electrons (collected by the spherical electrode S) divided by the total number of primary electrons at E_p.

All the secondary electron emitting materials have similar yield curves. With increasing primary electron energy, the yield increases and reaches a maximum

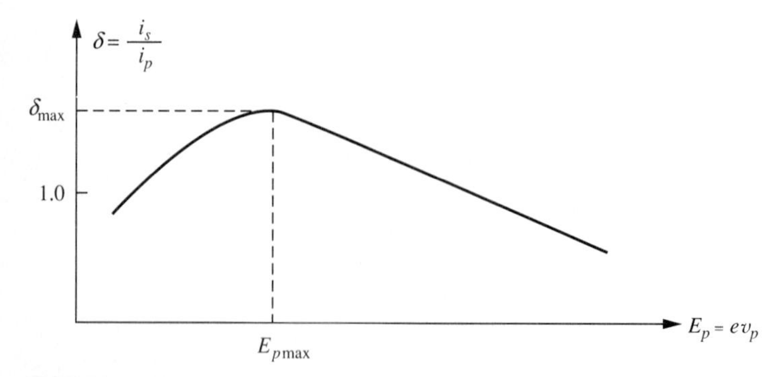

FIGURE 7.13
Typical yield curve of secondary electrons. $E_{p\max}$ is the corresponding primary electron energy yielding maximum secondary electrons.

value of $\delta_{\max}$ corresponding to $E_{p\max}$. As the primary energy is further increased, the yield again starts to decrease.

When primary electrons impinge on a solid surface, some of these electrons are reflected back by the surface potential barrier. Other electrons penetrate the barrier and enter the solid. They then move in arbitrary directions and at the same time suffer both elastic and inelastic collisions with the electrons inside the solid, in this interaction process lose energy. Some of these primary electrons will be directed back toward the surface and those with enough energy are re-emitted from the surface back into the vacuum. The electrons that were originally bound to the solid will gain energy through inelastic collision with the primary electrons and will move inside the solid in arbitrary directions. Some of these electrons, referred to as the true secondary electrons, will also reach the surface. They may either be reflected back to the solid or escape from the solid by overcoming the potential surface barrier and appear in the vacuum region as emitted secondary electrons.

The energy distribution of the secondary electrons are shown in Fig. 7.14. Here the secondary electron current is plotted against E/E_p, the ratio of the energy E of the emitted electrons to that of the energy E_p of the incoming primary electrons. The secondary electrons are divided into three regions. Region I is the true secondary electrons, Region II is the re-emitted primary electrons (referred to as rediffused electrons) from the solid, and Region III is the primary electrons elastically reflected back from the surface.

Some of the observed properties of the true secondary electrons:

The number of secondary electrons emitted per unit angle is greatest in the direction of normal to the emitting surface and decreases with increasing angle of incidence β as $\cos(\beta)$.

The time of liberation between the arrival of primary electrons and the emission of a secondary electron is essentially zero (10^{-12} s).

The velocity distribution of the true secondary electrons is similar to a Maxwell–Boltzmann distribution.

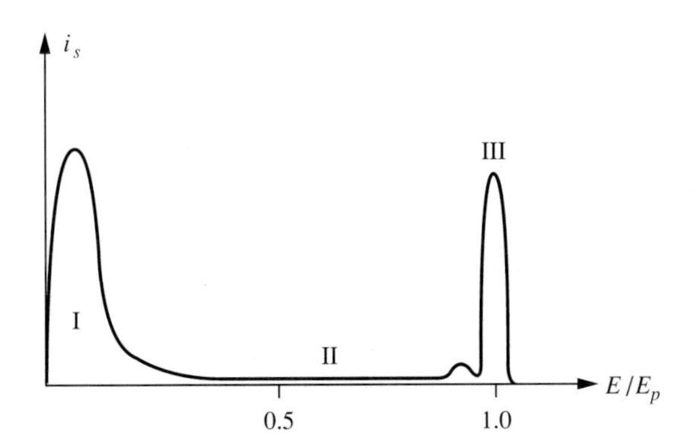

FIGURE 7.14
Secondary electron energy distribution. (I) True secondary electrons, (II) rediffused primary electrons, and (III) elastically reflected primary electrons.

The maximum yield from metals is between 0.7 and 1.7.

The maximum yield from semiconductors and insulators is between 1 and 20. The exchange of energy between the incoming primary electrons and the bound electrons in an insulator is more effective than the interaction with free electrons in a metal.

The surface conditions of the solid have a very marked effect on the secondary emission. The presence of even a monolayer of absorbed gas drastically changes the emission yield of the solid. The yield also depends on the surface roughness of the solid and decreases with increasing roughness.

When an ion strikes a solid, secondary electrons are emitted from the solid but not with much yield as in the case of electrons.

The secondary emission is actually a volume effect similar to the photoelectric effect. The yield depends largely on the band structure of the solid, and some of the conclusions derived for photoemission also apply to the secondary emission. High yields are obtained in substances with greater band gap energy. Table 7.2 lists the maximum yields from different solid materials.

TABLE 7.2
Maximum secondary electron yields from different solids (from Hachenberg and Braun, 1959)

	Element	δ_m	E_{pm}(eV)
Metals	Li	0.5	85
	Al	0.95	300
	Mo	1.25	375
	Ag	1.47	800
	W	1.35	650
Semiconductor elements (crystal)	Ge	1.2–1.4	400
	Si	1.1	250
	Se	1.35–1.40	400
Semiconductor compounds	Cu_2O	1.19–1.25	400
	PbS	1.2	500
	ZnS	1.8	350
Intermetallic compounds	$SbCs_3$	5–6.4	700
	$BiCs_3$	6–7	1000
	GeCs	7	700
Insulators	NaCl (layer)	6–6.8	600
	NaCl (crystal)	14	1200
	BeO	3.4	2000
	NaBr (single crystal)	24	1800
	SiO_2 (quartz)	2.4	400
	Mica	2.4	300–384

7.6 APPLICATIONS

Cathodes are surfaces that emit electrons when activated by external means such as thermal heat, photons, or charged particles. Physical mechanisms responsible for the emission of electrons through various external excitations were described in detail in the previous sections of this chapter. The practical applications and the material aspects of cathodes that incorporate one or a combination of these emission mechanisms will be given in the following sections.

7.6.1 Thermionic Cathodes

Thermionic emitters are either directly heated, by passing electrical current through the emitter element, or indirectly heated by the transfer of thermal energy from a separately heated filament to the emitter element. The basic design goal of a cathode element is to operate it with a maximum thermal yield, defined as the number of electrons emitted per heater power input. As can be seen from the Richardson equation (Eq. 7.10), this requires cathode materials with low work functions so that they can be operated at lower temperatures.

There are three major cathode types: pure metal cathodes, thin film cathodes, and dispenser cathodes.

1. *Pure metal cathodes* are generally used as directly heated cathodes. A planar or cylindrical-shaped metal piece is heated by passing an electrical current between the two ends. The materials commonly used for pure metal cathodes are tungsten and tantalum. These materials have $\mathcal{A} = 70$ A/cm^{2}K^2 and have very high operating temperatures. They also evaporate at these temperatures, thus limiting the operational life of these cathodes. These cathodes do not need any activation procedure and their containers can be brought to the atmospheric pressure and evacuated over and over again without adversely affecting the emission characteristic of the cathode.
2. *Thin film cathodes*, as the name implies, use a thin emitter film of a material on a pure metal base. The effective work function of the thin film emitter cathode is lower than the work functions of both the emitter and the base materials. The lower work function greatly enhances the emission characteristic of the cathode at the expense of reducing $\mathcal{A}_0$ to 4 A/cm^{2}K^2. Most of the thin film cathodes use tungsten as the base material. A suitable thin film is either deposited from the outside by vapor deposition or within the cathode by diffusion from the inside. Thoriated tungsten filaments are examples of thin film thermionic emitters. They operate around 1800–2200 K ($W_0 = 2.7$ eV).
3. *Dispenser cathodes* are cathodes on which a thin film of emitting material is produced at the surface and which is continuously replenished at the operating temperature of the cathode. Barium strontium carbonate is either used as a pellet or impregnated in a tungsten mesh. Heat breaks up the carbonate, and the metals diffuse to the surface to form the thin film. Any material that is evaporated from the surface is replenished by the same process. Barium

$$\mathcal{A}_0 = \frac{4\pi emk^2}{h^3} = 1.2 \times 10^6 \text{ A/m}^2\text{K}^2 = 120 \text{ A/cm}^2\text{K}^2$$

We can write J_{th} as strontium thin film has an effective work function of 1.67 eV, $\mathcal{A}_0 = 2.5$ A/cm^2 K^2, and operates around 1400 K. Dispenser cathodes are commonly used in present day vacuum-tube devices.

Figure 7.15 shows various types of thermionic cathode configurations. After sealing the glass envelopes, dispenser cathodes require an activation procedure usually achieved by raising and keeping the cathode temperature above the operating temperature of the cathode for a specified time. Bringing the cathode back to the atmospheric pressure generally poisons the cathode, and reactivating of the cathode becomes impossible. Properties of common thermionic cathodes are given in Table 7.3.

7.6.2 Photomultiplier Tubes

Photomultiplier tubes are photon detectors, and their operations are based both on the photoelectric effect and the secondary electron emission. A typical photomultiplier tube (Fig. 7.16) consists of a *photocathode*, secondary electron-emitting electrodes called *dynodes*, and an *anode*. Photons that are incident on the photocathode produce photoelectrons, and those that reach the inner surface of the cathode are emitted as photoelectrons. These electrons are then accelerated toward the first dynode under the application of a positive potential. The biasing of the dynodes, which are made up of secondary electron-emitting materials with very high yields, is such that the next dynode is at a higher positive potential than the previous one and the potential difference between each dynode is usually made

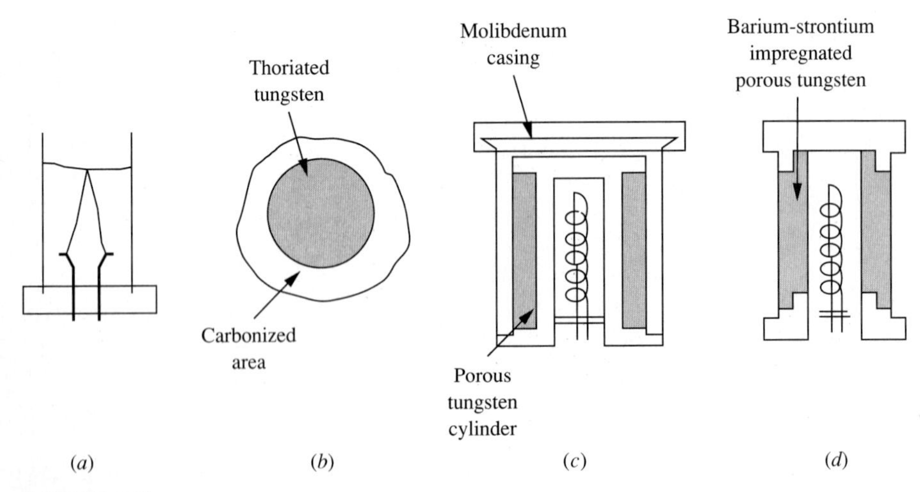

FIGURE 7.15
Various cathode configurations. (*a*) Directly heated cathode; (*b*) filament of a typical thoriated tungsten; dispenser cathodes using (*c*) separate barium-strontium cavity; and (*d*) impregnated porous tungsten (Eisenstein, 1948).

TABLE 7.3
Thermal emission data for present day cathodes (Philips cathodes) (Kohl, 1967)

		Impregnated			
Type of cathode	**L**	**Type A**	**Type B**	**Pressed cathode**	
Operating temperature T_0 (K)	1400	1400	1400	1400	1290
Electron work function W (eV)	1.67	1.53	1.67	1.7	
Emission density (pulsed) (A/cm^2)	5	1.2	5	2.5	0.9
Richardson constant $\mathcal{A}$(A/cm^2 deg^2)	2.5	0.14	2.5	2.4	
Activation temperature (K)	1520		1520	1470	
Time required for activation (minutes)			6	6	
Time required for aging (minutes)		150	150		
Life (hrs)		8000	15000	5000	
Rate of barium evaporation	high	low	low	med.	
Resistance to poisoning	good	good	very good	good	
Resistance to ion bombardment	good	good	good	good	
Available shapes (f = flat or shaped disk c = cylindrical)	f or c	f or c	f or c		
Maximum sizes		3/4 in O.D., 1/4 in long			

equal. As the photoelectrons strike the first dynode, the dynode emits secondary electrons. The number of the initial electrons are thus multiplied in number by the emitted secondary electrons which are then accelerated toward the second dynode where now these electrons produce secondary electrons. Finally, by the time all the electrons reach the final anode, their number becomes very large due to the electron multiplication processes in the dynodes. Electron multiplication of $10^5 - 10^6$ is typical in photomultiplier tubes.

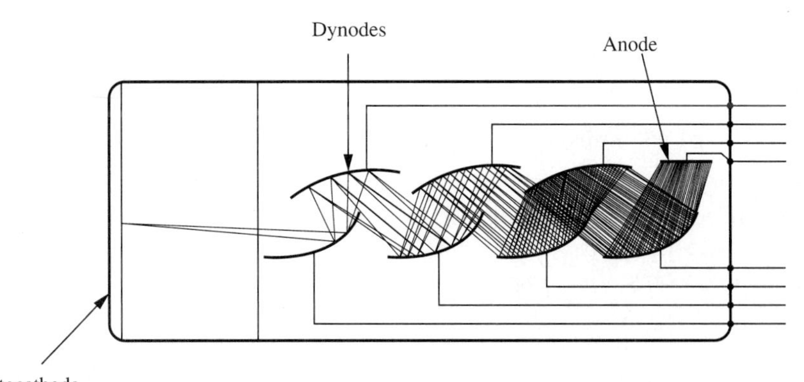

FIGURE 7.16
Photomultiplier tube. Photoelectrons are multiplied by the dynodes before being collected by the anode.

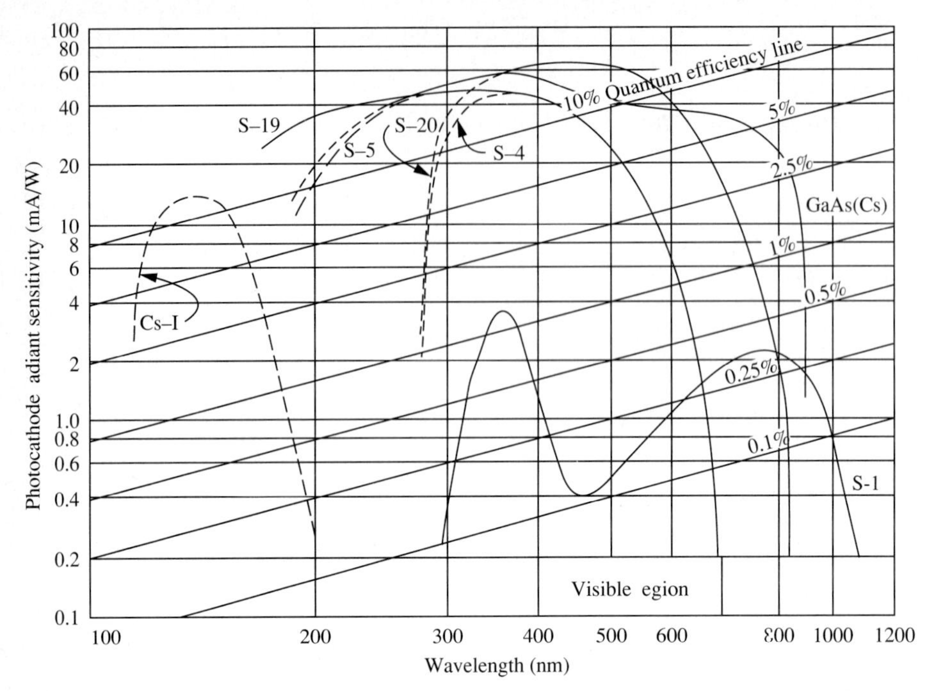

FIGURE 7.17
Spectral response of commercially available photocathodes.

Photomultiplier tubes are used as photon detectors with a large internal electronic gain. There are various types of photocathodes and their choice in a photomultiplier tube is dictated by the spectral response of the photocathode. The response of photomultipliers tubes to infrared spectra requires photocathodes with very small work functions. Using different materials for the photocathode, the response of the present day photomultipliers ranges from near infrared through visible light to ultraviolet spectra. (Fig. 7.17) When cooled down to low temperatures, photomultiplier tubes can be used as low-noise light detectors with sensitivities high enough that they can detect even the presence of a single photon.

7.6.3 TV Image Tubes

The operational principles of most of the present day television image pick-up tubes are also based on the photoelectric and secondary electron emission phenomena. Image orthicon and vidicon are examples of TV pick-up tubes. The essential components of an image orthicon are shown in Fig. 7.18. The image of an object is formed on the surface of a photocathode by a conventional optical system. Each point on the photocathode emits electrons proportional to the intensity of light falling onto that point. These electrons are then accelerated toward

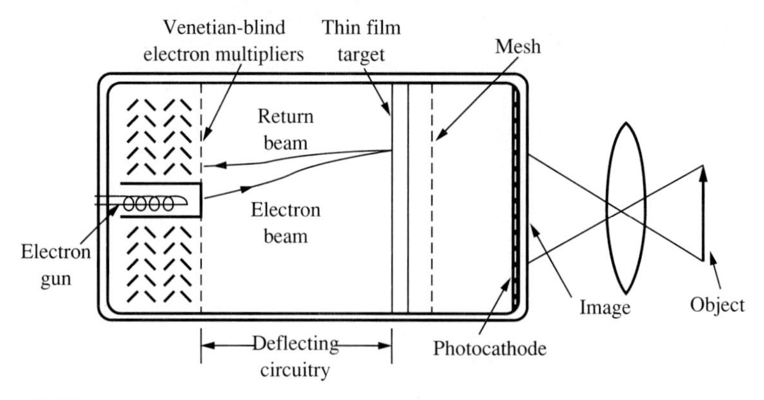

FIGURE 7.18
Essential components of an image orthicon TV tube.

a high resistive thin film secondary electron-emitting target. Secondary electrons that are emitted by the incident electrons are collected by a mesh screen near the target, leaving behind on the target a positively charged image of the optical image. The higher the number of photoelectrons, the higher the positive charge that builds up at that point on the target element. An electron beam produced by an electron gun and deflected by an external magnetic field is scanned across the target. The electron beam neutralizes the positive charge present on the target at a given location. The portion of the electron charge of the beam that is not used in neutralizing the positive charge at the target is reflected back toward the electron gun and impinges on venetian blind-type electron multipliers. The amount of charge collected by the anode is proportional to the charge that is reflected back from the target. The small current in the external circuit is equivalent to the neutralization of a large positive charge, which in turn means a high optical intensity at the photocathode. The final electrical signal amplitude at the anode is opposite to the brightness of the optical image formed on the photocathode.

Image orthicons are very sensitive electron tube devices that are used in industry and science for the detection of optical signals and images.

7.6.4 Field Emission Microscopy

Field emission microscopy is based on the high field emission phenomena and is used in the study of the surface composition of materials. The field emission microscope (Fig. 7.19) has a hemispherically pointed cathode tip and a hemispherical fluorescent screen (material that emits photons in proportion to the intensity of the electrons impinging on it) arranged around the emitter. The most important feature of the design is the perfection of the hemispherically pointed tip which has to be smooth to the lattice steps of atomic dimensions. Field emitted elec-

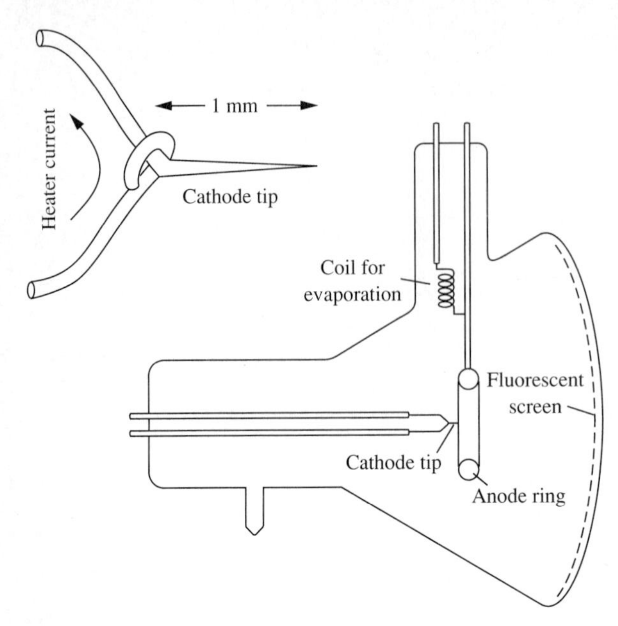

FIGURE 7.19
Schematic diagram of field emission microscope (Good and Muller, 1956).

trons leaving the tip surface with low initial energy start to follow the lines of forces essentially in radial directions. Since the tip radius is usually around 10^{-4} to 10^{-5} cm, the magnification is essentially the ratio of the screen distance to the tip radius and 10^5 to 10^6 magnification is easily obtained. A heater coil inside the evacuated envelope is used to evaporate surface films on the host cathode. This allows the structural study of absorbed surface films on materials. Through field emission microscopy, crystalline structure of metals are investigated.

FURTHER READING

Eisenstein, Albert S., "Oxide Coated Cathodes," *Advances in Electronics and Electron Physics*, Marton, L., ed., vol. I, Academic Press, 1948, 1–64.

Good, R. H., Jr., and Erwin W. Muller, "Field Emission," *Handbuch der Physik*, vol. XXI, 1956, 176–231.

Gorlich, P., "Recent Advances in Photoemission," *Advances in Electronics and Electron Physics*, Marton, L., ed., vol. IX, 1956, 1–31.

Hachenberg, O., and W. Braur, "Secondary Electron Emission From Solids," *Advances in Electronics and Electron Physics*, Marton, L., ed., vol. XI, Academic Press, 1959, 413–499.

Kohl, Walter J., *Handbook of Materials and Techniques for Vacuum Devices*, Reinhold, 1967, Chapter 16, 475–529.

vander Ziel, Aldert, *Solid-State Physical Electronics*, 3d ed., Prentice Hall, 1976, Chapters 7, 8, 9 and 10.

PROBLEMS

7.1. A monochromatic light of frequency ν is incident on a metal which has a work function of 2.6 eV.

(*a*) Find the minimum frequency necessary to emit photoelectrons from the metal.

(*b*) If the frequency is $\nu = 10^{15}$ Hz, find the maximum velocity of the ejected electrons.

7.2. Blue light $\lambda = 4000$ A (1A$= 10^{-10}$ m) is incident on a photocathode. If a potential difference of 0.8 V is needed to stop the current flow, find the work function of the cathode material.

7.3. Energy band diagrams of cesium and cesium oxide are shown in Fig. P.7.3 (Gorich, 1956). Find the maximum wavelength necessary to eject photoelectrons from both of these materials.

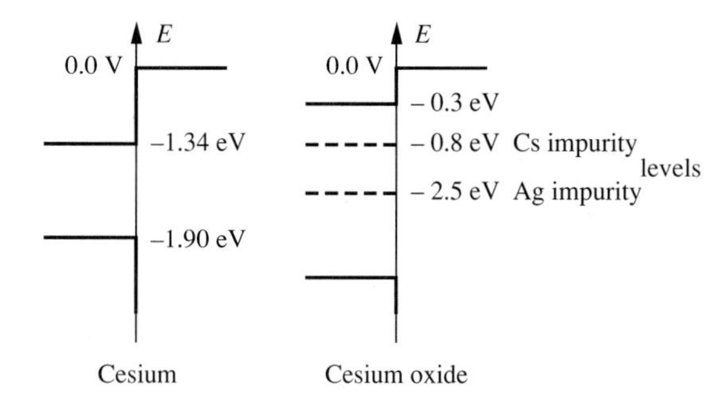

FIGURE P.7.3

7.4. A tungsten filament is heated to 1800 C. Find the current density generated by the filament.

7.5. If a metal plate is placed opposite to the cathode of Problem 7.4 at a distance of 2 mm and a potential difference of 10,000 V is applied, find the reduction in the work function of the tungsten and the corresponding increase in current density.

7.6. If the heat to the cathode of Problem 7.4 is removed, find the necessary electric field that can produce 1 mA/cm^2 of current density.

7.7. If the temperature dependence of the work function is given by $W = W_0 + \alpha T$, show that the effective Richardson constant can be written as

$$A_0 \exp(-\alpha/k)$$

CHAPTER 8

JUNCTIONS AND RELATED DEVICES

In the previous chapters, the physical characteristics of solids were given, with emphasis on the semiconductors. Additional parameters were also introduced, in order to understand some of the basic charge transport phenomena in solids. We saw that models based on the band theory of solids can help us to better understand many of these concepts. In all of these analyses, it was assumed that the solid was uniform in cross section and the charges were distributed uniformly within the solid, and that whatever happened in the solid occurred throughout the solid with the same probability. The results of nonuniform spatial charges were used in explaining the concepts of recombination time and diffusion length. All of these derivations were based on semiconductor samples whose overall physical uniformity was not violated.

In this chapter, we are going to look into what happens when two dissimilar materials are brought together and form a contact with one another. These contacts may be the physical contact of two dissimilar materials or they can be obtained by doping two different regions of a semiconductor with two different impurity atoms in the process of manufacturing the device. Therefore, in the sections that follow, the word "contact" may be used interchangeably for either actual diffused or grown junctions. A p-type or an n-type semiconductor may also be referred to as a different element—they may both be obtained from the same base material, but with different impurity atoms.

In this chapter, metal-metal contacts, metal-semiconductor contacts and semiconductor-semiconductor contacts will be studied. Interesting features of these contacts will also be given.

8.1 FERMI LEVEL IN JUNCTIONS

Although there are some basic properties that are common to all the junctions, many forms of junctions show completely different electrical characteristics. We will begin with the common characteristics of the junctions. The energy band diagram related to an overall junction is very useful in understanding the physical operation of a specific device that may incorporate one or more junctions in its implementation. One of the common consequences of bringing two dissimilar materials into contact with each other is the alignment of the Fermi energy levels in each of the elements to the same energy level after the junction is formed. As a result of this property, other energy levels associated with the elements forming the junction are placed relative to the common Fermi energy level. In all the subsequent analysis, the corrections necessary due to nonideal cases, such as the presence of surface states associated with various materials, will be neglected in considering the overall energy-band diagram of a junction.

Each material forming a junction may be doped with different impurity atoms and may have different charge concentrations. Depending on the type of material used, electrons partially or completely fill in the available energy states in various bands of the given material. In each element, the occupancy of levels is governed by the Fermi–Dirac distribution. At $T = 0$ K, the Fermi energy E_F is defined as the energy above which the probability of occupancy by an electron of an energy level E is zero. Consequently, the probability of occupancy of the energy E below E_F is one. At $T > 0$ K, the probability that an electron will occupy an energy state E is given by the Fermi–Dirac factor $f(E) = 1/\{1 + \exp[(E - E_F)/kT]\}$. The probability that an electron will not occupy an energy state E is similarly given by $[1 - f(E)]$. We should also keep in mind that, initially, the elements forming the junction are electrically neutral before the junction is established. With these basic concepts, we can analyze the general behavior of the contact problem of the two dissimilar materials.

We first consider two metals that have different electron concentrations. When the junction forms, electrons from the element with the higher concentration, say X, will diffuse toward the element Y with the lesser electron concentration. As this electron motion occurs, the electrons from X leave behind, near the vicinity of the junction, a positively charged region of space mostly made up of the immobile atoms of the solid. As the mobile charge flow continues into Y, because of the resulting space charge, an electric field which is directed toward the diffusing electrons is generated. The direction of this field is such that the resulting force is opposite to the direction of the diffusion of mobile electrons. Finally, as this built-in field amplitude increases, an equilibrium condition is reached such that further diffusion of mobile charge carriers stops. As a result of this electric field, an equivalent potential barrier in the vicinity of the contact is established.

Similar arguments apply to a p-n junction. As will be shown later in this chapter, an equivalent potential barrier will again be established across the junction, this time as a result of both the electron and hole diffusion. Therefore, whatever the two elements forming the junction are, as long as these two elements have different Fermi energy levels the resulting junction will have an internally generated built-in potential barrier. The barrier height will be dependent on the density of mobile charge carriers in the two materials since the magnitude of internally generated potential depends on the mechanism of charge diffusion and, thus, to the relative magnitude of these charges in the respective regions of the junction. The electrical characteristic of the resulting device largely depends on the extent and magnitude of this potential barrier. With the application of an external potential, equilibrium conditions can be violated and the magnitude of this barrier seen by the charges in the respective side of the junction can be changed.

We now consider two materials X and Y with $E_{Fx} > E_{Fy}$. When these materials are brought together to form a contact, we expect electrons from the material of higher concentration (X) to diffuse into the material with smaller electron concentration (Y). This diffusion can take place provided certain conditions are met. Suppose the same energy level E is in both of the materials. First of all, there should be an electron in X with energy E. This is proportional to the probability that an electron will occupy the state E in X. This can be written as $f_x(E)$. The second condition for the diffusion to take place is that there should be an available energy state E in Y for the electrons from X to diffuse into. This also means that there should be no electrons in Y at the energy state E. This condition can be interpreted as the probability that the energy state in Y is empty, that is, $[1 - f_y(E)]$. The total probability that electrons will diffuse from X to Y is the product of these two probabilities, that is, $f_x(E)[1 - f_y(E)]$. There is no reason that an electron in Y with energy E should not diffuse into the energy level E in X. The probability that this diffusion will take place requires similar conditions and is given by $f_y(E)[1 - f_x(E)]$. At equilibrium, these probabilities should be equal. We can thus write

$$f_x(E)[1 - f_y(E)] = f_y(E)[1 - f_x(E)] \tag{8.1}$$

Substituting the corresponding Fermi factors into Eq. 8.1, we find that the equality holds only if

$$\exp\left(\frac{E - E_{Fx}}{kT}\right) = \exp\left(\frac{E - E_{Fy}}{kT}\right)$$

is satisfied. This is true if

$$E_{Fx} = E_{Fy} \tag{8.2}$$

This result shows that when the system reaches an equilibrium, the Fermi energy levels of the two materials align to the same energy level.

8.2 METAL-METAL CONTACTS

The energy-band diagrams of two different isolated metals are drawn in Fig. 8.1*a* to show the typical important energy parameters associated with metals. Among these parameters, the work function W_m, the electron affinity χ_m, and the Fermi energy E_F will be different for each metal.

If two different metals with $W_2 > W_1$ are brought into contact, the electrons from the metal with a higher Fermi energy will diffuse into the other metal. Once equilibrium is reached, the two Fermi levels will align themselves to the same energy level as shown in Fig. 8.1*b*. This will be reflected as the lowering of the vacuum level of material 1 relative to material 2. The difference of these energy levels is given by $\Phi_c = W_2 - W_1$. The potential $V_c = \Phi_c/e$ corresponding to this barrier energy is called the contact potential.

Metal-metal contacts find use in thermocouples which are formed by bringing two dissimilar materials into contact with each other. The temperature dependence of the electromotive force is used to measure the temperature of the surrounding medium. The thermocouple effect strongly depends on the temperature dependence of charge carriers in a metal in a more complicated manner and this topic will not be covered here further.

8.3 METAL-SEMICONDUCTOR JUNCTIONS

As shown in Fig. 8.2*a*, a metal-semiconductor junction can be formed by depositing a metal over a highly polished and cleaned semiconductor wafer by various deposition techniques that will be discussed in Chapter 14. The lower contact should have ohmic characteristics (conduction of current as if through a normal resistor), in order not to mask the electrical characteristics of the actual metal-semiconductor junction.

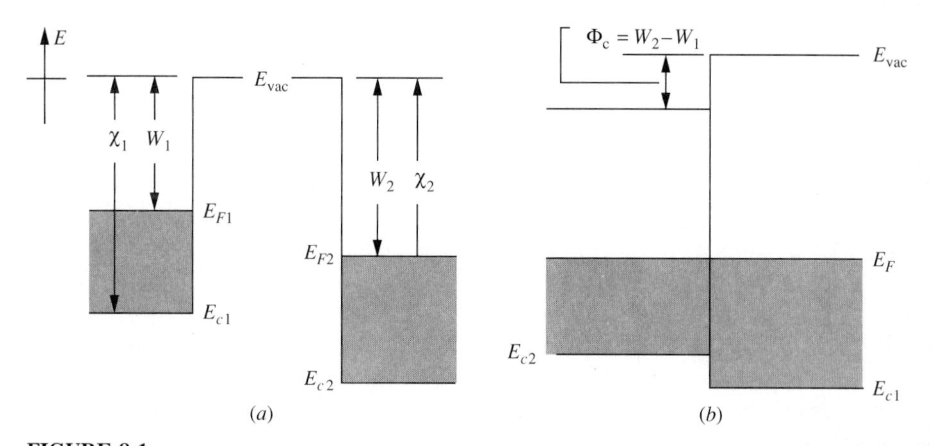

FIGURE 8.1
Energy-band diagram of two metals (*a*) before contact and (*b*) after contact ($W_2 > W_1$).

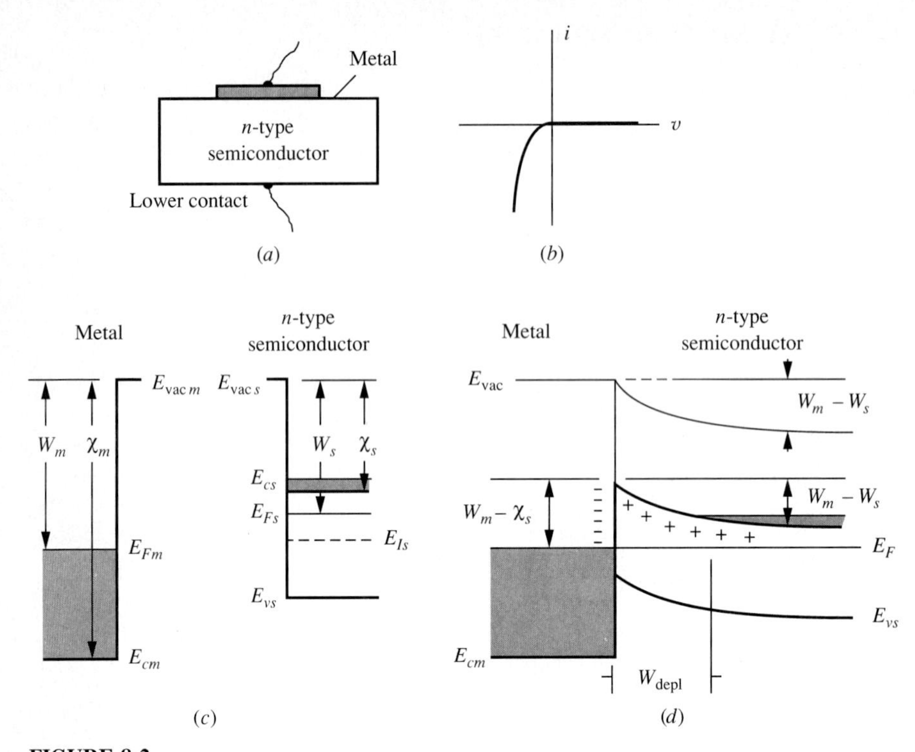

FIGURE 8.2
(*a*) Metal-semiconductor junction (*b*) i-v characteristics. Energy-band diagrams (*c*) isolated atoms and (*d*) after junction formation. $W_m > W_s$.

The metal semiconductor junctions show either a rectifying electrical characteristic (nonlinear current-voltage characteristics as a function of applied voltage) or an ohmic contact. Fundamental physical properties of a metal-semiconductor junction can be explained by referring to the associative energy levels of the metal and the semiconductor. Since the choice of the type of the metal as well as the type of the semiconductor determines the resulting electrical characteristics, we will first concentrate on a junction formed by a specific metal and an n-type semiconductor.

The ideal energy-band diagrams of an isolated metal and an isolated n-type semiconductor before contact are shown in Fig. 8.2*d*. W_m and W_s are the respective work functions of the metal and the semiconductor. For the present case, it will be assumed that $W_m > W_s$, that is, $E_{Fs} > E_{Fm}$. Upon contact of the two materials, electrons from the conduction band of the semiconductor will flow into the available energy states of the metal and, at the same time, leave behind near the contact area a positive immobile charge layer of unneutralized impurity atoms in the semiconductor. This charge imbalance produces an electric field directed away from the semiconductor toward the metal which, in turn, produces a force that starts to oppose the electron transport from the semiconductor into the metal. As the charge transfer continues, the magnitude of this opposing field increases until an equilibrium condition is reached where further electron diffusion across the junction is prevented. This equilibrium condition is the same as the lining up of the Fermi levels of the two materials forming the junction. The opposing electric field is reflected as a formation of a potential barrier across the junction.

The finite region extending from the beginning of the junction and into the semiconductor is depleted of mobile charges and only immobile ionized donor atoms reside in this region. This region is known as the *depletion layer*.

The resulting junction energy-band diagram of the corresponding metal-semiconductor junction is shown in Fig. 8.2*d*. The height of the potential barrier on the semiconductor side is given by $(W_m - W_s)$. This barrier is reflected as a gradual lowering of the vacuum-energy level of the semiconductor relative to the metal. This same potential barrier is reflected on the conduction band of the semiconductor. The potential barrier found on the metal side of the junction is equal to $(W_m - \chi_s)$. W_{depl} is the width of the depletion layer.

From our previous discussions on the Fermi level in semiconductors, we know that if the semiconductor has more electrons than holes, E_F lies above the intrinsic Fermi level and gets closer to the conduction band with increasing electron concentration. The opposite is true for the reduction of electrons in the semiconductor, that is, E_F moves further below the conduction band. Thus, as shown in Fig. 8.2*d,* the rise in the conduction-band energy relative to the common equilibrium Fermi energy E_F is consistent with the reduction of electrons from the semiconductor side of the junction as a result of the diffusion process.

When the current flow across the junction is considered, charge transport from the metal into the semiconductor or from the semiconductor into the metal will occur for those electrons that can overcome their respective energy barriers in their conduction bands. Those electrons that possess energies greater than their barrier energies will appear on the opposing sides of the junction and will thus be free to move toward the corresponding contact terminals if suitable external electric fields are applied. Only those electrons at the energy tails of the Fermi–Dirac distribution that have energies greater than the barrier potentials will be able to overcome the potential barriers of each side. If no external fields are applied, the two electron flows from the opposite sides will cancel each other and, thus, at equilibrium no external current will flow in the circuit. If the metal is taken as the reference point, an externally applied bias voltage modifies and, depending on the polarity, either reduces or increases the height of the potential barrier on the semiconductor side. This results in a nonlinear current voltage characteristic, as shown in Fig. 8.2*b*. The mechanisms and the corresponding voltage dependence of the nonlinear current flow across the junction will be explained in detail in the following sections. Note that the changes in the vacuum-energy levels do not play any role in the charge conduction processes with regard to the charge transport discussed here.

The energy-band diagram of a metal and a p-type semiconductor before and after contact are also shown in Fig. 8.3. For this case, we assume the work function of the metal to be smaller than that of the semiconductor, that is, $W_m < W_s$. This time, electron flow takes place from the metal into the semiconductor, and again, a

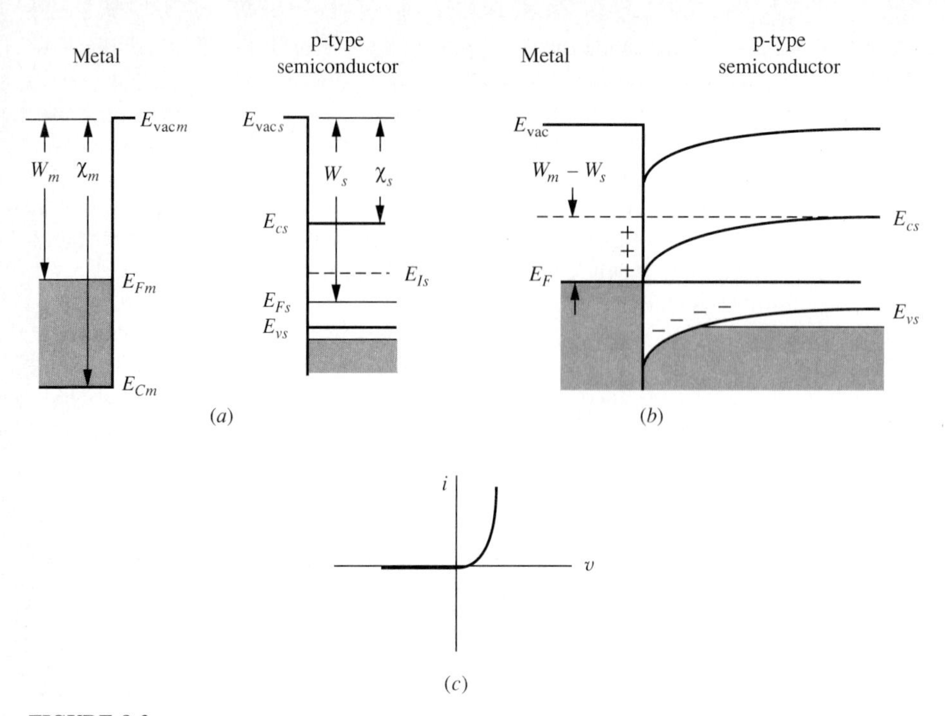

FIGURE 8.3
Energy-band diagram of a metal and a p-type semiconductor. (*a*) Isolated elements and (*b*) after junction formation. $W_m < W_s$ (*c*) *i*-*v* characteristics.

potential barrier forms between the two materials. Any electron on the metal side has to overcome the barrier height $W_m - W_s$ to appear on the semiconductor side. In Fig. 8.3*b* this is reflected as a barrier for the electrons to move from the metal into the conduction band of the semiconductor. But since the holes are the majority carriers in the semiconductor, there is the possibility of the holes moving from the semiconductor into the metal by the application of an external potential. If the semiconductor is made positive, the energy levels on the semiconductor side will be lowered, which implies that the energy barrier seen by the holes will be lowered. That means more holes can move into the metal side. We expect recombination of the holes with the majority carrier electrons in the metal. In order to compensate for the recombination in the metal, electrons should be supplied from the connecting terminal of the metal. Thus, the continuity of current in the circuit is maintained. If the semiconductor is made negative, the energy levels of the semiconductor will be raised, and thus, the holes on the metal side will be subjected to a higher energy barrier and very little current will flow. The resulting nonlinear current voltage characteristic of the junction for this case is shown in Fig. 8.3*c*.

When the conditions on the work functions of the metal and semiconductor given for the above two cases are reversed, the resulting contacts become ohmic contacts. When an n-type semiconductor is in contact with a metal where $W_m < W_s$ (as shown in Fig. 8.4*a*), the electrons from the metal now flow into the semiconductor, charging the surface of the semiconductor with more negative charges. Because of the increase in electrons on the semiconductor side, the conduction band of the semiconductor tilts downward, that is, the conduction-band edge gets closer to the Fermi levels. If an external potential is applied across the junction so that the semiconductor is positive, a potential equal to at least $W_s - W_m$ is required to initiate the flow of electrons from the metal to the semiconductor. This is reflected as a small discontinuity in the i–v characteristics of the junction as shown in Fig. 8.4*a*. When the semiconductor is made positive, the electrons in the conduction band can easily flow toward the metal. The behavior with this polarity is purely ohmic.

Similarly, when a p-type semiconductor is in contact with a metal such that $W_m > W_s$, the resulting band diagram forms (as shown in Fig. 8.4*b*). This time, the roles of the electrons are interchanged with those of the holes. When the semiconductor is made negative, the energy levels of the p-side are raised, implying reduction in the energy barrier for the holes in the metal. Beyond a minimum bias potential, the holes see no barrier and current flows easily. With reverse polarity, the energy levels decrease, implying higher energy for the holes on the semiconductor side, thus the holes can easily move into the metal. The resulting i–v characteristics for this case is shown in Fig. 8.4*b*.

When a heavily doped semiconductor is in contact with a metal, the resulting junction, irrespective of the relative work functions of the two materials, behaves

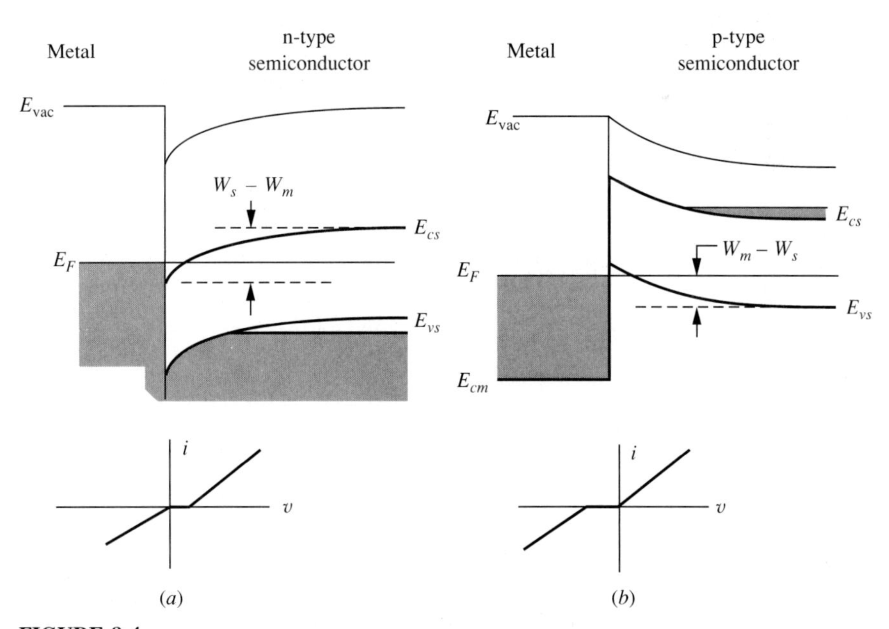

FIGURE 8.4
(*a*) Metal and n-type semiconductor contact, $W_s > W_m$, and (*b*) metal and p-type semiconductor contact, $W_m > W_s$. The resulting i-v characteristics are shown below each diagram.

as an ohmic contact, since the barrier that forms is over such a short distance that the electrons from both sides now have a high probability of tunneling through their respective junction barriers. Depending on the polarity of the applied potential, electrons either from the metal or the semiconductor can tunnel through the barrier to the opposing region. Thus, as shown in Fig. 8.5 for the case of metal n-type semiconductors, the junction behaves as close to a purely ohmic contact. The highly doped semiconductor is usually referred to as a *degenerate* semiconductor and is generally labeled an n^+-type. The Fermi level usually lies near the bottom, but inside, the conduction band. The electrical behavior of an n^+-type semiconductor is itself similar to a metal because of the large number of majority carriers present in the semiconductor. If there is a p-region in a device where ohmic contact is necessary, a p^+ region with a metal contact over it can also be used to make an indirect but effective ohmic contact to the original p-region.

In many cases, especially when it is required to bring leads from the semiconductor to the outside world, an ohmic contact to a semiconductor becomes essential. For the successful operation of the device, proper care should be taken in forming the junction so that they will not introduce any nonlinear effects that can change the characteristics of the actual device.

8.3.1 Schottky Barrier Diode

The rectifying metal-semiconductor contact is an integral part of many modern semiconductor devices. These junctions were originally studied in detail by Schottky and, therefore, are known as the Schottky barrier diodes.

In Fig. 8.6*a*, the energy-band diagram of a metal-semiconductor diode is redrawn with emphasis only on the barrier associated with the conduction bands of the two materials. If the electrons from each element can overcome the potential

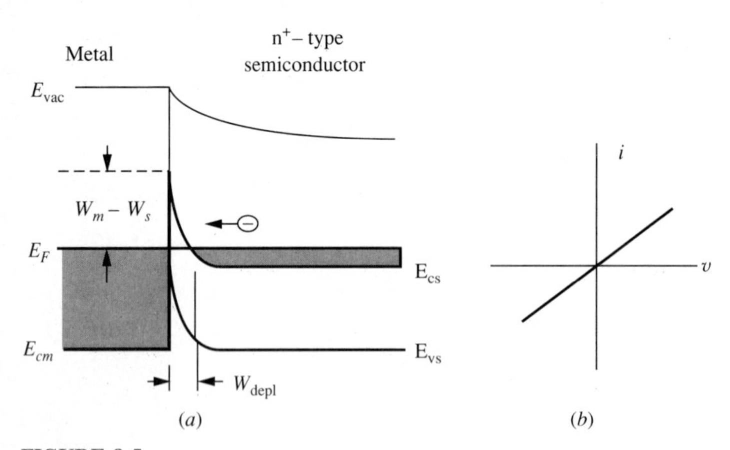

FIGURE 8.5
(*a*) A junction between a highly doped semiconductor and a metal results in an ohmic contact due to tunneling. (*b*) i-v characteristics.

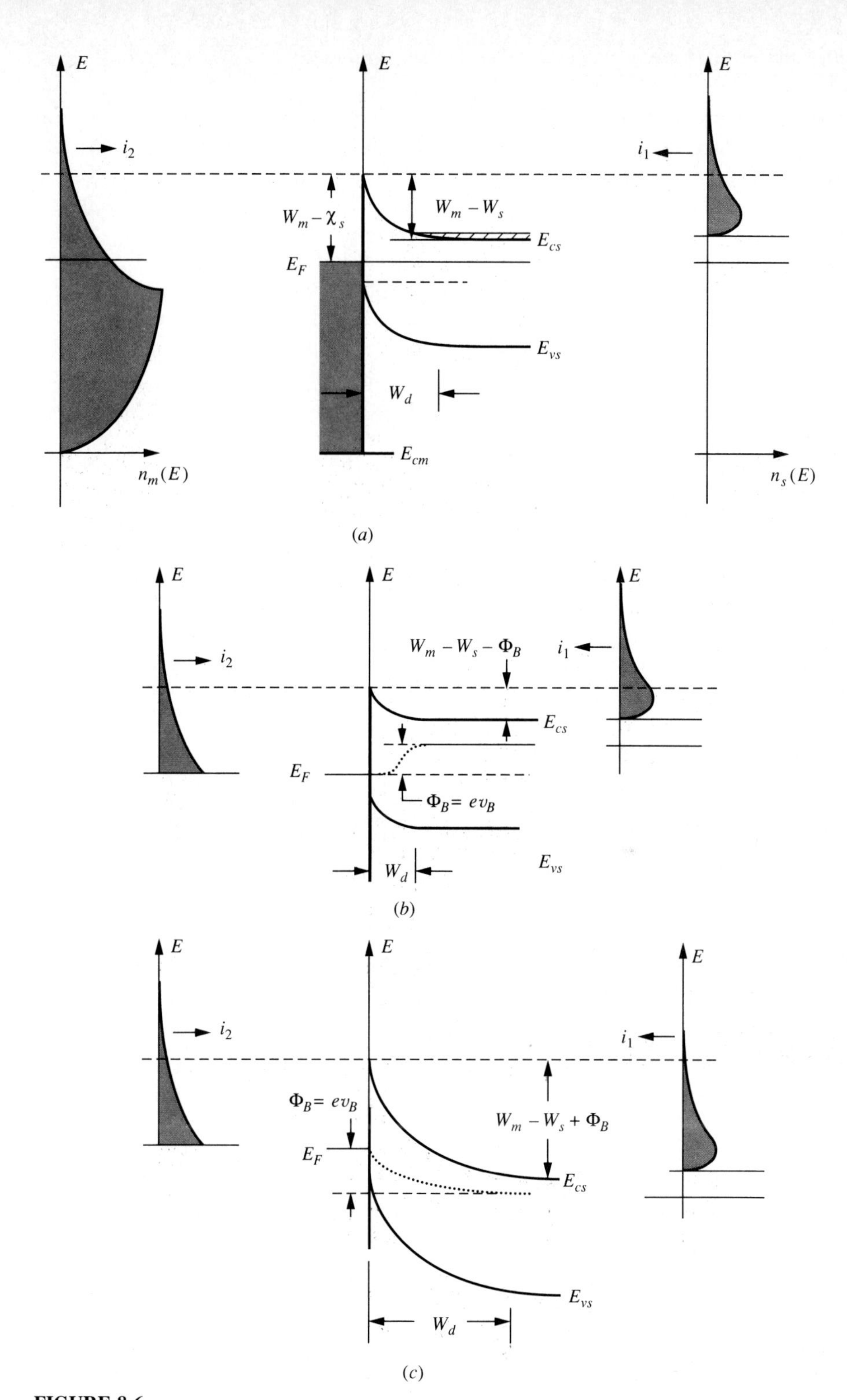

FIGURE 8.6
The energy-band diagrams and the electron densities as a function of energy in a metal-semiconductor junction with (*a*) no bias, (*b*) forward bias, and (*c*) reverse bias. The electron energy distributions are highly exaggerated to show their effects on currents.

barrier seen from their side of the junction, they can move into the conduction band of the other material and, by the application of an external field, can produce current in an external circuit. For studying the conduction mechanisms across the junctions, the barriers associated with the conduction-band electrons are of interest only for the case considered here. The changes in the vacuum-energy level will be useful if photoemission into the vacuum level is to be studied.

The energy dependence of the density of electrons in both the metal and the semiconductor is shown in Fig. 8.6*a*. Note that these distributions are highly exaggerated above the Fermi levels in order to emphasize their effects on the current flow across the junction. Electrons on each side with energies greater than E_F can overcome their respective potential barriers at the junction and produce the currents i_1 and i_2. At equilibrium, these currents are equal.

When a forward potential $\Phi_B = ev_B$ is applied across the junction, that is, the semiconductor is made negative relative the metal, the potential barrier for the electrons in the semiconductor is reduced. The barrier for the electrons in the metal does not change since the metal is at the reference potential and the potential on the metal is constant throughout its physical length. We are now ready to calculate the current imbalance in the diode resulting from this externally applied bias voltage v_B.

The calculation of the current is made by considering the electron flow from one material to the other as a consequence of only those electrons that can overcome the potential barrier of their respective sides of the junction. Any electron which possesses an energy E greater than the barrier height will be able to appear on the other side of the junction and will move freely in the second material. This problem can be treated in the same way as the thermionic emission discussed in Chapter 7. Here the electrons do not have to be emitted to the outside of the semiconductor—just overcoming the internal barrier is sufficient for their conductive motion.

The thermionic current density due to those electrons that can overcome a potential barrier is given by Eq. 7.10

$$J = \mathcal{A}^* T^3 \exp\left(\frac{-W'}{kT}\right) \tag{8.3}$$

where W' is the effective work function of the material and $\mathcal{A}^*$ is a modified Richardson constant that can take care of the effective masses of the electrons and other surface-related factors that were not considered in deriving Eq. 8.3.

Replacing the barrier height W' by $W_m - \chi_s$, we can write the current density flowing from the metal into the semiconductor as

$$J_2 = \mathcal{A}_2^* T^3 \exp\left(-\frac{W_m - \chi_s}{kT}\right) \tag{8.4}$$

Similarly, the current density flowing from the semiconductor to the metal is due to those electrons that overcome the potential barrier of $(W_m - W_s)$.

$$J_1 = \mathcal{A}_1^* T^3 \exp\left(-\frac{W_m - W_s}{kT}\right) \tag{8.5}$$

where constants $\mathcal{A}_1^*$ and $\mathcal{A}_2^*$ are taken to be different in order to take into account the small differences in the barrier heights of the two regions. At equilibrium these current densities should be equal. From this we find

$$J_1 = J_2$$

$$\mathcal{A}_1^* = \mathcal{A}_2^* \exp\left(\frac{\chi_s - W_s}{kT}\right) \tag{8.6}$$

When an external forward bias voltage v_B is applied across the junction, the barrier height on the semiconductor side decreases to $W_m - W_s - ev_B$ (Fig. 8.6*b*). The current flow from metal to semiconductor stays the same, but the current density from semiconductor to the metal is modified by

$$J_1 = \mathcal{A}_1^* T^3 \exp\left(-\frac{W_m - W_s - ev_B}{kT}\right) \tag{8.7}$$

The net current density is

$$J = J_1 - J_2$$

$$J = \left[\mathcal{A}_2^* T^3 \exp\left(-\frac{W_m - \chi_s}{kT}\right)\right]\left[\exp\left(\frac{ev_B}{kT}\right) - 1\right] \tag{8.9}$$

Multiplying by the cross sectional area of the junction, the diode current reduces to

$$i = I_0\left[\exp\left(\frac{ev_B}{kT}\right) - 1\right] \tag{8.10}$$

where I_0 is known as the *reverse saturation current* and is a function of the material parameters and the temperature. I_0 is given by

$$I_0 = A\,\mathcal{A}_2^* T^3 \exp\left(-\frac{W_m - \chi_s}{kT}\right) \tag{8.11}$$

where A is the cross-sectional area of the diode. Note that when $v_B = 0$, current i, given by Eq. 8.10, is zero.

If a reverse bias is applied by making the semiconductor more positive, the barrier height on the semiconductor side increases (Fig. 8.6*c*). Again, only Eq. 8.7 is modified by the new barrier $(W_m - W_s) + ev_B$, and the same equation, Eq. 8.10, is obtained with the difference that the sign of the exponent becomes negative. Equation 8.10 can thus be used to represent the complete current-voltage dependence in a Schottky diode if the proper sign is used for different polarities.

The Schottky barrier diode operates with the majority carriers. Therefore, in deriving Eq. 8.11, only electrons in the semiconductor are considered and the holes are neglected. Actually, the holes may have a small contribution to the

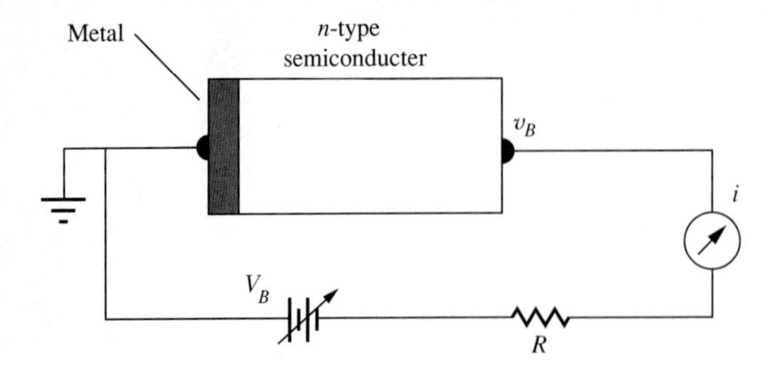

FIGURE 8.7
Basic circuit for studying a metal-semiconductor rectifying junction.

reverse saturation current I_0. On the other hand, the operation of a p-n junction diode, which will be discussed next, relies on the minority carrier flow across the junction of the semiconductors.

Figure 8.7 shows a typical circuit to study the electrical characteristic of a Schottky diode. A current-limiting resistor, an ammeter, and a variable-voltage source whose polarity can be changed are connected in series with the diode. The metal side of the diode is taken as the reference for the applied potential.

Example 8.1. A Schottky diode is made by depositing gold over an n-type silicon with a donor concentration of 5.0×10^{16} cm^{-3}. The diode area is 15 μm by 15 μm square. The work function for gold is $W_{\text{Au}} = 4.90$ eV and the work function for the given n-type Silicon is experimentally found to be $W_{\text{Si}} = 4.212$ eV. Find the resulting reverse saturation current of the diode at room temperature. Use 1.6×10^6 $(A\ \text{m}^{-2}\ K^{-1})$ for $\mathcal{A}_2^*$.

Solution. In order to find I_0, we need to know the electron affinity χ_{Si} of Silicon. This will be equal to

$$\chi_{\text{Si}} = W_{\text{Si}} - E_F$$

For 100 percent ionization of the donor atoms, the Fermi level for the given concentration is calculated to be 0.162 eV below the conduction band (see Example 6.1). Thus

$$\chi_{\text{Si}} = 4.212 - 0.162 = 4.05 \text{ eV}$$

The reverse saturation current density becomes

$$J_0 = (1.6 \times 10^6 \text{ A m}^{-2}\text{K}^{-1}) \times (300 \text{ K})^3 \exp\left[-\frac{(4.9 - 4.05) \text{ eV}}{0.026 \text{ eV}}\right]$$
$$= 0.27 \text{A m}^{-2}$$

Multiplying this by the junction area of 2.25×10^{-10} m^{-2} we find

$$I_0 = 6.16 \times 10^{-11} \text{ A}$$

8.4 SEMICONDUCTOR JUNCTIONS

A semiconductor junction is formed when two semiconductors of different characteristics are brought into contact with one another. If the same semiconductor element is used for both the p- and the n-regions, the resulting junction is called a *homojunction*. If two dissimilar semiconductors are used for the p- and n-regions, the resulting p-n junction is called a *heterojunction*. Homojunctions were the first practical junctions, and hence, they are generally referred to as p-n junctions.

8.4.1 Homojunctions (p-n Junctions)

A homojunction is generally manufactured by doping different portions of a semiconductor with different impurity elements using various processes such as diffusion, ion implantation, and so on. In practice, a chosen portion of a p-type semiconductor is doped with donor impurity atoms so as to produce the impurity profile shown in Fig. 8.8. Thus, a p-n junction is formed at the interface of the two regions. Similarly, acceptor atoms can be diffused into a p-type semiconductor. To form a junction, one can also epitaxially grow an n-layer or a p-layer over a p-type or an n-type substrate.

Energy-band diagrams of an n-type and a p-type semiconductor when they are isolated from each other are shown in Fig. 8.9, with their pertinent physical parameters. The two regions' parameters differ only in their Fermi energies or the corresponding work functions of each region.

We will consider a one-dimensional model for a p-n junction as shown in Fig. 8.10*a*. It will be assumed that the transition from the n-region into the p-region in the semiconductor is very sharp. This is referred to as an *abrupt junction* approximation. When the junction is initially formed, electrons that are the majority carriers in the n-type material diffuse into the p-region and combine with holes. Similarly, the holes from the p-side will diffuse into the n-region. These

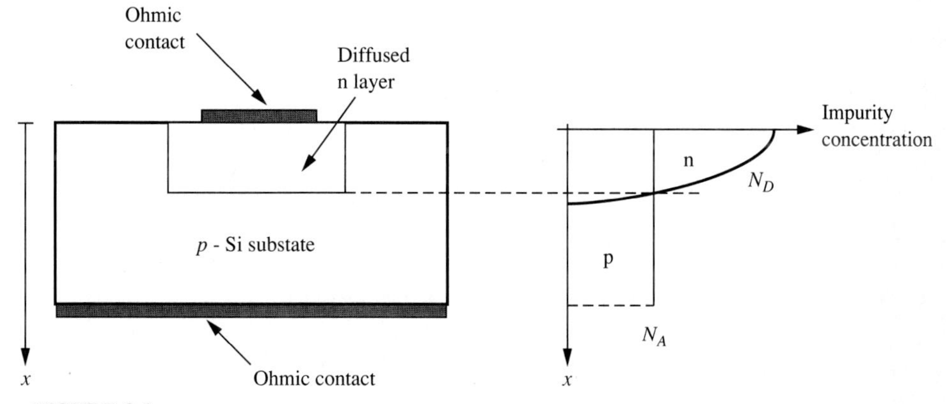

FIGURE 8.8
Actual implementation of a p-n homojunction and the resulting charge concentrations.

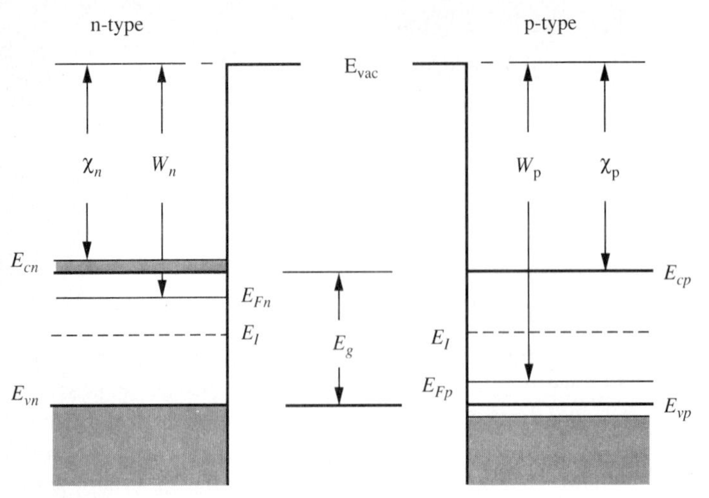

FIGURE 8.9
Isolated energy-band diagrams of n-type and p-type semiconductors.

diffusing mobile charges leave behind, near the junction, uncompensated immobile ionized impurity atoms (donors and acceptor ions in the respective regions), as shown in Fig. 8.10b. These immobile charges generate an internal electric field directed from the n- to the p-side (Fig. 8.10*c*) that reduces further diffusion of the mobile charges. An equilibrium is finally reached when this internal force completely balances the charge diffusion. As a result of this, a potential barrier is formed across the junction (Fig. 8.10*d*). The equations that relate the charge concentrations to the electric field and the electric field to the potential barrier are also given in the same figure. The resulting equilibrium energy-band diagram of the p-n junction, obtained by multiplying the potential by the electronic charge $-e$, is shown in Fig. 8.10*e*.

The barrier potential, sometimes referred to as the diode potential, is reflected as the lowering of the vacuum-energy level of the n-region relative to the p-side. With the exception of the doping levels or different E_F's, since both sides of the junction have the same semiconductor properties, the same shift in energy levels is reflected both on the conduction and the valence bands. The potential barrier at equilibrium is a result of the alignment of the Fermi levels throughout the semiconductor. eV_D is equal to the difference of the Fermi levels of the isolated elements, that is, $|E_{Fn} - E_{Fv}|$.

8.4.1.1 DIODE CURRENT AND BARRIER POTENTIAL. Before we derive the pertinent equations of a p-n junction, the magnitudes of the various charge concentrations will be discussed. Although electrons in the n-region are the majority carriers, only those electrons with energy greater than the potential barrier energy can move into the p-region. On the other hand, the minority hole carriers in the

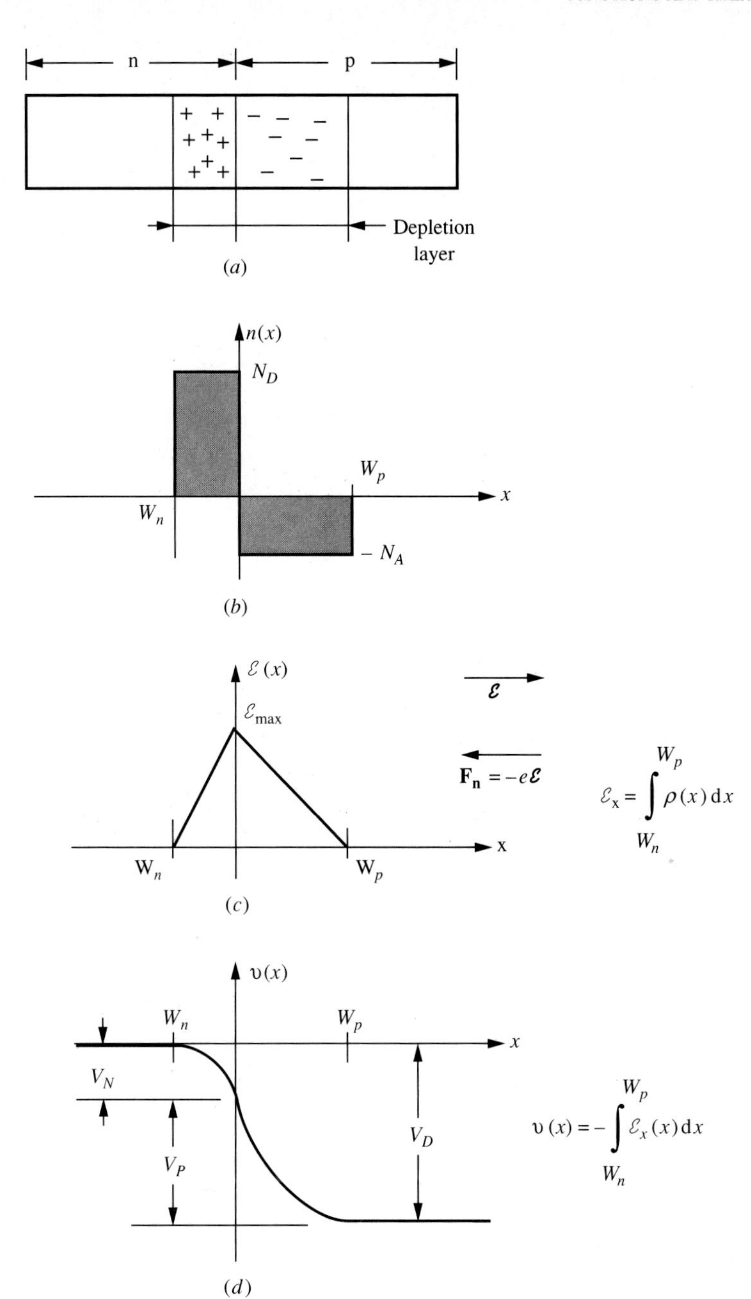

FIGURE 8.10
An abrupt p-n junction: (*a*) idealized physical model, (*b*) charge density, (*c*) internally generated electric field, (*d*) potential barrier.

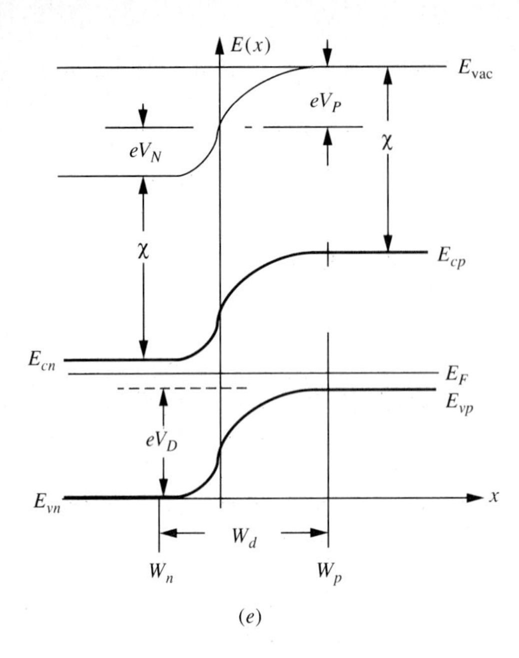

FIGURE 8.10
(e) Resulting energy-band diagram as a function of distance.

n-region whose concentration is very small compared to the electrons, but still finite, see an attractive potential at the junction and, irrespective of the potential height, move easily into the p-region. Similar charge transport from the p- to n-region takes place with the interchange of roles of the holes and electrons.

As shown in Fig. 8.11, four possible currents can flow across the junction at equilibrium. With no external bias, these cancel each other out, and no net current flows across the junction. We will presently show that because of the polarity of the internally established diode potential and as a result of the subsequent application of a reverse bias voltage, the resulting potential barriers will be increased in such a way that only minority carrier currents (electrons from p to n and holes from the n to p regions) will continue to flow across the junction. For practical purposes, these types of charge flow will remain constant irrespective of the magnitude of the reverse voltage (provided it does not exceed breakdown voltage of the junction) and will be the main factors contributing to the reverse-saturation current of the p-n junction diode. On the other hand, the fraction of the majority carriers overcoming the junction barrier are dependent on and controlled by the externally applied bias potential v_B. This bias potential lowers or raises the equivalent potential barrier height seen by the electrons on the n-side and the holes on the p-side, and thus controls the total current flow across the junction.

Let the equilibrium electron and hole densities of the charge carriers on the n- and p-sides away from the junction (in the bulk regions) be respectively n_{no}, p_{no}, n_{po}, p_{po}. Here n and p in the subscripts represent the type of the semiconductor and o stands for the equilibrium values. When the charge density and the subscript

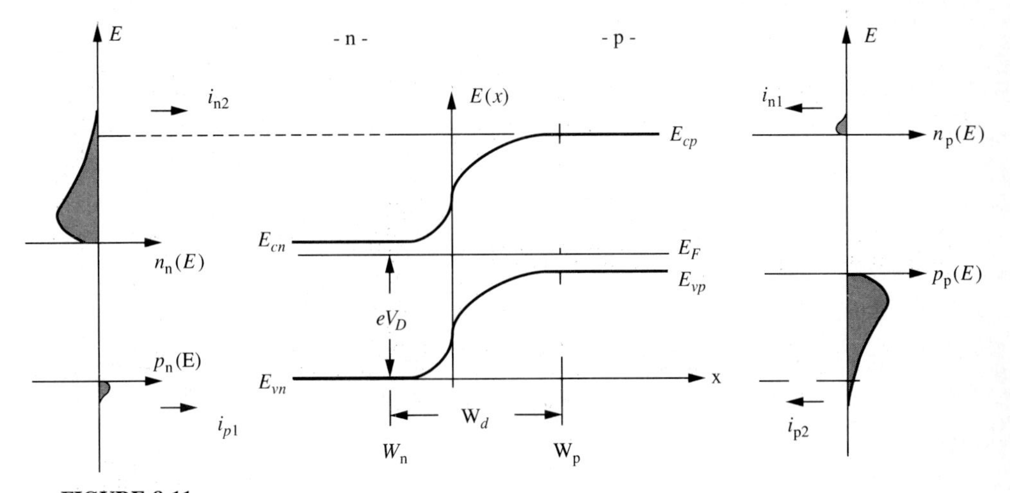

FIGURE 8.11
Equilibrium charge concentrations and the different currents flowing across a p-n junction. The charge density distributions are highly exaggerated to show their effects on current.

are the same, it represents the majority carriers, and when they are different, it represents the minority carriers. The various currents flowing at equilibrium with no external bias are shown in Fig. 8.11.

The electron flow from the p to n region is proportional to the electron density $n_{\rm po}$ and can be written as

$$\leftarrow \qquad i_{n1} = K_1 n_{\rm po} \tag{8.12}$$

where K_1 is a constant that depends on the physical processes involved in the charge transport of electrons across the junction. The electron contribution to the current flow from the n to p region is only by those electrons which can overcome the potential barrier eV_D. Since $E \geq \Phi_D = eV_D \gg kT$, this probability can be written in terms of the Boltzmann factor as

$$\rightarrow \qquad i_{n2} = K_1 n_{\rm no} \exp\left(-\frac{(eV_D)}{kT}\right) \tag{8.13}$$

Note that from Fig. 8.10*d*, V_D is actually a negative quantity. Here the sign of V_D is included in the Boltzmann factor, and V_D in Eq. 8.13 represents the magnitude of the diode potential. Similarly, the currents from the n- and p-regions due to holes can be written as

$$\rightarrow \qquad i_{p1} = K_2 p_{\rm no} \tag{8.14}$$

$$\leftarrow \qquad i_{p2} = K_2 p_{\rm po} \exp\left(-\frac{eV_D}{kT}\right) \tag{8.15}$$

where K_2 is another constant related to the hole transport.

At equilibrium

$$i_{n1} = i_{n2} \qquad \text{and} \qquad i_{p1} = i_{p2} \tag{8.16}$$

Using the above equations, we obtain

$$n_{po} = n_{no} \exp\left(-\frac{eV_D}{kT}\right) \tag{8.17}$$

and

$$p_{no} = p_{po} \exp\left(-\frac{eV_D}{kT}\right) \tag{8.18}$$

We can now calculate the potential barrier V_D by using these equations and the relations between various charge and doping concentrations. With the notation used here, Eq. 6.9 of Chapter 6 can be written in the form

$$p_{no}n_{no} = n_i^2 \qquad p_{po}n_{po} = n_i^2 \tag{8.19}$$

Assuming 100 percent ionization of the impurity ions, we can write

$$n_{no} = n_D^+ \approx N_D \qquad p_{po} = n_A^- \approx N_A \tag{8.20}$$

Substituting Eqs. 8.19 and 8.20 into Eqs. 8.17 or 8.18, the potential barrier can be written as

$$V_D = \left(\frac{kT}{e}\right) \ln\left(\frac{n_A^- n_D^+}{n_i^2}\right)$$

$$V_D = \left(\frac{kT}{e}\right) \ln\left(\frac{N_A N_D}{n_i^2}\right) \tag{8.21}$$

The potential barrier depends, therefore, on the impurity doping levels of the n- and p-regions and the semiconductor parameters through n_i. Since the product of the impurity doping levels is larger than n_i^2, Eq. 8.21 gives a positive V_D, which is the magnitude of the potential since the negative sign of V_D was taken care of in deriving Eq. 8.21.

To understand the effect of the externally applied bias potential on the resulting current flow, we refer to Fig. 8.10*d*. We keep the n-side as the reference point, The equivalent diode potential V_D on the p-side is negative relative to the n-side. If now a positive potential is applied to the p-side, the total potential difference across the junction will be the sum of the two potentials or $v_T = -V_D + v_B$. If, on the other hand, a negative potential is applied to the p-side, the total potential becomes $v_T = -V_D - v_B$. Multiplying these potential differences by $-e$, the potential energies for the two bias cases take the forms shown in Fig. 8.12. Thus, for the forward bias, the barrier height is reduced by $\Phi_B = -ev_B$, and for the reverse bias, it is increased by $\Phi_B = ev_B$.

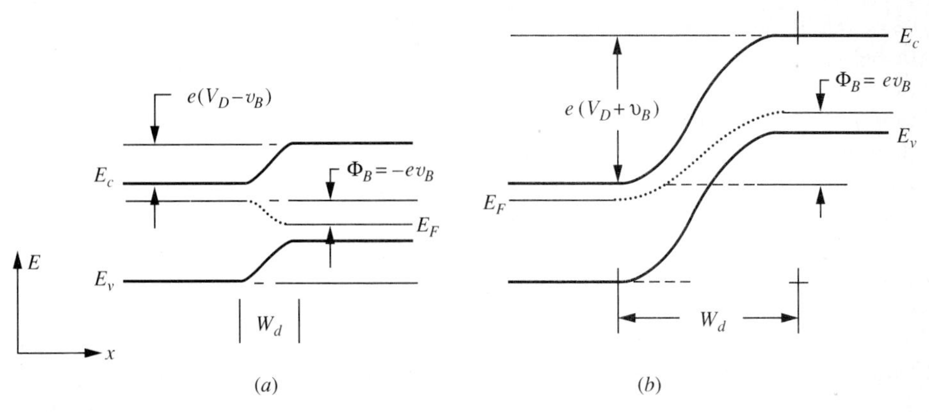

FIGURE 8.12
Variation of the potential barrier with applied bias voltage where $|\Phi_B| = |ev_B|$. (*a*) Forward bias and (*b*) reverse bias.

Referring to Fig. 8.12, under bias conditions, we do not expect currents given by Eqs. 8.12 and 8.14 to be changed. On the other hand, in Eqs. 8.13 and 8.15, the diode potential should be replaced by the equivalent potential barriers under the application of the given polarity of the bias potentials. Thus, the currents i_{n2} and i_{p2} are modified accordingly.

The total circuit current can be written as

$$i = (i_{\text{p2}} - i_{\text{p1}}) + (i_{\text{n2}} - i_{\text{n1}})$$

Replacing these currents by their respective values, including the changes in the potential barriers under forward bias conditions, the resulting total current can finally be written as

$$i = i_0 \left[\exp\left(+ \frac{ev_B}{kT} \right) - 1 \right] \tag{8.22}$$

where I_0 is the reverse-saturation current of the diode and is given by

$$I_0 = (K_1 n_{\text{po}} + K_2 p_{\text{no}}) \tag{8.23}$$

For the reverse bias, the barrier height is increased by ev_B and the total current becomes

$$i = I_0 \left[\exp\left(- \frac{ev_B}{kT} \right) - 1 \right] \tag{8.24}$$

This equation is exactly the same as Eq. 8.22, with v_B replaced by $-v_B$. Therefore, if we remember to use the right polarity (+ for positive and − for negative), we can use Eq. 8.22 for any bias condition. For the general use of Eq. 8.22, we can also remove the subscript *B* from the potential.

We next calculate the constants K_1 and K_2 in Eq. 8.23 by a detailed analysis of the charge transport across the junction. The equilibrium charge concentration of various carriers far away from the junction are shown in Fig. 8.13. Note that in this figure, no mobile charges are shown in the depletion region since the charges that enter this region are swept across by the fields in a very short time interval. Here, it will be assumed that the doping concentration in the two regions of the semiconductor are different, so that the corresponding bulk (regions far away from the junction) majority as well as the minority carrier densities are different. We also assume that the generation or recombination of the charge carriers while traversing the depletion layer is negligible.

We will consider the forward-bias case first and follow the flow of electrons from the n-side to the p-side of the junction. Because of the reduced potential barrier, large numbers of electrons (all with $E > ev_T = e(V_D - v_B)$) will have the necessary energy to overcome the barrier. The electrons which traverse the depletion layer and appear on the p-side of the depletion layer are said to be injected into the p-region. Next to the depletion layer, there will now be excess electrons on the p-side relative to the equilibrium minority carrier density n_{po}. In the analysis that follows, it will be assumed that the magnitude of the injected electron density is smaller than p_{po} but larger than the minority electron density n_{po} on the p-side. This assumption is referred to as the *low-level injection* model for the junction. With similar arguments, holes can be shown to be injected into the n-side from the p-side with similar concentration inequalities. Our analysis will be centered on the electron injection from the n to p side since the derivation that follows will be identical for the holes injected into the n-side from the p-side with appropriate changes in the corresponding parameters.

Electrons injected into the p-side will drift to the right by means of two processes: due to (*a*) the externally applied field and (*b*) diffusion away from the

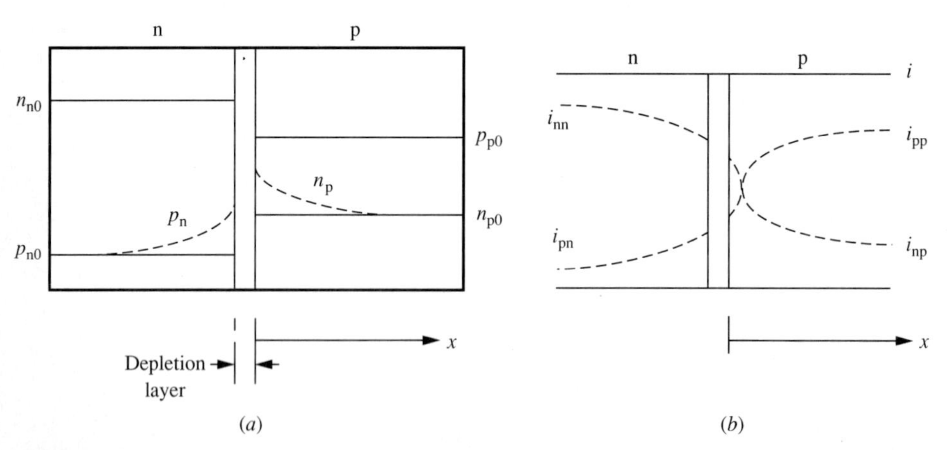

FIGURE 8.13
(*a*) Charge carrier variation near the junction and (*b*) resulting currents for injected charges. Total current i is made up of electron and hole currents.

junction because of the gradient of the electron concentration produced by the injection process. As the electrons diffuse away from the junction, they start to recombine with the majority holes on the p-side. As shown in Chapter 6, under these conditions the electron density decays exponentially with distance away from the junction, as shown in Fig. 8.14.

Using Eq. 8.17, the electrons that are injected into the n-region can be written as

$$n_p = n_{\mathrm{no}} \exp\left[-\frac{e(V_D - v_B)}{kT}\right] \tag{8.25}$$

The excess electron concentration n'_p on the p-side is then

$$n'_p = n_{\mathrm{no}} \exp\left[-\frac{e(V_D - v_B)}{kT}\right] - n_{\mathrm{po}} \tag{8.26}$$

But n_{no} and n_{po} are related by Eq. 8.17. Therefore n'_p reduces to

$$n'_p = n_{\mathrm{po}}\left[\exp\left(\frac{ev_B}{kT}\right) - 1\right] \tag{8.27}$$

The total current that flows across the junction and throughout the semiconductor is due to (*a*) the drift of the electrons and holes under the influence of the externally applied electric field and (*b*) the diffusion of the charges resulting from the injected-charge gradients. The electron and hole current densities can be calculated from the one-dimensional equations

$$J_n = en\mu_n \mathcal{E} + eD_n\left(\frac{dn}{dx}\right) \tag{8.28}$$

$$J_p = ep\mu_p \mathcal{E} - eD_p\left(\frac{dp}{dx}\right) \qquad (-.29)$$

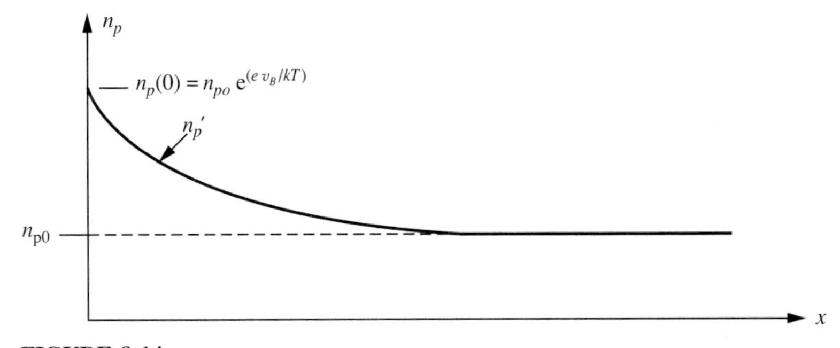

FIGURE 8.14
Variation of electron density on the p-side of the junction. n'_p is the excess electron density variation as a function of distance.

Far away from the junction only the drift terms (first terms of Eqs. 8.29 and 8.28) due to the majority holes and electrons on the respective sides will be dominant. On the two sides and near the junction, the minority electron and hole drifts can be neglected compared to the corresponding diffusion currents. Due to the continuity of the current throughout the semiconductor, the electron contribution to the total current on the n-side will be the same as the electron diffusion current on the p-side. Respective conditions will hold for the hole contribution to the current. Therefore, the total electron current flowing through the p-n junction will be the sum of these two diffusion currents (electron diffusion on the p-side and hole diffusion on the n-side) evaluated at the boundaries of the respective sides of the depletion layer. Once the electron contribution to the current is calculated, the hole contribution can be written immediately with proper parameter changes.

Similar to Eq. 6.22, and as shown in Fig. 8.14, the excess electron density variation on the p-side near the depletion layer as a function of distance can be written as

$$n'_p = n_p - n_{po}$$

$$n'_p = n'_p(0) \exp\left(-\frac{x}{L_n}\right) \tag{8.30}$$

Multiplying the diffusion term of Eq. 8.28 by the cross-sectional area A of the semiconductor, the electron current can be written as

$$i_n = AeD_n \left(\frac{dn'_p}{dx}\right)\bigg|_{x=0} \tag{8.31}$$

Using Eq. 8.27 for $n'_p(0)$, and performing the differentiation of n'_p with respect to x, we obtain the contribution of the electrons to the total current as

$$i_n = -A\left(\frac{eD_n n_{po}}{L_n}\right)\left[\exp\left(\frac{ev_B}{kT}\right) - 1\right] \tag{8.32}$$

Similarly, the contribution of the hole diffusion to the current calculated from the n-side gives

$$i_p = A\left(\frac{eD_p p_{no}}{L_p}\right)\left[\exp\left(\frac{ev_B}{kT}\right) - 1\right] \tag{8.33}$$

Combining Eqs. 8.32 and 8.33, we find for the total current

$$i = I_0\left[\exp\left(\frac{ev_B}{kT}\right) - 1\right] \tag{8.34}$$

where the reverse saturation current I_0 is given by

$$I_0 = Ae\left(\frac{D_n n_{po}}{L_n} + \frac{D_p p_{no}}{L_p}\right) \tag{8.35a}$$

Assuming again 100 percent impurity ionization, and using the equilibrium relations (Eqs. 8.19 and 8.20) for the charge densities, Eq. 8.35*a* can be put into the form

$$I_0 = Aen_i^2\left(\frac{D_n}{L_nN_A} + \frac{D_p}{L_pN_D}\right) \tag{8.35b}$$

If the applied bias to the junction is now reversed, exactly the same equation will be arrived at with v_B replaced by $-v_B$. For the reverse-bias case, when $e|-v_B| \gg kT, I = -I_0$. This is why I_0 is called the *reverse-saturation current* of the diode.

An important conclusion from these results is that the saturation current, in addition to being a function of temperature, is also dependent on the band gap energy of the semiconductor through n_i. The larger the band gap energy, the lower is the saturation current of the diode. Figure 8.15 shows the *i-v* characteristic of a typical p-n junction diode (v_B is replaced by v). The current-voltage characteristic of the metal semiconductor junction is also shown in the same figure for comparison, assuming that the I_0 for the Schottky diode is larger than that of p-n junction.

8.4.1.2 DEPLETION-LAYER WIDTH. The depletion-layer width can be calculated using the spatial variation of the immobile charges at the junction. Figure 8.16 shows the variation of these charges on the two sides of an abrupt junction. From the conservation of charge, the total charge on the two sides of the depletion layer should be equal. These are made up of the unneutralized acceptor ions on the p-side and the donor ions on the n-side. Their respective densities are

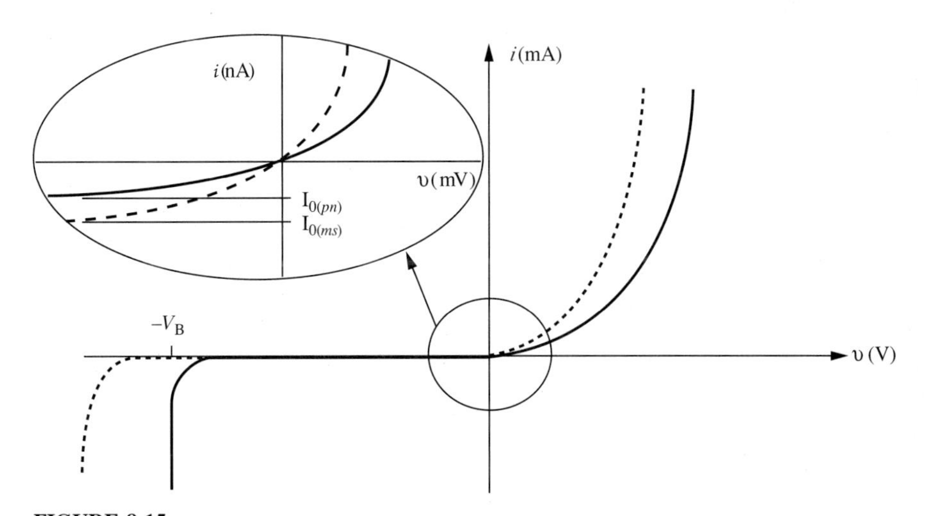

FIGURE 8.15
i-v characteristics of a p-n junction (solid lines) and a metal-semiconductor (dotted line) junction. I_0 and V_B are different for the two junctions as shown in the insert for I_0.

$n'_A \approx N_A$ and $n^+_D \approx N_D$. If the cross-sectional area of the n- and p-sides of the junction are equal, the equality of charge can be written as

$$eN_D(AW_n) = eN_A(AW_p) \tag{8.36}$$

where W_n and W_p are the extensions of depletion layer into the n- and p-sides respectively.

At a point in space, the charge density and the corresponding potential are related by the Poisson equation

$$\frac{d^2v}{dx^2} = -\frac{\rho}{\kappa_s\epsilon_o} \tag{8.37}$$

where κ_s is the dielectric constant of the semiconductor. For the n-region and uniform charge distribution, this becomes

$$\frac{d^2v}{dx^2} = -\frac{eN_D}{\kappa_s\epsilon_o}$$

Integrating with respect to x, we obtain

$$\frac{dv}{dx} = -\left(\frac{eN_D}{\kappa_s\epsilon_o}\right)x + C_1$$

The constant C_1 is calculated by setting the electric field $\mathscr{E}_x = -dv/dx = 0$ at $x = -W_n$ (see Fig. 8.10d). This gives

$$-\mathscr{E}_x = \frac{dv}{dx} = -\left(\frac{eN_D}{\kappa_s\epsilon_o}\right)(x + W_n) \tag{8.38}$$

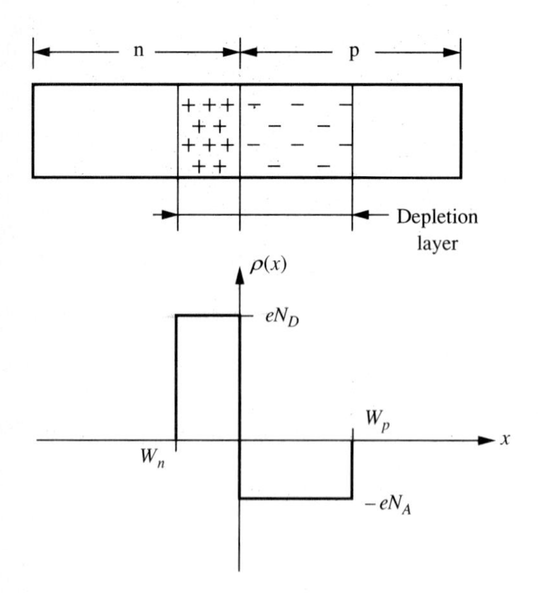

FIGURE 8.16
Unneutralized charge distribution across the junction. The depletion layer is assumed to be 100 percent ionized.

Integrating the potential with respect to x for the second time

$$v(x) = -\left(\frac{eN_D}{\kappa_s \epsilon_o}\right)\left[\left(\frac{x^2}{2}\right) + W_n x + C_2\right]$$

and using the condition that at $x = -W_n$, $v = 0$, we obtain

$$v(x) = -\left(\frac{eN_D}{\kappa_s \epsilon_o}\right)\left[\left(\frac{x^2}{2}\right) + W_n x + \left(\frac{W_n^2}{2}\right)\right]$$

$$v(x) = -\left(\frac{eN_D}{2\kappa_s \epsilon_o}\right)(x + W_n)^2 \tag{8.39}$$

At $x = 0$, if we let $v = -V_N$, we find for the magnitude of the potential contribution from the n-side to the diode potential

$$V_N = \left(\frac{eN_D}{\kappa_s \epsilon_o}\right) W_n^2 \tag{8.40}$$

Using a similar derivation, the contribution to the diode potential from the p-side can be written as

$$V_P = \left(\frac{eN_A}{2\kappa_s \epsilon_o}\right) W_p^2 \tag{8.41}$$

The magnitude of total barrier voltage becomes

$$V_D = V_N + V_P = \left(\frac{e}{2\kappa_s \epsilon_o}\right)(N_D W_n^2 + N_A W_p^2) \tag{8.42}$$

Using the charge neutrality condition given by Eq. 8.36, we can solve for W_n or W_p and obtain

$$W_n = \left[\left(\frac{2\kappa_s \epsilon_o V_D}{e}\right)\left(\frac{N_A}{N_A N_D + N_D^2}\right)\right]^{1/2} \tag{8.43}$$

$$W_p = \left[\left(\frac{2\kappa_s \epsilon_o V_D}{e}\right)\left(\frac{N_D}{N_D N_A + N_A^2}\right)\right]^{1/2} \tag{8.44}$$

Note that the signs of the respective potentials are taken into account in the derivation of these equations. Therefore, V_D is the magnitude of the total diode potential. The total depletion-layer width is given by the sum of these two equations

$$W_d = W_n + W_p \tag{8.45}$$

If one side of the junction is heavily doped with respect to the other side, that is, if we assume $N_D \gg N_A$ as shown in Fig. 8.16, the total depletion width can be approximated by

$$W_d \approx W_p = \left(\frac{2\kappa_s \epsilon_0 V_D}{eN_A} \right)^{1/2} \tag{8.46}$$

the depletion layer lies mostly on the lightly doped side, that is, the p-side of the junction under the given condition. If the total voltage across the junction is changed, the depletion layer changes accordingly. As can be seen from Eq. 8.46, the total depletion-layer width increases with increasing reverse bias, since the reverse bias increases the total barrier potential across the junction, that is, V_D becomes $(V_D + |v_B|)$. Similarly, the forward bias decreases W_d since V_D becomes $(V_D - v_B)$. From now on V_D in Eq. 8.46 will be replaced by v_T, the total barrier potential. Note also that the depletion-layer width is also dependent on the doping level of the semiconductor through N_A.

8.4.1.3 JUNCTION AND DIFFUSION CAPACITANCE. The depletion layer is free of mobile charges. The p-n junction can be thought of as a parallel plate capacitance with a cross-sectional area A and plate separation W_d separated with a material of dielectric constant, that of the semiconductor. The equivalent *junction capacitance* can be written as

$$C_j = \left(\frac{dQ}{dv} \right) = \left(\frac{\epsilon_s A}{W_d} \right)$$

$$C_j = \frac{\epsilon_s A (eN_A)^{1/2}}{(2\epsilon_s v_T)^{1/2}} \approx K_c (v_T)^{-1/2} \tag{8.47}$$

where Eq. 8.46 is used for W_d.

As the applied reverse-bias voltage magnitude increases, the junction capacitance decreases. A p-n junction can thus be used as a variable capacitor by varying the reverse-bias voltage across the diode. The diodes, specifically manufactured to act as voltage-controlled variable capacitors, are known as *varactor diodes*.

When the junction is forward biased, because of the large amount of injected charge, the junction capacitance discussed above is masked by a larger capacitance known as the *diffusion* or *diode capacitance*. With an increase (or decrease) of the forward bias, the injected-charge density in the vicinity of the depletion layer increases (or decreases) (Fig. 8.17) due to the additional (or reduced) charge injection. The change in this charge with bias voltage leads to the additional diffusion capacitance C_d.

The change in the diffused charge can be related to the change in the applied voltage by

$$dq_{\text{diff}} = \left(\frac{dq_{\text{diff}}}{dv} \right) dv \tag{8.48}$$

which allows us to define the diffusion capacitance as

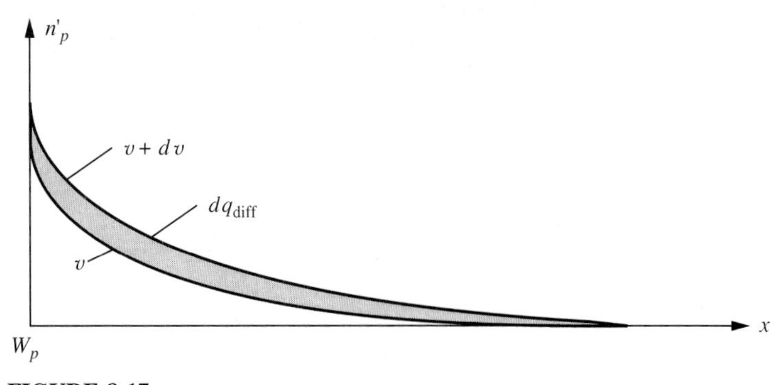

FIGURE 8.17
Injected excess charge leads to diffusion capacitance.

$$C_d = \left(\frac{dq_{\text{diff}}}{dv}\right) \tag{8.49}$$

where the subscript B is dropped from the bias voltage for convenience.

Using Eq. 8.30, the total diffused charge on the p-side can be calculated from the injected excess charge carriers as

$$q_{\text{diff}} = \int_0^\infty qAn'_p(0)\exp\left(-\frac{x}{L_n}\right)dx$$

$$q_{\text{diff}} = qAn'_p(0)\,L_n \tag{8.50}$$

Taking the derivative of Eq. 8.50 with respect to v, we obtain

$$C_d = qAL_n\left(\frac{dn'_p(0)}{dv}\right) \tag{8.51}$$

From Eq. 8.31, the diffusion current density for the electrons can be written as

$$J_{\text{diff}} = \frac{qD_n n'_p(0)}{L_n} \tag{8.52}$$

Multiplying both sides of this equation by A, and solving for $n'_p(0)$, we obtain

$$n'_p(0) = \left(\frac{L_n i_{\text{diff}}}{qD_nA}\right) \tag{8.53}$$

Taking the derivative of this expression with respect to v, and substituting into Eq. 8.51, we finally find for the diffusion capacitance

$$C_d = \frac{L_n^2(di_{\text{diff}}/dv)}{D_n} \tag{8.54}$$

Since the diffusion current is also equal to the total current, we can use Eq. 8.22 for i_{diff}, that is,

$$i_{\text{diff}} \equiv i = I_0 \left[\exp\left(\frac{ev}{kT}\right) - 1 \right]$$

Taking the derivative of i with respect to voltage, we find

$$\frac{di}{dv} = i\left(\frac{e}{kT}\right) \tag{8.55}$$

where it was assumed that $ev \gg kT$ in the diode equation, and only the exponential term was maintained in deriving Eq. 8.55.

Using $L_n^2 = D_n\tau_n$, C_d can finally be written as

$$C_d = \tau_n i \left(\frac{e}{kT}\right) \tag{8.56}$$

The equivalent circuit of a diode can be drawn as shown in Fig. 8.18. It is made up of two capacitances, C_j and C_d, in parallel with the dynamic resistance r_d of the diode evaluated at the operating point. In this model, the parasitic elements such as the lead inductances, stray capacitances, and bulk material resistances are neglected. When the diode is forward biased, C_d is 3–4 orders of magnitude larger than C_j. On the other hand, C_d drops from the model when the diode is reverse biased.

The time constant of the circuit in switching from the forward to the reverse bias can be approximated simply by

$$\tau \approx r_d C_d = \left(\frac{kT}{ei}\right)\left(\frac{ei}{kT}\right)\tau_n$$

$$\tau \approx \tau_n \tag{8.57}$$

The switching time constant of the diode is therefore in the order of the recombination time of the electrons on the p-side for a p-n junction with $N_D \gg N_A$.

When the polarity of the applied potential is switched suddenly from forward to reverse bias, the charge associated with C_d should be discharged through the external circuit before the system can come to the steady state in the reverse direction. If the polarity is then reversed back to the forward case, this time it

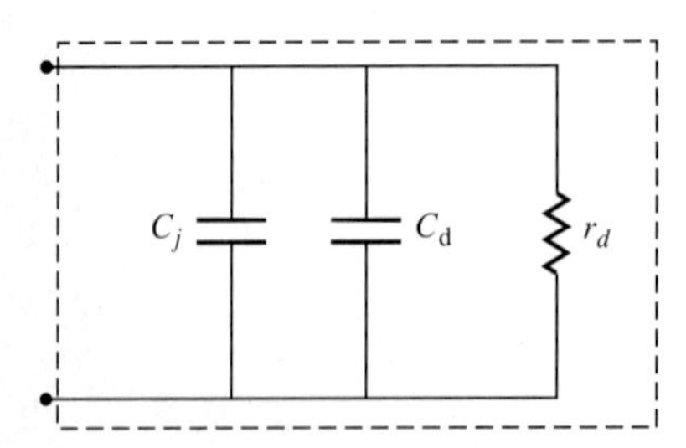

FIGURE 8.18
Small signal equivalent circuit of a diode. r_d is the dynamic resistance, with forward bias, $C_d >> C_j$, with reverse bias, $C_j >> C_d$. All parasitic and material effects away from the junction are neglected in this model.

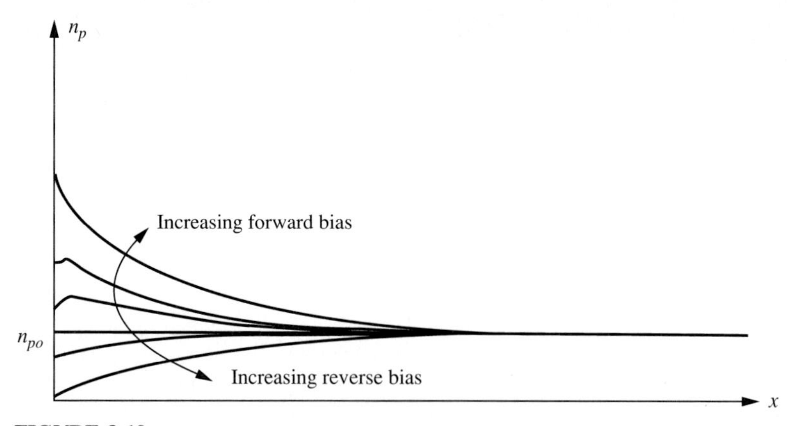

FIGURE 8.19
Change in the injected charge distribution when the polarity across the diode is reversed.

will again take a finite time until the diffused charge can build up on the p-side (see Fig. 8.19). The switching time constant for charging and discharging these charges before steady states can be reached is on the order of τ_n. Thus, the fastest time response of the diode is limited by the recombination time of the electrons on the p-side of the junction for the diode under discussion.

Example 8.2. A Si p-n junction has the following parameters:

$$N_D = 10^{17}\ \text{cm}^{-3}, \quad N_A = 10^{15}\ \text{cm}^{-3}, \quad \text{Area} = 50\ \mu\text{m} \times 50\ \mu\text{m}$$

Find

(*a*) the reverse-saturation current

(*b*) the depletion-layer widths and the corresponding junction capacitances at bias voltages of $v = -10$ V and $+0.75$ V

(*c*) the diffusion capacitance at $v = +0.75$ V ($T = 300$ K)

Solution.

(*a*) In order to calculate the reverse-saturation current of the diode using Eq. 8.35, we need various parameters associated with Si. From the tables in Chapter 5, we find

$$D_n = 38 \times 10^{-4}\ \text{m}^2\text{s}^{-1}, \quad \tau_n = 50\ \mu\text{s}, \quad D_p = 13 \times 10^{-4}\ \text{m}^2\text{s}^{-1}, \quad \tau_p = 10\ \mu\text{s}$$

$$n_i = 1.4 \times 10^{16}\ \text{m}^{-3}$$

The diffusion lengths are then equal to

$$L_n = (D_n\tau_n)^{1/2} = 4.36 \times 10^{-4}\ \text{m} \quad \text{and} \quad L_p = (D_p\tau_p)^{1/2} = 1.14 \times 10^{-4}\ \text{m}$$

I_0 then becomes

$$I_0 = (2.5 \times 10^{-9}\text{m}^2)(1.6 \times 10^{-19}\ \text{C})(1.4 \times 10^{16}\ \text{m}^{-3})^2$$
$$\times \left[\frac{38 \times 10^{-4}\ \text{m}^2\text{s}^{-1}}{(4.36 \times 10^{-4}\ \text{m})(10^{21}\ \text{m}^{-3})} + \frac{13 \times 10^{-4}\ \text{m}^2\text{s}^{-1}}{(1.14 \times 10^{-4}\ \text{m})(10^{23}\ \text{m}^{-3})} \right]$$
$$I_0 = 6.92 \times 10^{-16}\ \text{A}$$

(*b*) Before we can calculate the depletion width, we need the diode potential V_D. From Eq. 8.21

$$V_D = (0.026\ \text{eV}) \times \ln \left[\frac{(10^{21}\ \text{m}^{-3})(10^{23}\ \text{m}^{-3})}{(1.4 \times 10^{16}\ \text{m}^{-3})^2} \right]$$
$$V_D = 0.701\ \text{eV}$$

For the reverse bias of -10.0 V, the depletion layer W_d (from Eq. 8.46) is

$$W_d = \left[\frac{2 \times 11.8 \times 8.854 \times 10^{-12}\ \text{Fm}^{-1})(10.0 + 0.701\ \text{V})}{(1.6 \times 10^{-19}\ \text{C})\ (10^{21}\ \text{m}^{-3})} \right]^{1/2}$$
$$W_d = 3.74 \times 10^{-6}\ \text{m}$$

The corresponding junction capacitance is

$$C_j = \frac{(11.8 \times 8.854 \times 10^{-12}\ \text{Fm}^{-1})(2.5 \times 10^{-9}\ \text{m}^2)}{(3.74 \times 10^{-6}\ \text{m})}$$
$$C_j = 6.98 \times 10^{-14}\ \text{F}$$

Note that for $v = +0.75$, the v is greater than V_D, and W_d becomes imaginary. This indicates that the low-level injection approximation does not hold any more, and it may be that the depletion layer disappears completely.

(*c*) For the diffusion capacitance, we need the current through the diode at the operating bias potential. This is equal to

$$i = 6.92 \times 10^{-16}(\ \exp^{(0.75/0.026)} - 1)\ \text{A}$$
$$i = 2.33 \times 10^{-3}\ \text{A} = 2.33\ \text{mA}$$

The diffusion capacitance is then equal to

$$C_d = (50 \times 10^{-6}\ \text{s})(2.33 \times 10^{-3}\ \text{A})(0.026\ \text{eV})$$
$$C_d = 3.03 \times 10^{-9}\ \text{F} = 3.03\ \text{nF}$$

8.4.2 Heterojunctions

With the recent technological advances in compound semiconductor manufacturing processes, heterojunctions now play a very important role in semiconductor device implementation. With their higher energy band gaps and much higher electron mobilities, compound semiconductors are likely candidates in high-frequency device applications extending into the millimeter wavelengths. The technology for growing these semiconductors has reached such an advanced level that now even atomic layers of these semiconductors can easily be grown, one over the other. Certain recently developed techniques, such as molecular beam epitaxy (MBE) or metallo-organic chemical vapor deposition (MOCVD), allow growth of true abrupt heterojunctions. In this section we will present the unique properties of the heterojunctions and their major differences compared to the homojunctions, using the energy-band diagram of a heterojunction. Specific heterojunctions related to specific devices will be dealt with in later chapters.

To understand the properties and the resulting band diagram, we will consider a heterojunction made up of an n-type Ge and a p-type GaAs. The energy-band diagram of these semiconductors before contact and far away from each other is shown in Fig. 8.20. As can be seen from the diagram, almost all of the parameters associated with the two semiconductors have different values. Note that the two semiconductors in a heterojunction can both be mono-crystals (i.e., Ge-Si) or mono-compound crystals (i.e., Si-GaAs) and both compound crystals (i.e., GaAs-GaAlAs).

According to Fig. 8.20, in addition to the usual parameters that relate to a regular homojunction, there are two more important parameters: ΔE_c and ΔE_v. In terms of the energies of the two sides, these can be identified as

$$\Delta E_c = E_{c2} - E_{c1} \tag{8.54}$$

$$\Delta E_v = E_{v1} - E_{v2} \tag{8.55}$$

When a junction is formed by the two semiconductors, electrons from n-Ge and holes from p-GaAs diffuse across until the equilibrium diode potential

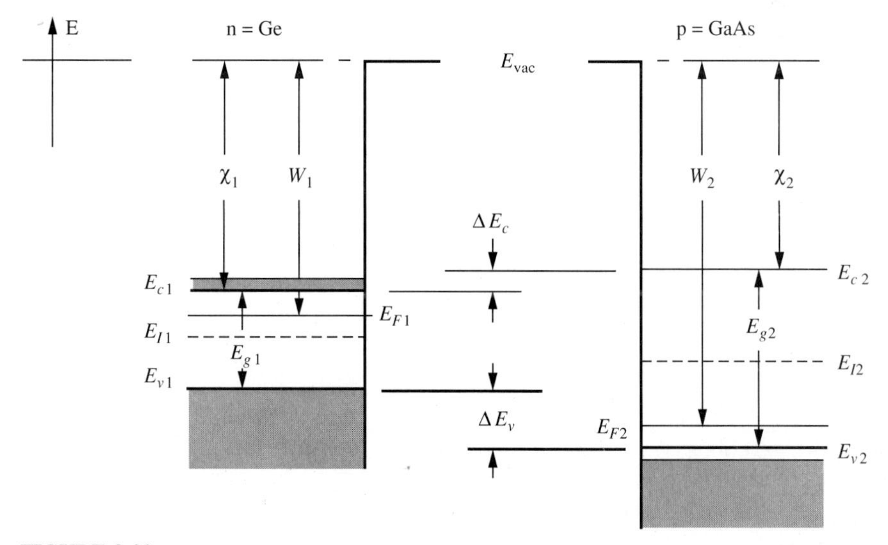

FIGURE 8.20
Energy-band diagrams of isolated n-type Ge and p-type GaAs semiconductors.

energy V_D is established. This also means the alignment of the Fermi energy levels of both sides to the same energy value. The barrier potential that forms due to the presence of immobile charges can again be calculated from the solution of the Poisson equation on both sides of the junction. This time, the solutions for the two sides differ from the homojunction case by the differing dielectric permittivities of the corresponding semiconductors. Using Fig. 8.21 and using similar arguments that led to the derivation of Eqs. 8.40 and 8.41, the solutions for the diode potentials for each region can be written as

$$v_1(x) = -\left(\frac{eN_D}{2\epsilon_1}\right)(x + W_n)^2 \tag{8.56}$$

$$v_2(x') = +\left(\frac{eN_A}{2\epsilon_2}\right)(x')^2 \tag{8.57}$$

where $v_2(x')$ is calculated for the shifted coordinate system shown in Fig. 8.21. The continuity of the displacement vector **D** requires that

$$\epsilon_1 \left(\frac{dv_1(x)}{dx}\right)\bigg|_{x=0} = \epsilon_2 \left(\frac{dv_2(x')}{dx'}\right)\bigg|_{x'=-W_p} \tag{8.58}$$

Substituting Eqs. 8.56 and 8.57 into Eq. 8.58, we find

$$eN_D W_n = eN_A W_p \tag{8.59}$$

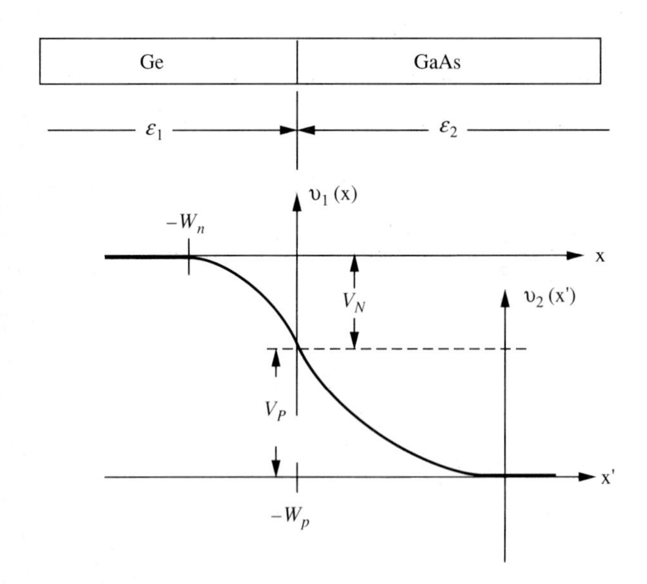

FIGURE 8.21
The diode potential variation across the heterojunction. The coordinate system and the dielectric permittivity ϵ for the n and p regions are different.

This is the same charge neutrality condition given by Eq. 8.36.

Equations 8.56 and 8.57 allow us to define the contributions of $\Phi_N = eV_N$ and $\Phi_P = eV_P$ to the diode potential from the respective sides of the junction. From the above equations

$$V_N = -v_1(x = 0) = \left(\frac{eN_D}{2\epsilon_1}\right) W_n^2 \tag{8.60}$$

$$V_P = v_2(x' = -W_p) = \left(\frac{eN_A}{2\epsilon_2}\right) W_p^2 \tag{8.61}$$

As can be seen from Fig. 8.21, these potentials are negative with respect to Ge side of the junction. The corresponding potential energies are reflected on the respective sides of the junction in the overall energy-band diagram of the heterojunction.

The overall energy-band diagram of an abrupt heterojunction is shown in Fig. 8.22. Here the diagram is drawn using the *affinity rule*. This rule assumes that the shift in the vacuum-energy levels is continuous and the n-side of the junction is displaced downward in energy by an amount equal to $\Phi_D =$

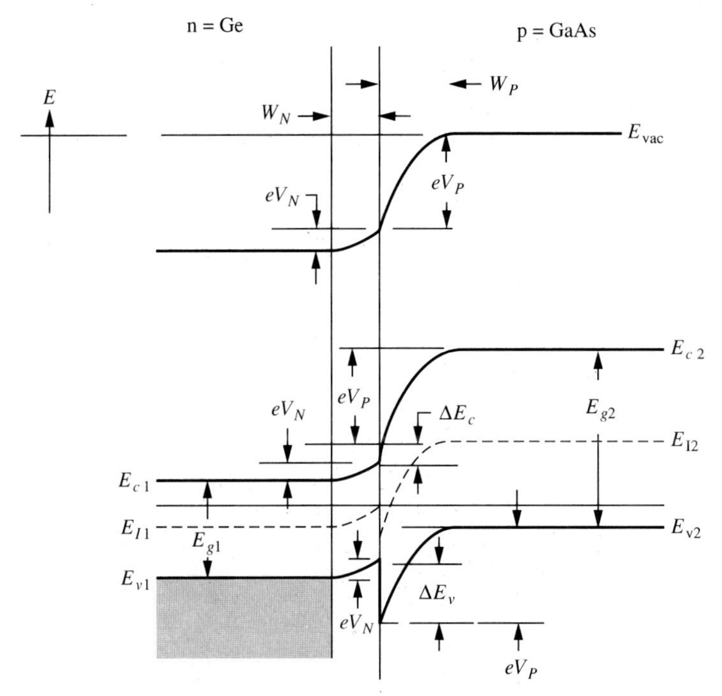

FIGURE 8.22
Equilibrium energy-band diagram of an n-type Ge and a p-type GaAs after the junction formation.

$-\Phi_N - \Phi_p = -(eV_N + eV_P)$. The same potential energies Φ_N and Φ_P are reflected in the corresponding conduction and valence bands of the respective sides of the junction. If the difference of the energies from the conduction bands to the vacuum levels are maintained, that is, keeping the electron affinities of both sides constant, there results a discontinuity of energy ΔE_c on the conduction band and ΔE_v on the valence band. For the example given in Fig. 8.22, ΔE_c is reflected as a jump discontinuity in the conduction band and as a spike equal to ΔE_v in the valence band of the semiconductor with higher band gap energy.

The simple model given here has to be modified when other heterojunctions are considered. For example, when a heterojunction is formed by two different intrinsic semiconductors (no intentional impurity doping), we don't expect the usual diode potential to form due to the lack of charge diffusion. But due to a mismatch in the crystal structure of the semiconductors or due to the different atomic positions and effective ionic charges facing each other in the binding process across the junction, an equivalent dipole layer forms across the junction, which is reflected again as an equivalent dipole potential V_d in the energy-band diagram. This dipole potential can also modify the actual diode potential that may arise from the charge diffusion from doped semiconductors. In general, the dipole potential can be included in the parameters ΔE_c and ΔE_v as

$$\Delta E_c = E_{c1} - E_{c2} - V_d \tag{8.62}$$

$$\Delta E_v = E_{v1} - E_{v2} + V_d \tag{8.63}$$

These values are calculated theoretically for many heterojunctions, and in many cases there is good agreement with the experimentally measured values, but there is still no unified theory which is applicable to all the heterojunctions.

The heterojunctions have many applications. They also have many distinct advantages over the homojunctions. These can be summarized as

1. Higher temperature characteristics due to the larger band gap energy of one or both of the semiconductors.
2. Lower saturation current because of larger band gaps of the semiconductors leading to very low n_i's and therefore, to very low minority carrier densities. Because of the unusual band barrier formations, the resulting barriers may further hinder the flow of minority carriers as can be seen from the spike in the valence band of Fig. 8.21.
3. As ternary compounds are formed, the change in the concentration of one of the constituent atoms modifies the band structure, the band gap energy and the index of refraction of the composite semiconductor. Thus multiple layers of these heterojunctions can be used to guide optical signals by proper choice of the element concentrations in different layers. These are useful in designing lasers and electrooptic components.
4. Multiple layers of the heterojunctions such as superlattices show unusual physical and electrical properties, which are used to create many unusual new electronic devices. Some of these devices will be discussed in later chapters.

8.5 AVALANCHE BREAKDOWN

We have shown that across a junction, a potential barrier was set up which increases with the applied reverse-bias voltage. Although generally small voltages are applied across the terminals of a junction, the voltage drops in the bulk material of the two elements away from the junction are usually very low because of the short physical lengths and the relatively large conductivity of these elements. Thus, most of the applied voltage appears across the junction.

The depletion-layer width increases with the applied reverse voltage as $W_d = K_1 v_T^{1/2} = K_1(V_D + |v|)^{1/2} \approx K_1|v|^{1/2} \text{for} |v| \gg V_D$. The maximum electric field can then be approximated by

$$\mathcal{E}_{\max} \approx \frac{|v|}{W_d} = K_1^{-1}|v|^{1/2} \tag{8.64}$$

where we see that the $\mathcal{E}_{\max}$ increases as the square root of the applied reverse-bias voltage. The charge carriers that cross the junction are the respective minority carriers that enter the junction from each side of the depletion layer, that is, electrons from p and holes from n sides.

Let us now follow the motion of the electrons as they enter the depletion region at $x = 0$ as shown in Fig. 8.23. In moving a distance dx an electron can gain energy from the electric field an amount $\Delta W = e\,\mathcal{E}_{\max} dx$. If the field magnitude is large, before leaving the depletion layer the electron can gain enough energy from the field so that it can make an ionizing collision with the valence electrons of the impurity or the host crystal atoms and produce additional electron-hole pairs inside the depletion layer. If the applied reverse bias is large enough, these newly generated electrons and holes move to the right and left, respectively, and can in turn make additional ionizing collisions. Thus, the charge density crossing the junction is multiplied in magnitude. This increase in charge is reflected as a sharp increase in the reverse saturation current flowing through the junction.

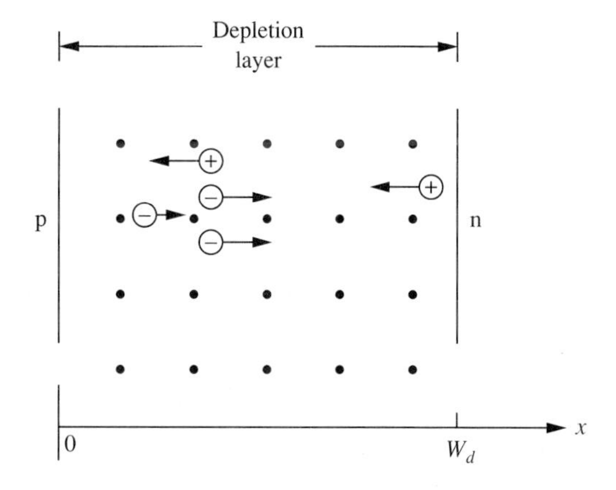

FIGURE 8.23
As the electrons and holes move through the depletion layer, they gain energy from the field and make ionizing collisions with the impurity and lattice atoms leading to creation of additional charge in the depletion region multiplying the charge crossing the depletion region.

Of course, the holes which enter the depletion layer at $x = W_d$ also generate additional electron-hole pairs and contribute to the charge multiplication. This multiplication can reach as high as $10^6 I_0$. The increase in charge due to the high fields appearing across the junction is known as *avalanche breakdown* and occurs in many solid-state devices which have, as a part of their operation, a reverse-biased junction. Although avalanche breakdown can be used in bipolar transistors to an advantage in designing fast electronic switching circuits, one generally avoids operating the device near its breakdown. On the other hand, some microwave devices, such as the Impatt diode, avalanche photodiode, and so on, are based on the avalanche breakdown properties of junctions for their operation.

In Fig. 8.13, the i-v characteristic of a p-n junction shows the reverse bias voltage where breakdown (or impact ionization) sets in. Depending on the semiconductor and the doping level, the reverse saturation current can be written as

$$I_r = I_0 \left[1 - \left(\frac{v}{V_B} \right)^n \right]^{-1} \tag{8.65}$$

where the exponent n varies between 1.5 and 4, depending on the semiconductor. V_B is the breakdown voltage.

For uniformly doped samples, the breakdown voltage increases with the semiconductor energy band gap ($V_B \propto E_g^{3/2}$) and decreases with increasing doping ($V_B \propto N^{-3/4}$). Impact ionization is a sensitive function of the applied field. If the field increases by a factor of 2–3, the ionization rate may increase five fold.

Avalanche breakdown can be used in some diodes by proper doping and profiling of the junction so that the breakdown voltage can be set to occur at a chosen voltage. These diodes are known as Zener diodes and are used in many electronic circuits, especially in power supplies as voltage regulators.

FURTHER READING

Frensley, W. R. and H. Kroemer, "Theory of Energy-Band Lineup at an Abrupt Semiconductor Heterojunction," *Phys. Rev. B*, vol. 16, 1977, 2642–2652.

Nussbaum, Allen, "The Theory of Semiconductor Junctions," in *Semiconductors and Semimetals*, R. E. Willardson and A. C. Beer, eds. vol. 15, Academic Press, New York, 1981, Chapter 2.

Pollmann, J. and A. Mazur, "Theory of Semiconductor Heterojunctions," *Thin Solid Films*, vol. 104, 1983, 257–276.

Sharma, B. L., "Ohmic Contacts to III–V Compound Semiconductors," in *Semiconductors and Semimetals*, R. E. Willardson and A. C. Beer, eds., vol. 15, Academic Press, New York, 1981, Chapter 1.

Sze, S. M.,*Physics of Semiconductor Devices*, 2d ed., Wiley, New York, 1981.

PROBLEMS

8.1. A Schottky barrier diode is made up from gold and n-type silicon. Find the reverse-saturation current of the diode if the diode area is 0.5 mm^2. $T = 300$ K. Explain why this current does not depend on the donor density of the n-side.

8.2. If the temperature of the Schottky diode in Problem 8.1 is increased by 50° C, find the increase in the saturation current density I_0.

8.3. A silicon p-n junction diode has donor and acceptor dopings of 10^{17} atoms/cm^3 and 10^{16} atoms/cm^3 respectively. Find the maximum value of the electric field at the junction. Note: $\mathscr{E}_{max}$ occurs at $x = 0$.

8.4. For the diode in Problem 8.3, find the saturation current I_0 if the area of the junction is 25 μm^2. (T = 25° C).

8.5. For the diode Problem 8.4, calculate the equilibrium Fermi energy for each region (assume 100 percent ionization). Calculate the diode potential from the Fermi levels and compare it with V_D calculated from Eq. 8.21.

8.6. Find the saturation current I_0 of a (a) germanium and (b) GaAs p-n homojunction if the conditions given in Problem 8.3 apply.

8.7. For Problem 8.3, find the total depletion-layer width of the junction. Compare this with W_n and W_p.

8.8. For the p-n junction in Problem 8.4, the physical lengths of the n- and p-regions are 100 μm each. If a voltage of 10 V reverse bias is applied across the contacts of the semiconductor, find the fields in each region, including the depletion layer.

8.9. A p-n junction has a saturation current $I_0 = 0.05 \times 10^{-6}$ A. Find an analytical expression that relates the change in current to the change in temperature. For a forward bias of 0.5 V, find this change for a temperature increase of 10 degrees from 300 K.

8.10. For Problem 8.4, find the junction and diffusion capacitances for a forward bias of 0.5 V and a reverse bias of 5 V.

8.11. A heterojunction is formed from a p-Ge and n-GaAs. The parameters corresponding to these semiconductors are given in the following table.

Parameter	p = Ge	n = GaAs
χ(eV)	4.13	4.07
$(E_c - E_F)$ eV	–	0.1
$(E_F - E_v)$ eV	0.1	–
E_g (eV)	0.70	1.45
N_D(m^{-3})	–	10^{22}
N_A(m^{-3})	3×10^{22}	–
κ_s	16	13.1

(a) Find the contributions to the diode potential from both sides of the junction. Is the sum equal to the difference in the Fermi levels?

(b) Draw the junction-band diagram assuming no additional dipole layer forms.

(c) Find the depletion-layer width of each side.

8.12. The charge distribution for a graded junction can be approximated by $\rho(x) = -ax(W_n \leq x \leq W_p)$ where a is the slope of the graded region as shown in Fig. P.8.12.

By integrating the Poisson equation, find the electric field, built in diode potential, depletion-layer widths on each side, and the total depletion-layer width.

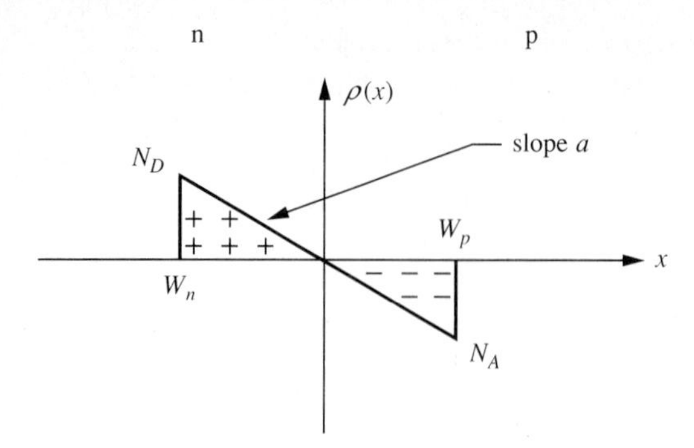

FIGURE P.8.12

8.13. For the metal-metal contact, although electron diffusion takes place from the metal with higher electron concentration to the metal with the lower concentration, and we expect a barrier to form for the electrons, the contact potential that formed at the vacuum interface was not reflected as a barrier for the electrons in the conduction band. Explain why this was neglected in drawing Fig. 8.1.

8.14. A point contact diode is made by bringing a very sharp metallic point (generally tungsten) in contact with a semiconductor. A very thin layer of oxide resides on the surface of the metal or the semiconductor before the contact. Draw an energy-band diagram for the diode and explain its operation in view of the discussions presented in this chapter. Why would you expect a higher frequency of operation for this diode? Hint: refer to the equivalent circuit model of the diode and estimate the magnitudes of the various parameters.

8.15. The acceptor and donor concentrations in a Ge p-n junction diode are 2×10^{15} cm^{-3} and 5×10^{17} cm^{-3} respectively. Find the extent of the depletion layers on both sides of the junction and estimate the error in using the depletion-layer approximation for $N_A \gg N_D$.

8.16. The actual current voltage characteristic of a diode can be approximated by

$$i = I_0 \left[\exp\left(\frac{ev}{nkT} \right) - 1 \right]$$

where n is known as the ideality factor ($n = 1$ for an ideal diode). Set up an experiment to find n and show how you will calculate n from that data.

8.17. Does diffusion capacitance exist in a Schottky barrier diode? Explain the advantages or disadvantages of using a Schottky diode in switching circuits relative to a p-n junction.

8.18. A Si p-n junction has impurity doping of $N_A = 10^{17}$ cm^{-3} and $N_D = 10^{15}$ cm^{-3}. A Schottky diode is made by depositing Au on n-type Si ($N_D = 10^{17}$ cm^{-3}). Calculate the ratio of the reverse saturation currents of the two diodes if they have the same cross-sectional area. Which diode will be useful in a rectifying circuit?

8.19. A heterojunction is made from intrinsic GaAs and intrinsic $Al_{0.7}Ga_{0.3}As$. Draw the band diagram of the resulting junction. Since both crystals are lattice matched, neglect any possible dipole potential. Assume that the dielectric constants of the two regions are equal.

$$E_g(\text{GaAs}) = 1.42 \text{ eV} \qquad E_g(\text{AlGaAs}) = 2.02 \text{ eV}$$

$$\Delta E_c = 0.48 \text{ eV} \qquad \Delta E_v = 0.12 \text{ eV}$$

8.20. A heterojunction is made from intrinsic GaAs and n-type $Al_{0.7}Ga_{.30}As$. Draw the band diagram of the resulting junction. Since both crystals are lattice matched, neglect any possible dipole potential. Assume that dielectric constants of the two regions are equal.

$$E_g(\text{GaAs}) = 1.42 \text{ eV} \qquad E_g(\text{AlGaAs}) = 2.02 \text{ eV}$$
$$\Delta E_c = 0.48 \text{ eV} \qquad \Delta E_v = 0.12 \text{ eV}$$
$$E_F(\text{n-AlGaAs}) = 0.2 \text{ eV below conduction band}$$

8.21. You are to design a p-n junction with an $N_D = 10^{17}$ cm^{-3} and $N_A 10^{15}$ cm^{-3}.
(*a*) What type of Si material will you start with? Explain.
(*b*) If the p-region is to be 2 μm thick, what kind of impurity atom will you pick? What kind of diffusion process will you use (i.e., temperature, time of diffusion, etc.)?
(*c*) If, instead of diffusion process, you were to use ion implantation, what kind of energy, dosage, time of implantation, number of different energy levels, and so on, would you have to use?

8.22. A p-n junction is to be used as a temperature sensor.
(*a*) Where will you bias the diode for maximum sensitivity?
(*b*) What will be the temperature sensitivity of this thermometer?

8.23. The hole contribution to the current flow in a Schottky diode was neglected in deriving Eq. 8.10. If the minority hole current is also included, what will be the modification to the reverse-saturation current? Will this have an important contribution to I_0?

8.24. Using the equations for the Fermi levels in n-type and p-type semiconductors, show that the built-in diode potential is equal to

$$eV_D = (E_{Fc} - E_{Fv})$$

and is the same as Eq. 8.21

8.25. A silicon p-n junction diode has donor and acceptor dopings of 5×10^{17} atoms/cm^3 and 2×10^{16} atoms/cm^3, respectively. The cross-sectional area of the diode is 50 μm $\times$ 50 μm. If $E_B = 6.0 \times 10^5$ V cm^{-1}, find the maximum voltage that can be applied across the diode before avalanche breakdown sets in at room temperature.

8.26. An ohmic contact is required for an n-region in a device. The n$^+$ region of Fig. P.8.26 provides an effective contact to the n region of the device. Draw an energy-band diagram and show that this is a good ohmic contact to the n-region.

8.27. A Schottky diode is made by depositing Au on n-type Si ($N_D = 10^{17}$ cm^{-3}). By calculating the depletion layers both on the metal side and the semiconductor side, show that practically the barrier width on the metal side is zero as shown in figures on metal-semiconductor junctions.

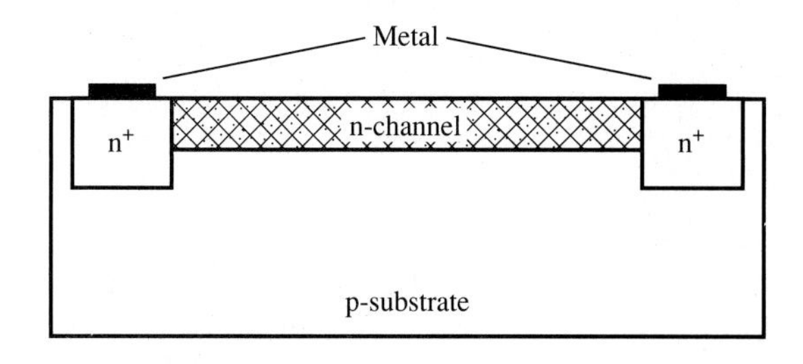

FIGURE P.8.26

CHAPTER 9

BIPOLAR JUNCTION TRANSISTORS (BJTs)

The general features of a p-n junction were discussed in Chapter 8. The resulting current-voltage characteristic is a nonlinear curve, where the current increases exponentially with increasing forward bias and reaches a constant value $-I_0$ with a decreasing reverse bias. The diode is a two-terminal or one-port device, and its characteristic can be used only for rectification or detection of an ac signal. Devices become interesting if there is a third terminal through which the current flowing through the other two terminals can be controlled.

In this chapter, the essential features of the bipolar junction (n-p-n or p-n-p) transistors (BJT) will be discussed. The control of the current flow is achieved by introducing an additional p-n junction next to an n-p junction. The first junction is used to control the reverse saturation current of a second reverse-biased junction. In this way, a control of the current is achieved.

9.1 PHYSICAL BASIS FOR BJT

The i-v characteristics of an n-p junction are shown in Fig. 9.1. The reverse saturation current I_0 is given by Eq. 8.35,

$$I_0 = Ae\left(\frac{D_n n_{\text{po}}}{L_n} + \frac{D_p p_{\text{no}}}{L_p}\right) \tag{9.1}$$

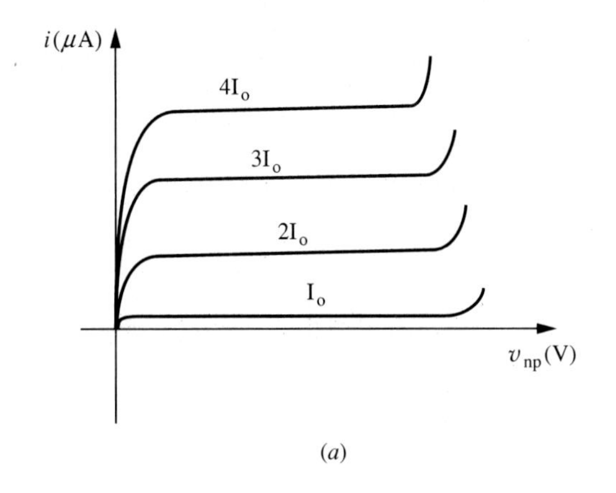

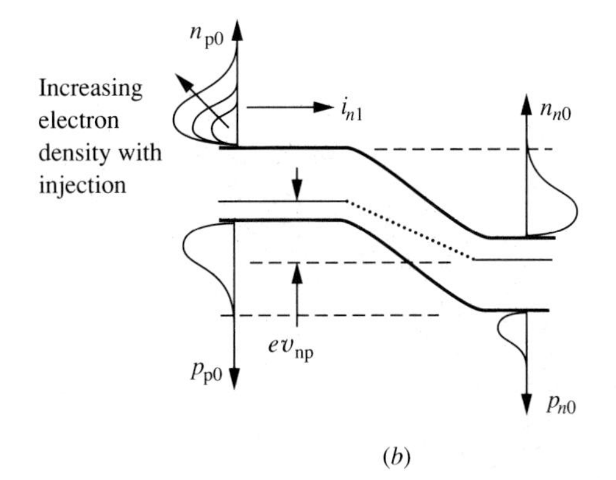

FIGURE 9.1
(*a*) Reverse current-voltage characteristics of a p-n junction diode with changing minority electron density on the p-side. (*b*) through i_{n1} increases with some means of injection of additional electrons onto the p-side.

and it depends especially on the minority carrier density of the electrons in the p-region and the minority carrier density of holes in the n-region. If one of these minority carrier densities is varied by some external means, the reverse saturation current through the diode will also be changed. For example, if n_{po} on the p-side is tripled, the I_0 will be tripled as shown in Fig. 9.1.

The minority carrier density n_{po} in the p-region of the diode is manipulated by introducing a second semiconductor junction in series with the first junction, as shown in Fig. 9.2. Instead of physically connecting the two diodes back to back, one of the elements, p-region for the n-p-n, (n-region for p-n-p) is made common to both of the diodes on the same semiconductor chip. A short length of the p-region (or the n-region) is necessary, otherwise, as we shall see later, if it were larger than the diffusion length of the electrons in the p-region (or the diffusion length of holes in the n-region), the control of I_0 in the second diode would not have been possible. The schematic representations of bipolar junction transistors are shown in Fig. 9.2 for (*a*) an n-p-n transistor and for (*b*) a p-n-p configuration. The arrows indicate the direction of the conventional electrical current.

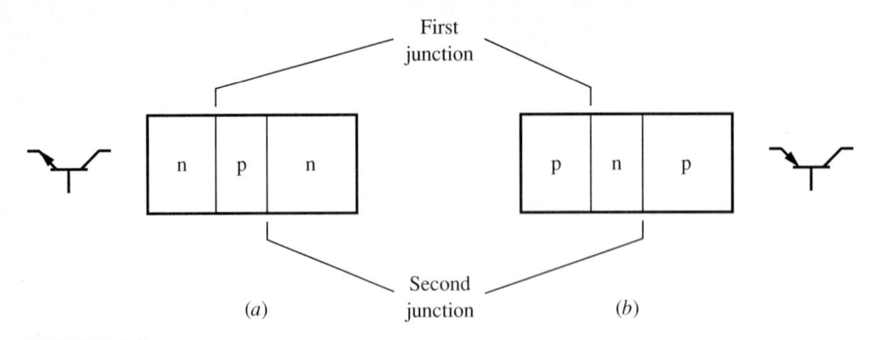

FIGURE 9.2
(*a*) n-p-n and (*b*) p-n-p transistor simple physical configurations and corresponding schematic representations.

9.2 BIPOLAR JUNCTION TRANSISTOR

A typical constructional detail of an n-p-n transistor is shown in Fig. 9.3. Over a highly conducting silicon wafer, an n-layer of desired donor concentration and thickness is grown. Next, a mask is used to diffuse the base p-region. A second set of masks is then used to diffuse the n^+-emitter layer. Proper oxide films separate the corresponding regions of the transistor from each other so that necessary ohmic contacts can be deposited over each of the respective regions.

An idealized model of an n-p-n BJT is shown in Fig. 9.4*a*. This may be seen as the uniform region of AA' of Fig. 9.3. The respective elements of the transistor are known as the *emitter*, the *base*, and the *collector*. Although in an actual device the effective cross-sectional areas of the emitter-base junction (EBJ) and the collector-base junction (CBJ) are different, the uniform cross-sectional model depicted in Fig. 9.4*a* helps simplify the understanding and analysis of the various charge transport mechanisms and the associated current flowing through various

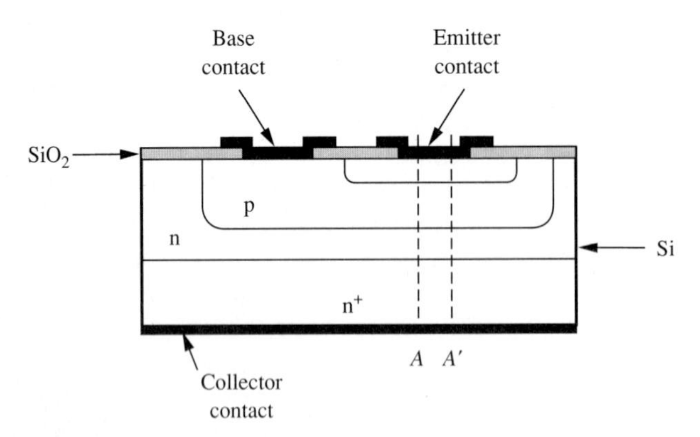

FIGURE 9.3
Cross section of a typical modern n-p-n silicon transistor. n's and p are replaced by p's and n for a p-n-p transistor. The base region can be made symmetric, that is, a mirror image of the left side to the right of AA' with the two bases connected together.

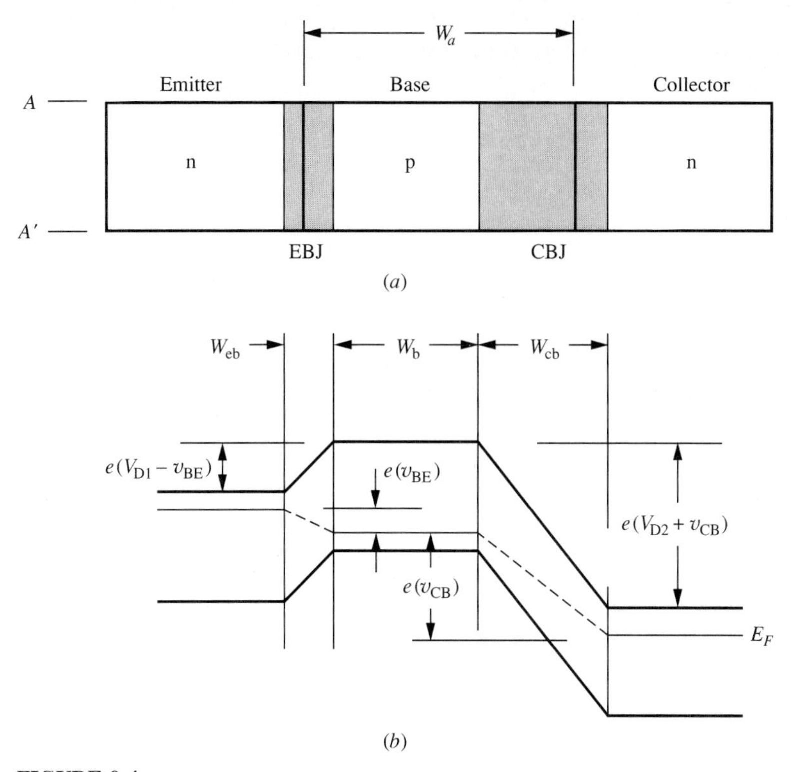

FIGURE 9.4
(*a*) Physical layout of a uniform cross-sectional n-p-n transistor corresponding to cross section AA' of Fig. 9.3, and (*b*) energy-band diagram with bias for the active mode of operation.

regions of the transistor. In the following discussions, we will further assume that both junctions are abrupt junctions and that they are physically separated by a distance W_a. The physical lengths of the emitter and the collector are not as important as the base width (width is used instead of base length) of the transistor.

The following convention is used in identifying the polarities of the voltages applied across the device terminals. The initial subscript is the positive terminal, and the second is the negative terminal. The v_{BE} implies that the base is positive relative to the emitter, or that the emitter is at negative potential with respect to the base. Similarly, $-v_{EB}$ means again the emitter is negative with respect to the base.*

*The notation used with transistor representations is as follows: a small letter with a capital letter subscript represents the total voltage, a capital letter with a capital letter subscript represents the dc component of voltage, and a small letter with a small letter subscript represents the ac component of voltage. For example, the total base collector voltage can be the sum of an ac voltage v_{bc} superimposed on a dc voltage V_{BC}, that is, $v_{BC} = V_{BC} + v_{bc}$. This notation also applies for currents.

TABLE 9.1
Modes of operation of BJT. The voltage magnitudes are for an n-p-n transistor.

Mode of operation	EBJ bias	CBJ bias
Active	Forward ($v_{BE} > 0$)	Reverse ($v_{CB} > 0$)
Cutoff	Reverse ($v_{BE} < 0$)	Reverse ($v_{CB} > 0$)
Saturation	Forward ($v_{BE} > 0$)	Forward ($v_{CB} < 0$)

The operation of the BJT occurs with three distinct modes, as given in Table 9.1. Although all modes of operation are important, in the subsequent discussions of various current components active-mode bias conditions will be used. Application to other modes can easily be obtained by using the appropriate terminal voltages.

Since the EBJ is forward-biased, in the active mode the depletion-layer width, labeled as W_{eb}, is very short compared to the reverse-biased collector-base junction depletion-layer width W_{cb}. Because of the presence of two depletion-layer widths, the effective base width W_b of the transistor is shorter than the physical length of the base p-region. The equivalent energy-band diagram of a homojunction (same semiconductor material throughout) transistor with the proper energy shifts corresponding to the active mode of operation is shown in Fig. 9.4*b*.

The equations related to a bipolar junction transistor will be derived for an n-p-n transistor. Similar derivations can be used for a p-n-p transistor by changing the electrons into holes and by reversing the polarities of the respective bias voltages at each of the junctions. Various currents that flow across the junctions are shown in Fig. 9.5. (Alphabetical symbols in the figure refer to the various current components that will be identified with various charge flows among different

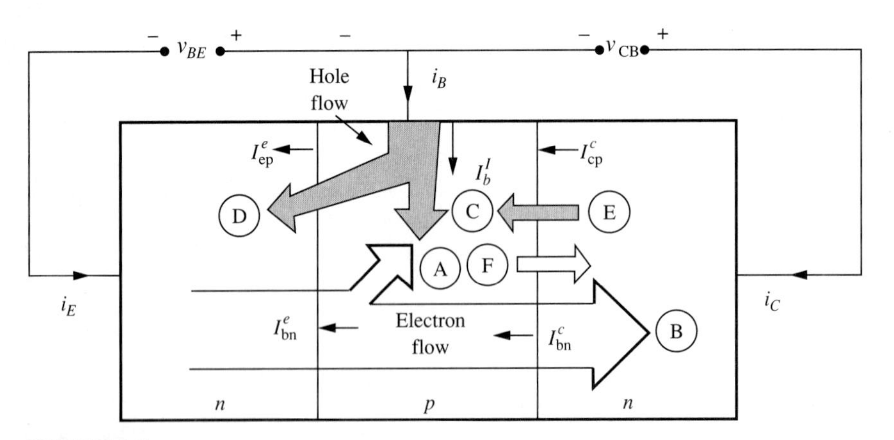

FIGURE 9.5
Various charges and currents flowing in an n-p-n transistor under dc bias conditions. Inside the transistor, the directions of the current arrows are in the direction of the conventional current and also indicate the locations where they are calculated.

regions of the BJT.) The main current is made up of the electron flow from the emitter to the collector. As the electrons are injected across the emitter-base junction into the base region by the applied forward bias, they diffuse in the base region toward the base-collector junction. Some of these electrons are lost through recombination with the holes in the base region (current A), but those that reach the collector-base boundary are swept across the CBJ junction by the collector-base bias potential (current B). The holes that recombine with some of the injected electrons at the base region are supplied by the holes flowing into the base region (current C). Holes are also injected (referred to as back-injection) from the p-side into the n-side (current D). This is small relative to the electron injection from the n- to p-side because of the relative doping levels of the two sides of the junction. There are also the minority hole and electron currents through the CBJ (currents E and F) which are important when the transistor is at cut-off.

The overall efficiency of the transistor is governed by the relative magnitudes of all these currents. The best transistor performance is obtained by transferring all of the injected electrons from the emitter into the collector. The emitter-base voltage or the base current is used to control the total electron current reaching the collector.

9.3 n-p-n HOMOJUNCTION TRANSISTOR

Where applicable, and with proper modifications, many of the results derived in Chapter 8 for a p-n junction will be used in the analyses of the transistor performance. The following helpful assumptions will be made:

1. As shown in Fig. 9.4*a*, a BJT with a uniform cross section will be used. This will allow us to carry only a simplified one-dimensional analysis.

2. It will be assumed that the injected-electron density into the base region is large compared to the equilibrium minority carrier electron density in the base region, but much smaller than the majority hole density at the base region. This is referred to as low-level injection.

3. Electric fields exist only in the depletion layers. There are no voltage drops across the bulk materials, that is, no fields exist at the emitter, base, and collector regions. This allows us to set the applied terminal voltages equal to those that appear across the respective junctions of the transistor. If the resistivities of the bulk materials are not negligible, these will introduce spatially varying voltages across the junctions and introduce nonlinearities into the analysis. This is especially true for high-frequency BJTs that require very small base widths where spatially varying fields cannot be neglected.

4. The applied voltages are assumed to be steady-state voltages. Even if there are small temporal changes in these voltages, they occur in time intervals long enough to reach steady-state conditions.

Figure 9.6*a* shows the various bulk equilibrium carrier densities away from the junctions. It also shows some of the minor changes in the notation for the

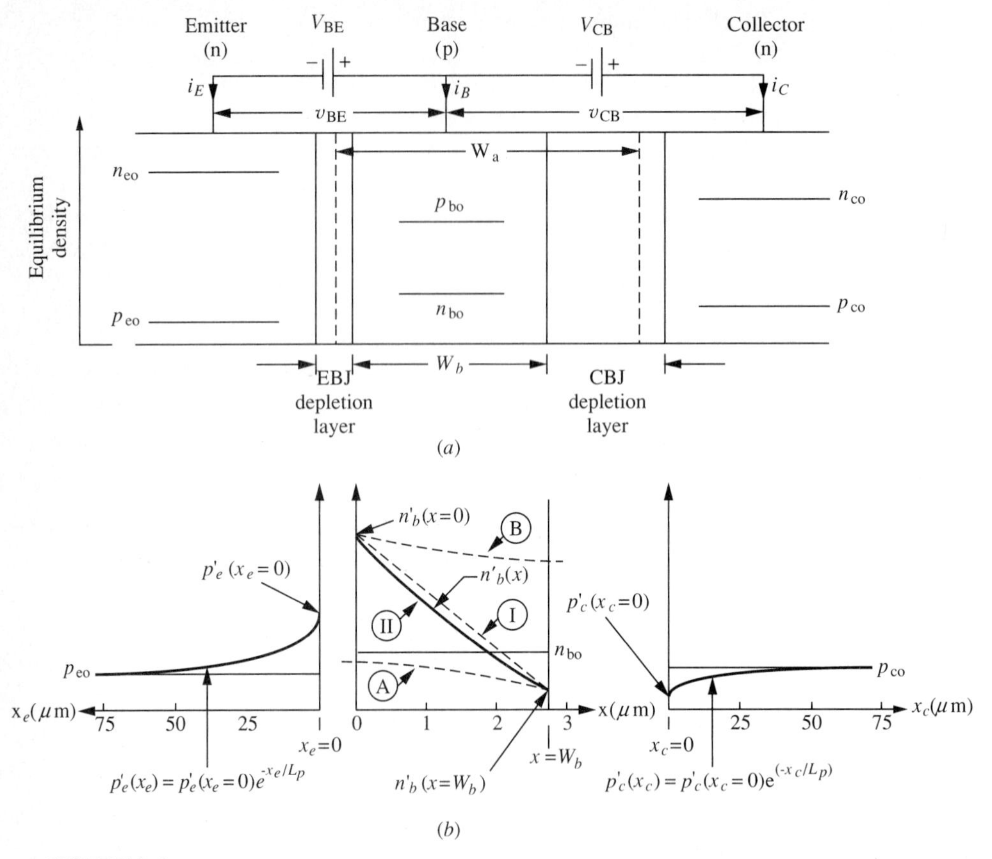

FIGURE 9.6
(*a*) Equilibrium carrier densities and (*b*) excess charge density variations in various regions of the BJT. The excess carrier densities at the junction boundaries are also shown. Curve I is an ideal approximation of the actual excess electron distribution in the base region of Curve II. Curves A and B would be the distribution of excess electrons if they were to diffuse in bulk materials under their given biases. The coordinate scales are different in various regions. In (*a*) the terminal voltages are taken to be different than the battery voltages for generality.

various regions of the BJT compared to the notation used for the p-n junction in Chapter 8. Here, one should keep in mind that the emitter is an n-type, the base is a p-type, and the collector is an n-type. Some of the subscripts n, p, and n in the parameters of the various regions are replaced by the element subscripts e, b, and c.

Figure 9.6*b* also shows the excess minority charge density variations in the emitter, the base, and the collector. Distance variables are labeled as x_e, x, and x_c for the respective regions. In the figure, the coordinate scales x_e and x_c are an order of magnitude larger than the base coordinate x.

With the given bias voltages, magnitudes of the excess charge carrier densities at the respective junction boundaries can be written as (see Eq. 8.27)

$$p'_e(x_e = 0) = p_{eo}\,[\exp(ev_{BE}/kT) - 1] \quad \text{at emitter side of EBJ} \tag{9.2a}$$

$$n'_b(x = 0) = n_{bo}\,[\exp(ev_{BE}/kT) - 1] \quad \text{at base side of EBJ} \tag{9.2b}$$

$$n'_b(x = W_b) = n_{bo}\,[\exp(-ev_{CB}/kT) - 1] \quad \text{at base side of CBJ} \tag{9.2c}$$

$$p'_c(x_c = 0) = p_{co}\,[\exp(-ev_{CB}/kT) - 1] \quad \text{at collector side of CBJ} \tag{9.2d}$$

Since the physical lengths of the emitter and collector regions are assumed to be long compared to hole diffusion length L_p, the excess hole carriers decay exponentially away from the respective junctions. Note that as v_{BE} becomes greater than a few kT, the exponential terms dominate in the first two equations. When v_{CB} becomes greater than a few kT, the excess carrier densities become negative at two sides of the CBJ. This implies that at $x = W_b$, the injected-electron density falls below n_{bo}, and the total minority electron density becomes zero. In other words, no electrons can be stored at this boundary. Any electron that appears on the base side is swept across the CBJ immediately by the reverse bias of the CBJ.

For all BJTs, the base width is very short compared to the electron diffusion length in the base, that is, $W_b << L_b$. Some of the electrons, though relatively few, actually recombine with the holes in the base region. If the loss of electrons through recombination with the holes is taken into account, the excess electron density variation in the base region will follow Curve II in Fig. 9.6*b*. In many simplifying calculations, the $n'_b(x)$ can be approximated by a straight line (Curve I), as shown by the dotted line in the same figure. If the excess electron densities injected into the base from the emitter and the collector junctions were to decay exponentially due to diffusion processes similar to the decay of holes in the emitter and base regions, these electrons would have followed Curves A and B as shown by the dashed curves. If the base physical width W_a were long compared to L_b in the base region, the exponential decays A and B would have been the dominant factors in determining the currents flowing through the base.

Referring to Fig. 9.6, we are now ready to calculate various currents flowing in an n-p-n transistor. In order to calculate the contributions of various charges on the resulting terminal currents, we use, in view of assumption (3) above, only the diffusion components of currents given by Eqs. 8.28 and 8.29.

The emitter current i_E is made up of two injected currents on the EBJ. These diffusion current components can be calculated from

$$I^e_{\text{bn}} = eA_bD_b \left.\frac{dn'_b(x)}{dx}\right|_{x=0} \tag{9.3a}$$

$$I^e_{\text{ep}} = -eA_{\text{be}}D_e \left.\frac{dp'_e(x_e)}{dx_e}\right|_{x_e=0} \tag{9.3b}$$

In these equations, superscript e signifies that the current components are for the emitter. The first subscript is the region where the current is to be calculated, and the second subscript shows whether the current is due to holes (p) or electrons (n) in that region. These current components are shown in Fig 9.5. D_b and D_e

are the respective diffusion constants of the electrons and holes in the base and emitter.

It should be noted that in the one-dimensional model given in Fig. 9.4, the transistor has a uniform cross section throughout. In deriving the current equations, different subscripts will be used for the base-emitter area A_{be} when currents on the emitter side are calculated, for the collector-base area A_{bc} when currents on the collector side are calculated, and for the effective base area A_{b} when currents at the base region are calculated. This is done purposely to remind the reader that, as can be seen from Fig. 9.3, in an actual device these areas are considerably different from each other and play important roles in determining the actual terminal currents.

Equation 9.3*b* is the current due to the back-injected holes into the emitter. The exponentially decaying excess hole density at the emitter gives rise to the same diffusion current as given by Eq. 8.33. With the present notation, this current is given by

$$I_{\mathrm{ep}}^{e} = -A_{\mathrm{be}}\left(\frac{eD_e p_{\mathrm{eo}}}{L_e}\right)\left[\exp\left(\frac{ev_{\mathrm{BE}}}{kT}\right)-1\right] \tag{9.4}$$

To calculate Eq. 9.3*a*, we have to know the excess electron density gradient at the base side of the EBJ. For this, we have to calculate spatial variation of the excess electron density in the base region, taking into account the loss of electrons through recombination with the holes while traversing the base region. The excess charge density variation is obtained from the solution of Eq. 6.21. Written for the electrons, it becomes

$$\frac{d^2 n_b'(x)}{dx^2} - \frac{n_b'(x)}{L_b^2} = 0 \tag{9.5}$$

Solution of Eq. 9.5 gives

$$n_b'(x) = A\,\exp\left(-\frac{x}{L_b}\right) + B\,\exp\left(\frac{x}{L_b}\right) \tag{9.6}$$

Since the base width $W_b << L_b$, both terms have to be maintained. In order to determine the constants A and B, we have to apply the boundary conditions

$$n_b'(x=0) = n_{\mathrm{bo}}\left[\exp\left(\frac{ev_{\mathrm{BE}}}{kT}\right) - 1\right] \tag{9.7a}$$

$$n_b'(x=W_b) = n_{\mathrm{bo}}\left[\exp\left(-\frac{ev_{\mathrm{CB}}}{kT}\right) - 1\right] \tag{9.7b}$$

Substituting these into Eq. 9.6, we obtain two equations for A and B in the form

$$A + B = n_{\mathrm{bo}}\left[\exp\left(\frac{ev_{\mathrm{BE}}}{kT}\right) - 1\right] \tag{9.8a}$$

$$A\exp\left(-\frac{W_b}{L_b}\right) + B\exp\left(\frac{W_b}{L_b}\right) = n_{\mathrm{bo}}\left[\exp\left(-\frac{ev_{\mathrm{CB}}}{kT}\right) - 1\right] \tag{9.8b}$$

Solving for A and B and substituting back into Eq. 9.6, we obtain, after some manipulations,

$$n_b'(x) = \frac{n_{\mathrm{bo}}}{\sinh(W_b/L_b)}\left\{\sinh\left(\frac{W_b - x}{L_b}\right)\left[\exp\left(\frac{ev_{\mathrm{BE}}}{kT}\right) - 1\right] + \sinh\left(\frac{x}{L_b}\right)\left[\exp\left(-\frac{ev_{\mathrm{CB}}}{kT}\right) - 1\right]\right\} \tag{9.9}$$

It is easy to verify that Eq. 9.9 satisfies the given boundary conditions. In order to calculate the currents, we need the gradient of Eq. 9.9. Taking the derivative of Eq. 9.9, we obtain

$$\frac{dn_b'(x)}{dx} = -\frac{n_{\mathrm{bo}}}{L_b\sinh(W_b/L_b)}\left\{\cosh\left(\frac{W_b - x}{L_b}\right)\left[\exp\left(\frac{ev_{\mathrm{BE}}}{kT}\right) - 1\right] - \cosh\left(\frac{x}{L_b}\right)\left[\exp\left(-\frac{ev_{\mathrm{CB}}}{kT}\right) - 1\right]\right\} \tag{9.10}$$

The general form of the electron current in the base region then becomes

$$I_{\mathrm{bn}} = -\frac{eA_bD_bn_{\mathrm{bo}}}{L_b\sinh(W_b/L_b)}\left\{\cosh\left(\frac{W_b - x}{L_b}\right)\left[\exp\left(\frac{ev_{\mathrm{BE}}}{kT}\right) - 1\right] - \cosh\left(\frac{x}{L_b}\right)\left[\exp\left(-\frac{ev_{\mathrm{CB}}}{kT}\right) - 1\right]\right\} \tag{9.11}$$

The contribution of electrons to the emitter current is obtained by setting $x = 0$ in Eq. 9.11. This leads to

$$I_{\mathrm{bn}}^e = -\frac{eA_bD_bn_{bo}}{L_b\sinh(W_b/L_b)}\left\{\cosh\left(\frac{W_b}{L_b}\right)\left[\exp\left(\frac{ev_{\mathrm{BE}}}{kT}\right) - 1\right] - \left[\exp\left(-\frac{ev_{\mathrm{CB}}}{kT}\right) - 1\right]\right\} \tag{9.12}$$

The total emitter current is then given by the sum of Eq. 9.12 and 9.4.

$$i_E = -\left\{\frac{eA_bD_bn_{\mathrm{bo}}}{L_b}\coth\left(\frac{W_b}{L_b}\right) + \left(\frac{eA_{be}D_ep_{\mathrm{eo}}}{L_e}\right)\right\}\left[\exp\left(\frac{ev_{\mathrm{BE}}}{kT}\right) - 1\right] + \frac{eA_bD_bn_{\mathrm{bo}}}{L_b\sinh\left(\frac{W_b}{L_b}\right)}\left[\exp\left(-\frac{ev_{\mathrm{CB}}}{kT}\right) - 1\right] \tag{9.13}$$

Note that the emitter current given by Eq. 9.13 is in the $+x$ direction, which is the same direction as the transistor emitter terminal current as shown in Fig. 9.5.

In a similar way, the components of the collector current can be calculated from

$$I_{\text{bn}}^{c} = eA_bD_b \left.\frac{dn_b'(x)}{dx}\right|_{x=W_b} \tag{9.14a}$$

$$I_{\text{cp}}^{c} = eA_{\text{cb}}D_p \left.\frac{dp_c'(x_c)}{dx_c}\right|_{x_c=0} \tag{9.14b}$$

The hole contribution to current from the collector side is given by

$$I_{\text{cp}}^{c} = -A_{\text{cb}}\left(\frac{eD_c p_{\text{co}}}{L_c}\right)\left[\exp\left(-\frac{ev_{\text{CB}}}{kT}\right) - 1\right] \tag{9.15}$$

The electron contribution to the collector current is found by setting $x = W_b$ in Eq. 9.11. This yields

$$I_{\text{bn}}^{c} = -\frac{eA_bD_bn_{\text{bo}}}{L_b\sinh(W_b/L_b)}\left[\exp\left(\frac{ev_{\text{BE}}}{kT}\right) - 1\right] + \frac{eA_bD_bn_{\text{bo}}}{L_b}\coth\left(\frac{W_b}{L_b}\right)\left[\exp\left(-\frac{ev_{\text{CB}}}{kT}\right) - 1\right] \tag{9.16}$$

The currents given by Eqs. 9.15 and 9.16 are in the $+x$ direction and are opposite to the terminal collector current i_C. The total terminal collector current can thus be written as the negative of the sum of Eqs. 9.15 and 9.16. This gives

$$-i_c = \left[\frac{eA_{cb}D_cp_{\text{co}}}{L_c} + \frac{eA_bD_bn_{\text{bo}}}{L_b}\coth\left(\frac{W_b}{L_b}\right)\right]\left[\exp\left(-\frac{ev_{\text{CB}}}{kT}\right) - 1\right] - \frac{eA_bD_bn_{\text{bo}}}{L_b\sinh(W_b/L_b)}\left[\exp\left(\frac{ev_{\text{BE}}}{kT}\right) - 1\right] \tag{9.17}$$

If we are to calculate the gradient of the excess electron density distribution for the case where $x << L_b$ and $W_b << L_b$, we can use the approximations

$$\sinh(y) = y + \frac{y^3}{3!} + \frac{y^5}{5!} + \ldots \tag{9.18a}$$

$$\cosh(y) = 1 + \frac{y^2}{2!} + \frac{y^4}{4!} + \ldots \tag{9.18b}$$

Using the first-order approximation, we find that the slope given by Eq. 9.10 reduces to

$$\frac{dn_b'(x)}{dx} \approx \frac{-n_{\text{bo}}\left[\exp\left(\frac{ev_{\text{BE}}}{kT}\right) - 1\right] + n_{\text{bo}}\left[\exp\left(-\frac{ev_{\text{CB}}}{kT}\right) - 1\right]}{W_b} \tag{9.19}$$

$$= \frac{n_b'(x = W_b) - n_b'(x = 0)}{W_b}$$

This is the equation of the slope of the straight line with end points given by $n_b'(x = W_b)$ and $n_b'(x = 0)$. If we had used Eq. 9.19 in Eqs. 9.3*a* and 9.13*a*, we would have obtained the same electron contributions to the emitter and collector currents. The reduction of electrons in the base region due to recombination with holes should then have to be taken into account considering the recombination process in the base region (see Problem 9.4).

9.4 EBERS–MOLL MODEL

The equations derived for a BJT can be used to model a BJT over a very wide range of voltages. The currents given by Eqs. 9.13 and 9.17 can be written as

$$i_E = -I_{\mathrm{ES}}\left[\exp\left(\frac{ev_{\mathrm{BE}}}{kT}\right) - 1\right] + \alpha_R I_{\mathrm{CS}}\left[\exp\left(-\frac{ev_{\mathrm{CB}}}{kT}\right) - 1\right] \tag{9.20}$$

and

$$i_C = \alpha_F I_{\mathrm{ES}}\left[\exp\left(\frac{ev_{\mathrm{BE}}}{kT}\right) - 1\right] - I_{\mathrm{CS}}\left[\exp\left(-\frac{ev_{\mathrm{CB}}}{kT}\right) - 1\right] \tag{9.21}$$

where

$$I_{\mathrm{ES}} = \frac{eA_b D_b n_{\mathrm{bo}}}{L_b}\coth\left(\frac{W_b}{L_b}\right) + \frac{eA_{\mathrm{be}} D_e p_{\mathrm{eo}}}{L_e} \tag{9.22a}$$

$$I_{\mathrm{CS}} = \frac{eA_{\mathrm{cb}} D_c p_{\mathrm{co}}}{L_c} + \frac{eA_b D_b n_{\mathrm{bo}}}{L_b}\coth\left(\frac{W_b}{L_b}\right) \tag{9.22b}$$

$$\alpha_F I_{\mathrm{ES}} = \alpha_R I_{\mathrm{CS}} = \frac{eA_b D_b n_{\mathrm{bo}}}{L_b \sinh(W_b/L_b)} \tag{9.22c}$$

Eqs. 9.20 and 9.21 can be put in a different form in terms of two functions of only v_{BE} and v_{CB}. These become

$$i_E = I_{\mathrm{EO}}\left[\exp\left(\frac{ev_{\mathrm{BE}}}{kT}\right) - 1\right] - \alpha_R i_C \tag{9.23a}$$

$$i_C = -I_{\mathrm{CO}}\left[\exp\left(-\frac{ev_{\mathrm{CB}}}{kT}\right) - 1\right] - \alpha_F i_E \tag{9.23b}$$

with

$$I_{\mathrm{CO}} = (1 - \alpha_R \alpha_F) I_{\mathrm{CS}} \tag{9.24a}$$

$$I_{\mathrm{EO}} = (1 - \alpha_R \alpha_F) I_{\mathrm{ES}} \tag{9.24b}$$

In these sets of equations, I_{CS} and I_{ES} are the equivalent emitter and collector open curcuit saturation currents obtained when one of the other diodes is

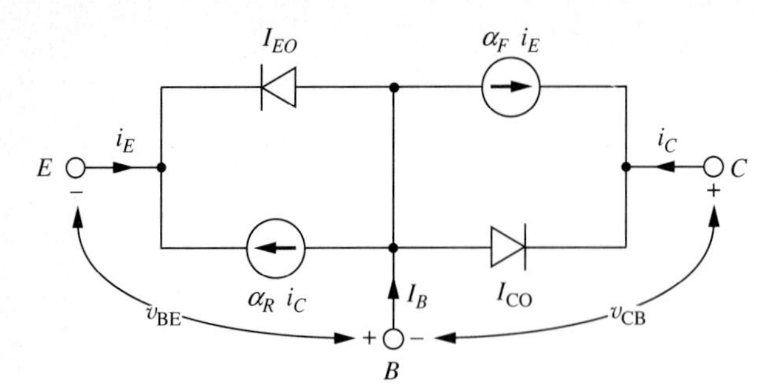

FIGURE 9.7
Ebers–Moll model for an n-p-n BJT.

short circuited. The constants α_R and α_F are known respectively as the common base static *reverse* and *forward* gains of the transistor. As we will show in the next section, a good transistor has an α_F very close to but less than one. α_R is considerably smaller than α_F.

Equations 9.23*a* and 9.23*b* represent two diodes with equivalent reverse saturation currents I_{EO} and I_{CO} parallel to two current sources. The resulting model is known as the Ebers–Moll model and is shown in Fig. 9.7 for an n-p-n transistor.

9.5 TRANSISTOR CURRENT GAIN α_0

The overall dc current amplification of a bipolar junction transistor is measured by the dc forward current gain α_0. It is defined as the ratio of the total collector current to that current initiated at the emitter. The current amplification can at best be equal to one and depends on three parameters. These are (*a*) γ_e, the emitter injection efficiency, (*b*) β_t, base transport factor, and (*c*) γ_c, the collector efficiency. We will consider the active mode of operation of the BJT. For this, we will also take $ev_{\text{BE}} >> kT$ and $ev_{\text{CB}} >> kT$ and the exponential terms in the current equations will be approximated by $[\exp(ev_{\text{BE}}/kT) - 1] \approx \exp(ev_{\text{BE}}/kT)$ and $[\exp(-ev_{\text{CB}}/kT) - 1] \approx -1$.

9.5.1 Emitter Injection Efficiency γ_e

Emitter efficiency γ_e is defined as the ratio of the electron current due to the electrons injected from the emitter into the base to that of the total emitter current.

$$\gamma_e = \frac{I^e_{\text{bn}}}{i_E} = \frac{I^e_{\text{bn}}}{I^e_{\text{bn}} + I^e_{\text{ep}}}$$

$$\gamma_e = \frac{1}{1 + \left(I^e_{\text{ep}}/I^e_{\text{bn}}\right)} \tag{9.25}$$

where I_{ep}^{e} is the back-injected hole current, given by Eq. 9.4. With the voltage approximations assumed above

$$I_{\text{ep}}^{e} \approx -A_{\text{be}}\left(\frac{eD_e p_{\text{eo}}}{L_e}\right)\exp\left(\frac{ev_{\text{BE}}}{kT}\right) \tag{9.26a}$$

The injected-electron current is obtained from Eq. 9.12.

$$I_{\text{bn}}^{e} \approx -\frac{eA_b D_b n_{\text{bo}}}{L_b \tanh(W_b/L_b)}\exp\left(\frac{ev_{\text{BE}}}{kT}\right) \tag{9.26b}$$

Note that under the approximations made here, the additional constant term in Eq. 9.12 is neglected because of the dominant exponential term.

Finally, the emitter efficiency becomes

$$\gamma_e = \frac{1}{1 + (A_{\text{be}}p_{\text{eo}}D_e L_b/A_b n_{\text{bo}}D_b L_e)\tanh(W_b/L_b)} \tag{9.27}$$

If the base width is small compared to the electron diffusion length, $\tanh(W_b/L_b)$ can be replaced by (W_b/L_b) (see Eq. 9.18). The emitter efficiency becomes

$$\gamma_e \approx \frac{1}{1 + (A_{\text{be}}p_{\text{eo}}D_e W_b/A_b n_{\text{bo}}D_b L_e)} \approx 1 - \left(\frac{A_{\text{be}}p_{\text{eo}}D_e W_b}{A_b n_{\text{bo}}D_b L_e}\right) \tag{9.28}$$

The emitter efficiency approaches one if (*a*) the base width W_b is very small compared to electron diffusion length L_b in the base, and (*b*) if the emitter is heavily doped relative to the base so that the equilibrium hole minority carrier density p_{eo} in the emitter is smaller than the equilibrium minority carrier electron density n_{bo} in the base region. These conditions can easily be met in the implementation of the BJT by heavily doping the emitter relative to the base. As will be shown in Chapter 12, a small base width necessary for a large γ_e is also necessary to increase the high-frequency response of the transistor.

Equation 9.27 also shows the reason why two ordinary diodes cannot be physically connected back-to-back to obtain transistor action. Because of the large lengths of the p-regions of the two diodes and the connecting leads (in this case, $W_b >> L_b$), no injected charge will survive the long equivalent length of the base region and appear at the adjacent diode junction.

9.5.2 Base Transport Factor β_t

Base transport factor β_t is defined as the ratio of electron current reaching the base-collector junction to the electron current injected into the base of the base-emitter junction, that is,

$$\beta_t = \frac{I_{\text{bn}}^{c}}{I_{\text{bn}}^{e}} \tag{9.29}$$

These currents are given by Eqs. 9.16 and 9.12. The base transport factor becomes—neglecting the constant term with respect to the term with exp (ev_{BE}/kT)—

$$\beta_t \approx \frac{1}{\cosh\left(W_b/L_b\right)} \tag{9.30}$$

Again, for a base width that is small compared to the electron diffusion length, Eq. 9.30 becomes

$$\beta_t \approx 1 - \left(\frac{W_b^2}{2L_b^2}\right) \tag{9.31}$$

This also indicates that a short base width is necessary for a base transport factor to be as close to one as possible.

9.5.3 Collector Efficiency γ_c

Collector efficiency is the ratio of the electron current that reaches the collector to the total base-collector current. The current that reaches the collector junction is already given by Eq. 9.16. This electron current crosses the base-collector junction without almost any loss. Because of the high reverse voltage normally applied to the CBJ, the electric field across this junction is very high. As the electrons traverse the junction, there may be additional electrons created by impact ionization of the impurity atoms in the depletion layer, thus increasing the overall electron current flowing from the base to the collector. Overall, the collector efficiency can be taken to be equal to unity.

Assuming a uniform cross section for all the regions, the dc common emitter current gain parameter α_o can finally be written as

$$\alpha_o = \gamma_e \beta_t \gamma_c$$

$$\alpha_o = \left[1 - \left(\frac{A_{\mathrm{be}}\, p_{\mathrm{eo}} D_e W_b}{A_b n_{\mathrm{bo}} D_b L_e}\right)\right]\left[1 - \left(\frac{W_b^2}{2L_b^2}\right)\right] \tag{9.32a}$$

$$\alpha_o \approx 1 - \left(\frac{W_b^2}{2L_b^2}\right) \tag{9.32b}$$

Although the emitter efficiency correction is a first-order term, because of the higher impurity doping of the emitter, the second-order correction term of the base transport factor dominates and is greater than the first-order correction term of γ_e.

Example 9.1. The following parameters are given for a Si n-p-n junction transistor:

$W_b = 2.5\ \mu\mathrm{m}$
$N_{\mathrm{De}} = 10^{18}\ \mathrm{cm}^{-3}$ (emitter doping)

$$\begin{aligned}
N_{\text{Ab}} &= 10^{15}\ \text{cm}^{-3}\ \text{(base doping)} \\
L_b &= 43.6\ \mu\text{m} \\
L_e &= 26.0\ \mu\text{m} \\
D_b &= 38\ \text{cm}^2/\text{s} \\
D_e &= 13\ \text{cm}^2/\text{s} \\
A_{\text{be}} &= A_b = A = 100\ \mu\text{m}^2
\end{aligned}$$

Calculate α_0 of the transistor.

Solution. In order to calculate γ_e, we have to calculate the ratio $p_{\text{eo}}/n_{\text{bo}}$, which is

$$\frac{p_{\text{eo}}}{n_{\text{bo}}} = \frac{(N_{\text{De}}/n_i^2)}{(N_{\text{Ab}}/n_i^2)} = \frac{N_{\text{De}}}{N_{\text{Ab}}} = 10^{-3}$$

The emitter efficiency is

$$\begin{aligned}
\gamma_e &= 1 - \frac{p_{\text{eo}} D_e W_b}{n_{\text{bo}} D_b L_e} = 1 - \frac{(10^{-3})(13\ \text{cm}^2/s)(2.5\ \mu\text{m})}{(38\ \text{cm}^2/s)(26.0\ \mu\text{m})} \\
\gamma_e &= 1 - 0.0000329 = 0.99997
\end{aligned}$$

The base transport factor is

$$\begin{aligned}
\beta_t &= 1 - \frac{W_b^2}{2L_b^2} = 1 - \frac{(2.5\ \mu\text{m})^2}{2(43.6\ \mu\text{m})^2} \\
\beta_t &= 1 - 0.00164 = 0.998
\end{aligned}$$

The common base current gain is then equal to

$$\begin{aligned}
\alpha_o &= (1 - 0.0000329)(1 - 0.00164) \\
\alpha_o &= 1 - 0.00164 - 0.0000329 \\
\alpha_o &= 1 - 0.00167 \approx 0.998 = \beta_t
\end{aligned}$$

Thus, the base transport factor determines the current gain, provided the emitter is heavily doped relative to the base.

Using Eq. 9.23, in the active mode and in the common-base configuration shown in Fig. 9.8a, the emitter and collector currents are related by

$$i_C = -\alpha_0 i_E + I_{CO} \tag{9.33}$$

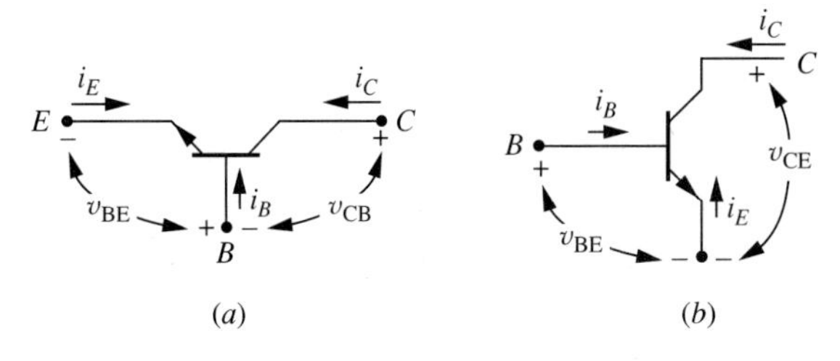

FIGURE 9.8
(a) Common-base and (b) common-emitter transistor configurations.

Here $\alpha_0 = \alpha_F$. The base current is than equal to

$$i_B = (1 - \alpha_0) i_E + I_{CO} \tag{9.34}$$

In the cutoff mode of operation, that is, when $v_{BE} < 0$ and $v_{CB} > 0$, the only current that flows is the reverse saturation current of the CBJ which is equal to I_{CO} of Eq. 9.24.

A widely used transistor configuration known as the *common emitter* is shown in Fig. 9.8*b*. The ratio of the collector current to the base current is known as the β_0 of the transistor. From the ratio of Eq. 9.33 to Eq. 9.34, neglecting I_{CO}, β_0 is

$$\beta_o = \frac{\alpha_0}{1 - \alpha_0} \tag{9.35}$$

β_0 becomes large when α_o is close to one. Again, this shows the importance in the transistor design of obtaining an α_o as close to one as possible.

In the common-emitter configuration, the base current becomes the controlling factor for the collector current. Instead of the collector-base supply, a battery is now connected across the collector-emitter terminals. The base current can be varied by connecting a battery across the base-emitter terminals. As the base current is increased (or holes are introduced into the base region in an n-p-n BJT) electrons have to be injected into the base region from the emitter to maintain the charge neutrality in the base region. If τ_t, the time within which electrons traverse the base width, is short compared to τ_b, the electron recombination time in the base, for each hole many electrons have to be injected into the base region before that hole recombines and disappears from the base. Thus, we expect a larger emitter and consequently a correspondingly larger collector current to flow for small base currents. We can thus manipulate large collector currents with smaller base currents. That is why β_o, given by Eq. 9.35, is much larger than unity.

In the common-emitter configuration, the transistor will operate in the active mode as long as the collector-emitter voltage v_{CE} is greater than v_{BE}. When v_{CE} becomes less than v_{BE}, the collector-base junction becomes forward biased, and the transistor moves into the saturation mode of operation. This mode becomes important in switching applications of the BJT.

The current-voltage characteristics of an n-p-n transistor are shown in Fig. 9.9 for the common-base and the common-emitter configurations. In the active regions, the dotted lines represent the intrinsic transistor characteristics where the collector currents are constant and independent of v_{CB} or v_{CE}.

9.5.4 Deviations from Ideal Operation

In an intrinsic transistor, the effective base width W_b, the distance between the two depletion layers of the transistor, is assumed to be constant and independent of v_{CB}. Actually, the depletion-layer widths depend on the square root of the applied voltages across the respective junctions (Eq. 8.45). Since the base-emitter junction is normally forward biased, the emitter-base depletion width W_{eb} is very short.

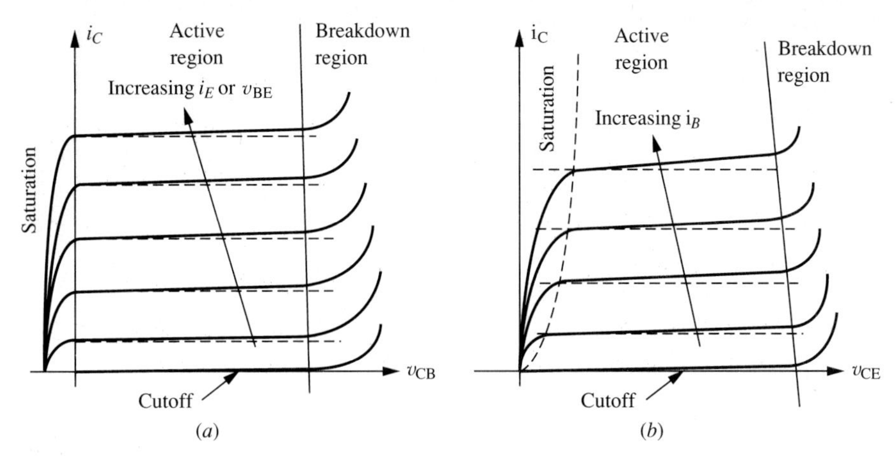

FIGURE 9.9
Current-voltage characteristics of a typical n-p-n transistor for (*a*) common-base and (*b*) common-emitter configurations. Dotted lines in the active regions are for an intrinsic transistor where i_C is assumed to be independent of v_{CB} or V_{CE}.

On the other hand, since the base-collector junction is usually reverse biased, the effective base width is determined by the extent of the collector-base depletion-layer width W_{bc} into the base region. Since W_{bc} depends on the square root of v_{CB}, and increases with increasing v_{CB}, the collector current given by Eq. 9.17 increases with increasing v_{CB} due to an increase in the slope of the excess electron density in the base region. The actual i-v characteristics of the transistor are not horizontal and current increases with v_{CB} as shown by the solid lines in Fig. 9.9.

If the applied collector-base voltage is large enough, it is possible that the collector-base depletion region may extend all the way into the emitter-base depletion layer and the transistor action stops. The collector-base voltage where this occurs is called the *punch-through* voltage and can be calculated from Eq. 8.45 by finding the corresponding junction voltage v_{CB}, which will bring about the punch-through condition. Neglecting the base-emitter depletion layer, the punch-through voltage is obtained when the depletion layer W_{bc} of the collector-base junction becomes equal to W_a, the physical length of the base. This yields

$$V_{\text{pt}} \approx \frac{eN_A W_a^2}{2\kappa_b \epsilon_o} \tag{9.36}$$

Note that because of the large electric field at the base-collector junction, avalanche breakdown of the charge carriers occurs before the punch-through voltage is reached and a sharp increase in the collector current is observed. These breakdown regions are shown in Fig. 9.9.

If an ac signal is superimposed into a fixed dc bias voltage of the base-collector junction, the effective base width of the transistor is accordingly modulated by the variation of this ac signal. Modulation of the base width is known as

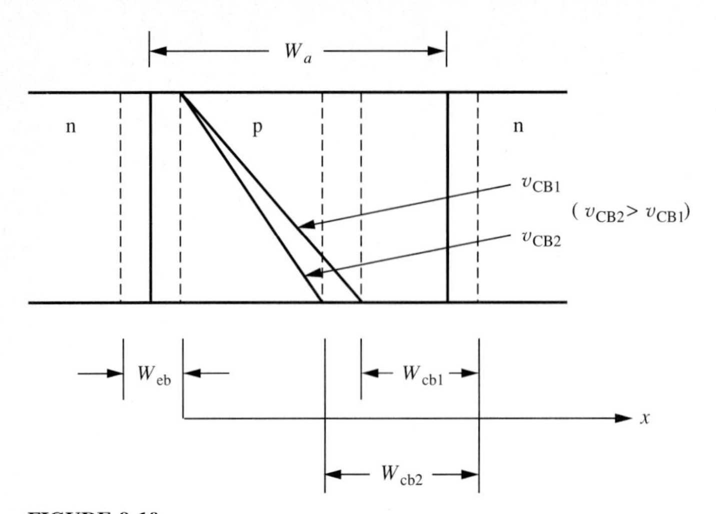

FIGURE 9.10
Base width modulation due to the Early effect. As the collector-base voltage increases, the base width decreases.

the *Early effect*. Figure 9.10 shows the equivalent base width for two values of v_{CB}.

Base-width modulation also effects the current gain α_0 of the transistor, since both the emitter efficiency and the base transport factor depend on the base width of the transistor. At low voltages, the change in these parameters is not noticeable, but as the base-collector voltage is increased, both of these parameters and, therefore, the transistor current gain increases.

9.6 SMALL-SIGNAL EQUIVALENT CIRCUIT OF A BIPOLAR TRANSISTOR

The low-frequency small-signal equivalent circuit for the common-base bipolar transistor configuration is shown in Fig. 9.11. The currents and voltages represented here are the ac components. r_e and r_b are the series resistances of the emitter and the base bulk materials, and $\alpha_0 i_e$ is the current source representing the effect of the coupling of current from the emitter to the collector. α_0 is the actual low-frequency common-base current gain of the transistor. r_c is the collector resistance resulting from the base-width modulation of the transistor and is equal to the slope of the *i-v* curve at the operating point of the BJT. The feedback current source in parallel with the emitter resistance arises also from the base-width modulation. g_{eff} is a *transconductance* defined by

$$g_{\text{eff}} = \left. \frac{\partial i_e}{\partial v_c} \right|_{v_e = \text{constant}} \tag{9.37}$$

In low-frequency applications, g_{eff} can generally be neglected. In the equivalent circuit given in Fig. 9.11, the capacitive effects are also completely neglected.

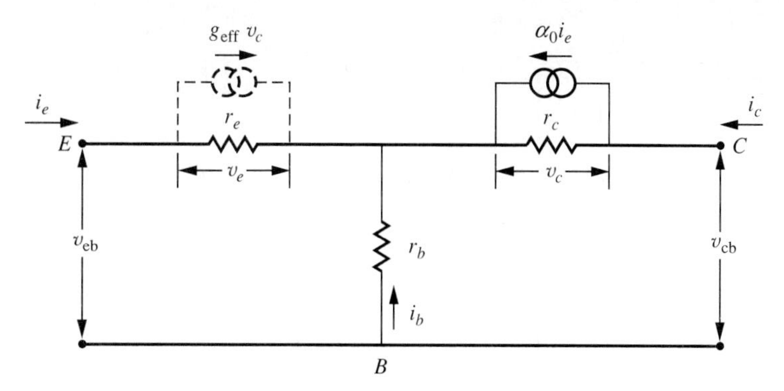

FIGURE 9.11
Small-signal low-frequency equivalent circuit of a common-base BJT.

They will be included when high-frequency effects are considered in more detail in Chapter 12.

9.7 SWITCHING CHARACTERISTICS OF BJTs

To study its switching characteristics, a BJT transistor is connected in a circuit, as shown in Fig. 9.12. A square wave input voltage is applied to the base of the transistor in series with the current-limiting resistor R_{in}. Output voltage is taken from the collector. R_L is the load resistor.

Assume that the transistor is initially in the cutoff mode, that is, the input voltage is such that $v_{\text{in}} = -V_1$ ($v_{\text{BE}} < 0$) (see Fig. 9.13). In this mode, the excess charge in the base region is very small (Curve 3 in Fig. 9.14) and only the I_{CO} flows through the circuit. Since this is negligibly small, the voltage drop across the load resistor is practically zero, and thus $v_{\text{out}} \approx V_{\text{CC}}$.

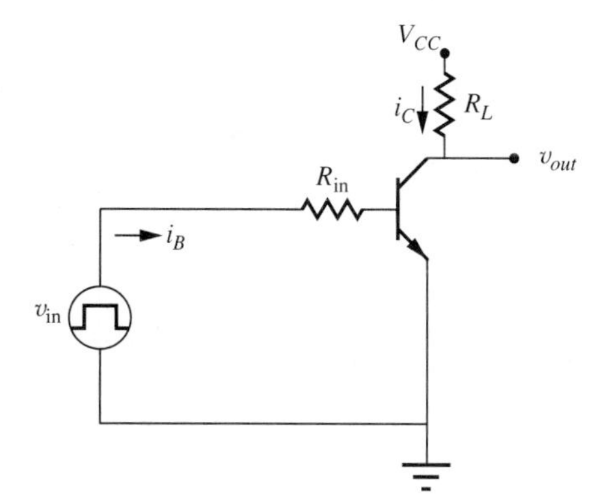

FIGURE 9.12
Switching circuit for observing the transient response of the transistor.

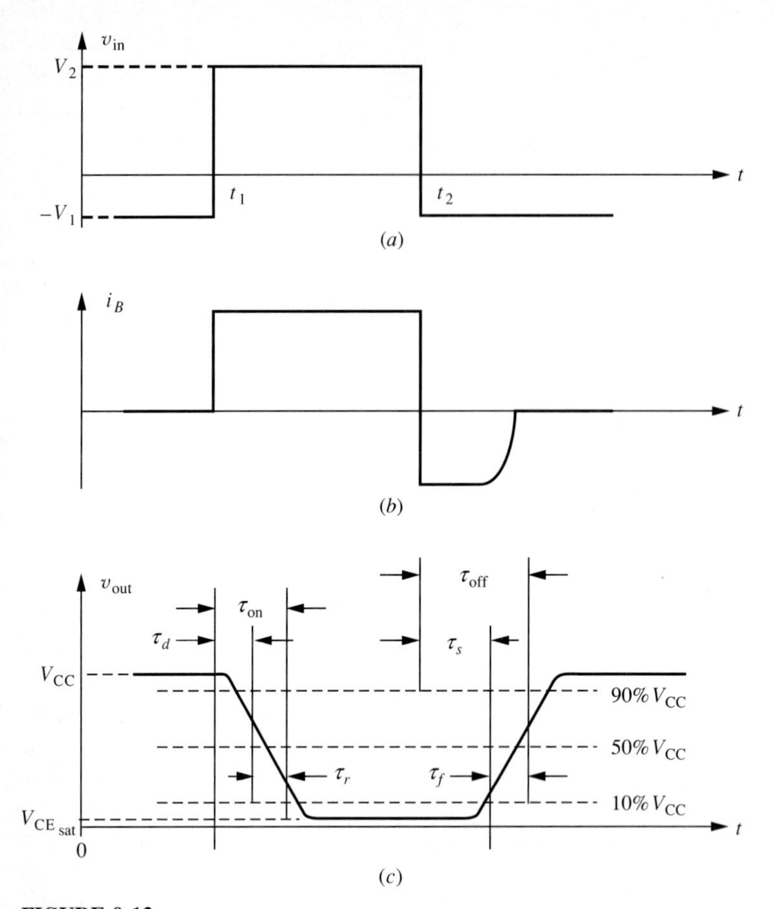

FIGURE 9.13
Switching characteristics of an n-p-n transistor: (*a*) v_{in}, (*b*) i_b, and (*c*) v_{out}. The rise and fall times are measured between 0.90 v_{CC} and 0.10 v_{CC}.

If v_{in} switches to $+V_2$, the base current starts to flow immediately. Since V_2 is larger than v_{BE}, a relatively high base current flows and forces the emitter to inject a large number of electrons into the base region. The resulting collector current increases and with it the voltage drop across R_L also increases. This drop becomes so large that it forces the v_{CE} to fall below v_{BE}. This means that the CBJ becomes forward biased. Consequently, additional electrons are injected into the base region from the collector. The excess electron density on the CBJ side becomes greater than the equilibrium minority carrier electron density n_{bo} in the base region. Although injection of electrons to the base occurs from both of the junctions, since $v_{BE} > v_{CB}$, still more electrons are injected into the base from the emitter than the collector. The variation of the excess electron density in the

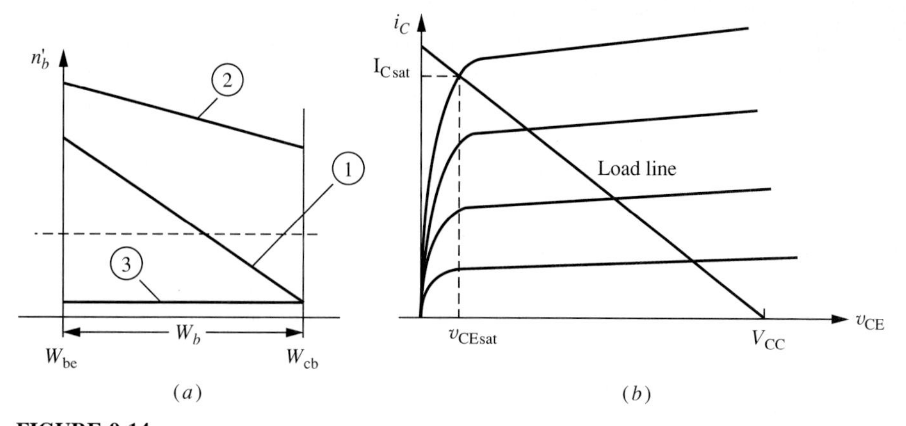

FIGURE 9.14
(*a*) Excess electron density in the base region for (1) active, (2) saturation, and (3) cutoff modes. Although W_b is shown to be the same for all these conditions, in an acutal transistor W_b changes with applied voltages, especially with v_{CE}. (*b*) Parameters associated with switching.

base is shown by Curve 2 in Fig. 9.14. For the currents to reach steady-state conditions, excess charge equal to the area under Curve 2 (Fig. 9.14*a*) should be supplied to the initially almost empty base region. In reaching the saturation mode, the transistor goes from the cutoff region through the active region to the saturation region following the load line of Fig. 12.14*b*. It takes a while for the base excess electron density to reach n_{bo}, and during this time interval, no collector current flows. This generates the delay time τ_d in Fig. 9.13. As the transistor goes through the active region and reaches the saturation mode, the collector current increases from I_{CO} to I_{Csat}. The final collector voltage drops down to V_{CEsat}. This occurs in a time interval τ_r called the *rise time* of the transistor.

When the input voltage is switched back to $-V_1$, the base current cannot become zero instantaneously. At this moment, there are still a large number of excess electrons in the base region. These are to be removed from the base before steady state (back to Curve 3 in Fig. 9.14) can be reached. The electron diffusion into the CBJ continues while recombination of electrons is still taking place at the base region. For this recombination to continue, holes still have to be supplied to the base. Since the polarity of the base voltage is reversed by the application of $-V_1$, continued supply of holes into the base implies a reversal of the base current direction at $t = t_2$. Thus i_b becomes negative at $t = t_2$. It takes τ_s seconds for the collector current to decrease until the CBJ again becomes reverse biased. During τ_s, the transistor continues to operate in the saturation mode. Finally, while the rest of thc electrons are removed, it takes τ_f seconds for the collector current to drop back to the cutoff value I_{CBO}. τ_s and τ_f are known as the *saturation* and *fall times*. The τ_{on} and τ_{off} times of the transistor are the sums of $(\tau_d + \tau_r)$ and $(\tau_s + \tau_f)$, respectively.

9.8 HETEROJUNCTION BJT

In studying the physical and electrical characteristics of a BJT, it was assumed that all elements were made from the same semiconductor material. With the advances made in the manufacture of compound semiconductor devices, heterojunction bipolar transistors (HBJT) are now being manufactured with distinct advantages over homojunction BJTs.

A typical energy-band diagram of a single HBJT is shown in Fig. 9.15. Here a wide band gap emitter (GaAlAs) is used in the implementation of the transistor. Although here both the base and the collector are made from the same material (GaAs), it is also possible to make the collector from a wide band gap material. In a double heterojunction bipolar transistor (DHJBT), the collector

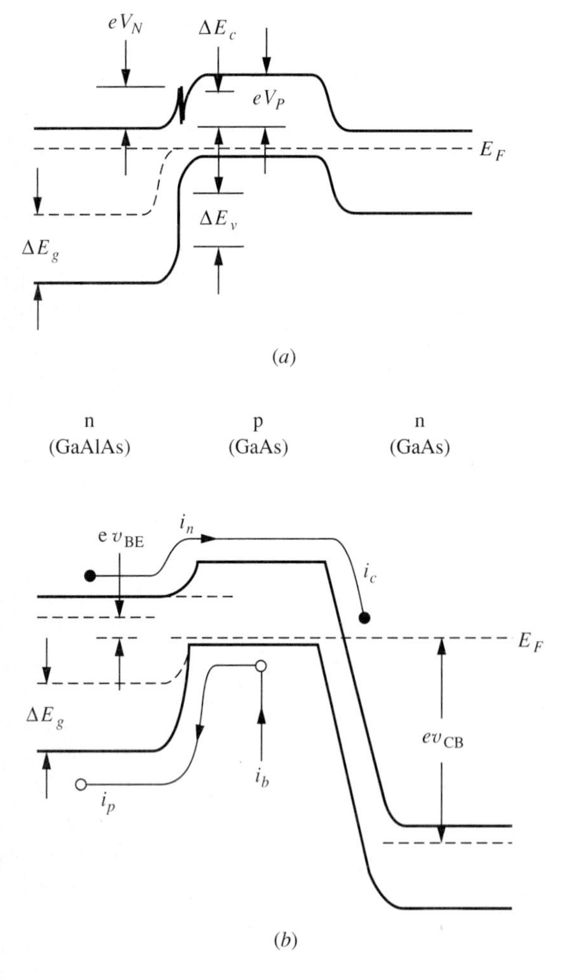

FIGURE 9.15
Energy-band diagram of an n-p-n single heterojunction bipolar transistor with a wide band gap emitter (*a*) with no bias and (*b*) with bias showing various current components and the hole-repelling effect of the additional energy gap in the emitter. (Part (*b*) from Herbert Kroemer, "Heterostructure Bipolar Transistors and Intergrated Circuits," *Proc. IEEE*, vol. 70, 1982, 13–25, © 1982 IEEE.)

is also made up from n-type GaAlAs with a different Al molar concentration. In either case, use of a wide band gap emitter is essential for the operation of an HBJT.

In an HBJT, electrons on the emitter side have to overcome a lower built-in barrier potential to be injected into the base. On the other hand, the holes have to overcome a larger barrier potential on the base side to be injected into the emitter. Thus, the back-injected hole current is considerably suppressed. From Eq. 9.25, reduction of I_{ep}^{e} implies an increase in γ_e and consequently an increase in α_{o}. Therefore, almost all of the reduction to α_{o} comes through the recombination of holes in the base. The wide band gap emitter and the increased doping at the base region also (*a*) reduce the offset and saturation voltages associated with the transistor, (*b*) decrease the depletion junction capacitances, and (*c*) decrease the base resistance. As will be shown in Chapter 12, the last two parameters affect the high-frequency response of the transistor. Lower values mean an increase in the limit of the high frequency of operation.

FURTHER READING

Bar-Lev, Adir, *Semiconductor and Electronic Devices*, 2d ed., Prentice-Hall, Englewood Cliffs, N. J., 1984.

Dekker, A. J., *Solid-State Physics*, Prentice-Hall, Englewood Cliffs, N. J., 1961.

Kroemer, Herbert, "Heterostructure Bipolar Transistors and Integrated Circuits," *Proc. IEEE*, vol. 70, 1982, 13–25.

Marty, A., J. Jamai, J.P. Vannel, N. Fabre, J.P. Bailbe, N. Duhamel, C. Dubon-Chevallier, J. Tasselli, "Fabrication and dc Characterization of GaAlAs/GaAs Double Heterojunction Bipolar Transistors," *Solid-State Electronics*, vol. 31, 1988, 1375–1382.

Neudeck, Geold W., "The Bipolar Junction Transistors," in *Modular Series in Solid-State Devices*, vol. 3, G. W. Neudeck and R. F. Pierret, eds., Addison-Wesley, Reading, Mass., 1983.

Spenke, E., *Electronic Semiconductors*, McGraw-Hill, New York, 1958.

Streetman, Ben G., *Solid-State Electronic Devices*, 3d ed., Prentice-Hall, Englewood Cliffs, N. J., 1990.

PROBLEMS

9.1. Derive Eq. 9.9.

9.2. Derive Eq. 9.19.

9.3. Derive Eq. 9.23.

9.4. Consider the excess base electron distribution to be given by the straight-line approximation of Fig. 9.6*b*, Curve I. In order to take into account the recombination with the holes, assume that the additional base current can be written as

$$I_{\text{bp}}^{b} = \frac{N_b}{\tau_n} \tag{P. 9.1}$$

where N_b is the total electron charge in the base region, and $1/\tau_n$ is the probability per unit time that the electrons recombine with the holes in the base. By taking the area under the straight-line approximation to be equal to the total electron density N_b, show that the contribution to the base current due to recombination can be written as

$$I_{\text{bp}}^{b} = -\frac{1}{2}\frac{eA_bW_b}{\tau_n}\left\{n_{\text{bo}}\left[\exp\left(\frac{ev_{\text{BE}}}{kT}\right)-1\right]+n_{\text{bo}}\left[\exp\left(-\frac{ev_{\text{CB}}}{kT}\right)-1\right]\right\} \quad \text{(P. 9.2)}$$

9.5. Show that if the slope of the excess electron density at the base region of an n-p-n transistor were approximated by a straight line, the electron current in the base region would be given by

$$I_{\text{bn}}^{e} = -\frac{eA_{\text{be}}D_b}{W_b}n_{\text{bo}}\left[\exp\left(-\frac{ev_{\text{CB}}}{kT}\right)-1\right]+\frac{eA_{\text{be}}D_b}{W_b}n_{\text{bo}}\left[\exp\left(\frac{ev_{\text{BE}}}{kT}\right)-1\right] \quad \text{(P. 9.3)}$$

9.6. With the currents given in Problems 9.4 and 9.5, assume that the base transport factor can now be written as

$$\beta_t \equiv \frac{I_{\text{bn}}^{c}}{I_{\text{bn}}^{e}} = \frac{I_{\text{bn}}^{e}-I_{\text{bp}}^{b}}{I_{\text{bn}}^{e}} = 1-\frac{I_{\text{bp}}^{b}}{I_{\text{bn}}^{e}} \quad \text{(P. 9.4)}$$

Show that one obtains Eq. 9.31 again. If the emitter and collector efficiencies are taken to be equal to one, show that the common-emitter current gain β_{o} can be written as

$$\beta_{\text{o}} \approx \frac{1}{2}\frac{W_b^2}{D_b\tau_n} \quad \text{(P. 9.5)}$$

9.7. If we define β_{o} as the ratio of the electron recombination time τ_n to the transit time τ_t of the electrons to traverse the base region, show that from Problem 9.6, the transit time in the base region can be written as

$$\tau_t \approx \frac{1}{2}\frac{W_b^2}{D_b} \quad \text{(P. 9.6)}$$

9.8. Derive the Ebers–Moll model for a p-n-p transistor by making proper changes in the voltages given by Equations 6.23.

9.9. The following parameters of a silicon n-p-n transistor are given

$N_{\text{De}} = 5\times10^{17}$ atoms/cm^3 (emitter doping)
$N_{\text{Ab}} = 1\times10^{15}$ atoms/cm^3 (base doping)
$N_{\text{Dc}} = 4\times10^{15}$ atoms/cm^3 (collector doping)
$W_a = 5\ \mu\text{m}$ $\quad A = 50\ \mu\text{m}^2$ (constant area)
$v_{\text{CB}} = 10$ V, use constants from Example 9.1

Find
(*a*) the emitter efficiency of the transistor
(*b*) the base transport factor
(*c*) the current gain of the transistor (assume $\gamma_c = 1$)
(*d*) the β_{o} of the common emitter configuration
(*e*) the punch-through voltage of the transistor

9.10. Show that for the transistor given in Problem 9.9, the correction due to emitter efficiency can be neglected compared to the correction due to the base transport factor in the calculation of α_0.

9.11. Find an expression for the gain of the transistor whose small-signal model is given in Fig. 9.11. Neglect g_{eff} and assume $r_c = \infty$.

9.12. An HBJT has a p-type GaAs base and an n-type $\text{Ga}_x\text{Al}_{1-x}\text{As}$ as an emitter as shown in Fig. 9.14. $eV_N = 0.25$ eV, $eV_P = 0.45$ eV, $\Delta E_c = 0.25$ eV, and $\Delta E_v = 0.45$ eV. Find the ratio of the $p_{\text{eo}}/n_{\text{bo}}$ and show that the emitter injection efficiency is practically equal to 1.0 for this HBJT.

CHAPTER 10

JUNCTION FIELD-EFFECT TRANSISTORS (JFETs)

There is another class of widely used transistor configurations, known as the field-effect transistor (FET), which has different operational as well as different i-v characteristics. Its operation depends on the fields that exist between various regions of the transistor, and it is a voltage-controlled device, in contrast to the current-controlled bipolar transistor. The input impedances of the FETs are also very high. Unlike bipolar transistors, where minority carriers determine the electrical characteristics, field-effect transistors are majority carrier devices. There are many variations of the field-effect transistor. In this chapter, we are going to restrict our study to the junction field-effect transistors (JFETs) and the Metal-Semiconductor or Schottky barrier field-effect transistors (MESFET). Another very important variation of an FET, the MOSFET, will be discussed in Chapter 11. The physical basis for the operation of field-effect transistors is the same, but the actual implementation of the transistor as well as the resulting electrical characteristics show marked differences.

10.1 VOLTAGE-CONTROLLED RESISTOR

Figure 10.1 shows a uniform bar of a semiconductor. The resistance of the bar, which has a cross-sectional area of $A = (hd)$ and a total length of L, is given by

$$R = \rho\left(\frac{L}{hd}\right) \tag{10.1}$$

where $\rho = 1/\sigma$ is the resistivity of the sample bar.

If one of the dimensional parameters of the bar is changed, the resistance of the sample will change accordingly. If a constant voltage is applied across the sample, the current flowing through the bar will decrease or increase with increasing or decreasing resistance. A physical change of one of the parameters will not be practical if such a variable resistor is to be implemented as a useful electronic device. On the other hand, if one of the dimensional parameters of the resistor is changed by an electrical means, the current flowing in the sample can then be electrically controlled. If the height of the bar in Fig. 10.1 is changed from h_1 to h_2, the resistance of the bar will increase, and thus the resulting current flowing through the bar will decrease for the same applied voltage.

In a junction field-effect transistor, the resistance of a semiconductor sample is varied by an externally applied voltage, as shown in Fig. 10.2. Although the actual device is implemented in a different and more practical way, the model demonstrates the essential properties of the device and, because of the symmetry, allows a simple mathematical derivation of the current voltage characteristic of the device.

An n-type semiconductor bar of length L, width d, and height $2h$ has two reverse-biased junctions formed at the upper and lower walls. These junctions can both be either p^+-n junctions or metal-semiconductor junctions. A voltage v_{DS} is applied across the main semiconductor bar. The electrode through which

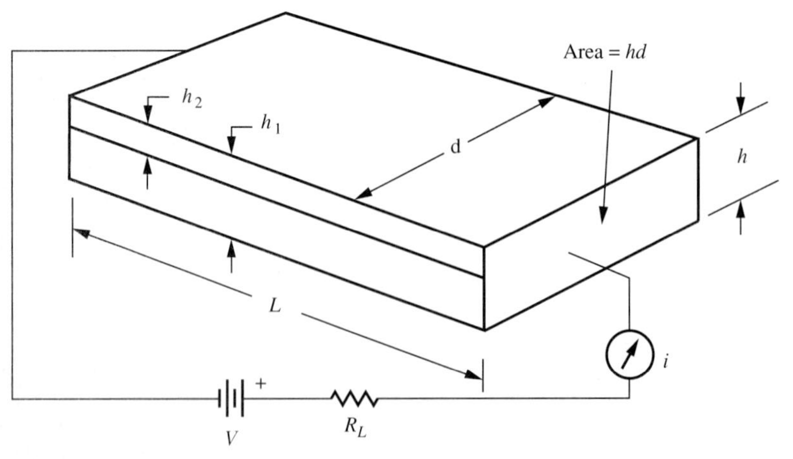

FIGURE 10.1
Resistance of a semiconductor bar. Changing h changes the resistance of the sample.

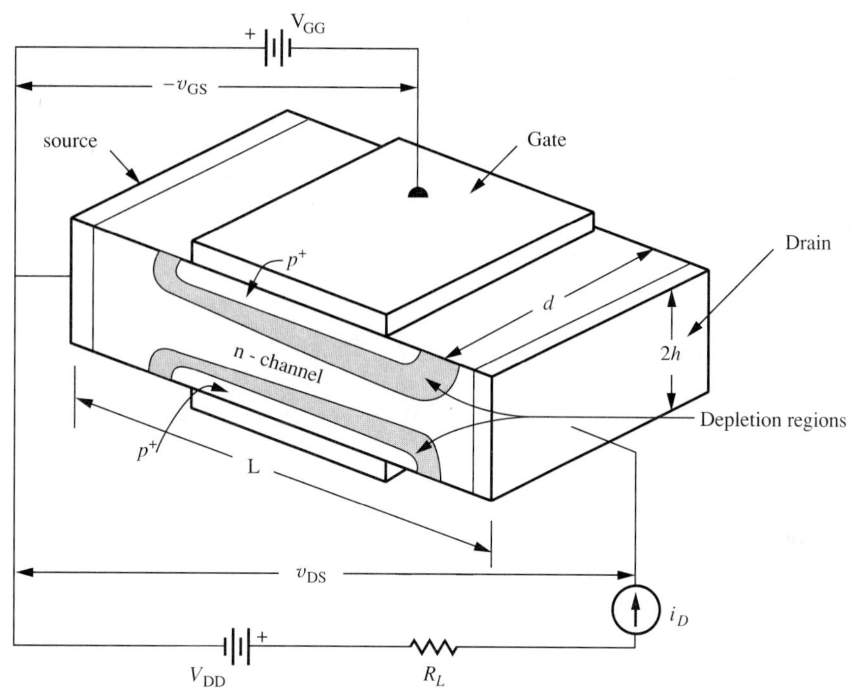

FIGURE 10.2
Essential components of a junction field-effect transistor (JFET).

the electrons originate is called the *source* and the place where they leave the bar is called the *drain*. The upper and lower junctions, which are tied together, are called the *gate*. The region from the source to the drain, through which the electrical current flows, is called the *channel*. For the transistor shown in Fig. 10.2, the channel is an *n-channel* since the electrons that are the majority carriers constitute the charge carriers in an n-type semiconductor.

A voltage $-v_{GS}$ is applied between the gate and the source. The polarity of this voltage is such that the gate is negative with respect to the source—that is, the base-channel junctions are reverse biased. We know from Chapter 8 that a reverse-biased junction has a large depletion layer containing no mobile charges, and that the depth of the depletion layer is proportional to the square root of the voltage that appears across the junction. Figure 10.3*a* shows the depletion layers associated with the two junctions when both the gate-source and the drain-source voltages are zero. The depletion layers are due to the usual un-biased diode depletion layers and are practically uniform across the gate.

When the source and the drain are tied together, that is, $v_{DS} = 0.0$ V, the resulting depletion-layer depths and the reduction in the channel height with various gate-to-source voltages is shown in Fig. 10.3*b*. At $v_{GS} = -3.0$ V, the channel is almost closed—that is, the depletion layers of the two junctions almost touch each other. If v_{GS} is further reduced beyond -3.0 V, the channel is completely closed;

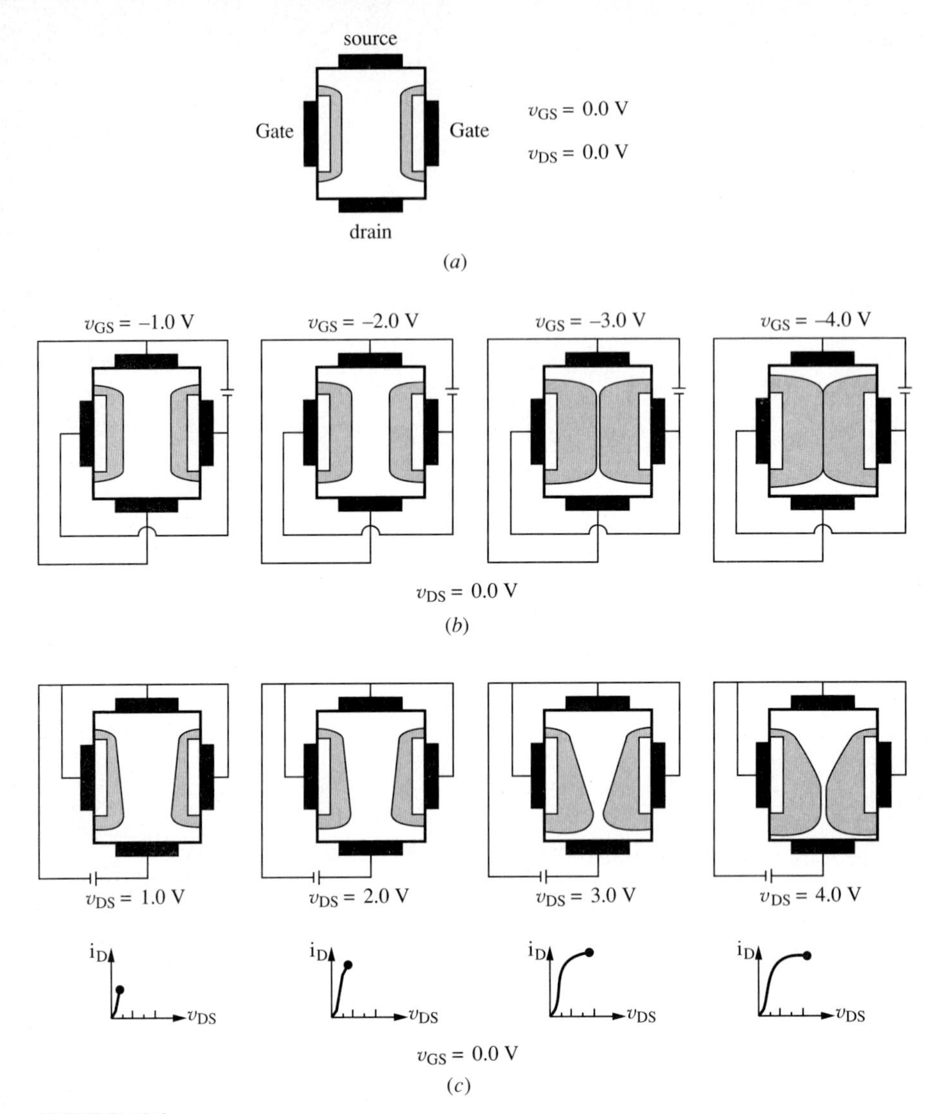

FIGURE 10.3
Operational characteristics of a JFET with $V_P = -3.0$ V. (*a*) With no bias. Depletion layers are the usual diode depletion layers. (*b*) Channel height for various gate-source voltages with $v_{DS} = 0$. Beyond $v_{GS} = -3.0$ V, the channel is closed, the transistor at cutoff. (*c*) Current increases with increasing v_{DS} for no gate bias. The current saturates at 3.0 V.

in other words, the two depletion layers overlap. For the case $v_{GS} = -4.0\ V$ shown in Fig. 10.3*b*, no current can flow through the channel, with the exception of the reverse saturation current of the diodes which, for practical purposes, can be neglected.

To see the effect of the drain-to-source voltage, initially the two gates are connected to the source and the v_{DS} is gradually increased (Fig. 10.3*c*). When a small v_{DS} is applied between the source and the drain, a current flows through the sample from the source to the drain. This current, labeled i_D, is expected to increase with increasing v_{DS}. For a given v_{DS}, the voltage along the channel increases toward the drain because of the $i_D R$ voltage drop where R is the channel resistance. Therefore, the reverse voltage seen by the gate-channel junction increases toward the drain. Consequently, the depletion layer becomes wider toward the drain end relative to the source end.

At low voltages, the current increases linearly with voltage. With a further increase in v_{DS}, a reduction in the source-to-drain current i_D is observed, due to the increase in the resistance of the channel because of the reduction in the effective height of the channel. As v_{DS} is further increased, the voltage difference between the channel and the gate increases proportionately, and the depletion layers from the two junctions approach each other at the drain end of the transistor, at the same time reducing the effective channel height. Thus a further reduction in current is observed.

As the v_{DS} is further increased, the changes in i_D with respect to v_{DS} show a peculiar behavior with an additional interesting phenomenon taking place. To start with, the height of the channel is physically fixed at $2h$. If, with increasing voltage, the two depletion layers of the junctions extend into the channel so that their widths near the drain end become equal to h, the channel is said to be *pinched off*. At pinch-off, since the channel height is practically zero, one expects the current flow from the source to the drain to stop. This does not occur. The current continues to flow and practically stays constant with increasing drain-to-source voltage v_{DS}. This means that even beyond pinch-off, provided the channel was open to start with, the channel stays open.

If now a negative voltage is applied to the gate relative to the source, i_D will again increase with increasing v_{DS}. This time, the increase in current will be smaller than the case of $v_{GS} = 0$. At any point along the channel, the reverse bias appearing across the gate and the channel is larger than the case when $v_{GS} = 0$ because of the additional applied voltage between the gate and the source. The higher reverse voltage across the channel reduces the effective height of the bar in a correspondingly larger proportion. This time the pinch-off condition and the corresponding current saturation are reached at a lower value of v_{DS}. As shown in Fig. 10.4, a nonlinear current voltage characteristic with lower currents is thus observed for various fixed values of v_{GS}.

The combination of voltages v_{GS} and v_{DS} for which pinch-off occurs is called the pinch-off voltage V_P (V_p is a negative voltage for an n-channel). If the initial v_{GS} is large enough to produce pinch-off to start with, the channel cannot be opened and stays closed, irrespective of v_{DS}, since the depletion regions extend uniformly all the way from the source to the drain. The transistor is now in the cutoff mode of operation. It should also be mentioned that when $-v_{GS}$ is greater than V_P ($|v_{GS}| < |V_p|$) and there is a finite drain current flowing through the channel for a given v_{DS}, if now the $-v_{GS}$ is decreased (made more negative for an n-channel) and becomes less than V_P, the drain current again stops flowing through the channel. The effect of v_{GS} is such that when the $-v_{GS}$ is equal to or less than V_P, the channel is closed and additional v_{DS} cannot keep the channel open.

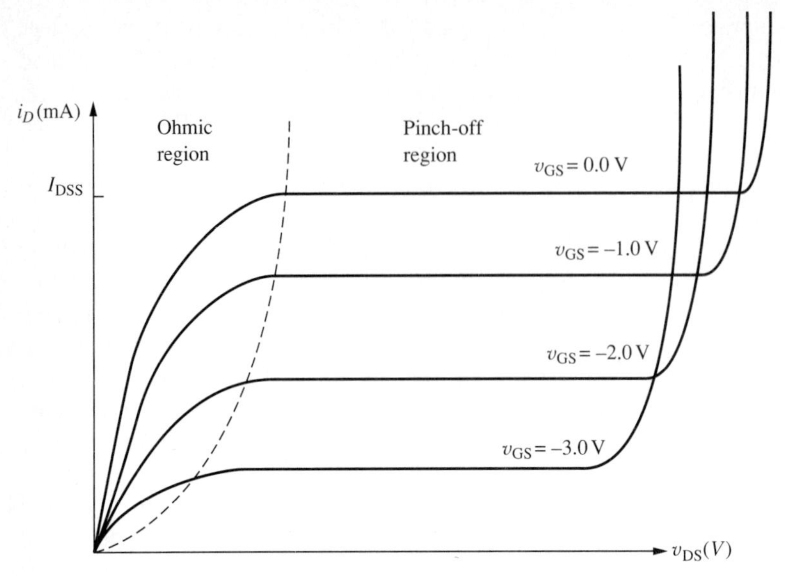

FIGURE 10.4
Current-voltage characteristics of a typical field effect transistor ($V_P = -4$ V).

For a given v_{GS} greater than V_P, the drain current is controlled by the drain-to-source voltage v_{DS}, up to pinch-off, but beyond pinch-off, v_{DS} has very little control over the current. The region of operation before pinch-off is referred to as the *ohmic region* or the *triode region*. The active region of the FET is the *pinch-off* region.

For a given open channel v_{GS}, if v_{DS} is increased further in the saturation region, the very high electric fields that are set up in the channel increase the energy of the electrons going through the channel and ionize the valence electrons of the donor ions so that the breakdown condition similar to the avalanche breakdown of BJTs occurs. This is observed by the rapid increase in the drain current, as shown in Fig. 10.4.

Actual operation of the FET is more complicated than the simplified picture given above. In order to understand the dependence of the resulting drain current on the applied drain-to-source voltage, we have to consider the electric field dependence of the drift velocity in a given semiconductor. At low electric fields, the drift velocity follows the linear relation $v = \mu\, \mathscr{E}$, but as the field magnitude increases, the velocity deviates from the linear relation as was shown in Fig. 6.7, which is redrawn here as Fig. 10.5. The drift velocity in Si starts to saturate at higher fields and becomes constant at the value v_{sL} beyond a critical field $\mathscr{E}_c$. In GaAs, the initial increase in drift velocity is faster than in Si because of the higher low-field electron mobility and reaches a peak value of v_p at $\mathscr{E}_p$. As the electric field is further increased, velocity starts to decrease and eventually reaches a saturation velocity which is not much different than the saturation velocity of Si.

With the typical voltages applied across the terminals of an FET, it is very easy to produce electric fields which can exceed $\mathscr{E}_c$ both in Si and GaAs. For ex-

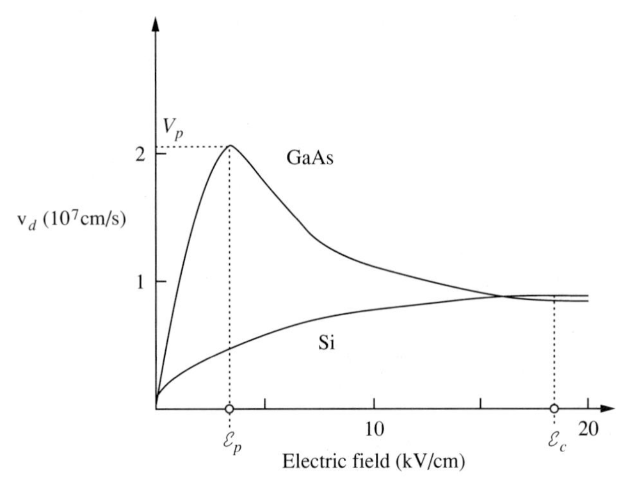

FIGURE 10.5
Electric field dependence of drift velocity in Si and GaAs for electrons.

ample, if 5 V is applied across the source-drain terminals which, for low-frequency transistors, may have a source-drain separation of 5 μm, one can easily generate average fields greater than 10 kV/cm. In today's high-frequency transistors, the source-to-drain distances are in the submicron range where critical field $\mathscr{E}_c$ can easily be exceeded. Considering the differences in the dependence of the drift velocity on the electric field, we expect the $i_D - v_{DS}$ characteristics to differ among FETs made from various semiconductor materials.

Typical constructional details of an n-channel FET are shown in Fig. 10.6. A thin buffer layer (with the highest resistivity) is grown over a semi-insulating substrate. An active n-layer is then grown over this buffer layer. Either a thin p^+-layer is diffused into the n-layer to form the p^+-n junction of a Junction Field-Effect Transistor (JFET) or a rectifying metallic contact is deposited over the n-layer to form a Schottky barrier for a Metal Semiconductor FET (MESFET). Ohmic electrode contacts are then deposited over the source, the gate (for JFET), and the drain locations.

In the analysis of the FET, we will consider a one-dimensional model, that is, as shown in Fig. 10.6, all the parameters will vary only as a function of the distance x measured from the source toward the drain.

For a given set of v_{DS} and v_{GS}, we expect the same current to flow throughout the length of the FET. This is given by

$$\begin{aligned} I &= AJ_x \\ I &= A(x)\rho(x)v(x) \\ I &= edh(x)n(x)v(x) \end{aligned} \tag{10.2}$$

where e is the electronic charge, d is the depth of the channel (into the paper in Fig. 10.6), $h(x)$ is the height of the channel, $n(x)$ is the charged-particle density (electrons for an n-channel), and $v(x)$ is the velocity of the charged particles at a distance x from the source.

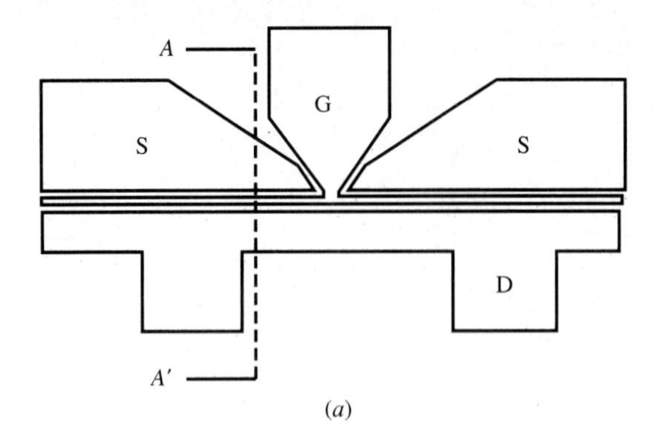

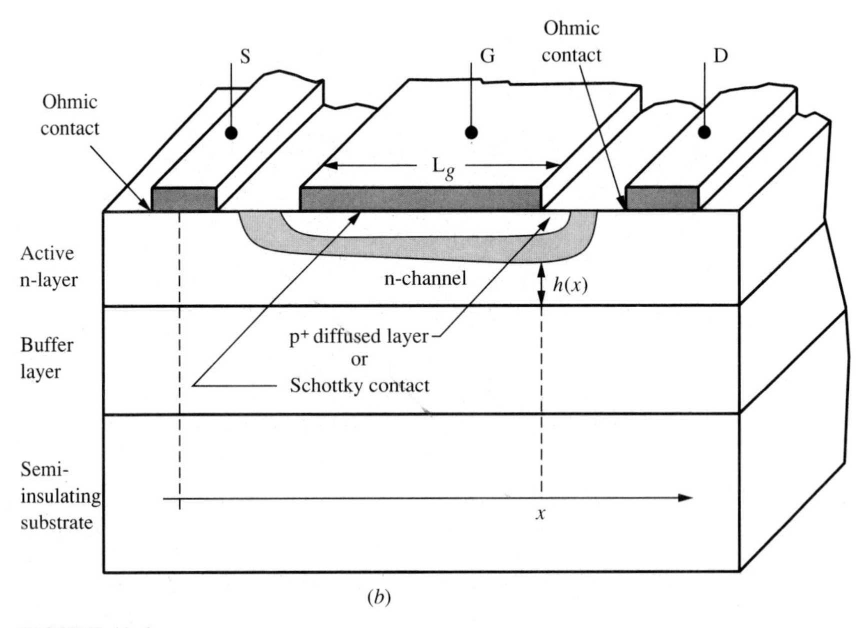

FIGURE 10.6
Constructional details of an n-channel field-effect transistor (FET). *(a)* top view and *(b)* cross section AA'. Diffusion of a p^+-layer produces a junction FET (JFET) or forming a Schottky contact produces a metal semiconductor FET (MESFET). L_g is referred to as the gate-width rather than gate-length.

We will first consider a silicon FET. Figure 10.7*a* shows a FET without any gate electrodes. Current flows from source to drain. Since the channel is uniform, the electric field is expected to be uniform throughout. At low drain-to-source voltages v_{DS}, the resulting electric field is low, and we expect a linear increase in current with increasing voltage following $v = \mu\mathcal{E}$. But as v_{DS} is increased further, the drift velocity does not increase as fast as the electric field and the velocity starts to saturate (see Fig. 10.5). Actually, since all the parameters in Eq. 10.2 stay constant except $v(x)$, the current increase follows closely the velocity dependence of the electrons. Once the field reaches $\mathcal{E}_c$, the current saturates and stays practically constant.

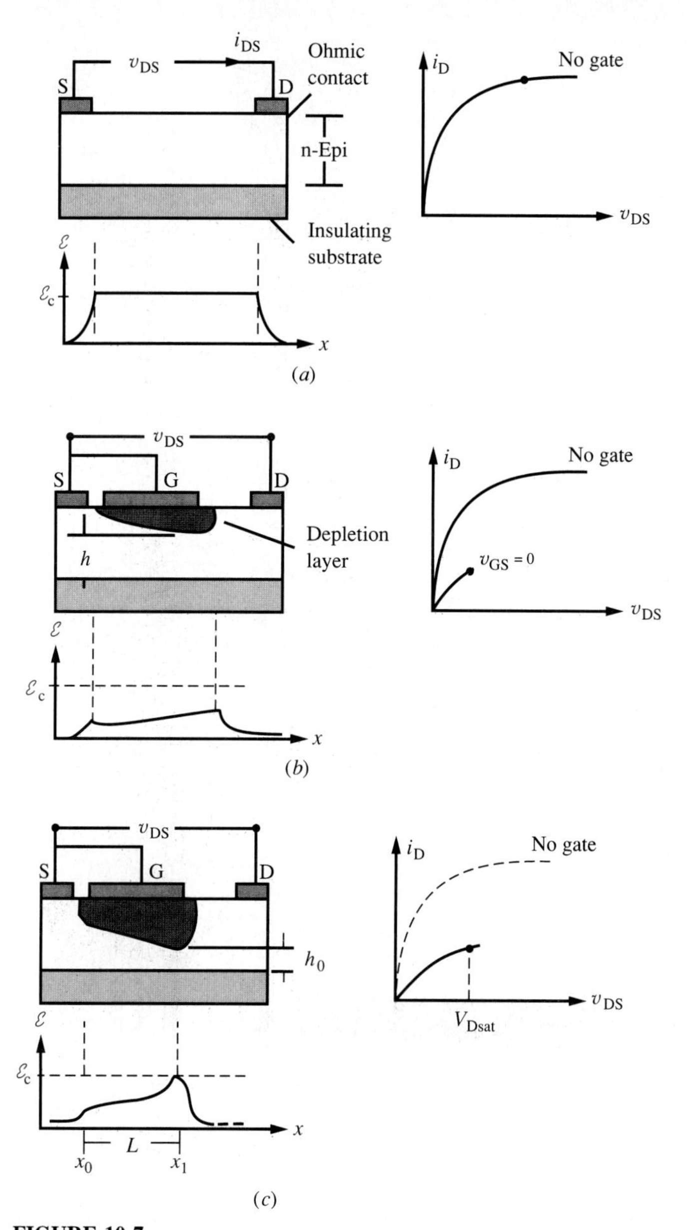

FIGURE 10.7
(*a*) Shows i-v characteristics of an n-type Si layer with two ohmic contacts. The current saturates because electrons reach a maximum drfit velocity at the critical field $\mathcal{E}_c$. In (*b*)–(*d*) the current is controlled by the depletion under a Schottky gate, shorted to the source. In (*c*), the current starts to saturate at V_{Dsat}, and (*d*) shows formation of a stationary dipole layer for $v_{DS} > V_{\text{Dsat}}$. (*e*) illustrates the condition for negative bias (Charles A. Liechti, "Microwave Field-effect Transistors–1976," *IEEE Trans. MTT,* vol. MTT-24, 1976, 279–300, © 1976 IEEE). (*figure continues on next page*)

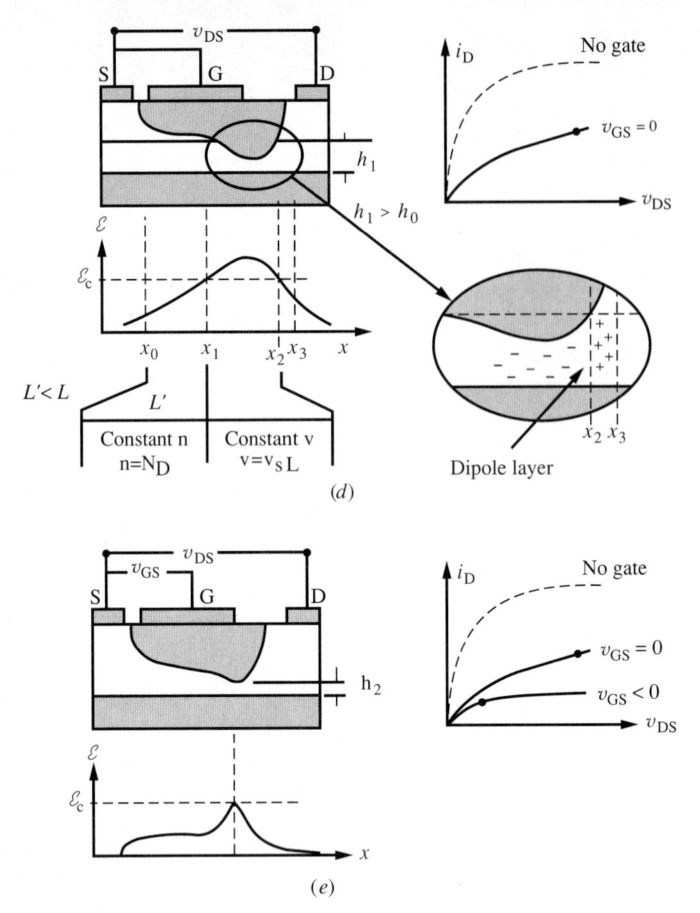

FIGURE 10.7
(*d*) and (*e*)

If either a p^+-n junction or a Schottky barrier is introduced between the source and the drain, any reverse-bias voltage that will appear across the barrier will produce a depletion layer whose extent into the n-region will be determined by the square root of the potential difference appearing between the channel and the gate electrode at that point. The depletion-layer depth will be practically free of mobile charges and electrons will be restricted to flow in the constricted channel with reduced height $h(x)$. Initially, the gate electrode will be assumed to be connected to the source, that is, $v_{GS} = 0$ V. Whenever a current flows in the channel, the voltage drop increases toward the drain due to the ohmic voltage drop. Thus, the reverse-bias voltage that appears across the junction increases as we move toward the drain. Consequently, the depletion-layer depth also increases toward the drain as shown in Fig. 10.7*b*. The minimum channel height occurs near the drain end of the gate. The reduced channel height increases the equivalent resistance of the channel and thus, compared to the no-gate electrode case, less current flows through the channel for a given v_{DS}. Under these conditions, current saturation occurs at a lower value of v_{DS} with reduced current. The saturation current follows the drift velocity versus $\mathscr{E}$ field dependence given in Fig. 10.5, provided the field does not exceed the critical $\mathscr{E}_c$ (Fig. 10.7*c*).

The line integral of the electric field through the channel is equal to the voltage difference between the source and the drain. If v_{DS} is further increased, the depletion layer extends more into the channel and the average field in the channel also increases. Point x_1, where the critical field occurs, moves toward the source (Fig. 10.7d). Thus, we expect a smaller potential to appear at x_1. This means that the channel height h_1 is larger at that point compared to h_0 of Fig. 10.7*c*. As long as the electric field is below $\mathscr{E}_c$, the electron density stays equal to the donor concentration N_D. Since h decreases beyond x_1, it means that more charge has to be injected into the velocity saturation region. Between x_1 and x_2, the electron velocity stays constant at v_{sL}. Since $h(x)$ is narrower than h_1 in this region, more electrons should be injected into this region at x_1 to increase the electron density in order to keep the current constant. Therefore, a negative charge accumulation layer forms in this region.

At x_2, the channel height is again equal to h_1. To keep the current constant beyond x_2, a positive charge layer should form beyond x_2. The additional field generated by the dipole layer between x_2 and x_3, which is in the same direction as the applied field, keeps the electric field above $\mathscr{E}_c$ and thus the velocity stays at v_{sL} up to the point x_3. The positive charge implies a reduction in the majority electron density in that region. The overall effect of dipole layer is a slight increase in the drain current with increasing v_{DS}, as shown in Fig. 10.7*d*.

If a negative potential is now applied to the gate with respect to the source, the total voltage appearing across the gate and the channel at a given distance x is the sum of v_{GS} and the channel voltage which is larger than the case when $v_{GS} = 0$. To start, with the additional v_{GS}, the channel height is already reduced for the same applied drain-to-source voltage. The current is less and the saturation occurs at a lower v_{DS} as shown in Fig. 10.7*e*.

GaAs MESFET operation follows similar behavior with one major difference. The electric field variation along the channel is shown in Fig. 10.8. Since the field exceeds the peak field $\mathscr{E}_p$ between x_1 and x_3 and reaches a maximum at x_2, the electron drift velocity where it is a maximum at $\mathscr{E}_p$ drops down to v_{sL} at x_2. In order to keep the current constant, additional electrons have to be injected into the region between x_1 and x_2 because of the reduced-channel cross section. But this time, since the electrons slow down in this region, more charge is accumulated in this region compared to the case of the Si FET. Beyond x_2, the channel height starts to increase and at the same time the drift velocity of the electrons also increases, and thus a relatively higher positive space charge is created between x_2 and x_3.

It may seem at first that since the saturation velocities v_{sL} for both Si and GaAs are not too much different from each other, the figure of merit of an FET, which is the maximum frequency of operation and which is inversely proportional to the transit time of the charge carriers through the channel (see Chapter 12), will

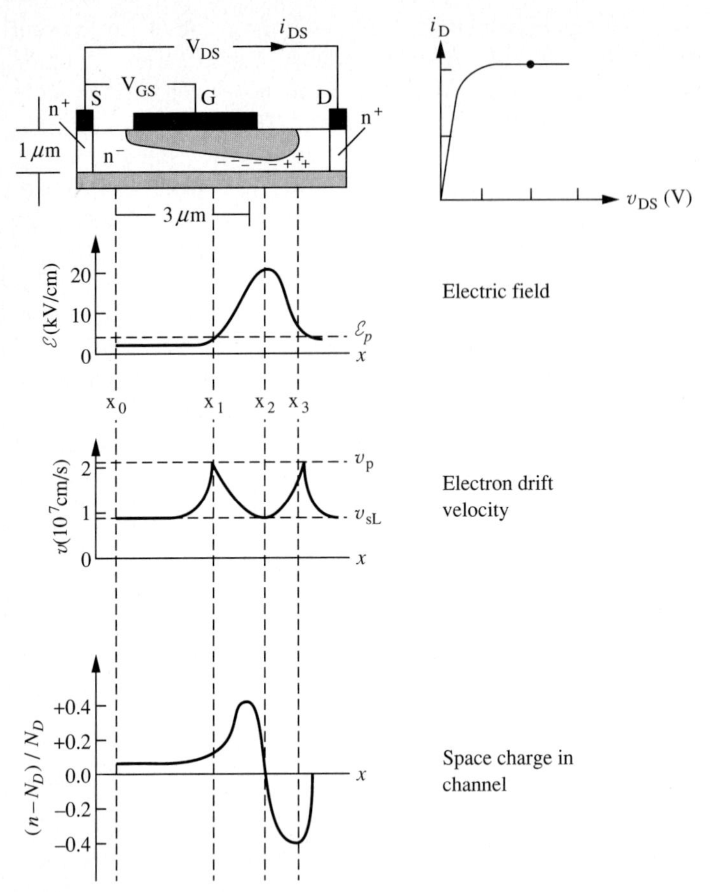

FIGURE 10.8
The channel cross section, electric field, electron drift velocity, and space-charge distribution in the channel for a GaAs MESFET opeated in the current-saturated mode. Proceeding from x_1 to x_2, the channel cross section becomes narrower and, in addition, electrons "slow down." To preserve current continuity, a heavy electron accumulation has to form. The opposite occurs between x_2 and x_3 (Charles A. Liechti, "Microwave Field-effect Transistors–1976," *IEEE Trans. MTT,* vol. MTT-24, 1976, 279–300, © 1976 IEEE).

not be too much different from each other. Although electrons in GaAs possess a higher velocity at fields lower than $\mathscr{E}_c$, this may at first suggest that there is no advantage of using GaAs over silicon in implementing an FET. This is not true for short gate-length devices.

Due to the very short submicron gate lengths, electrons entering the channel may not have too many relaxation collisions. This means they will not have enough time to reach their saturation velocity before reaching the drain end of the channel. Since their low-field mobility is higher, the electrons overshoot their saturation velocity and arrive at the drain in a shorter time interval. This phenomenon reduces the transit time and thus improves the response time of GaAs and other compound semiconductor MESFETs (such as InP, which exhibit similar peak drift velocities) over Si for very high-frequency microwave applications.

It should also be mentioned that it is easier to implement a MESFET than a JFET because of the reduction in the manufacturing steps for the MESFET. Diffusion of the p^+-layer is eliminated in a MESFET.

10.2 DRAIN CURRENT AND PINCH-OFF VOLTAGE

In order to understand some of the parameter dependencies of the drain current on the applied voltages, a simple model will be used to derive the current-voltage characteristics of an FET in the triode region. The FET shown in Fig. 10.6 can be thought to be the symmetric half of the JFET shown in Fig. 10.2.

Fig. 10.9 shows the various dimensional parameters of the JFET transistor. The following assumptions are made.

1. The total distance from the source to the drain is approximately equal to the depletion-layer width L.
2. At any point x, the depletion-layer depth is $w(x)$ and varies linearly with the distance variable x. The depletion-layer depths near the source and the drain are respectively W_s and W_d.
3. The mobility of electrons is constant throughout the channel and is independent of the applied fields.
4. In writing down the depletion-layer depths, the internal diode potentials at the p^+-n junctions are neglected, that is, it is assumed that both v_{GS} and v_{DS} are greater than V_d. (See Problem 10.5).

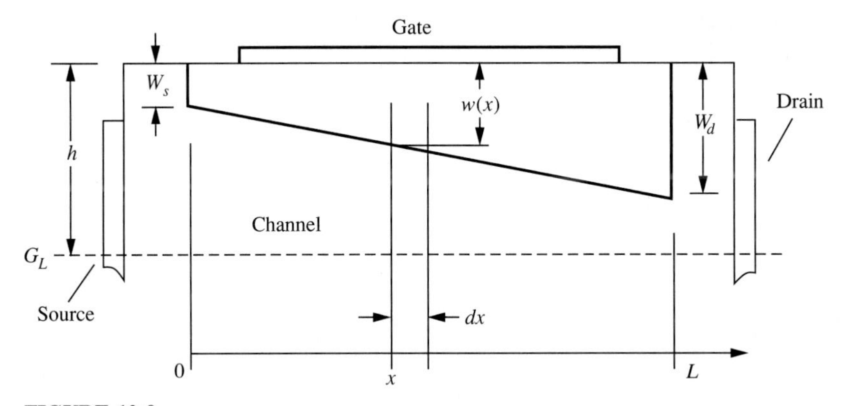

FIGURE 10.9
Parameters used in calculating the drain current.

Near x, an incremental distance dx is considered. The resistance of a sample of length dx, depth d, and height $h - w(x)$ is given by

$$dR(x) = \rho \frac{dx}{2d[h - w(x)]} \tag{10.3}$$

where ρ is the resistivity of the semiconductor. For the FET shown in Fig. 10.6, the factor 2 is neglected.

The potential that appears between the gate and the channel at position x is

$$v(x) = i_D R(x) + v_{\text{SG}} \tag{10.4}$$

The derivative of v with respect to x can be written as

$$\frac{dv}{dx} = \frac{dv}{dw}\frac{dw}{dx} \tag{10.5}$$

The relation between the depletion layer and the voltage appearing across the junction is given by Eq. 8.46 (with $N_{\text{A}} >> N_{\text{D}}$ so that most of the depletion layer resides on the n-side of the junction)

$$w^2(x) = \frac{2\kappa_s \epsilon_0 v(x)}{eN_D} \tag{10.6}$$

Using Eq. 10.4, the derivative of v with respect to x gives

$$\frac{dv(x)}{dx} = i_D \frac{dR(x)}{dx} \tag{10.7}$$

Similarly, the derivative of Eq. 10.6 with respect to w gives

$$\frac{dv(x)}{dw} = \left(\frac{eN_D}{\kappa_s \epsilon_o}\right) w(x) \tag{10.8}$$

Substituting Eq. 10.7 and 10.8 into Eq. 10.5 and manipulating terms, and integrating both sides, we obtain

$$\frac{i_D \rho}{2d} \int_0^{\text{L}} dx = \frac{eN_D}{\kappa_s \epsilon_o} \int_{W_s}^{W_d} [(h - w)w] dw \tag{10.9}$$

where W_s and W_d are the depletion-layer depths near the source and the drain and are equal to

$$W_s^2 = \frac{2\kappa_s \epsilon_0}{eN_D} v_{\text{SG}} \tag{10.10}$$

and

$$W_d^2 = \frac{2\kappa_s \epsilon_0}{eN_D} (v_{\text{DS}} + v_{\text{SG}}) \tag{10.11}$$

Integrating both sides of Eq. 10.9

$$\frac{i_D \rho}{2d} L = \frac{e N_D}{\kappa_s \epsilon_o} \left(h \frac{W_d^2 - W_s^2}{2} - \frac{W_d^3 - W_s^3}{3} \right) \tag{10.12}$$

We now solve this equation for i_D and write it in the following form

$$i_D = \frac{2hd}{\rho L} \frac{e N_D h^2}{2 \kappa_s \epsilon_o} \left[\left(\frac{W_d}{h} \right)^2 - \left(\frac{W_s}{h} \right)^2 - \frac{2}{3} \left(\frac{W_d}{h} \right)^3 + \frac{2}{3} \left(\frac{W_s}{h} \right)^3 \right] \tag{10.13}$$

The pinch-off voltage V_P occurs when the depletion-layer depth at the drain side is equal to h, that is,

$$V_P = -\frac{e N_D}{2 \kappa_s \epsilon_o} h^2 \qquad \text{(for n-channel)} \tag{10.14a}$$

$$V_P = \frac{e N_A}{2 \kappa_s \epsilon_o} h^2 \qquad \text{(for p-channel)} \tag{10.14b}$$

Substituting Eqs. 10.10, 10.11 and 10.14a into Eq. 10.13, we obtain

$$i_D = -G_0 V_P \left[\left(\frac{v_{DS}}{-V_P} \right) - \frac{2}{3} \left(\frac{v_{DS} + v_{SG}}{-V_P} \right)^{3/2} + \frac{2}{3} \left(\frac{v_{SG}}{-V_P} \right)^{3/2} \right] \tag{10.15}$$

where G_0 is an equivalent conductance of the channel and is given by $G_0 = (2hd/\rho L)$.

The pinch-off voltage is also equal to the maximum drain-to-gate voltage that appears at the drain side of the channel when

$$V_P = V_{GS} + V_{SD} = -(V_{DS} + V_{SG}) \tag{10.16}$$

where V_{GS} and V_{SD} are the values of v_{GS} and v_{SD} when the pinch-off condition is satisfied. Substituting this into Eq. 10.15, the drain current at saturation becomes

$$I_{DD} = -G_0 V_P \left[\left(\frac{V_{DS}}{-V_P} \right) - \frac{2}{3} + \frac{2}{3} \left(\frac{V_{GS}}{V_P} \right)^{3/2} \right] \tag{10.17}$$

Equations 10.15 and 10.17 are also valid for a p-channel JFET. The calculations given here represent the voltage dependence of the drain current in the ohmic region. Although some drastic simplifying assumptions are made in deriving Eq. 10.15, the resulting current is not too much off for FETs with long channel widths. There are extensive derivations which take into account the electric field dependence of the drift velocity. These calculations show that the resulting current magnitudes are smaller and for a given v_{GS}, saturation of current occurs at a smaller v_{DS}. But the essential characteristics of an FET do not differ much from those given in Fig. 10.4.

The saturation drain current at $V_{GS} = 0$ V is known as I_{DSS} and can also be calculated from Eq. 10.17. For any JFET, the pinch-off voltage can be calculated from Eq. 10.14a or 10.14b. For a given gate-to-source voltage v_{GS}, the drain-to-source voltage where pinch-off occurs can be calculated using Eq. 10.16. If, to start with $|V_{GS}|$ is greater than $|V_P|$, then the channel is already closed and no current flows through the channel.

Example 10.1. An n-channel JFET has $G_0 = 2.0$ mA/V and $V_P = -4.0$ V. Find
(*a*) The drain current for $v_{GS} = -1.0$ V and $v_{DS} = 2.0$ V.
(*b*) The current at pinch-off for $v_{GS} = -1.0$ V.
(*c*) I_{DSS} for the given transistor.

Solution

(*a*) From Eq. 10.15, the drain current is

$$i_D = -(2.0\text{mA V}^{-1})(-4.0V)\left[\frac{(2.0 \text{ V})}{-(-4.0 \text{ V})} - \frac{2}{3}\left[\frac{(2.0 \text{ V} + 1.0 \text{ V})}{-(-4.0 \text{ V})}\right]^{3/2} + \frac{2}{3}\left[\frac{(1.0 \text{ V})}{-(-4.0 \text{ V})}\right]^{3/2}\right]$$

$$i_D = 1.21 \text{ (mA)}$$

Note that we have used $+1.0$ V for v_{SG} which is equal to $-v_{GS}$.

(*b*) To calculate drain current at pinch-off, we need the drain-to-source voltage for the given gate-to-source voltage. This is calculated from Eq. 10.16 as

$$V_{DS} = -V_P - V_{SG} = -V_P + V_{GS}$$

$$V_{DS} = -(-4.0 \text{ V}) + (-1.0 \text{ V}) = 3.0 \text{ V}$$

Then I_{DD} from Eq. 10.17 is

$$I_{DD} = -(2.0\text{mA V}^{-1}(-4.0 \text{ V})\left\{\left[\frac{3.0 \text{ V}}{-(-4.0 \text{ V})}\right] - \frac{2}{3} + \frac{2}{3}\left[\frac{-1.0 \text{ V}}{-(-4.0 \text{ V})}\right]^{3/2}\right\}$$

$$I_{DD} = 1.33 \text{ mA}$$

(*c*) I_{DSS} can be calculated from Eq. 10.17 by setting $V_{GS} = 0$ V. The V_{DS} is then equal to $+4.0$ V. This gives

$$I_{DSS} = -(2.0 \text{ mA V}^{-1})(-4.0 \text{ } V)\left\{\left[\frac{4.0 \text{ V}}{-(-4.0 \text{ V})}\right] - \frac{2}{3}\right\}$$

$$I_{DSS} = 2.67 \text{ mA}$$

Beyond pinch-off, the drain current can approximately be assumed to be independent of the drain-to-source voltage. The drain current in the pinch-off region can simply be written as

$$i_D = I_{DSS}\left(1 - \frac{v_{GS}}{V_P}\right)^2 \tag{10.18}$$

Figure 10.10 shows the variation of this drain current as a function of gate-to-source voltage for an n-channel JEFT.

Actually, as we have explained before, the drain current does not stay constant beyond pinch-off and increases slightly with increasing drain current. This modifies the $i_D - v_{DS}$ characteristics of an FET as shown in Fig. 10.11.

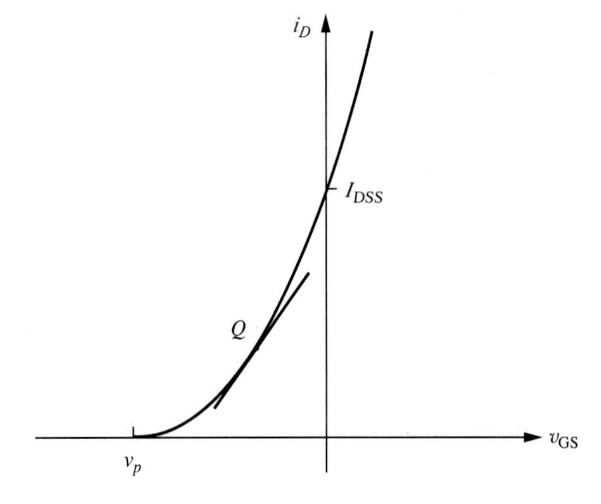

FIGURE 10.10
Drain-current transfer characteristics. The Slope at the operating point Q gives the transconductance of the transistor.

For a finite slope in the $i_D - v_{DS}$ characteristics, Eq. 10.18 can be modified as

$$i_D = I_{DSS}\left(1 - \frac{v_{GS}}{V_P}\right)^2\left(1 + \frac{v_{DS}}{V_A}\right) \tag{10.19}$$

where V_A is a negative voltage where practically all the slopes of the various v_{GS} curves intersect as shown in Fig. 10.11. V_A is typically greater than 100 V for many JFETs used in integrated circuits. Although the ratio v_{DS}/V_A is very small, it still modifies the transfer characteristics of an ideal JFET.

In an amplifier circuit, a field-effect transistor is usually operated in its pinch-off region where the current increase versus v_{DS} is almost constant (Fig. 10.11). The slope of the drain current versus v_{GS} at the operating point of the transistor (point Q in Fig. 10.10) is an indication of the voltage gain that one can get from an FET. The parameter that represents this gain capability of the transistor is the *transconductance* which is defined by

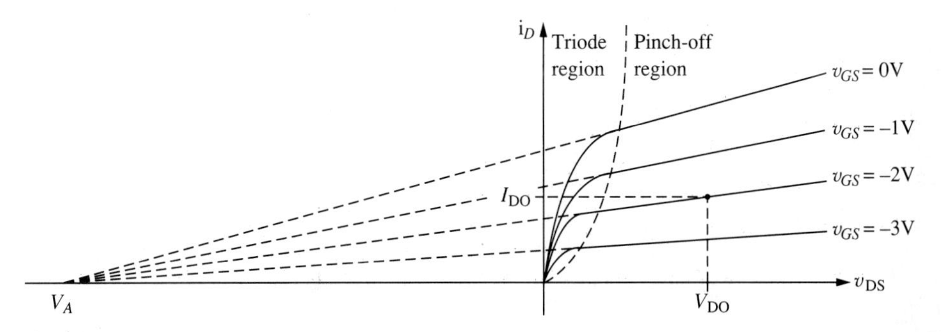

FIGURE 10.11
The drain current in an actual JFET is not constant beyond pinch-off. V_A is a voltage where all slopes of various v_{GS} curves meet.

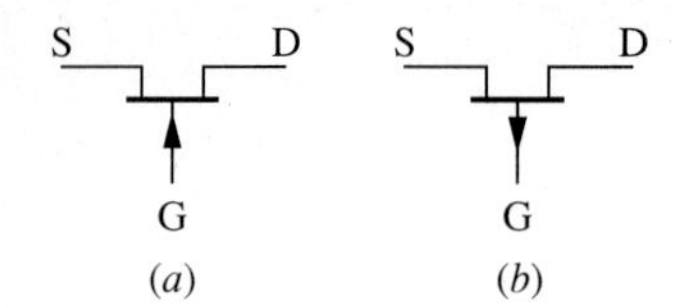

FIGURE 10.12
Schematic representation of (*a*) n-channel JFET and (*b*) p-channel JFET. Arrow shows the usual forward current direction of the p-n junction if it were biased in the forward direction.

$$g_m = \left.\frac{\partial i_{\mathrm{D}}}{\partial v_{\mathrm{GS}}}\right|_{v_{\mathrm{DS}} = \text{ Constant}} \tag{10.20}$$

The schematic representations of an n-channel and a p-channel field-effect transistor are shown in Fig. 10.12.

The small-signal low-frequency equivalent circuit of the JFET is shown in Fig. 10.13. Here the the output resistor in parallel with the current source arises because of the finite slope of the drain current with increasing drain-to-source voltage. If V_A is very large, r_0 can be approximated by

$$r_0 \approx \frac{V_A}{I_{\mathrm{DO}}} \tag{10.21}$$

Here, as shown in Fig. 10.11, I_{DO} is the drain current at the operating point. For the ideal transistor, this resistance will be infinite.

Another major difference between a bipolar transistor and the FET is that the FET has a very high input impedance compared to the bipolar transistor. This high impedance arises because of the reverse-biased gate-junction where the only current that flows is the reverse saturation current of the diodes, and this can be made very small by choosing the proper parameters in the manufacture of the transistor. The r_g in Fig. 10.13 is the input resistance. Therefore for practical purposes it can be assumed to be infinite.

The definitions of r_g and r_0 are

$$r_g = \left.\frac{\partial v_{\mathrm{GS}}}{\partial i_G}\right|_{v_{\mathrm{DS}} = \text{ Constant}} \qquad r_0 = \left.\frac{\partial v_{\mathrm{DS}}}{\partial i_D}\right|_{v_{\mathrm{GS}} = \text{ Constant}}$$

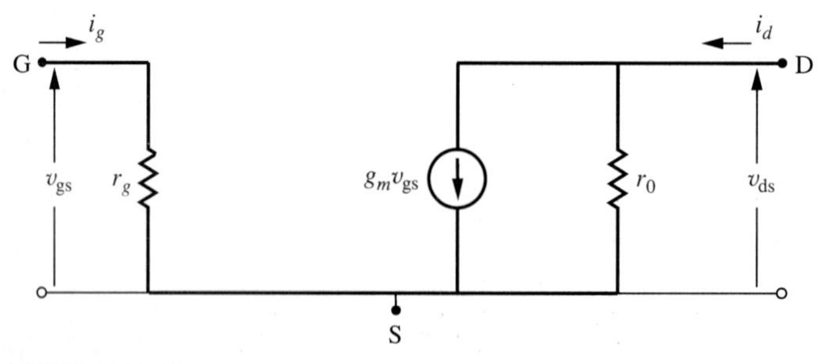

FIGURE 10.13
Small-signal low-frequency equivalent circuit of a JFET operating in the common-source configuration.

FURTHER READING

Lehovec, K., and R. Zuleeg, "Voltage-Current Characteristics of GaAs J-FETs in the Hot Electron Range," *Solid State Electr.*, vol. 13, 1970, 1415–1426.

Liechti, Charles A., "Microwave Field-effect Transistors—1976," *IEEE Trans. MTT*, vol. MTT-24, 1976, 279–300.

Pierret, Robert F., "Modular Series on Solid-State Devices," *Field Effect Devices*, vol. IV, Addison-Wesley, Reading, Mass. 1983.

Ruch, Jacques G., "Electron Dynamics in Short Channel Field-Effect Transistors," *IEEE Trans. Elect. Dev.*, vol. ED-19, 1972, 652–654.

Sedra, Adel S., and Kenneth C. Smith, *Microelectronic Circuits*, 2d ed., Holt, Reinhart and Winston, New York, 1987.

Streetman, Ben G., *Solid-State Electronic Devices*, 2d ed., Prentice Hall, 1980.

Sze, S. M., *Physics of Semiconductor Devices*, 2d ed., Wiley, New York, 1981.

Wada, Toshimi, and Jeffrey Frey, "Physical Basis of Short Channel MESFET Operation," *IEEE Trans. Elect. Dev.*, vol. ED-26, 1979, 476–489.

PROBLEMS

10.1. The following parameters of a p-channel ($N_A = 2 \times 10^{15}$ atoms/cm^3) silicon FET are given

$$\begin{aligned} N_D &= 2 \times 10^{18} \text{ donors/cm}^3 \\ L &= 50\ \mu\text{m} \\ 2h &= 5\ \mu\text{m} \\ d &= 25\ \mu\text{m} \end{aligned}$$

(*a*) Find the pinch-off voltage.

(*b*) Find the drain current for $v_{GS} = 1.2$ V, $v_{DS} = -1.5$ V.

(*c*) Find I_{DSS} when $v_{GS} = 0$ V.

(*d*) Plot the drain current versus gate-source voltage for the given transistor.

10.2. For the transistor in Problem 10.1, find the parameters of the small-signal equivalent circuit of the transistor at $v_{GS} = 1.2$ V and calculate the low-frequency voltage gain of the transistor. Assume $r_0 = \infty$.

10.3. From the discussions given in this chapter, explain how you would manufacture a transistor with a high transconductance.

10.4. For the transistor given in Problem 10.1, find the current in the triode region when $v_{GS} = +1.5$ V and $v_{DS} = -2.0$ V. Find also the current at which pinch-off occurs.

10.5. Derive the drain current of an FET in the triode region including the diode potential V_D in the derivation of the drain current.

CHAPTER 11

METAL OXIDE SEMICONDUCTOR TRANSISTORS (MOSFETs)

Another important class of transistors that finds wide application in the communication and computer industry is the metal oxide semiconductor (MOS) transistor, sometimes also referred to as the Metal Insulator Semiconductor (MIS) transistor. In addition to ease of manufacturability, they have unusual electrical characteristics. These transistors are used widely in the computer industry, especially in implementing integrated circuitry. There are basic similarities between an MOS transistor and a JFET, but there are also noticeable differences in physical and electrical characteristics. Both are field-controlled majority carrier devices. As switching elements, bipolar or FET transistors are in their saturation regions when they are in their conducting states and draw appreciable current from the external power supply. On the other hand, if MOS transistors are used in their complementary configuration, called CMOS, they draw current only while switching from on-to-off or off-to-on states, thus consuming considerably less power.

To illustrate the operation of an MOS transistor, we will introduce the many pertinent parameters unique to an MOS transistor, beginning with the properties of a metal insulator semiconductor (MIS) capacitor. If the insulator is an oxide, the capacitor is referred to as an MOS capacitor. Then we will introduce MOS transistors (MOSFET or MISFET), and explain an important offspring of the MOS capacitor, the charge-coupled device (CCD).

11.1 MIS CAPACITOR

The energy-band diagram of an isolated p-type semiconductor near the vacuum interface is shown in Fig. 11.1. The band energies of an idealized crystal structure extend uniformly all the way to the vacuum interface (Fig. 11.1.*a*). In an actual crystal, there exist additional charge states on or near the surface. These may be due to changes in the crystallographic structure, irregularities in the crystal, adsorbed gases, and/or impurities at or near the surface. The charge carriers filling these so-called *surface states* shift the energy levels up or down relative to the equilibrium bulk material Fermi level, depending on the sign of these charges. This phenomenon is called *band bending*, and the extent of this band bending is identified by a surface potential v_{ss}. The magnitude of v_{ss} depends on the density of the surface states. Figures 11.1*b* and *c* show the potential variation near the surface of the semiconductor when the surface states trap negative or positive surface charges for a p-type semiconductor.* The energy difference between the actual and intrinsic Fermi levels of the bulk semiconductor is identified as ev_b, that is, $ev_b = E_I - E_F$. The magnitude of v_b relative to the effective surface potential determines the resulting electrical characteristic of the MIS capacitor when external voltages are applied.

An MIS capacitor is made by depositing a thin insulating layer (SiO_2 or Si_3N_4) of thickness d_{ox} over a p-type semiconductor (generally silicon), as shown in Fig. 11.2. Metal electrodes are then deposited over the oxide as well as onto

* Negative charge implies that the equilibrium Fermi energy will be closer to the conduction band E_c. Thus, if the interface or surface charge density is negative, the band edges near the interface will bend downward to bring the conduction band closer to the Fermi level, which under equilibrium conditions should be the same throughout. For positive charges, the bands bend upward, that is, the valence band near the interface moves closer to the Fermi level.

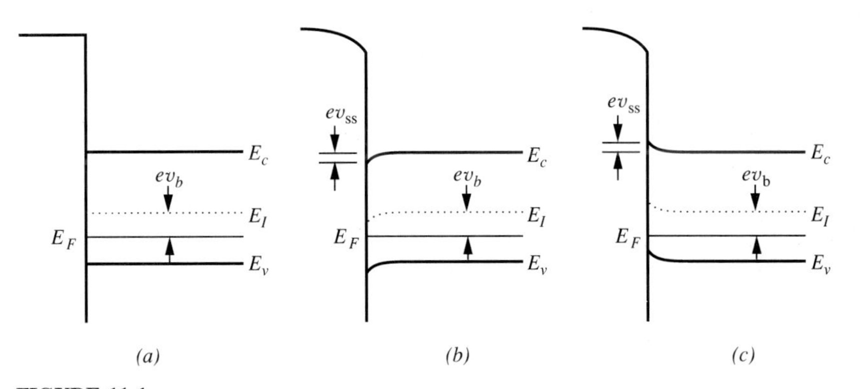

FIGURE 11.1
Energy-band diagram of an isolated p-type semiconductor. (*a*) Ideal case. Presence of (*b*) negative and (*c*) positive surface charges generate band bending and surface potential v_{ss} forms.

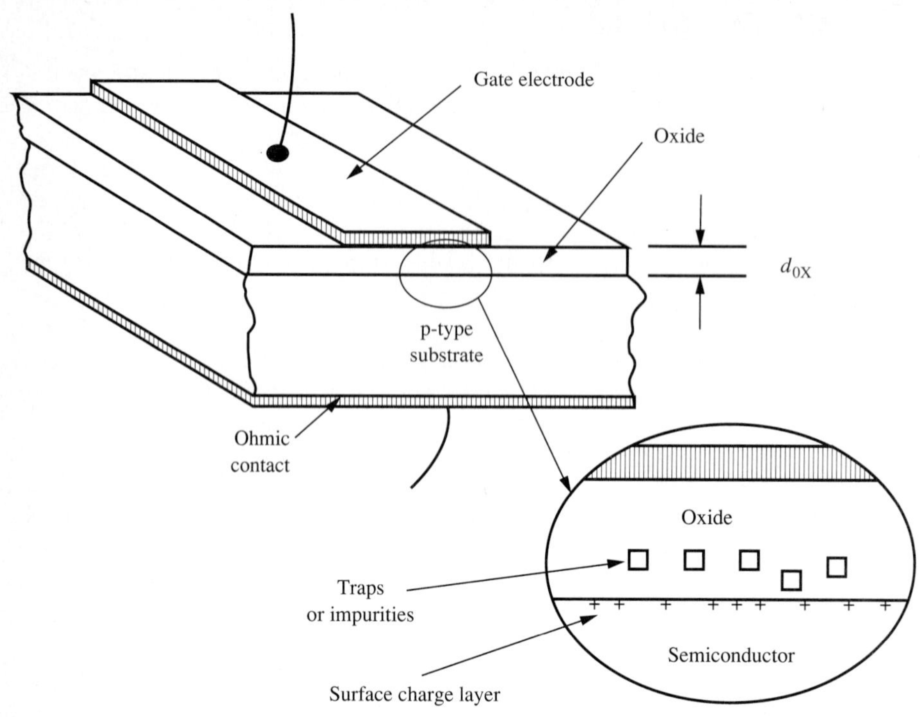

FIGURE 11.2
MIS capacitor. An insulator layer (generally SiO_2 or Si_3N_4) of thickness d_{ox} is grown over Si. The enlarged figure shows the actual interface between the oxide, which may contain impurity traps, and the semiconductor.

the bottom of the semiconductor substrate. An enlarged view of the oxide and the region near the semiconductor surface is also shown in this figure. In addition to the surface states of the semiconductor itself, an additional surface-charge layer is generated at the interface due to the lattice mismatch between the uncompensated Si atoms and the oxide. During the growth of the oxide, unintentional impurity atoms may also be deposited within the oxide. To a large extent, careful manufacturing processes can reduce the magnitude of the impurity levels and the interface-charge densities.

The ideal energy-band diagrams and the important identifying parameters for the metal, the insulator, and the p-type semiconductor are shown in Fig. 11.3*a* for the isolated elements. When the capacitor-forming process is completed (Fig. 11.3*b*), as expected, the Fermi levels of the materials align themselves to the same level. Instead of the resulting vacuum levels of the capacitor, it suffices to consider those energy levels of the metal and the semiconductor relative to the conduction band energy of the oxide. If the work functions of the metal and the semiconductor are the same, the vacuum levels, and so the conduction level of the insulator, are flat. In general, the semiconductor, whether it is a p-type or an n-type, has a work

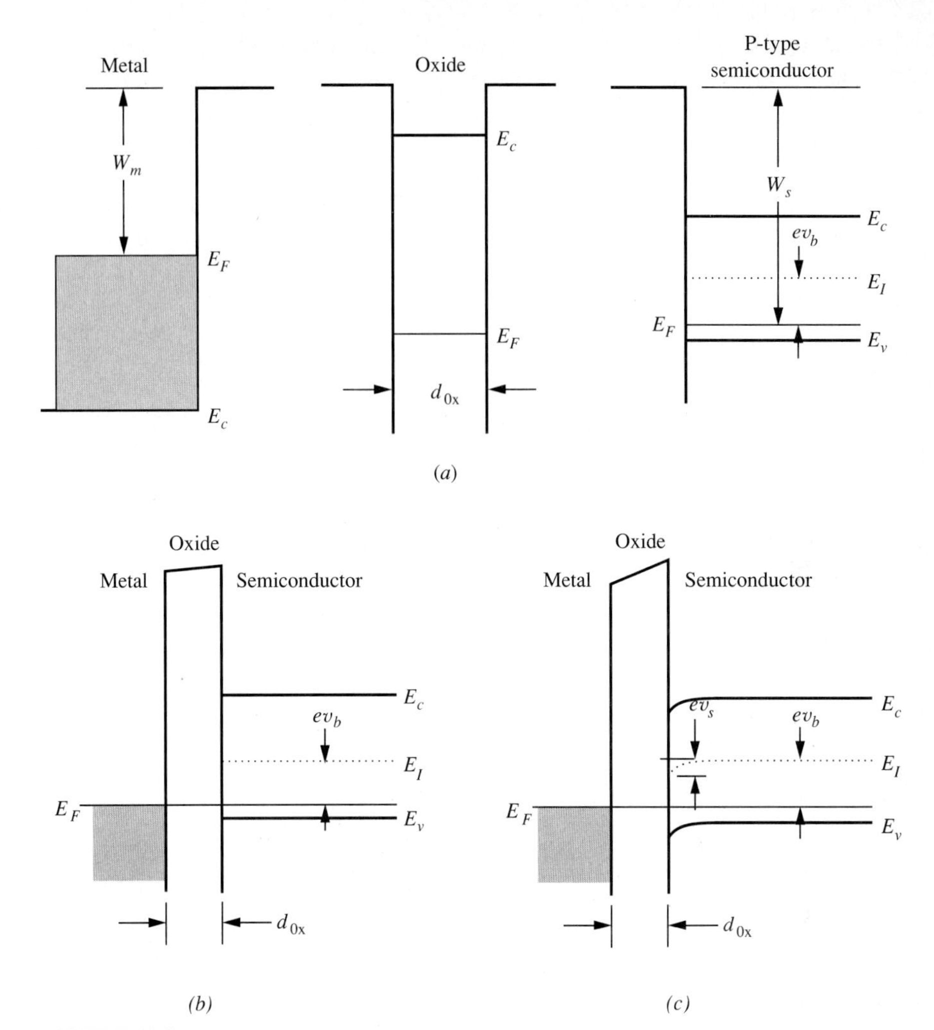

FIGURE 11.3
Energy-band diagram of an MIS capacitor. (*a*) Energies associated with the ideal elements. The resulting band diagram (*b*) for the ideal capacitor and (*c*) when oxide traps store positive charge. No surface potential v_s is formed for the ideal case.

function higher than that of the metal (usually Al), and the oxide conduction band slopes upward toward the semiconductor (Fig. 11.3*b*).

A potential barrier between the metal and the semiconductor is formed for the electrons in the metal and the holes in the semiconductor. This barrier height is determined by the energy difference between the conduction band and the Fermi level of the insulator. The barrier width is equal to the oxide thickness d_{ox} which is large enough to prevent any tunneling of charge carriers across the barrier. In an actual MIS capacitor, the presence of the trapped surface and the oxide charges produces an interface energy band bending, and thus produces the equivalent interface (surface) potential v_S. This is shown in Fig. 11.3*c*. Note that v_s is different than the v_{ss} that was shown in Fig. 11.1, since v_s now includes the effects of charges trapped in the oxide and possible crystallographic mismatches between the insulator and the semiconductor.

Compared to an ordinary capacitor, the MIS capacitor shows unusual electrical characteristics for different polarities of the applied voltage. In the discussions that will follow, a p-type semiconductor substrate will be used to explain the resulting electrical characteristics of the capacitor. Also, an ideal MIS capacitor will be considered. Extension to a nonideal case and also to an n-type semiconductor substrate can easily be made by including the effects of the surface potential and by interchanging the roles of holes and electrons. The electrode over the oxide layer is called the *gate*, and the externally applied voltages are referenced with respect to the substrate electrode which has an ohmic contact to the substrate.

As long as applied gate voltages are smaller than the dielectric breakdown voltage of the insulator, they are never capable of supplying enough energy to the electrons in the metal or the holes in the semiconductor to overcome the oxide barrier. Thus, whatever the polarity and magnitude of the applied gate voltage is, there will be no steady-state current flow through the capacitor. Therefore, with the exception of the initial transient currents as a result of charge alignment to balance the applied fields, all carriers in the semiconductor are in thermal equilibrium. The relation $np = n_i^2$ is satisfied everywhere. Also, the Fermi level in the semiconductor extends all the way up to the interface. Thus, if there are any shifts in energy, all the interface energy levels will shift relative to this Fermi level.

If a negative potential is applied to the gate electrode, we know from simple Coulomb theory that the gate electrode will be negatively charged. Consequently, the lower plate in contact with the semiconductor will be positively charged. The negative charge at the gate will attract the mobile holes in the bulk semiconductor toward the semiconductor-oxide interface. Since the semiconductor is a p-type material to start with, where holes are majority carriers, there will now be more holes than the bulk-hole density p_0 on the semiconductor near the interface. The hole density is said to be *enhanced* or an *accumulation* of holes takes place at the surface within a few hundred Angstroms.

Figure 11.4*a* shows the capacitor in the accumulation mode. More and more holes are attracted toward the interface, and the electric field lines, as a result of the applied gate voltage, start at the accumulated holes and end up at the negatively charged gate electrode. The interface behaves as if the reference contact from the bottom of the semiconductor is shifted to the interface. Thus, the overall capacitance of the MIS structure behaves as a parallel plate capacitor whose plates are separated by the oxide thickness d_{ox}.

The energy-band diagram for the accumulation mode of an MIS capacitor is given in Fig. 11.4*b*. Note that the applied gate potential v_G (equivalent to a potential energy $\Phi_G = ev_G$) shifts the energy levels of the gate side of the capacitor upward, relative to the substrate side. As the hole density at the interface

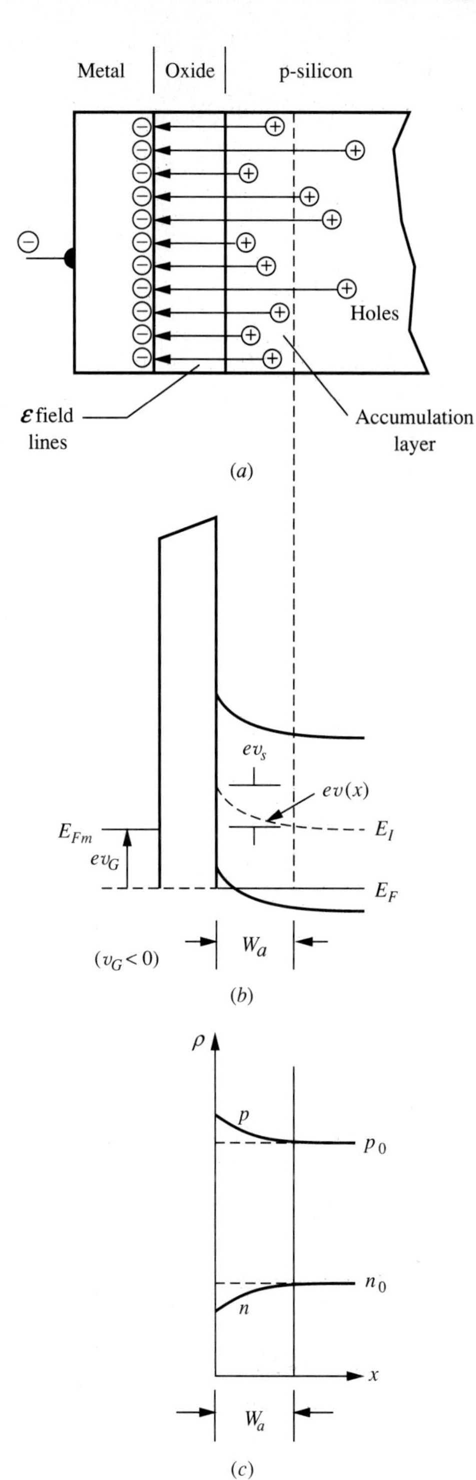

FIGURE 11.4
Accumulation mode of the MIS capacitor. (*a*) Negative potential on the gate attracts the holes toward the interface. Hole enhancement (accumulation) takes place. More and more electric fields end up at the interface. (*b*) Energy-band diagram. The bands bend up. The valence band at the interface gets closer to the Fermi level. (*c*) Charge distribution in the vicinity of the interface. Hole density increases above and electron density decreases below their respective equilibrium values.

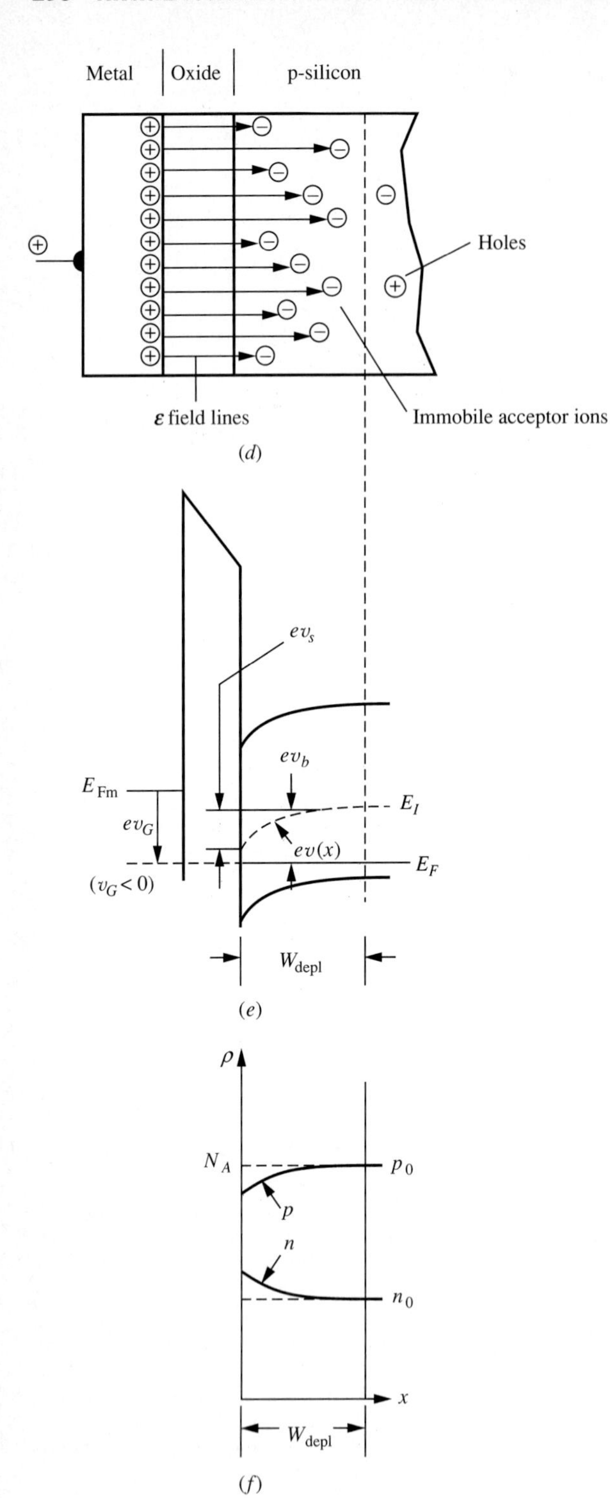

FIGURE 11.4 (*continued*)
Depletion mode of the MIS capacitor. (*d*) Holes are pushed away from the interface. Depletion layer forms. Electric field lines end up on immobile acceptor ions. (*e*) Energy-band diagram. As the hole density decreases, electron density increases at the interface. The valence band gets farther away from the Fermi level near the interface. (*f*) Charge distribution away from the interface. Hole density decreases below and electron density increases above their respective equilibrium values.

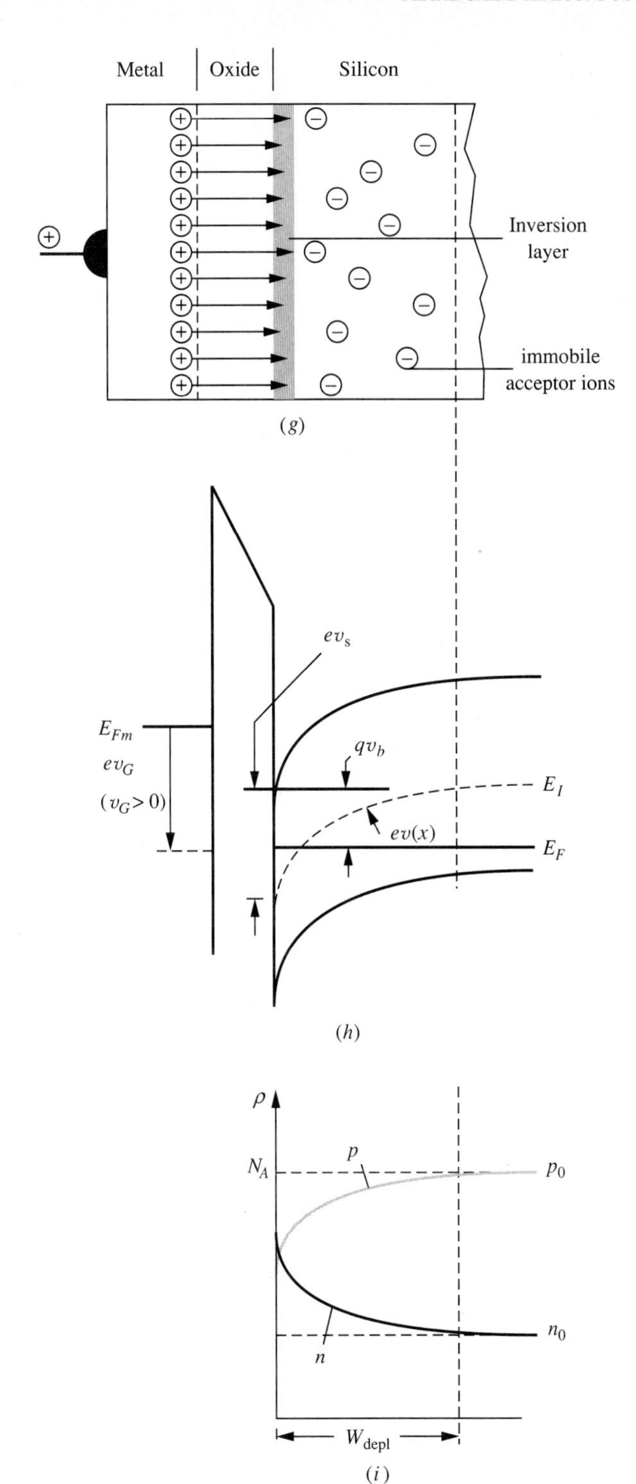

FIGURE 11.4 (*continued*) Inversion mode of the MIS capacitor. (*g*) As the gate voltage is made more positive, more holes are pushed away from the interface. Electron density increases at the interface. (*h*) Energy-band diagram for $v_b < v_s < 2v_b$. The intrinsic Fermi level drops below the actual Fermi level. The interface becomes n-type. Mobile electron density is still very low. (*i*) The electron density distribution for $v_b < v_s < 2v_b$. At the depletion layer, the negative charge is still due to the immobile acceptor ions.

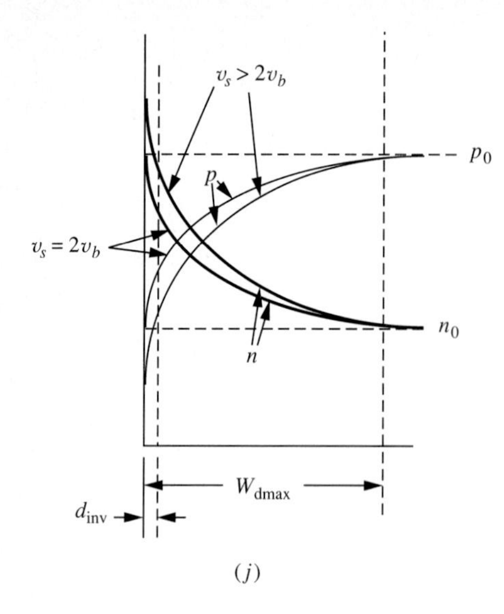

FIGURE 11.4 (*continued*)
(*j*) Electron density for $v_s = 2v_b$ and $v_s > 2v_b$. For the latter case, deep (strong) inversion takes place since now the electron density at the interface is greater than the bulk equilibrium hole density.

increases, the interface behaves more as a p^+-type semiconductor. Thus, the valence band moves upward and gets closer to the Fermi level. The resulting charge distribution near the interface is shown in Fig. 11.4*c*. At the interface, the hole density exceeds the bulk-hole density p_0, and the electron density falls below n_0.

If the polarity of the potential is now reversed and the gate electrode is made positive, the positively charged gate electrode will push away the holes near the interface in the semiconductor. The hole density near the interface will be reduced, and the interface will be *depleted* of mobile hole carriers. The holes are still the majority carriers in this region but their concentration near the interface is reduced from their bulk value p_0, where p_0 is practically equal to the acceptor impurity atom density N_A at room temperature. As a result of the attractive potential at the gate electrode, electrons are attracted toward the interface and thus their number density is increased accordingly near the interface. But the density of the electrons is still very low relative to p_0. The reduction of holes at the interface is compensated by the immobile acceptor ions near the interface. Thus, most of the electric field lines end up close to the interface on the negatively charged immobile acceptor ions, as shown in Fig. 11.4*d*. Since the interface is composed mostly of the immobile ions, this mode of operation is known as the *depletion* mode.

The equivalent energy-band diagram for the depletion mode is shown in Fig. 11.4*e*. A depletion layer of thickness W_{depl}, extending into the semiconductor, is generated. The depth of the depletion layer depends on the applied gate voltage and on the division of this voltage between the oxide and the semiconductor. Since beyond the depletion region the bulk of the semiconductor is neutral, the voltage variation in the semiconductor will be determined mostly by the charge distribution near the interface. Thus, the resulting voltage variation through the depletion layer follows the curve $v(x)$, which reaches a maximum value of v_s at the interface. Since, in this case, there is a reduction of holes near the interface, the valence band moves away from the E_F, that is, the interface behaves as a p^--type semiconductor. The depletion mode is maintained as long as the surface potential is less than v_b. The corresponding hole and electron density variations are shown in Fig. 11.4*f*. Near the interface the immobile ion density, which has a value of N_A, is still dominant.

If the applied positive gate voltage is increased further, a new phenomenon occurs at the interface, as shown in Fig. 11.4*g*. As more and more holes are pushed away from the interface, the band bending increases further. Consequently, as more electrons are attracted toward the interface, the valence-band edge moves away (less and less p-type), and at the same time the conduction-band edge moves closer to the Fermi level. When v_s becomes greater than v_b, as shown in Fig. 11.4*h*, E_I shifts below the Fermi level. Now the conduction band lies closer to E_F and makes the interface take on properties of an n-type semiconductor. In this way, the interface changes its electrical characteristics. If there are no externally injected electrons, the electrons that appear at the interface are still very few compared to the acceptor ion density. Almost all of the electric field lines still end up at the negative immobile ions. Thus, the depletion layer, made up of the immobile charges, still prevails but extends further into the semiconductor because of the higher gate voltages. This condition is maintained as long as v_s lies between $v_b < v_s < 2v_b$. Although the electrons now become more numerous than the holes at the interface, and the surface electrical characteristic is thus *inverted*, the interface still maintains the general properties of the depletion mode of operation.

As v_s gets closer to $2v_b$, the interface becomes more and more n-type, and the depletion layer extends further into the semiconductor. If the positive gate voltage is further increased so that $v_s = 2v_b$, the electron density at the interface becomes equal to that of the bulk-hole density. Thus, the sign of the majority charge carriers at the interface is reversed and the *strong* (or *deep*) *inversion* of the charge takes place. The electrons that were minority carriers become the majority carriers at the interface. At this point, the depletion layer also reaches its maximum value W_{dmax}, as shown in Fig. 11.4*j*. If the gate voltage is further increased, electrons at the interface start to intercept the electric field lines and prevent further increase of the depletion layer. Thus, the depletion layer remains practically constant at W_{dmax}, with further increase in the gate potential. The immobile acceptor atoms still dominate the depletion region and the electrical characteristics of this region, excluding the inversion layer, are still determined by these immobile acceptor ions.

The behavior of the MIS capacitor can be verified by terminal capacitance measurements. The equivalent capacitance of the system is made up of two capacitors in series: one due to the oxide layer and the other due to the capacitance between the interface and the bulk semiconductor. The latter capacitance is referred to as the *depletion-layer capacitance*. The capacitance of the oxide layer C_{ox} is the same as that of a parallel plate capacitor with a dielectric layer that has the same dielectric constant as the oxide layer. For a given potential barrier, the capacitance of the depletion layer is calculated from the regular depletion-layer formula, provided that the surface layer potential maximum v_s at the interface is known. The total capacitance is then

$$\frac{1}{C_t} = \frac{1}{C_{\text{ox}}} + \frac{1}{C_{\text{depl}}} \tag{11.1}$$

The external capacitance measurement shows the unique features of the MIS capacitor remarkably well. Two types of measurements are considered for the ideal capacitor, as shown in Fig. 11.5. If the measurements are made at low frequencies, the capacitance variation of the MIS capacitor as a function of the externally applied gate voltage follows the curve shown in Fig. 11.5*a*. If rapidly varying sinusoidal voltages are applied, the resulting *C*-v measurements follow the curve shown in Fig. 11.5*b*.

Whatever the mode of the MIS capacitor, holes and electrons are still referred to respectively as the majority and minority carriers for the p-type semiconductor. Except for the transient rise or fall times associated with capacitors, capacitance measurements are generally based on a small amplitude ac signal superimposed on a dc bias voltage. The incremental charge variations in response to an ac signal allow the measurement of the equivalent capacitance of the capacitor. As long as the frequency of the ac signal is less than $1/\tau_D$, where τ_D is the dielectric relaxation time, the majority carriers (holes) can easily follow the variations in the applied signal. On the other hand, the minority carriers (electrons) can follow ac variations only provided the frequency of the applied signal is below $1/\tau_{\text{inv}}$, where τ_{inv} is the response time of the inverted electrons to a disturbance. Thus, the capacitance measurements are expected to be frequency dependent. τ_{inv} depends on the detailed analysis of the minority carrier motion through the depletion layer. The deep impurity traps associated within the band gap of the semiconductor determine the response time of the minority carriers within this layer.

As the gate voltage is made negative, the overall capacitance increases very little and eventually settles at a constant value close to that of C_{ox} for the ideal capacitor. The capacitance at $v = 0$ V corresponds to where v_s is zero. This is

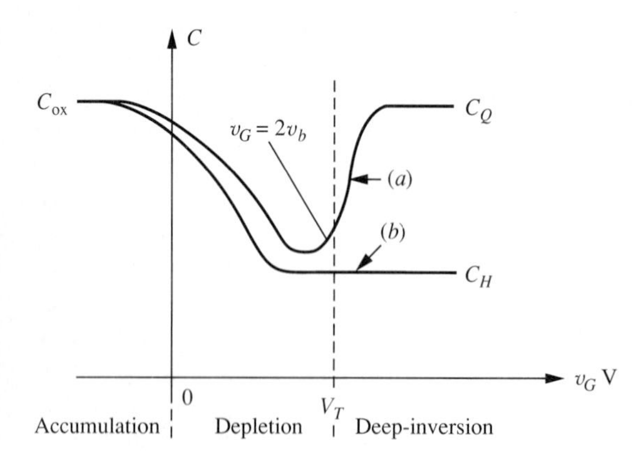

FIGURE 11.5
Capacitance of an ideal MOS capacitor as a function of the applied gate voltage. (*a*) Low-frequency and (*b*) high-frequency measurements (p-type substrate).

usually referred to as the *flat-band* case. If positive voltages are applied to the gate, the depletion layer near the surface of the semiconductor begins to form and introduces the additional capacitance C_{depl}. Since the value of the series combination of the two capacitors is smaller than the smaller of the two capacitances, the measured capacitance decreases. It reaches a minimum value when the depletion capacitance reaches a minimum value.

At frequencies lower than $1/\tau_{\text{inv}}$, the minority carriers at the inversion layer follow the ac signal variations. As the inverted number density of the electrons increases at the interface with increasing gate voltage, more and more of the inverted electrons mask the incremental charge variations associated with the depletion layer. This behaves as if the contact at the semiconductor substrate has shifted to the interface. Eventually, the overall measured equivalent capacitance approaches that of the capacitance of the oxide, as shown in Fig. 11.5*a*.

At high frequencies, that is, $f > 1/\tau_{\text{inv}}$, the inverted electrons cannot follow the rapid variations in the applied ac signal. The inverted layer behaves as if it is transparent to the ac signal. The capacitance thus remains at the minimum value, which is the series combination of the oxide and the minimum depletion-layer capacitors. Since the depletion layer stays constant with increasing inverted charge, the equivalent capacitor remains constant for increasing gate voltage as shown in Fig. 11.5*b*.

For nonideal MIS capacitors, there may be band bending at the interface and thus, to start with, a finite equivalent surface potential v_s is already set up at the interface even with no applied gate potential. Depending on the sign of the surface-charge layer, either a positive or negative external potential can produce a flat-band case corresponding to $v_s = 0$ V. Figure 11.6 shows the effect of the oxide charges on the *C*-v characteristics of the MIS capacitor. In Fig. 11.6I, the case of trapped positive oxide charges ($Q_0 > 0$) is shown. Because of the presence of positive charge at the oxide close to the interface, holes are pushed away from the interface and the semiconductor surface becomes slightly p^--type, as in Fig. 11.6I*a*. This results in a downward band bending near the interface so that the valence band moves away from E_F (Fig. 11.6I*b*). In order to flatten the bands, the E_F of the metal should be raised (or that of the semiconductor should be lowered). This requires a negative potential at the gate (Fig. 11.6I*c*). This can also be visualized in terms of the electric field lines. Because of the presence of the positive charge at the oxide layer, some of the electric field lines that are supposed to link the charges at the gate electrode and the semiconductor end up at the positive charges in the oxide layer rather than the ones in the semiconductor. More negative voltage on the gate is required to effect the holes in the semiconductor. Thus, the whole *C*-v curve shifts toward negative voltages. It requires a negative voltage to flatten the bands to reach the flat-band case. The magnitude of the shift in the voltage v_{FB} is an indication of the magnitude of the oxide charge Q_0.

Similar arguments apply to the case where the oxide layer traps negative charges. Under this case (Fig. 11.6II), the *C*-v curve shifts toward positive voltages.

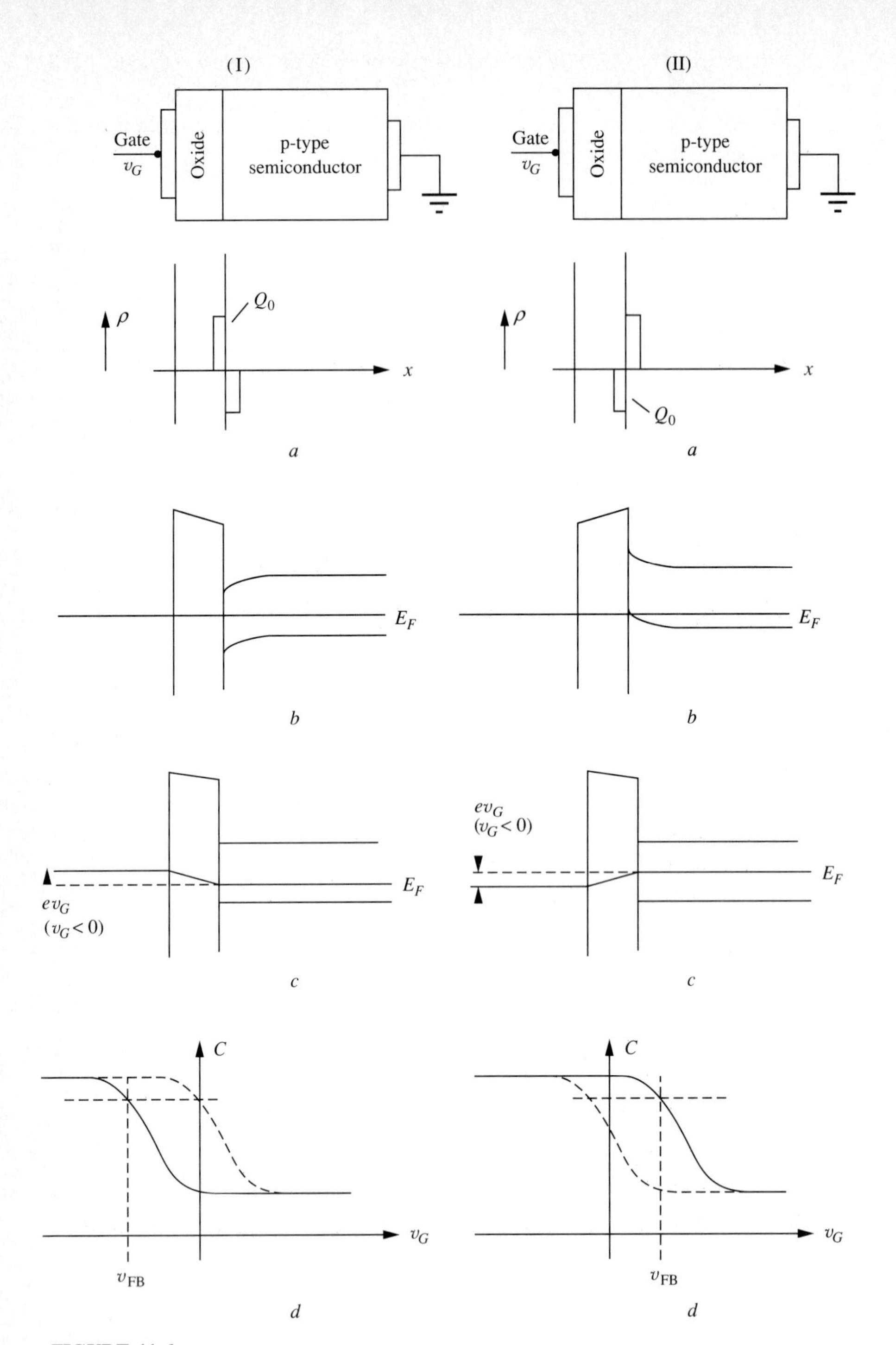

FIGURE 11.6
The changes in the C-v characteristics of an MIS capacitor with the presence of trapped charge Q_0 in the oxide layer. (I) with positive charge and (II) with negative charge. (*a*) Charge density near the interface, (*b*) equivalent energy-band diagram, (*c*) required v_G to obtain flat band, and (*d*) corresponding shift in C-v measurement relative to the ideal case (dotted line).

In addition to determining the polarity of the oxide charges, C-v measurements can also be used as an experimental technique to determine the doping profile of the impurity atoms, the interface trap densities, and trapped oxide charges.

An analysis of an ideal MIS capacitor in the depletion mode will illustrate some of the pertinent parameters of the capacitor. For this case, the corresponding charges, potential drops, and resulting capacitances are shown in Fig. 11.7.

We will assume that by the application of the gate voltage v_G, charge depletion in the semiconductor interface is produced. We will also assume that within a very short length (a few hundred angstroms) inverted negative charge is also present at the semiconductor interface. This charge may include an externally introduced charge (such as through photoelectric effect or through injection from a nearby source) near the surface of the semiconductor. The depletion layer is made up of the negatively charged immobile acceptor atoms, which will be assumed to be constant throughout the depletion layer.

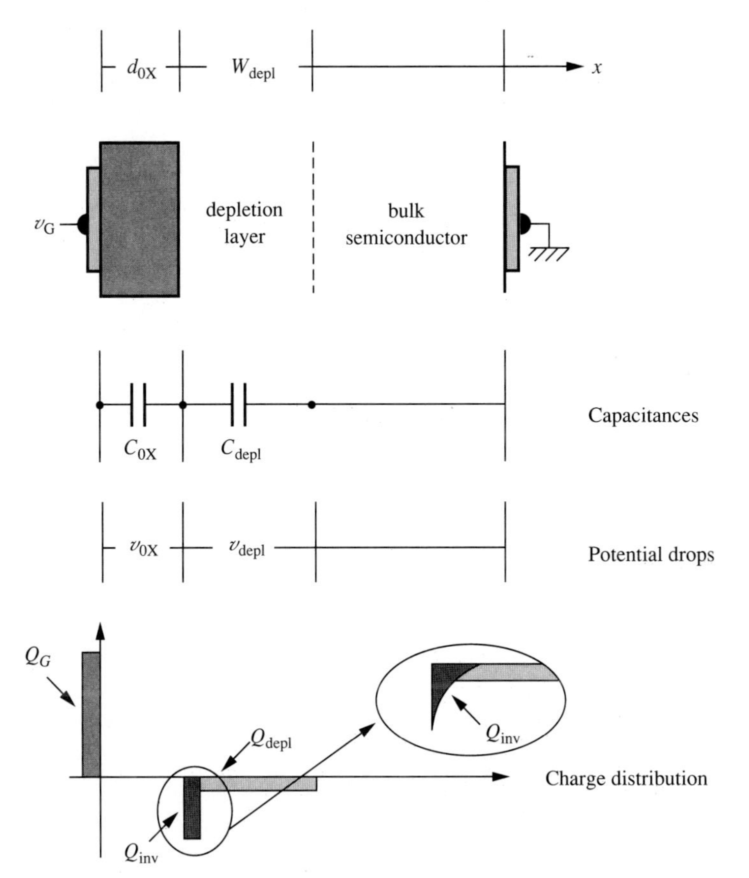

FIGURE 11.7
Capacitance, potential drops, and charge distributions in an MIS capacitor in strong depletion. Insert shows the actual inverted charge density variation at the interface.

The charge neutrality for the whole system can be written as

$$Q_G + Q_{\text{inv}} + Q_{\text{depl}} = 0 \tag{11.2}$$

where Q_G, Q_{inv}, and Q_{depl} are the corresponding charges per unit area at the gate, in the inversion layer, and in the depletion layer. Note that away from the interface the bulk semiconductor is electrically neutral and its presence is excluded from the discussion.

The applied potential is equal to the sum of the voltage drops across the corresponding capacitors.

$$v_G = v_{\text{ox}} + v_{\text{depl}} \tag{11.3}$$

The voltage across the oxide capacitor is given by

$$v_{\text{ox}} = \frac{Q_G}{C_{\text{ox}}} \tag{11.4}$$

where C_{ox} is the capacitance per unit area of the oxide layer and is given by

$$C_{\text{ox}} = \frac{\epsilon_{\text{ox}}}{d_{\text{ox}}}$$

The depletion-layer charge per unit area is simply

$$Q_{\text{depl}} = -eN_A W_{\text{depl}}$$

$$Q_{\text{depl}} = -\sqrt{2eN_A\epsilon_s v_s} \tag{11.5}$$

where for the depletion layer the usual depletion-width formula (Eq. 8.46) is used.

Combining Eq. 11.2, 11.3, 11.4, and 11.5 yields

$$v_G = \frac{Q_G}{C_{\text{ox}}} + v_s = \frac{-(Q_{\text{inv}} + Q_{\text{depl}})}{C_{\text{ox}}} + v_s$$

$$v_G = -\frac{Q_{\text{inv}}}{C_{\text{ox}}} + \frac{\sqrt{2eN_A\epsilon_s v_s}}{C_{\text{ox}}} + v_s \tag{11.6}$$

Leaving the square root term on the right, moving the other terms to the left, and squaring both sides gives

$$\left(v_G + \frac{Q_{\text{inv}}}{C_{\text{ox}}} - v_s\right)^2 = 2\frac{eN_A\epsilon_s}{C_{\text{ox}}^2}v_s = 2v_0 v_s$$

where $v_0 \equiv (eN_A\epsilon_s/C_{\text{ox}}^2)$. Expanding the squared term, and solving for v_s, we obtain

$$v_s = v_G + \frac{Q_{\text{inv}}}{C_{\text{ox}}} + v_0 - \sqrt{2\left(v_G + \frac{Q_{\text{inv}}}{C_{\text{ox}}}\right)v_0 + v_0^2} \tag{11.7}$$

In general v_0 is small and can be neglected. Thus

$$v_s \approx v_G + \frac{Q_{\text{inv}}}{C_{\text{ox}}}$$

For a given gate voltage, since Q_{inv} is negative, the surface potential decreases with increasing charge. Note also that since C_{ox} decreases with increasing oxide thickness, v_s also decreases with increasing oxide thickness. If the MOS capacitor is not ideal, v_G is replaced by $v_G - v_{\text{FB}}$, where v_{FB} is the flat-band voltage necessary to bring the initially band-bended energy levels to the flat-band level.

If, due to thermal generation or some external means of charge injection, the charge Q_{inv} in the well is allowed to increase, v_s is reduced, and the depletion region collapses. The time for thermal generation depends on the way the capacitance is prepared and can be greater than a few hundredths of a second. Thus, as long as the pulse duration of the applied voltage is short and there are no other injected charges, the interface will remain depleted during this interval.

Example 11.1. Find v_s in an ideal MOS Si-SiO_2 capacitor if $d_{\text{ox}} = 0.1\,\mu\text{m}$ and $N_A = 10^{15}\text{cm}^{-3}$ when the applied gate potential is 10 V and no inverted charge is present. $\kappa_r(SiO_2) = 3.9$.

Solution. Using the given parameters

$$C_{\text{ox}} = \frac{\epsilon_{\text{ox}}}{d_{\text{ox}}}$$
$$= \frac{3.9 \times 8.854 \times 10^{-12}(\text{Fm}^{-1})}{0.1 \times 10^{-6}\text{m}} = 3.453 \times 10^{-4}(\text{Fm}^{-2})$$

and

$$v_0 = \frac{eN_A\epsilon_s}{C_{\text{ox}}^2}$$
$$= \frac{(1.6 \times 10^{-19}\text{C})(10^{21}\text{m}^{-3})(11.8 \times 8.854 \times 10^{-12}\text{Fm}^{-1})}{3.435 \times 10^{-4}\text{Fm}^{-2}}$$
$$= 0.14 \text{ V}$$

The surface potential for $v_G = 10$ V is

$$v_s = v_G + v_0 - \sqrt{2v_Gv_0 + v_0^2}$$
$$= 10 + 0.14 - \sqrt{2 \times 10 \times 0.14 + (0.14)^2}$$
$$= 8.46 \text{ V}$$

If there is an injected surface charge density of -10^2 nC/cm^2, the use of

$$\frac{Q_{\text{inv}}}{C_{\text{ox}}} = \frac{-100 \times 10^{-9}\ \text{C/cm}^2 \times 10^4\ \text{cm}^2/\text{m}^2}{3.453 \times 10^{-4}} = -2.90\ \text{V}$$

modifies the surface potential as, using Eq. 11.7,

$$\begin{aligned} v_s &= 10 - 2.90 + 0.14 - \sqrt{2(10 - 2.90)0.14 + (0.14)^2} \\ &= 5.82\ \text{V} \end{aligned}$$

11.2 MOS FIELD-EFFECT TRANSISTOR

A metal oxide semiconductor transistor has two different versions. These are known as the (a) Depletion and (b) Enhancement MOSFETs.

11.2.1 Depletion-Type MOSFET

We will start with the transistor configuration shown in Fig. 11.8*a*. Two n^+ contact regions (heavily doped n-regions) are diffused into a p-type semiconductor substrate. A thin layer of donor atoms is then diffused between the n^+-regions to form an n-layer. Next, an oxide layer (or any other suitable insulator) is grown over the n-layer, and electrical contacts are deposited over the oxide and the n^+-regions. These electrodes are labeled as the *source*, the *gate*, and the *drain*. The source and the drain can be interchanged if the device is symmetrically manufactured. The thin n-layer is known as the *channel* and for an n-doped layer, it is called an *n-channel*. Note that an electrical contact is also made to the bottom of the substrate, which is generally connected to the source electrode of the transistor. All the voltages are applied relative to the source.

The electrical symbol for the n-channel MOS transistor is shown in Fig. 11.9*a*. The arrow shows the usual forward current direction of the p-n junction formed between the channel and the substrate. The arrow pointing into the channel signifies an n-channel device.

As shown in Fig. 11.8*b*, because of the opposite doping levels between various regions of the transistor, p-n junctions are formed across these respective regions with varying degrees of depletion-layer widths. As a result of the applied drain-source voltage v_{DS}, the combination of the applied gate-source voltage v_{GS} and the increasing voltage drop along the channel toward the drain end modifies and changes proportionately these depletion-layer widths. Thus, similar to a junction field effect transistor, v_{DS} and v_{GS} consequently control and reduce the channel cross section toward the drain end of the transistor. The depletion layers associated with the channel are instrumental in determining the electrical characteristic of the transistor.

The current-voltage characteristic of an MOS transistor for the present configuration is shown in Fig. 11.9*b*. Note that practically no current flows through

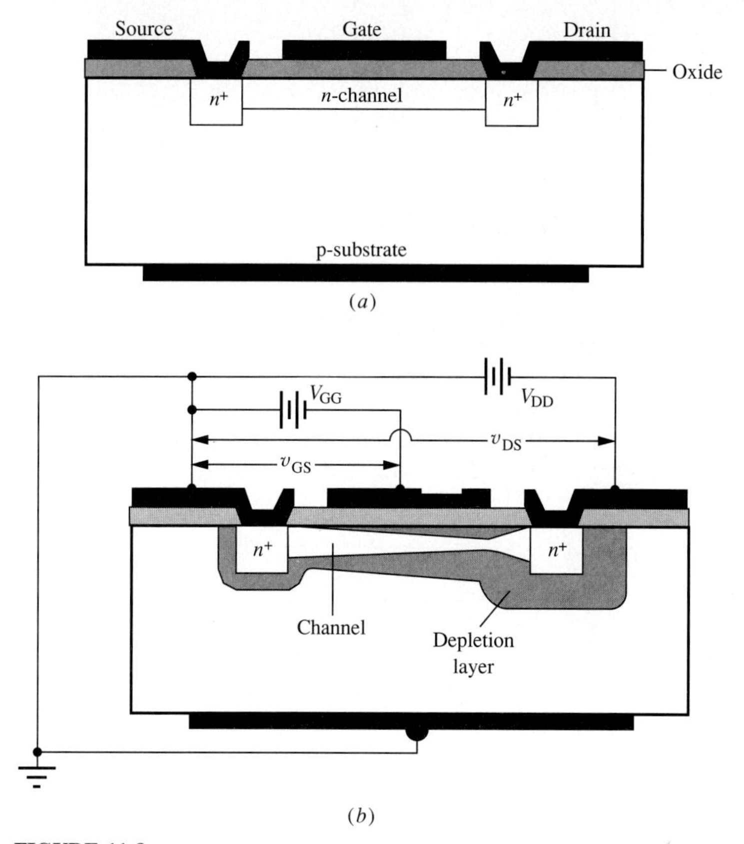

FIGURE 11.8
(*a*) Constructional details of a depletion-mode MOS transistor. (*b*) Relative depletion layers under operating conditions. For generality, the total terminal voltages are shown to be different from the battery voltages.

the substrate since the source-substrate junction is reverse biased for the positive values of v_{DS}.

For the case of $v_{GS} = 0.0$ V, the current initially increases linearly with v_{DS} because of the almost uniform channel cross section. As v_{DS} is increased further, two depletion layers control the flow of electrons from the source to the drain: the usual reverse-biased depletion layer between the n-channel and the substrate, and the depletion layer generated in the channel at the oxide-semiconductor interface. As a result of the increasing voltage drop toward the drain end of the channel, the increasing reverse bias between the channel and the substrate increases the corresponding channel-substrate depletion-layer width in the channel toward the drain end. As with the depletion of charge in an MIS capacitor, the same voltage drop creates a depletion layer between the gate oxide and the channel, whose width also increases toward the drain. The overall effect of the changes in these

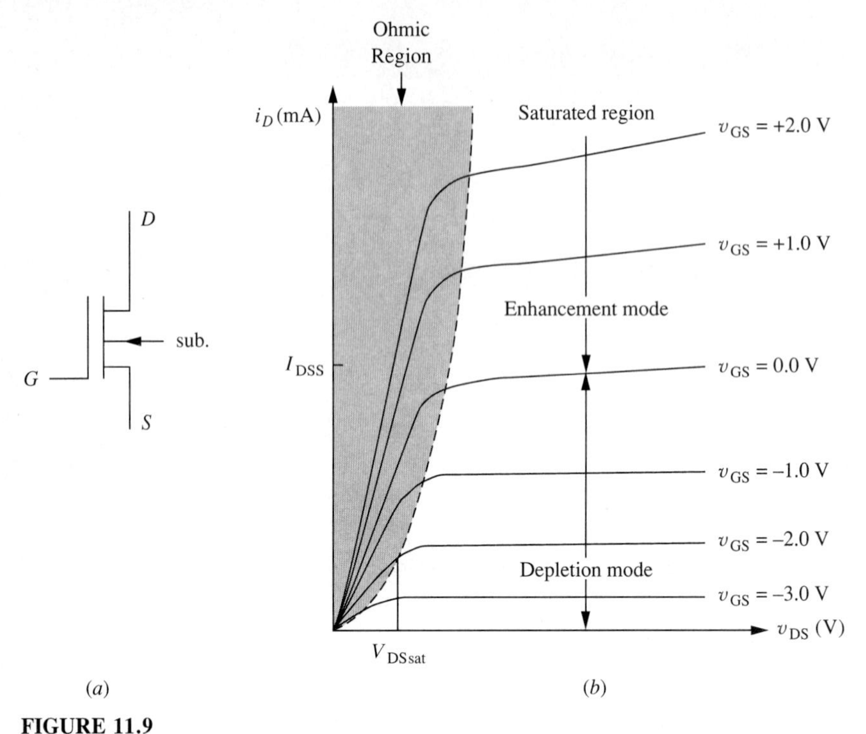

FIGURE 11.9
(*a*) Electrical symbol and (*b*) *i*-*v* characteristics of an n-channel depletion-mode MOS transistor. $V_P = -4.0$ V. A typical V_{DSsat} is shown for $v_{GS} = -2.0$ V.

depletion layers is a reduction in the channel cross section toward the drain end of the transistor. Thus, with increasing drain-to-source voltage, the increase in drain current starts to deviate from the usual linear voltage dependence on v_{DS}. As v_{DS} is increased further, both of these depletion layers extend more into the channel, and eventually the channel becomes *pinched-off* similar to the pinch-off in a junction field-effect transistor. Further increase of the v_{DS} has very little effect on the current and the transistor remains in saturation.

If now a negative potential v_{GS} is applied to the gate electrode to start with, there will be increases in the two channel depletion-layer widths in proportion to the applied v_{GS}. The transistor will conduct less current, a lower v_{DS} will be required to bring the channel to the pinch-off condition, and consequently it will reach saturation at a lower value of v_{DS}. If, to start with, v_{GS} is large enough so that it can produce channel pinch-off by itself, the channel will remain closed and no current will flow through the channel irrespective of v_{DS}. The transistor will be at cutoff.

For a given source-to-gate voltage v_{GS}, the pinch-off voltage V_P is related to the saturation drain-to-source voltage V_{DSsat} when

$$V_P = -(V_{DSsat} + v_{SG}) \tag{11.8}$$

Example 11.2. For a transistor when $v_{GS} = -2.0$ V, find V_{DSsat} if $V_P = -4.0$ V for an n-channel depletion mode MOSFET.

Solution. From Eq. 11.8, using $v_{SG} = -v_{GS} = -(-2.0)$ V,

$$-4.0 \text{ V} = -(V_{DSsat} + 2.0 \text{ V})$$

we find

$$V_{DSsat} = 2.0 \text{ V}$$

This is the value of v_{DS} where the pinch-off will occur for the given gate-to-source voltage. V_{DSsat} corresponding to this example is shown in Fig. 11.9.

Another interesting property of this MOSFET device is shown in Fig. 11.9*b*. If positive voltages are applied to the gate electrode, transistor action still continues to be achieved in the device. The effect of positive v_{GS} is to enhance the electron density at the channel, thus increasing the current flow. But since the reverse bias still appears across the substrate-channel junction, it requires a higher v_{DS} to bring the transistor into pinch-off and thus to saturation. Operation with positive gate potentials is referred to as the *enhanced* mode of operation. For negative gate voltages, it is known as the *depletion* mode of operation. Although both modes of operation are possible, this type of MOS transistor configuration is generally known as the *depletion mode* transistor.

In the ohmic region, the depletion mode MOS transistor has drain current versus v_{DS} characteristics similar to that of a JFET.

A depletion mode p-channel MOS transistor can be made by replacing all n's by p's and p's by n's, as shown in Fig. 11.10*a*. The polarity of the applied voltages are also reversed. Hole flow through the channel produces the drain current of the transistor. The *i-v* characteristic of the p-channel MOS transistor and corresponding electrical symbol are also shown in Fig. 11.10.

11.2.2 Enhancement-Type MOSFET

Figures 11.11, 11.12, and 11.13 show different but most widely used versions of the MOS transistor. These are known as *enhancement-type* MOSFETs, and they are also easier to implement because of one less diffusion process during the manufacturing stage of the transistor. In this version, there is no externally diffused channel between the source and the drain. With no gate voltage, the configuration is equivalent to two back-to-back diodes. As long as $v_{GS} = 0$ V, whatever the positive value of v_{DS} is, the transistor does not conduct any current because of the reverse bias on the diode at the drain-substrate end (closed channel) provided the reverse bias on the substrate-drain diode is smaller than a punch-through voltage.

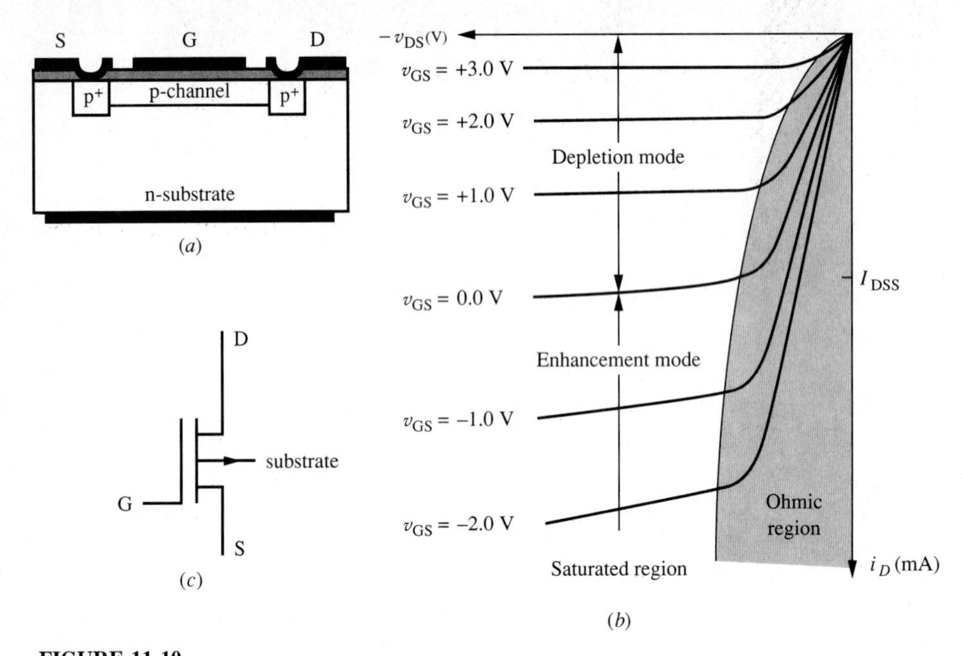

FIGURE 11.10
(*a*) Details of a p-channel depletion-mode MOS transistor, (*b*) its equivalent *i*-*v* characteristics, and (*c*) its electrical symbol.

If a positive voltage is applied to the gate, and if this voltage is greater than the threshold voltage v_T (also referred to as the *turn-on voltage*) for the inversion layer to be generated at the semiconductor interface near the oxide layer, an n-channel will be opened between the source and the drain. A current can now flow between the source and the drain terminals when a positive voltage is applied between the drain and source terminals. In Fig. 11.11*A*, the resulting depletion layer and the n-channel are shown. The channel depth is highly exaggerated in this and subsequent figures. The depth is usually within a few hundred angstroms wide.

As usual, the two terminal voltages control the flow of the electrons in this transistor. If only $v_{GS} > V_T$ is present and $v_{DS} = 0$, the potential seen by the channel, which is labeled as gate-channel potential v_{ch}, will be uniform as a result of substrate being connected to the source electrode. Thus, the channel depth will also be uniform. If now finite $v_{DS} > 0$ is applied across the source-drain electrodes, a changing channel voltage will appear between the induced channel and the substrate. The magnitude of v_{ch} will decrease toward the drain end of the transistor because of the $i_D R_{ch}$ (R_{ch} is the equivalent channel resisitance) voltage drop through the channel. At any point between the source and the drain, as long as v_{ch} is greater than the threshold voltage V_T, the channel will remain open (see Fig. 11.11*b*I). Since the semiconductor sees a decreasing potential close to its

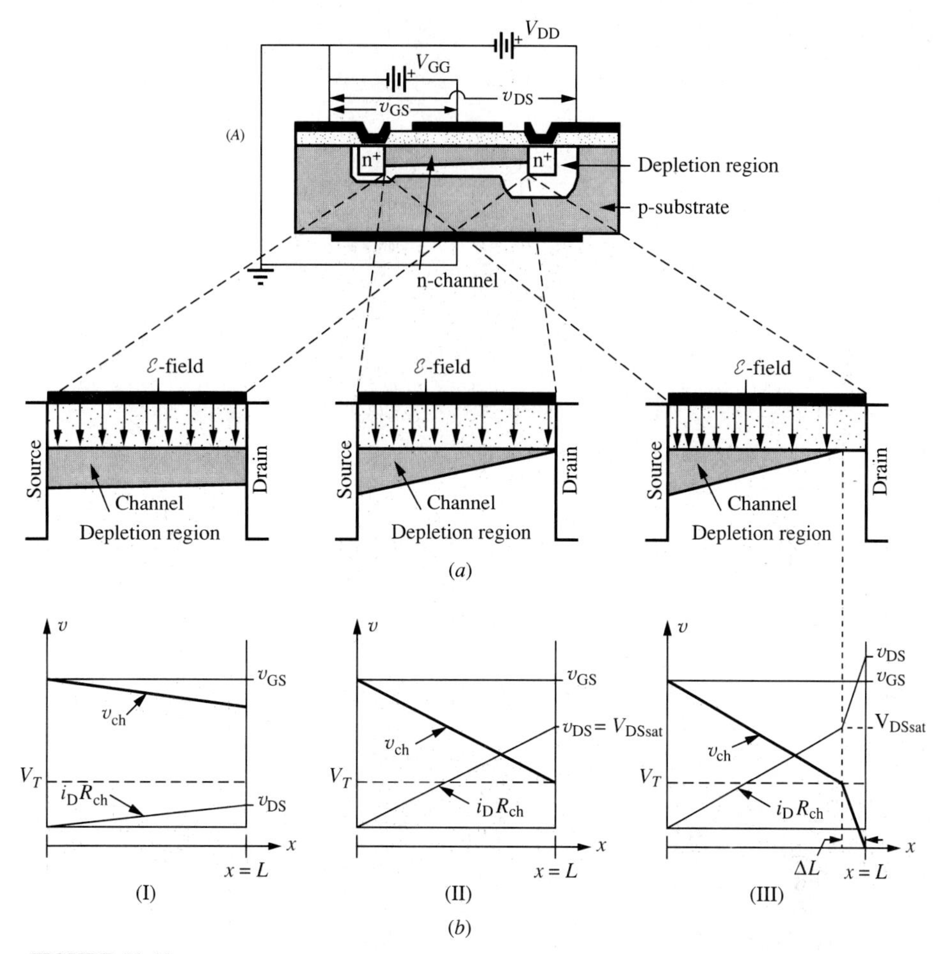

FIGURE 11.11
n-channel enhancement-mode MOSFET. (*A*) Constructional details. The channel and depletion regions are shown for small drain-source voltage $v_{DS} > V_T$. (*a*) Channel width (highly exaggerated) and electric field lines at the oxide. (*b*) Effective channel voltage, $v_{ch} = v_{GS} - i_D R_{ch}$. Characteristics for (I) small v_{DS}, (II) at pinch-off, and (III) at saturation.

surface toward the drain end, the channel depth will also proportionately decrease toward this end as shown in Fig. 11.11*a*I. Thus, for small values of v_{DS}, we do not expect the channel length to change too much from the source to the drain. This results in a linear increase in drain current as a function of v_{DS} for a given v_{GS}.

As we increase v_{DS}, we expect a larger current to flow through the channel. But because of the $i_D R_{ch}$ drop along the channel, the channel voltage $v_{ch} = v_{GS} - i_D R_{ch}$ decreases further toward the drain end of the transistor. As a result

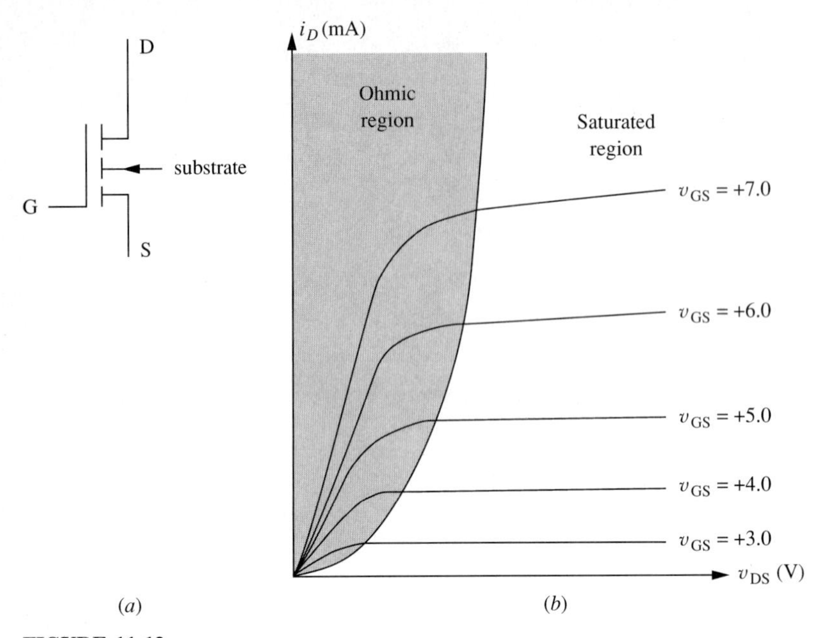

FIGURE 11.12
(*a*) Electrical symbol and (*b*) *i*-*v* characteristics of an n-channel enhancement-mode MOSFET. V_T = 2.0 V.

of this, the induced channel depth gets smaller toward the drain end. As long as $v_{ch} > V_T$, there will be an induced channel open throughout the region from source to drain. Since the equivalent induced-channel depth is decreasing, we expect a reduction in drain current for increasing v_{DS}. This decrease will continue until the channel voltage becomes equal to V_T at the drain end (see Fig. 11.11*b*II). At this point, the induced-channel depth at the drain end of the transistor will be pinched-off (see Fig. 11.11*a*II). The drain-source voltage at which the channel pinches off is labeled V_{DSsat}. Thus, for a given v_{GS}, the drain current reaches saturation. The condition for drain saturation voltage can be written as

$$V_{DSsat} = v_{GS} - V_T$$

As v_{DS} is increased beyond V_{DSsat}, the voltage near the drain end falls below the threshold voltage V_T, and no channel is induced within a length of ΔL (Fig. 11.11*a*III). For long gate lengths, ΔL is very small compared to the effective gate length L. Since a large voltage drop appears across this region (Fig. 11.11*b*III), the high electric fields keep the current flowing close to the surface of the semiconductor. Thus, the saturated constant drain current continues to flow in the transistor. If the gate length is short, the saturation current does not stay constant, but increases slightly with increasing v_{DS}.

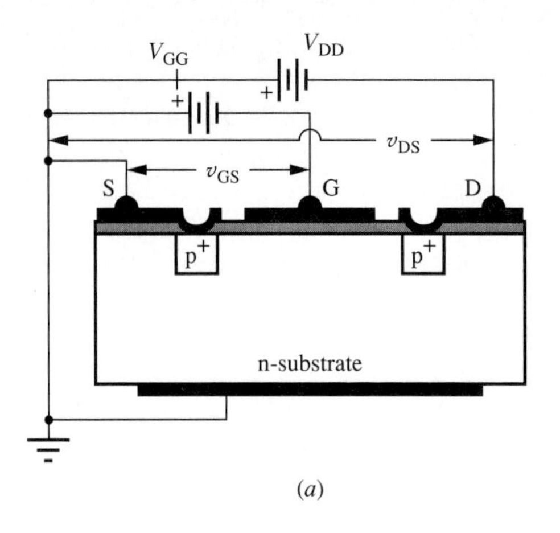

(a)

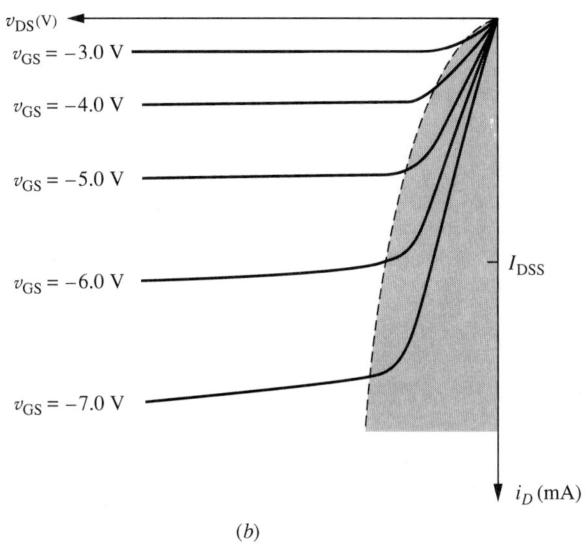

(b)

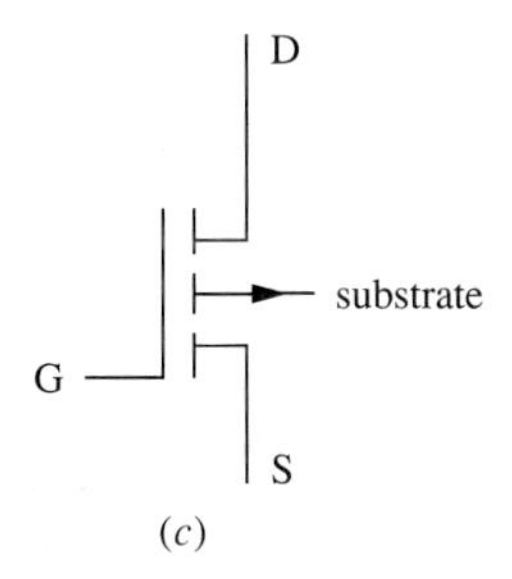

(c)

FIGURE 11.13
Enhancement-mode p-channel MOS transistor (a) $V_T = -2.0$ V, (b) its i-v characteristics, and (c) its electrical symbol.

The current-voltage characteristic and the corresponding electrical symbol of an enhancement-mode n-channel MOS transistor is shown in Fig. 11.12. In the symbol, the substrate is drawn as broken lines to indicate that the path between the source and the drain does not exist under zero-bias conditions.

An enhancement-mode p-channel MOS transistor, its i-v characteristics, and its electrical symbol are shown in Fig. 11.13.

The threshold voltage of an enhancement-mode MOS transistor can be calculated using Eq. 11.6.

$$v_G = \frac{Q_{\text{inv}} + \sqrt{2eN_A\epsilon_s v_s}}{C_{\text{ox}}} + v_s \tag{11.9}$$

The threshold voltage is determined when $v_s = 2v_b$. The corresponding gate voltage is known as the threshold voltage V_T (or the turn-on voltage) and is given by

$$V_T = 2v_b + \frac{\sqrt{2eN_A\epsilon_s(2v_b)}}{C_{\text{ox}}} - \frac{Q_{\text{inv}}}{C_{\text{ox}}} \tag{11.10}$$

Example 11.3. For an n-channel (p-type substrate) silicon enhancement-mode MOS transistor using SiO_2 as an oxide, find the threshold voltage if $N_A = 10^{17}\text{cm}^{-3}$, $d_{\text{ox}} = 0.1\mu\text{m}$, and κ_r (SiO_2) $= 3.9$. Assume $Q_{\text{inv}} \approx 0$.

Solution. For Si,

$$E_I = E_V + \frac{E_G}{2} = E_V + \frac{1.07\text{ eV}}{2} = E_V + 0.535\text{ eV}$$

From Eq. 6.4*b*, the Fermi level for the p-type Si semiconductor is ($p = N_A^- = N_A$)

$$E_F = E_v + \frac{kT}{e}\ln\left(\frac{N_v}{N_A}\right)$$

$$= E_v + 0.026\ln\left(\frac{2.5\times 10^{25}}{10^{23}}\right)$$

$$= E_v + 0.144\text{ eV}$$

The potential v_b is

$$ev_b = E_I - E_F$$

$$= (E_V + 0.535) - (E_V + 0.144)$$

$$ev_b = 0.391\text{ eV}$$

$$v_b = 0.391\text{ V}$$

Also

$$C_{\text{ox}} = \frac{\epsilon_{\text{ox}}}{d_{\text{ox}}} = 3.453\times 10^{-4}\text{F/m}^2$$

Finally

$$V_T = 2(0.391) + \frac{\sqrt{2(1.6 \times 10^{-19}\text{C})(10^{23})(11.8 \times 8.854 \times 10^{-12}\text{F/m})(2 \times 0.391)}}{3.453 \times 10^{-4}\text{F/m}^2}$$

$$V_T = 5.46 \text{ V}$$

For a nonideal capacitor, since the equivalent gate voltage is $v'_G = v_G - v_{FB}$ (v_{FB} is the flat-band voltage, voltage necessary to produce the flat-band case at the interface), the threshold voltage becomes

$$V_T = (5.46 + v_{FB}) \text{ V}$$

In an actual MOS transistor, the substrate is generally lightly doped, so that the Fermi level is initially close to the intrinsic Fermi level. The oxide charges trapped during the manufacturing stage are generally positive, thus the flat-band voltage is negative, that is, the conduction band is already bent downward toward the Fermi level. Therefore, in some cases, it may require considerably lower gate voltages (100 mV or so) to obtain threshold conditions.

In discussing the MOS capacitor, it was shown that the response time of the capacitor measurements were dependent on the frequency of the gate signal. In those measurements, τ_{inv} was the time necessary for charges to cross the depletion layer so that measurement could be made. Here, in a transistor, the charge motion is through the movement of electrons from the source to the drain. Thus, the injected charges do not have the same frequency limitations. The frequency responses of MOSFETs are indeed very high and are used in many high-speed digital applications. Their frequency response is related to the transit time of charge carriers through the channel. This will be discussed in detail in Chapter 12.

The drain current in the saturation region of an n-channel depletion-mode MOS transistor can be written as

$$i_D = I_{DSS}\left(1 - \frac{v_{GS}}{V_P}\right)^2$$

$$i_D = \frac{I_{DSS}}{V_P^2}(V_P - v_{GS})^2 \qquad v_{GS} > V_P \tag{11.11a}$$

where I_{DSS} is the saturation current when $v_{GS} = 0$ V and V_P is the pinch-off voltage.

The drain current for the enhancement-mode MOSFET can also be written as

$$i_D = kV_T\left(\frac{v_{GS}}{V_T} - 1\right)^2$$

$$i_D = k(v_{GS} - V_T)^2 \qquad v_{GS} > V_T \tag{11.11b}$$

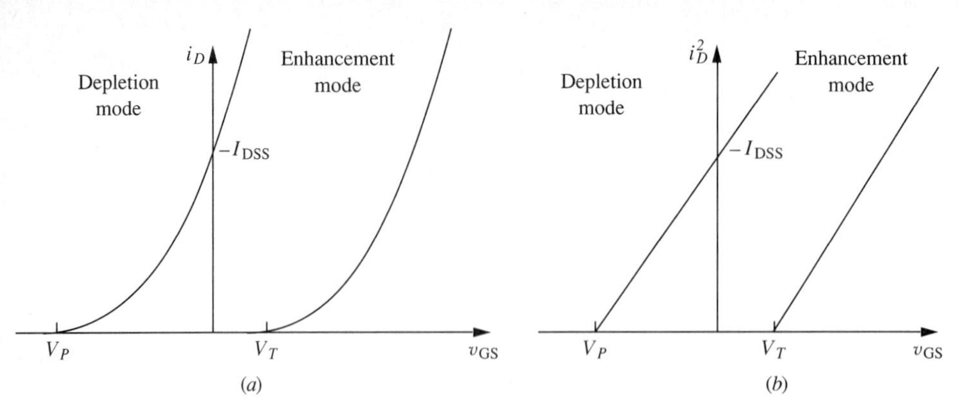

FIGURE 11.14
Current transfer characteristics of n-channel depletion-mode and enhancement-mode MOS transistors. (*a*) i_D versus v_{GS} and (*b*) i_D^2 versus v_{GS}.

with the constraint of

$$v_{DS} > v_{GS} - V_T$$

The constant k is given by

$$k = \frac{\mu_e \epsilon_{ox}}{2d_{ox}} \frac{W}{L} \tag{11.12}$$

where μ_e is the electron mobility, W is the channel width (into the paper), L is the channel length, d_{ox} is the oxide thickness, and ϵ_{ox} is the oxide dielectric permittivity. Plots of Eqs. 11.11*a* and *b* give the saturation current transfer characteristics of depletion- and enhancement-mode MOSFETs, as shown in Fig. 11.14, where (*a*) i_D versus v_{GS} and (*b*) i_D^2 versus v_{GS} are plotted. In Eq. 11.11, the slight increase in current with v_{DS} is neglected.

The small-signal equivalent circuit of a MOSFET is similar to the equivalent circuit of the JFET (Fig. 10.8), except that the MOS transistors have practically infinite dc input resistance because of the dielectric layer between the gate electrode and the channel. Thus, r_g is neglected in Fig. 10.8.

11.3 COMPLEMENTARY PAIR MOS TRANSISTOR CONFIGURATION (CMOS)

When a single transistor is used as a switching element, in the off state, it operates at cutoff and practically no current flows through the transistor. On the other hand, in the on state, the transistor is saturated and the full saturation current flows through the transistor. To conserve power and to reduce heat dissipation, this saturation current should be reduced as much as possible. If enhancement-mode n-channel and p-channel MOS transistors are connected in series, it is possible in switching circuits to reduce the power consumption from the power supply to a minimum. This scheme is called the Complementary MOS (CMOS) configuration.

The constructional details of the CMOS are shown in Fig. 11.15. Both the n- and p-channel MOS transistors are fabricated on the same chip. The drains and the gates of the two transistors are connected together. The symbol and the corresponding voltage transfer characteristics of a CMOS inverter for a power supply of 10 V is shown in Fig. 11.16.

When the input voltage is zero, transistor Q_2 is at cutoff. Although Q_1 is in its conducting state—that is, the channel is open to conduct current—no current flows through it since Q_2 is at cutoff. There is no dc path to the ground, and the voltage drop across Q_1 is zero. Thus, the output voltage taken between the source of Q_1 and the drain of Q_2 is nearly equal to V_{DD}. When v_{in} exceeds the threshold voltage of Q_2, Q_2 starts to conduct, but because of the potential that appears at the drain of Q_2, Q_2 is immediately forced into its saturation state. During this time, Q_1 is still in its unsaturated state, and the output voltage starts to decrease due to the increase in the current flowing through Q_1. As v_{in} increases further, and the output reaches a point where V_{DD} divides equally between the two transistors, both transistors are forced to operate at their saturation regions where the maximum drain current flows. As the input voltage continues to increase, the

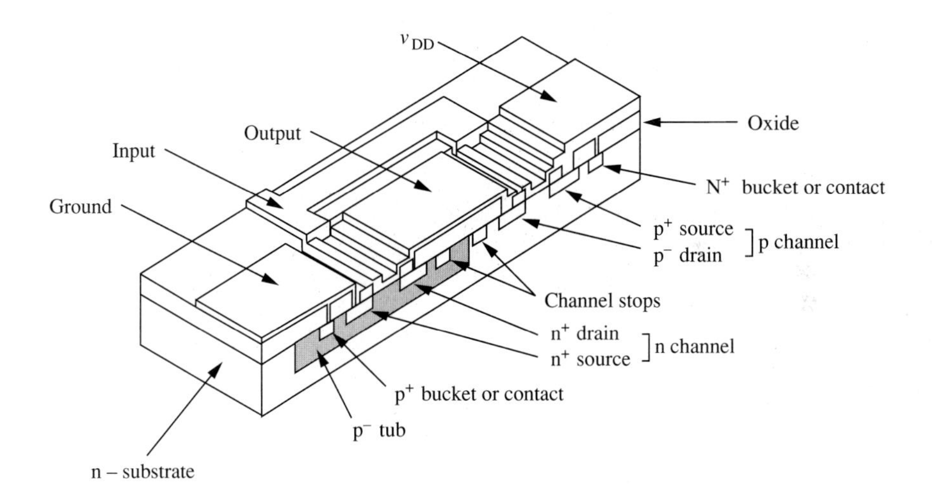

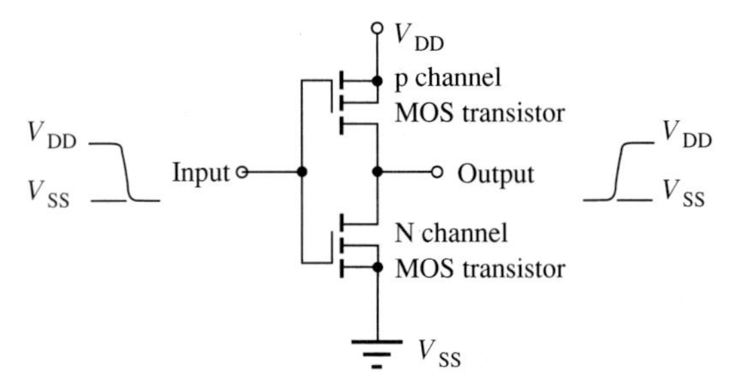

FIGURE 11.15
Constructional details of a CMOS inverter and its electrical connections (*McMOS Handbook*, 1972).

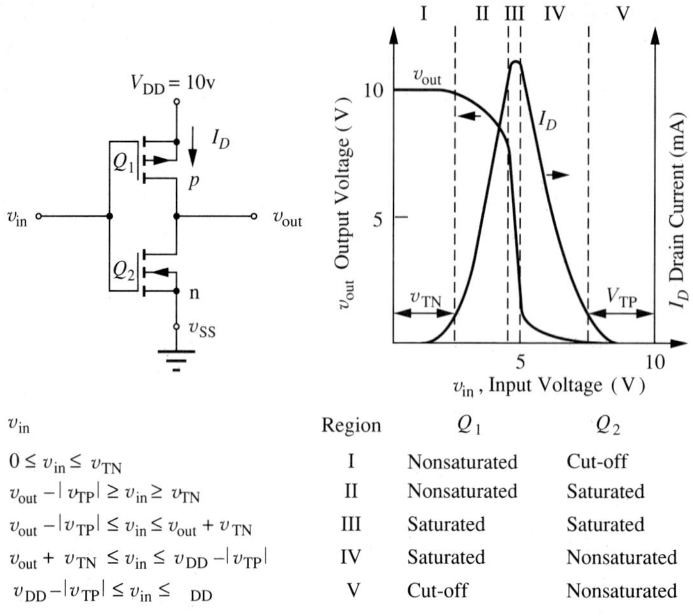

v_{in}	Region	Q_1	Q_2
$0 \le v_{in} \le v_{TN}$	I	Nonsaturated	Cut-off
$v_{out} - \lvert v_{TP} \rvert \ge v_{in} \ge v_{TN}$	II	Nonsaturated	Saturated
$v_{out} - \lvert v_{TP} \rvert \le v_{in} \le v_{out} + v_{TN}$	III	Saturated	Saturated
$v_{out} + v_{TN} \le v_{in} \le v_{DD} - \lvert v_{TP} \rvert$	IV	Saturated	Nonsaturated
$v_{DD} - \lvert v_{TP} \rvert \le v_{in} \le$ DD	V	Cut-off	Nonsaturated

FIGURE 11.16
Voltage transfer characteristic of the CMOS inverter and the total drain current as a function of the input voltage (*McMOS Handbook*, 1972).

voltage between the gate and the drain of Q_1 drops; Q_1 approaches its cutoff condition and stops conducting. Although this time the conditions are right for Q_2 to conduct, there is no dc path from V_{DD} to ground. The voltage drop across Q_2 is zero. Therefore, v_{out} is also zero. Figure 11.16 shows the current that flows in the CMOS during this transition stage. As one can see from this figure, the CMOS does not draw any current from the power supply at its on and off states. Power consumption occurs only during the transition states of the transistor, which occur in a relatively short time compared to the pulse duration.

11.4 CHARGE-COUPLED DEVICES (CCD)

Another very important application area of the MIS concept is the Charge-Coupled Device (CCD). With CCDs, one can make imaging devices, shift registers, and logic circuitry.

The CCD concept is based on storing charges (minority carriers) in a potential well generated in an MIS capacitor. Then, by manipulating potentials, this charge is transferred from one well into the adjacent potential wells. Figure 11.17 demonstrates the concept of charge transfer. Here two imaginary buckets have a common flexible wall (barrier) whose height can be controlled by some external means. The bucket on the right is partially filled with some fluid (charge) (Fig. 11.17*a*). The bucket on the left is empty. If the bucket on the left is lowered and,

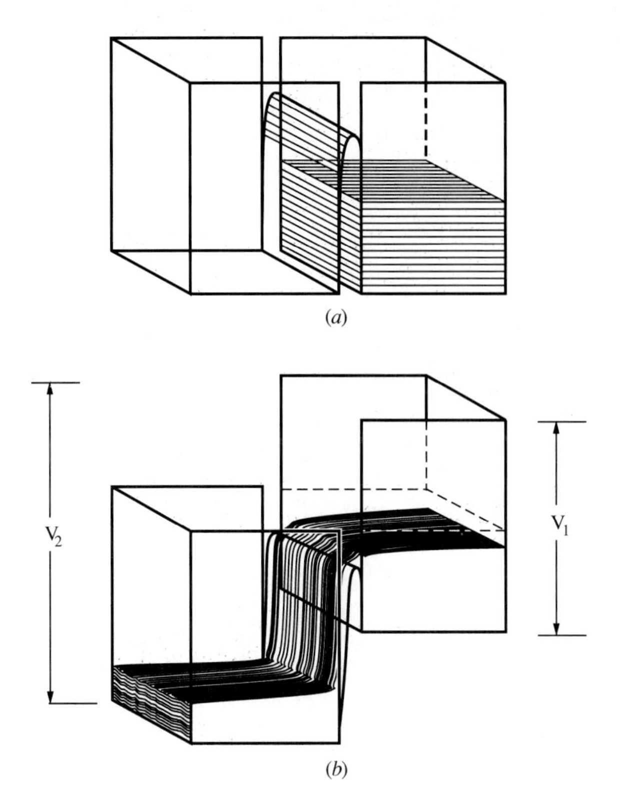

FIGURE 11.17
Bucket concept related to CCD. (*a*) Initially, the bucket on the right is full, the one on left is empty, and the common barrier wall is high enough not to spill any fluid to the left. (*b*) Lowering the left bucket and at the same time lowering the side wall (barrier) causes the fluid to spill from the right bucket to the left. $|V_2| > |V_1|$.

at the same time, the common wall is also lowered below the level of the fluid, the fluid in the right bucket will start to spill over into the bucket on the left (Fig. 11.17*b*). If the barrier seen by the bucket on the right is completely removed by lowering the bucket on the left further down, all the fluid from the right bucket will be transferred to the one on the left.

The bucket is the equivalent of a potential well generated in an MIS capacitor when a pulsed voltage produces deep inversion. In this mode of operation, it will be assumed that the time of thermally generated electron density build-up at the depletion layer is long compared to the duration of the applied voltage pulse (this is easily satisfied for practical purposes in a SiO_2-Si MOS capacitor). Following the application of a pulsed voltage, the interface stays in the deep-inversion mode but with no mobile charges, as shown in the energy-band diagram of Fig. 11.18*a*. If additional charge (for the case shown, electrons) is introduced into the layer by an external means, the surface potential changes in proportion to the injected charge as shown in Fig. 11.18*b*, but the well is still maintained, provided the injected charge is not excessive.

Charge injection takes place either electronically, by transferring charge from a neighboring device, or by optical illumination. Note that in this configuration, the charge motion takes place on the semiconductor surface next to insulator. Thus, this type of CCD is known as a *surface transfer CCD*.

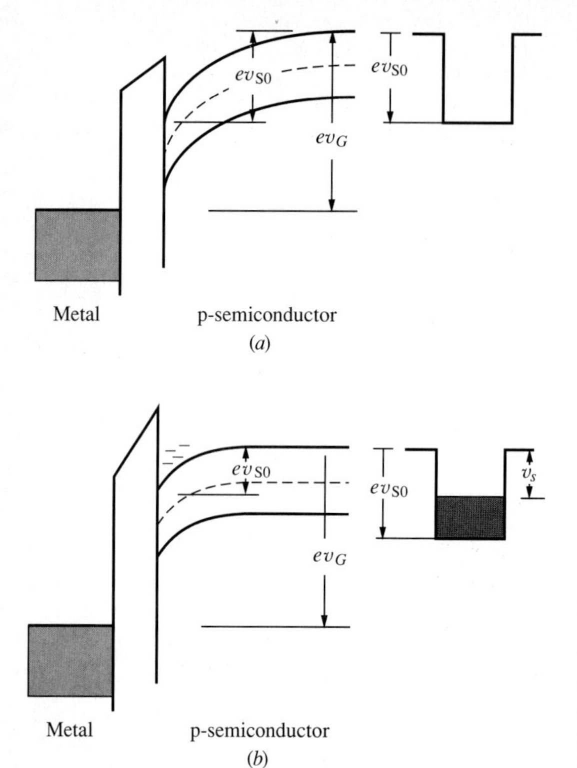

FIGURE 11.18
(*a*) Formation of a potential well when the MOS capacitor is pulsed. The well is represented as a bucket of potential. (*b*) Electrons generated either optically or by injection fill the well and at the same time lower v_s.

The actual CCD is made up of a series of MOS capacitors placed next to each other on a linear or planar array. In its simplest form, a typical surface transfer charge-coupled device is shown in Fig. 11.19*a*. A uniform oxide layer is deposited over a p-type semiconductor. For the configuration given here, a set of equally spaced gate electrodes is deposited over the oxide. Each electrode is connected to the adjacent third electrode, that is, first to fourth, second to fifth, and so on. Each set of three electrodes forms one cell bit. This configuration is known as a three-phase system. The voltage waveforms that are applied to the electrodes within a one-bit cell are shown in Fig. 11.19*c*. For the CCD shown, it will be assumed that charges are injected either photoelectrically or by an input diode-gate combination. An output diode finally collects the charge and produces the current in the external circuitry. The analysis will be restricted to a unit cell.

Figure 11.19*b* shows the sequence of charge transfer from well to well under the application of the phase potentials v_1, v_2 and v_3 at different time intervals t_1, t_2, t_3, and t_4. The sequence of events will be analyzed starting at time t_1 when phase v_1 is at a high potential. Consequently, there is a potential well under the v_1 electrode. If charges are now introduced into this deeply inverted layer, these charges will be collected in the potential well under the v_1 electrode. At time t_2 when phase v_2 is made high, a potential well is also created under the v_2 electrode. Since the two gates are so close together, the common wall of the

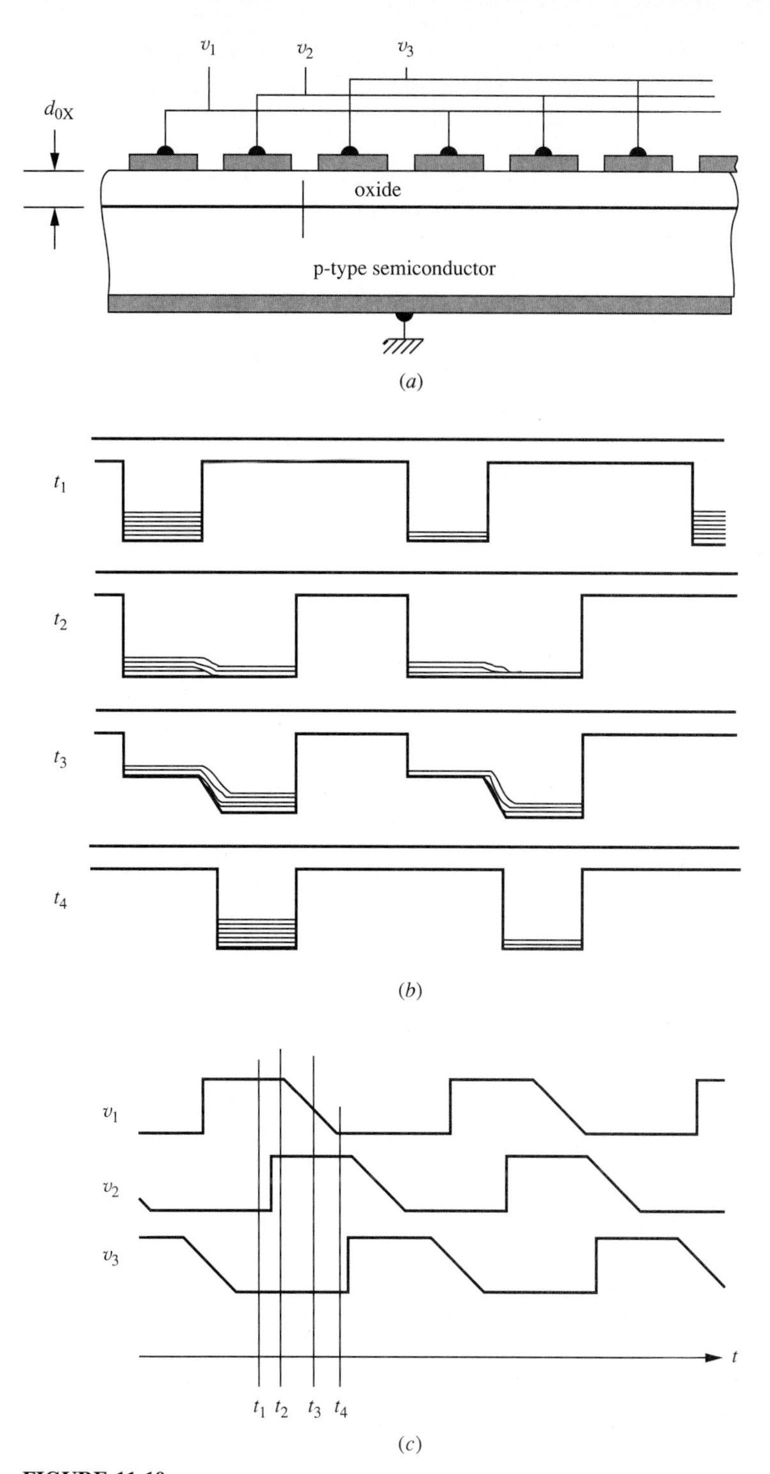

FIGURE 11.19
(*a*) The charge-coupled device (CCD). (*b*) The way charge is transferred from one well to the other as a result of (*c*) phase voltages v_1, v_2, and v_3.

potential barriers collapses. The charge in the first potential well spills over into the second. At time t_3, the potential v_1 is reduced, consequently reducing the potential well under the v_1 electrode. This allows remaining electrons from the v_1 well to completely transfer into the v_2 well. At time t_4 the well under the v_1 electrode is reduced to its minimum possible value and the charge now resides in the well under the v_2 electrode. The sequence is continued in the same manner by transferring the charge at the v_2 well to the well under the v_3 electrode. The process continues by transferring the charge to the next unit cell of the CCD device. The charge is finally collected by an output diode which produces a pulsed current in the external circuit whose amplitude depends on the amount of charge collected.

One of the major application areas of CCD devices is in imaging. Using a linear or a planar array of CCDs, optical imagery can be implemented. Either through the top where the gate electrodes are located or through back illumination, electron hole-pairs are generated in the semiconductor whose local number densities are proportional to the intensity of the incident light at a given point in the semiconductor. A control gate then transfers these collected charges into the CCD shift registers where the electrons are finally transferred to the output circuitry. Between each gating operation, the photoelectrically generated electrons are integrated (accumulated) under the photogates. This allows enough charge to be collected under each well before being transferred to the CCD registers. A state-of-the-art imaging linear array using GaAs is shown in Fig. 11.20.

There is a major problem associated with surface channel CCD devices. When electrons are transferred from one well to the other, the electrons fill the adjacent wells in a very short time. At the same time, these electrons also fill the surface states that reside at the interface. These states, created during the manufacturing process, are a result of the lattice mismatch between Si and SiO_2. When it is time for these electrons to empty a well, it takes a longer time for these charges to empty these states. If the clock pulse duration is shorter than the emptying time of the surface states, some electrons will be left behind. These will then be transferred with the next batch of electrons in the subsequent clock pulse. This phenomenon gives rise to erroneous results, especially when the CCD

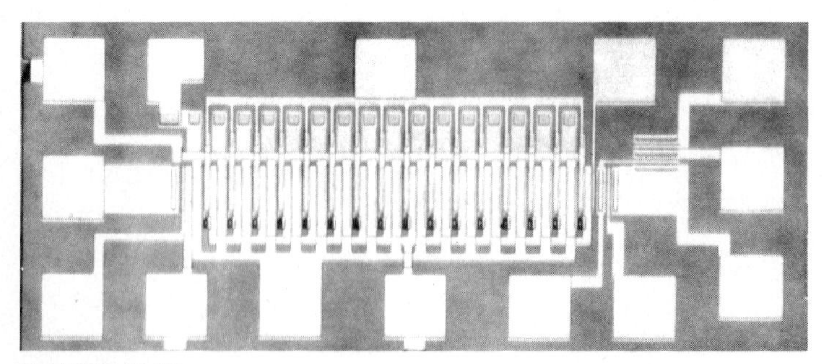

FIGURE 11.20
A state-of-the-art GaAs CCD linear array (Kosel *et al.*, 1988).

is used for optical imaging. Therefore, under these conditions, the pulses do not represent the true intensity of the optical signal.

A modified version of the CCD, which is referred to as the Buried Channel CCD (BCCD), allows charge transfer to take place away from the surface. As shown in Fig. 11.21, BCCD uses a lightly diffused n-layer over the regular p-type semiconductor. An n^+-region is also diffused into the n-layer to make an ohmic contact. A positive-bias voltage is then applied to the n^+-contact relative to the common bottom electrode, and thus, the main p-n junction is depleted of mobile charges ($W_{\mathrm{p-n}}$ in Fig. 11.21). A negative gate voltage is applied to the gate electrode above the oxide. This produces the deep inversion layer of the usual MIS capacitor with a depletion layer extending to W_{MOS}, as shown in Fig. 11.21. The bias voltage V_B can be adjusted so that the two depletion layers meet at a certain distance away from the oxide layer in the n-layer. The voltage necessary to produce this condition is known as the pinch-off voltage. The region where this occurs does not change when the bias voltage is further increased. This region is also the minimum-energy region. Thus, any external charge (electrons) introduced into the system will reside in this minimum-energy region. This becomes the location of the equivalent potential well in the BCCD. The charges are transferred from one well to the next one away from the surface, thus avoiding the surface-state traps. In this case, there is complete transfer of charges between the wells with each applied pulse sequence. Buried channel CCDs are also implemented in Ga-As, where the transfer speeds are an order of magnitude faster because of the higher electron mobilities.

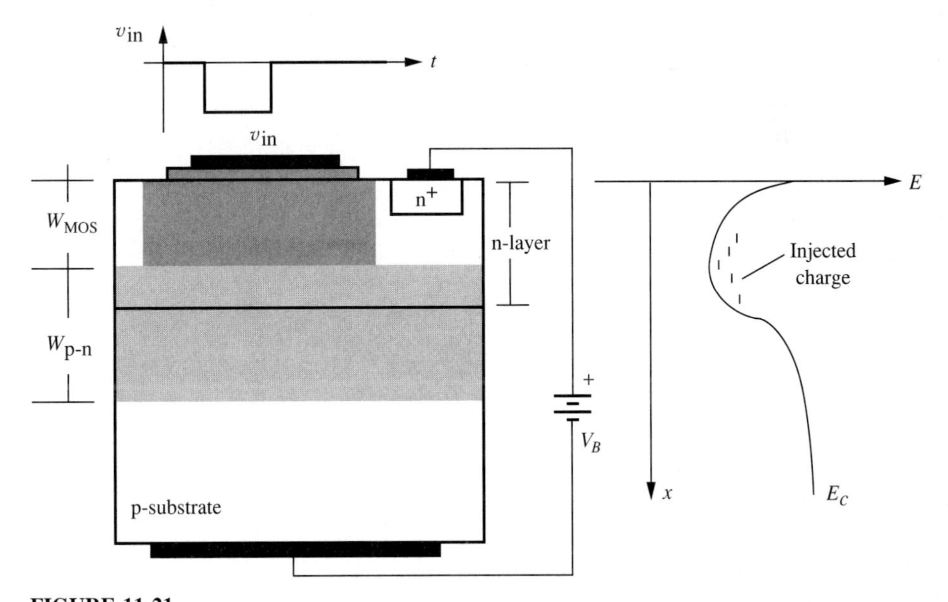

FIGURE 11.21
Buried Channel Charge-Coupled Device (BCCD). $W_{\mathrm{p-n}}$ and W_{MOS} are the respective depletion-layer widths due to the bias potential V_B and the MOS capacitor under the application of the pulsed voltage v_{in} (Beynon and Lamb, 1980).

FURTHER READING

Allen, P. E., and D. R. Holberg, *CMOS Analog Circuit Design,* Holt, Reinhart and Winston, New York, 1987.

Barbe, D. F., "Imaging Devices Using Charge-Coupled Concept," *Proc. IEEE,* vol. 63, 1975, 38.

Beynon, J. D. E. "The Basic Principles of Charge-Coupled Devices," *Microelectronics,* vol. 7, 1975, 7.

Beynon, J. D. E., and D. R. Lamb, *Charge-coupled Devices and Their Applications,* McGraw-Hill, UK, 1980.

Boyle, W. S., and G. E. Smith, "Charge-Coupled Semiconductor Devices," *Bell System Tech. Journ.,* 1970, 587.

Gray, P. E., and C. L. Searle, *Electronic Principles, Physics, Models and Circuits,* Wiley, New York, 1969.

Kosel, P. B., N. Bozorgebrahimi, L. Bechter and R. E. Poore, "Photodetectors for high speed image scanners on GaAs," SPIE, vol. 994, *Optoelectronic Materials, Devices, Packaging and Interconnects. II,* 1988, 108–116.

McMOS Handbook, 1st Ed., Motorola Inc., 1974.

Richman, Paul, *MOS Field-Effect Transistors and Integrated Circuits,* Wiley, New York, 1973.

Sze, S. M., *Physics of Semiconductor Devices,* 2nd ed., Wiley, New York, 1981.

Walden, R. W., R. H. Krambeck, R. J. Strain, J. McKenna, N. L. Schryer, and G. E. Smith, "The Buried Charge-Coupled Devices," *Bell System Tech. Journ.,* vol. 51, 1972, 1635.

PROBLEMS

11.1. Explain whether you will use an MOS capacitor as a voltage-controlled capacitor in an electronic circuit. What will be its advantages and disadvantages?

11.2. An MIS capacitor is made by depositing Si_3N_4 on GaAs. The thickness of Si_3N_4 is 0.15 μm. Find the equivalent capacitance of the MIS capacitor at a bias of (*a*) -2 V and (*b*) 10 V. Assume ideal conditions for GaAs and a cross-sectional area of 50 μm^2. Sketch the low- and high-frequency *C-v* curves. $\kappa_r(Si_3N_4) = 7.5$.

11.3. An n-channel depletion-mode MOSFET is experimentally found to have the following parameters: $V_P = -4.0$ V and $I_{DSS} = 4.0$ mA. Find the transconductance of the MOSFET at an operating point of $v_{GS} = -1.5$ V.

11.4. Experiments show that the transfer characteristic of a MOSFET depends heavily on temperature. From the definition of V_T (Eq. 11.10), discuss which parameters will contribute to this temperature dependence. Will V_T increase or decrease with increasing temperature?

11.5. Based on the operational characteristic of a depletion-mode MOSFET, as long as the manufacturing parameters, such as the doping level of the channel and the physical dimensions, are the same, do you expect any differences in V_T of an n-channel or p-channel transistor at dc conditions?

11.6. In an enhancement-mode n-channel transistor, if v_{GS} is zero and $v_{DS} = 10$ V, what will be the resulting drain current if the respective diode currents are given by $i = 10^{-9}\{\exp(v/0.026) - 1\}$ A where v is the voltage appearing across the corresponding diode?

11.7. If low-frequency and high-frequency measurements of the *C-v* curves of an MIS diode are made, show that the depletion-layer width can be calculated from

$$W = A\epsilon_s\left(\frac{1}{C_H} - \frac{1}{C_{ox}}\right)$$

11.8. If the low- and high-frequency capacitance measurements of an MIS capacitor give $C_{\text{ox}} = 190$ pF and $C_H = 75$ pF and the gate electrode area is 100 μm^2, find
(*a*) the depletion-layer width
(*b*) the thickness of the insulator layer if silicon nitride is used as a dielectric

11.9. If you make a capacitor as shown in Fig. P.11.9, what kind of capacitance measurements will you expect as a function of gate voltage? Sketch and explain.

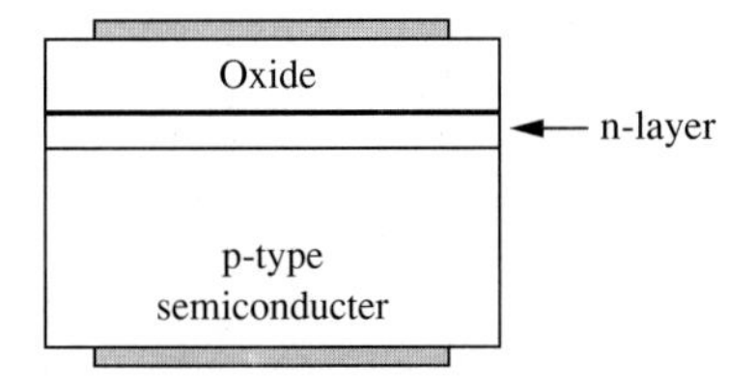

FIGURE P.11.9

11.10. An enhancement-mode MOSFET shown in Fig. 11.11 has $N_D = 10^{19}\text{cm}^{-3}$ (for n^+-regions) and $N_A = 10^{16}\text{cm}^{-3}$ (for p-substrate). If the equivalent areas of the n^+-regions are 50 μm^2, and L = 5μm, find the punch-through voltage for the channel. Assume that the diodes are ideal and all voltages appear across the junctions of the two diodes.

CHAPTER 12

HIGH-FREQUENCY SOLID-STATE DEVICES

When the frequency of operation of an electronic device is increased, certain parameters, such as the circuit effects and the transit time of charge carriers, become important factors that degrade the high-frequency performance of the device. The circuit effects associated with the package leads connecting the chip with the outside world can be compensated relatively easily by using passive circuit elements. On the other hand, intrinsic effects associated with the solid-state devices that we have studied thus far, such as the transit time of the charge carriers, become the most crucial internal parameters that limit the high-frequency performance of these devices.

When a charge carrier is injected either into the base from the emitter of a BJT or into the channel from the source in an FET, it takes a certain amount of time for these charges to travel to the collector or to the drain ends of the respective devices. As will be shown in this chapter, there are various parameters that control this transit time. If the signal does not reach the collector or the drain in a time interval that is short compared to the period of the applied signal, the output current or voltage will lag in phase behind the input signal. This degrades the high-frequency response of the device especially when the transit time becomes comparable to or larger than the period of the applied signal. The short transit time is also essential to the faster switching of a transistor from its on-to-off (and vice-versa) states. The parameters that govern the high-frequency operation of a microwave transistor also controls the switching time of the transistor.

There are now new novel transistor configurations that push the high-frequency operation of these transistors to their theoretical limits. There are also bulk-effect semiconductors as well as other ingenious two-terminal junction devices that extend the operation of solid-state devices to very high frequencies. Unfortunately, because of the very small submicron dimensions required for high-frequency operation, the power output of the present individual solid-state devices is relatively low.

In this chapter, the frequency limitations of transistors will first be analyzed, then the present state of very high-frequency transistors will be given. We will follow by introducing some important two-terminal devices.

12.1 FREQUENCY DEPENDENCE OF POWER GAIN AND NOISE

The frequency dependent gain of a transistor can be written as

$$G(f) = \frac{G_0}{\left[1 + G_0 (f/f_{\max})^4\right]^{1/2}} \tag{12.1}$$

where G_0 is the low-frequency gain of the transistor of a given configuration.

At high frequencies

$$G(f) \approx \left(\frac{f_{\max}}{f}\right)^2 \tag{12.2}$$

and the gain falls off at 6 dB per octave (as the frequency doubles).

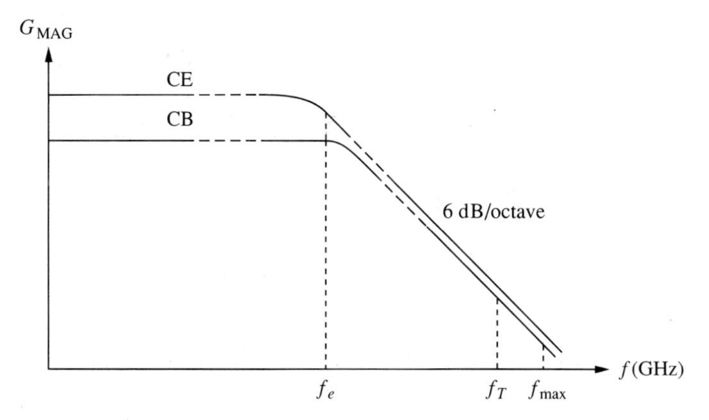

FIGURE 12.1
Maximum available gain (MAG) versus frequency of common-emitter (CE) and common-base (CB) configurations. The dashed lines indicate possible potentially unstable regions of the transistor.

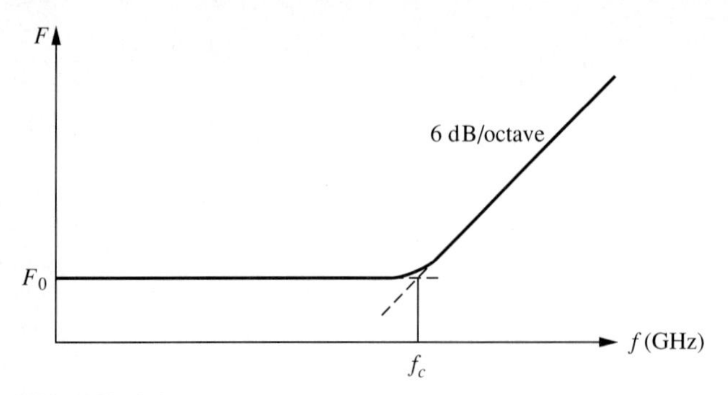

FIGURE 12.2
Noise figure of a microwave transistor as a function of frequency.

The gain function is usually discontinuous, and there are regions of potential instabilities.* Figure 12.1 shows typical gains as a function of frequency in the stable and unstable regions of operation for the common-base and common-emitter configurations of a typical bipolar junction transistor. Here f_{max} is the frequency where the gain is unity; f_T, known as the transit-time cutoff frequency, is discussed in detail later. f_e is the frequency at which gain drops to 3 dB below the low-frequency gain. FETs also show similar gain-frequency dependence.

In a similar way, the noise figure (see Appendix A) of a microwave transistor can be written as

$$F = F_0\left[1 + \left(\frac{f}{f_c}\right)^2\right] \tag{12.3}$$

where F_0 is the low-frequency noise figure of the transistor. It is constant up to a critical frequency f_c and increases 6 dB per octave above f_c. F_0 depends largely on h_{fe}, referred to as the common-emitter forward current gain, of the transistor and decreases with increasing h_{fe}. A typical noise-figure variation of a transistor as a function of frequency is shown in Fig. 12.2.

12.2 TRANSIT TIME EFFECTS IN BIPOLAR TRANSISTORS

At the high-frequency end of the spectrum ($f > f_e$ in Fig. 12.1), when a bipolar junction transistor is used in the common-emitter configuration, it has the following two-port simplified admittance parameters:

$$Re(Y_{\text{in}}) = \frac{1}{r'_b}$$

$$Re(Y_{\text{out}}) = \omega C_C$$

* By choosing proper external circuit elements, the transistor can still be operated in these potentially unstable regions by compensating for these instabilities with passive elements.

and the current gain

$$|h_{fe}| = \frac{\omega_T}{\omega}$$

where r'_b is the equivalent base resistor seen at the base terminal of the transistor, usually known as the *base-spreading resistance;* C_C is the collector capacitance; $\omega_T = 2\pi f_T = 1/\tau_T$; and h_{fe} is the current gain of the transistor in the common-emitter configuration at low frequencies. In this representation, any feedback is either neglected or assumed to be neutralized by a lossless network.

The equivalent circuit of the transistor that depicts these parameters is shown in Fig. 12.3. The gain of this two-port network is given by

$$G \approx \frac{\omega_T}{\omega^2 r'_b C_C} \tag{12.4}$$

At $f = f_{max}$, the gain is unity. Setting $G = 1.0$ and $\omega_{max} = 2\pi f_{max}$, we solve Eq. 12.3 for f_{max} and obtain

$$f_{max} = \sqrt{\frac{f_T}{2\pi r'_b C_C}} \tag{12.5}$$

The three critical parameters that determine the limits of the high-frequency operation of a bipolar junction transistor are thus f_T, r'_b, and C_C.

The gain and noise figure of a transistor both vary in the microwave region due to the frequency dependence of a number of elements. These in turn can best be characterized in terms of a number of characteristic or *cutoff* frequencies.

If a signal is applied to the emitter of a bipolar transistor in the common-base configuration, it will encounter four principal regions of delay or attenuation. Figure 12.4 shows a simplified model of the transistor that helps to identify these delays.

After the carriers have entered the emitter, the following successive phenomena take place:

1. The emitter-base capacitance shunts the active emitter region. This is equivalent to an R-C charging delay time which is given by

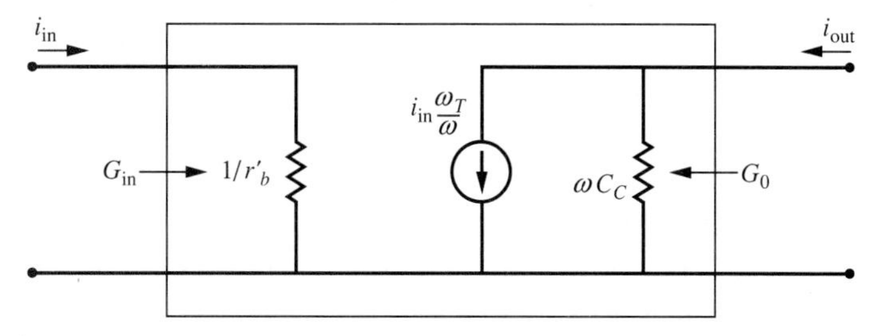

FIGURE 12.3
Simplified high-frequency equivalent circuit of a bipolar transistor.

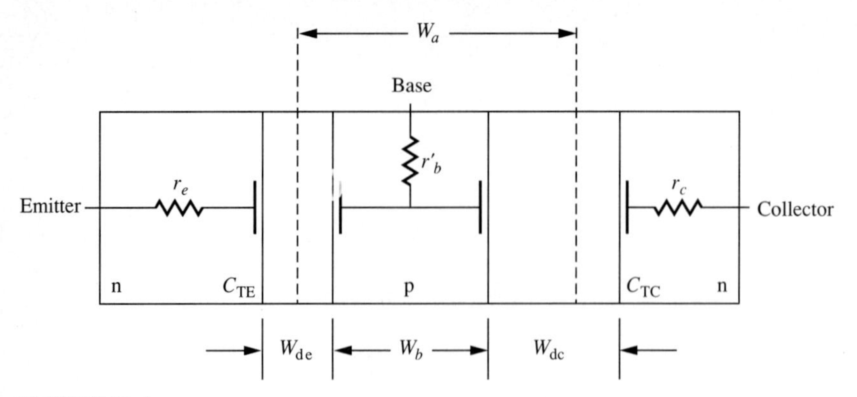

FIGURE 12.4
The parameters that control the transit time of electrons in an n-p-n bipolar transistor.

$$\tau_e = r_e C_{TE} \tag{12.6}$$

where r_e is the emitter resistance (sometimes referred to as the space-charge resistance), and C_{TE} is the emitter-base transition capacitance.

2. Carriers must cross the base region W_b through a combination of diffusion and drift. During this transition, some of the carriers will be lost due to recombination and the signal will be attenuated. Base transit time is given by

$$\tau_b = \frac{W_b^2}{[2D(0.8 + 0.46\eta)]} \quad \text{with drift} \tag{12.7a}$$

$$\tau_b = \frac{W_b^2}{\zeta D} \quad \text{without drift} \tag{12.7b}$$

where $\eta = [\mathcal{E}_b W_b/(kT/e)]$, $\mathcal{E}_b$ is the electric field at the base region, and ζ is a parameter that depends on doping profile of the base region. $\zeta = 2$ for a uniform doped base (see Problem 9.7).

3. Carriers next cross the collector-base depletion layer W_{dc} under the influence of an electric field. This leads to the collector depletion-layer time which is given by

$$\tau_{dl} = \frac{W_{dc}}{2v_{sL}} \tag{12.8}$$

where v_{sL} is the scatter-limited velocity of the electrons. For microwave transistors, generally $W_{dc} >> W_b$.

4. If there is any appreciable resistance between the collector depletion layer and the external collector terminal, a final $R - C$ delay time is encountered in the form

$$\tau_c = r'_c C_C \tag{12.9}$$

C_C includes the BCJ transition capacitance C_{TC}. The total time delay between the emitter and the collector will be the sum of these four delays and can be written as

$$\tau_{ec} = \tau_e + \tau_b + \tau_{dl} + \tau_c \tag{12.10}$$

or the total transit time in the base region becomes

$$\tau_{ec} = r_e C_{TE} + \frac{W_b^2}{[2D(0.8 + 0.46\eta)]} + \frac{W_{dc}}{2v_{sL}} + r_c' C_C \quad \text{(with drift)} \tag{12.10a}$$

$$\tau_{ec} = r_e C_{TE} + \frac{W_b^2}{\zeta D} + \frac{W_{dc}}{2v_{sL}} + r_c' C_C \quad \text{(without drift)} \tag{12.10b}$$

Time constants can be more conveniently expressed in terms of characteristic frequencies. Let

$$\tau = \frac{1}{\omega}$$

then, defining total transit time from the emitter to the collector as

$$\tau_T = \frac{1}{\omega_T} \equiv \tau_{ec}$$

we can write

$$\tau_T = \frac{1}{\omega_T} = \frac{1}{\omega_e} + \frac{1}{\omega_b} + \frac{1}{\omega_{dl}} + \frac{1}{\omega_c} \tag{12.11}$$

where all ω's are obtained from the reciprocal of the time delays given above. We also note that in the simplified time delays that are introduced here we have neglected many of the contributions from the parasitic elements. During the design stage, one can reduce C_{TE} and C_C by manipulating the geometry of the transistor. Since τ_{dl} is the shortest of the time delays, a short gate-length transistor is highly desirable to reduce τ_b below τ_e and τ_c, so that the f_T of the transistor can be increased to the highest value possible.

The model used thus far in calculating the transit time can only predict the theoretical maximum frequency, f_{Tmax}, that can be achieved for a BJT transistor. On the other hand, the parasitic effects as well as the other intrinsic coupling parameters, such as the collector-base capacitance C_{CB}, introduce degradation to the performance of the BJT and make a simplified transit time analysis very difficult. It is more convenient to introduce a complete small-signal equivalent circuit model for the BJT that allows using network theory to predict the actual high-frequency performance of the transistor. Figure 12.5*a* shows the location of possible circuit elements that results from the physical layout of the transistor. The corresponding common-emitter equivalent circuit of a BJT transistor treating the relatively long base region as a distributed *RC* circuit is given in Fig. 12.5*b*.

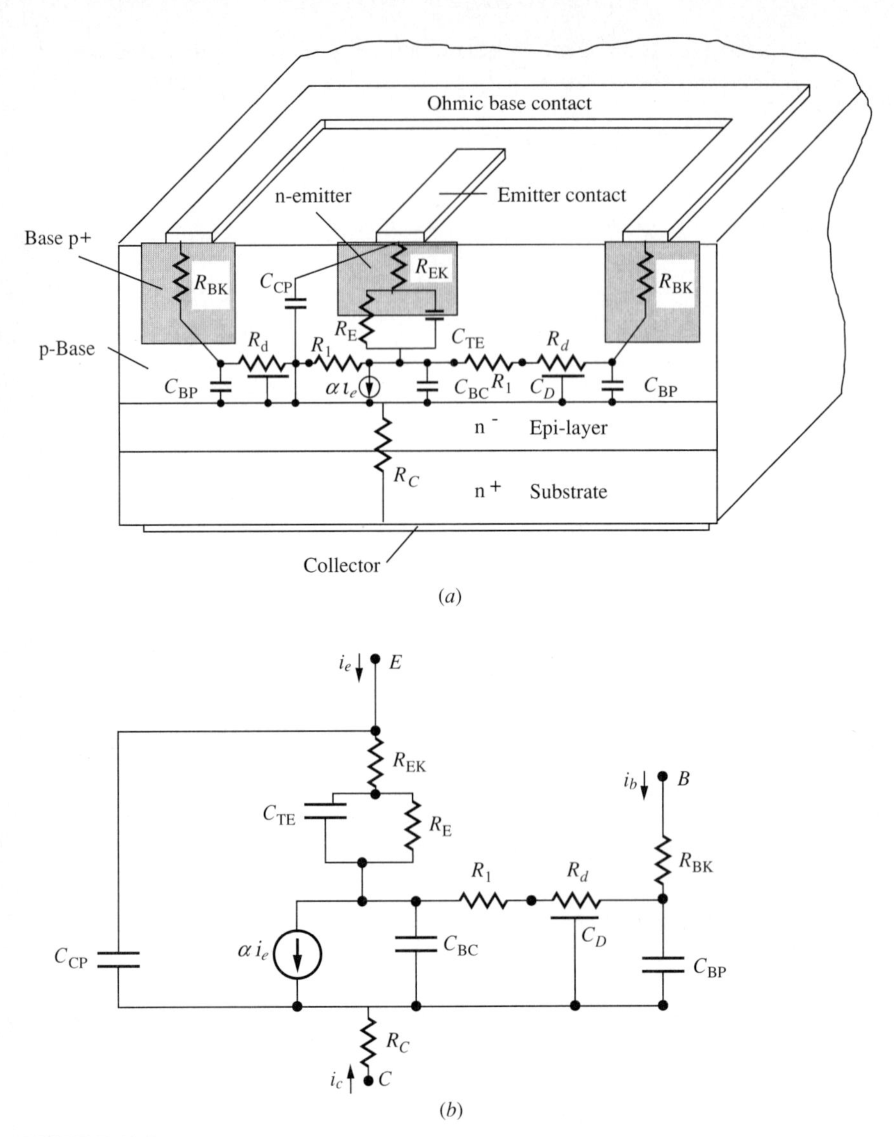

FIGURE 12.5
The bipolar junction transistor. (*a*) Location of parameters. (*b*) Equivalent circuit with distributed base (Bahl and Bahartia, 1988).

Proper combination of R_{BK}, R_d, and R_1 make up for the effective base-spreading resistor of the transistor.

The noise figure of the bipolar transistor depends on three principal noise sources. These are shot noise in the emitter, shot noise in the collector, and thermal noise in the base resistance.

12.3 TRANSIT TIME EFFECTS IN FET's

In a field-effect transistor, the total gate-to-drain transit time can be written as a sum of

1. τ_{gc}, the gate-channel junction charging time;
2. τ_d, the dielectric relaxation time of the charge distribution in the channel; and
3. τ_t, the carrier transit time through the channel.

The regions of an FET that control these delay times are shown in Fig. 12.6.

When a signal v_s is applied to the gate, due to the gate-channel capacitance the signal will be delayed by an amount given by

$$\tau_{gc} = r_c C_{GC} \tag{12.12}$$

where r_c is the equivalent channel resistance, and C_{GC} is the equivalent gate-channel capacitance. The gate-channel equivalent circuit may be represented as a distributed nonuniform transmission line. Since the channel resistance is concentrated in the narrow channel portion near the drain, especially for large signal operation beyond the saturation point, τ_{gc} can be approximately written by the following relation

$$\tau_{gc} = 0.05 r_c C_{GS} + 0.55 r_c C_{GD} \tag{12.13}$$

where C_{GS} and C_{GD} are the gate-to-source and gate-to-drain capacitances.

Once the depletion layer is charged, the applied voltage appears across the channel. The evolution of this second stage takes place through two simultaneous processes: charge distribution by dielectric relaxation and carrier transit time through the channel. The dielectric relaxation is given by

$$\tau_d = \frac{\kappa_s \epsilon_0}{en\mu} \tag{12.14}$$

where n is the carrier concentration, and μ is the mobility of the charge carriers.

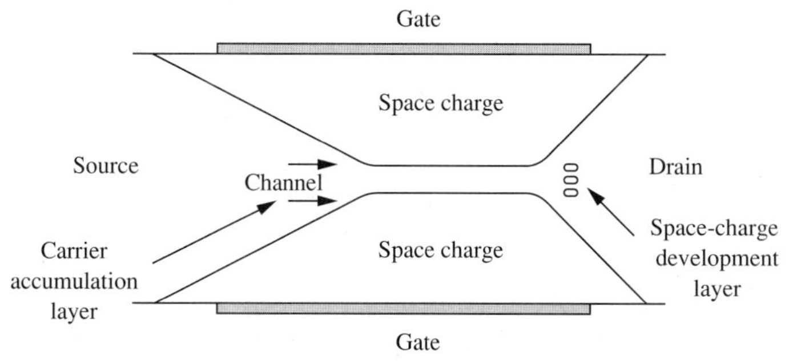

FIGURE 12.6
Delay regions in an FET.

For microwave FETs at high operating-voltage levels, the effective length of the channel is $\geq 0.5L$, where L is the actual gate length. The transit time through the channel can then be written as

$$\tau_t = \frac{0.5L}{v_{\mathrm{sL}}} \tag{12.15}$$

Finally, the longest of the time delays

$$\tau_{gd} = \tau_{gc} + \tau_t \tag{12.16a}$$

or

$$\tau_{\mathrm{gd}} = \tau_{\mathrm{gc}} + \tau_d \tag{12.16b}$$

is chosen for the total gate-drain time delay.

For complete analysis, and for both the common-source and common-gate configurations, other resistive and capacitive parasitic elements should also be included in Eq. 12.16. These additional parameters increase τ_{gd}. Junction FETs have higher parasitic capacitance and resistance values.

The frequency f_T becomes

$$f_T = \frac{1}{2\pi\tau_{\mathrm{gd}}} \tag{12.17}$$

As can be seen from these simplified arguments, the mobility, the saturated velocity of the charged carriers, and the gate length L play very important roles in the high-frequency performance of an FET.

A small-signal equivalent circuit of an FET can be drawn by referring to Fig. 12.7. In Fig. 12.7*a*, the geometric locations of the respective equivalent circuit elements are indicated. The corresponding equivalent circuit is shown in Fig. 12.7*b*. The transconductance g_m is defined by

$$g_m = \frac{di_D}{dv_{\mathrm{GS}}}\Big|_{v_{\mathrm{DS}}=\text{constant}} \tag{12.18}$$

To have a better high-frequency performance, the transistor should also have a good performance at low frequencies.

The noise figure in GaAs FETs is intrinsic to the device due to the parasitic resistances. Intrinsic noise arises from two sources. One is due to the thermal or Johnson noise produced by the ohmic section of the channel, and the other is due to the diffusion noise in the velocity-saturated section of the channel region. This diffusion noise can be dominant for short gate length devices.

With present day computers, the equivalent circuit models of transistors can be simulated with ease by using commercially available advanced-level computer software programs. Measurements made of the transistor parameters can even be used to predict the magnitudes of some of the parameters used in the representation of the equivalent circuit.

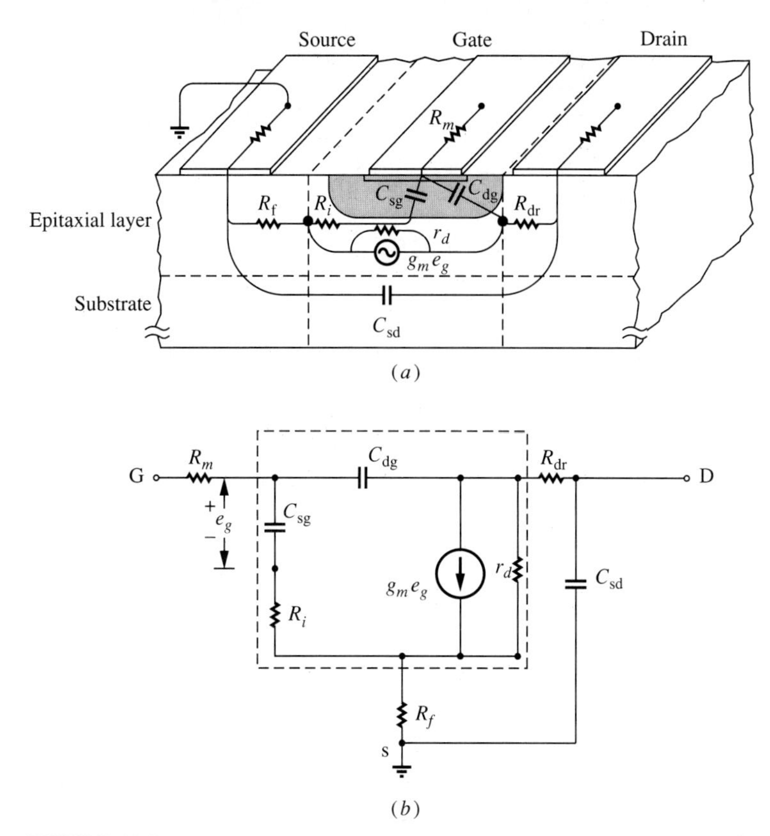

FIGURE 12.7
(a) Schematic of a junction FET (JFET) where the circuit elements of the equivalent circuit *(b)* appears (Pucel, *et. al.,* 1975).

12.4 SCHOTTKY BARRIER FET (MESFET)

If the p-n junction of the gate of a JFET is replaced by a Schottky barrier, the resulting transistor has electrical characteristics superior to JFETs, especially in field-effect transistors made from compound semiconductors. A schematic drawing of a typical MESFET (MEtal Semiconductor FET) is shown in Fig. 12.8. The configuration is easily implemented with existing planar integrated circuit technology with minimum processing steps.

The following are the advantages of Schottky barrier gate transistors over JFETs.

1. On many wide band gap compound semiconductors, good p-n junctions are difficult to form, often having high reverse-saturation currents. On the other hand, a properly made Schottky barrier has close to the theoretical reverse-saturation current, which is extremely low.

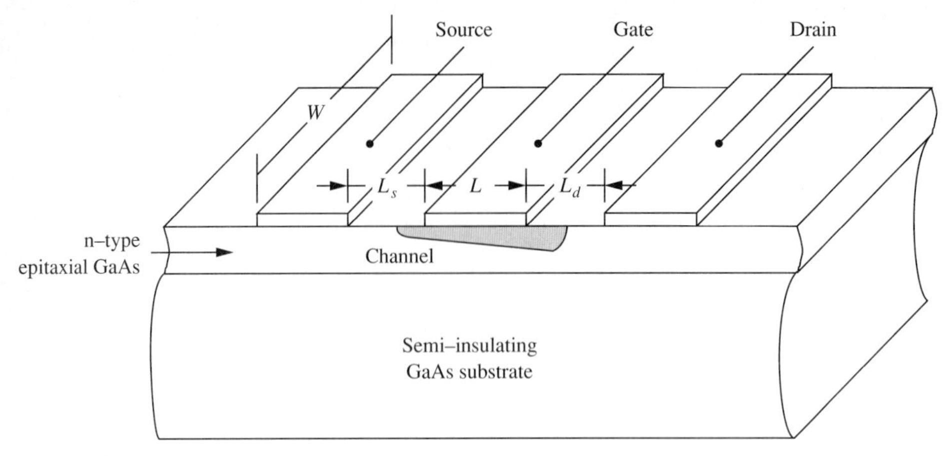

FIGURE 12.8
Schematic of a Schottky barrier Field-Effect Transistor (MESFET).

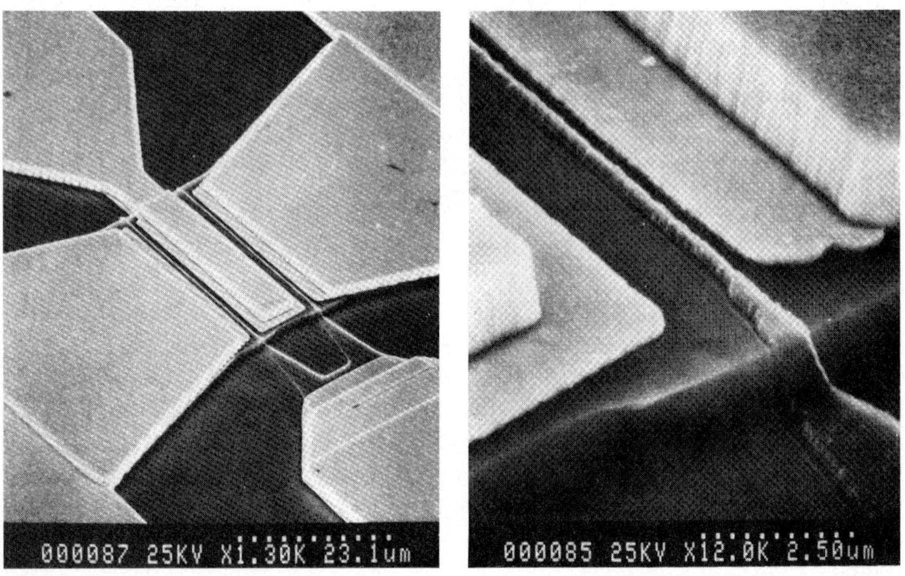

FIGURE 12.9
The device geometry and channel structure of an FET. The device geometry consists of two 0.25× 30 μm gate fingers in an interdigitized structure with the source metallization pattern surrounding the area. The gate is offset toward the source in the channel to reduce the source parasitic resistance (Watkins, 1983).

2. The energy barriers associated with the wide band gap semiconductors are quite large, hence, a Schottky barrier can be biased in the forward direction by a considerable amount to narrow the depletion layer without excessive gate current. Even if the gate current flows, this appears as a hot electron current into the gate; thus there are no storage effects to interfere with the high-frequency operation of the device.

Figure 12.9 is an enlarged photograph of the gate structure of a typical state-of-the-art GaAs low-noise MESFET.

12.5 MODULATION-DOPED TRANSISTORS (MODFET OR HEMT)

In a semiconductor, the principle scattering sources of electrons are lattice atoms (phonons) at high temperatures and ionized impurity atoms (donors) at low temperatures. If the electrons are forced to move in an undoped semiconductor at low temperatures, the mobility of the carriers will be highly increased due to the lack of impurity atoms and reduced number of phonons. This is accomplished by a process known as *modulation doping*, which is obtained by selective doping of barrier layers in heterostructures, alternating undoped and doped layers. At the heterojunction, electrons from the impurity donors in the doped region diffuse into the undoped region. Since there are no majority holes in the undoped region, the field set up by the immobile donor atoms and the electrons that moved into the undoped region forms a barrier at the junction, as shown in Fig. 12.10. Since the undoped layer now has excess electrons (n-type behavior), band bending occurs on the undoped region in the vicinity of the junction. The resulting change in the energy profile of the junction thus creates a potential well in the undoped region near the junction. The electrons transferred from the donor atoms then reside in this well. The well can also possess discrete energy levels, as shown in Fig. 12.10.

The transistors based on modulation doping are known as MODFETs (MOdulation Doped FETs) or HEMTs (High Electron Mobility Transistors).

The best example of modulation doping is the heterojunction formed by Si-doped AlGaAs and the undoped GaAs layers. The resulting modulation-doped single-layer interface is shown in Fig. 12.10. Electrons from the doped AlGaAs spill into the undoped GaAs, then reside in the well at the interface on the GaAs side. Under the application of an external field, these electrons move in the undoped GaAs layer with minimal impurity scattering. The resulting enhancement in the mobility of the electrons in modulation-doped GaAs as a function of temperature with years as a parameter is shown in Fig. 12.11. The increase in mobility through the years can be attributed to the advances made in compound semiconductor layer growth technology and the controlled deposition of very pure and thin intrinsic layers. The electrons residing in the conduction band of the undoped GaAs layer are referred to as *two-dimensional electron gas* since they can move

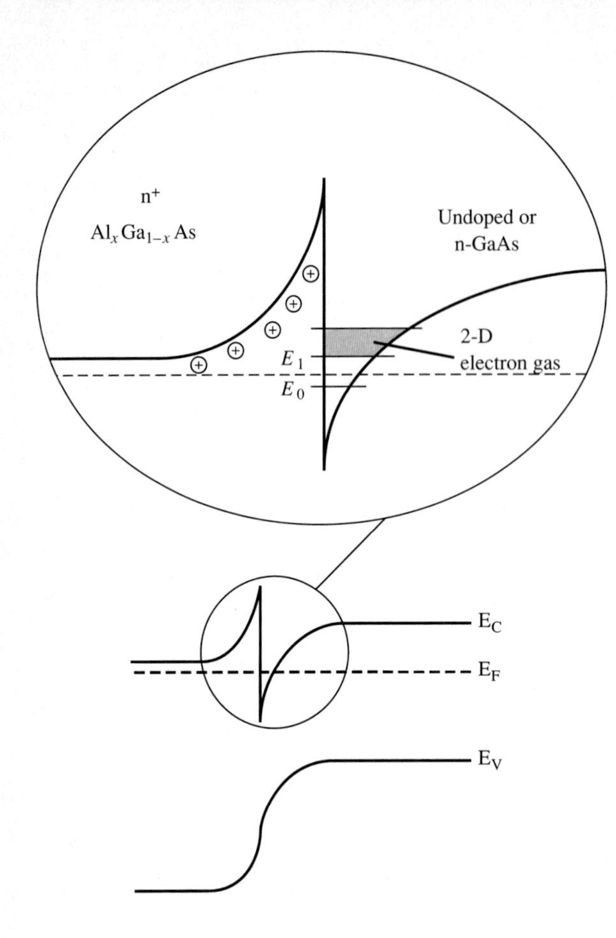

FIGURE 12.10
Formation of a two-dimensional electron gas at a heterojunction. GaAs is generally undoped or slightly n-type. Inset shows the discrete resonance levels that form on the GaAs side. The 2-D electron gas has a very high mobility.

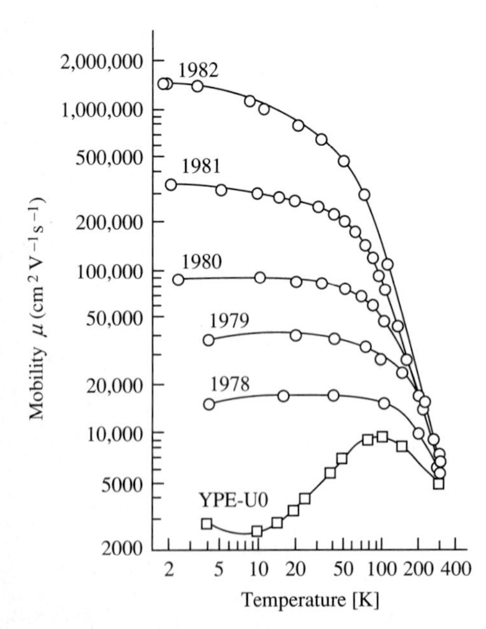

FIGURE 12.11
Mobility of electrons in GaAs as a function of temperature with years as a parameter (Tsang, 1985).

easily, with negligible scattering in the two-dimensional plane perpendicular to the direction of the junction.

The constructional details of a typical HEMT are shown in Fig. 12.12. The very thin undoped AlGaAs barrier layer grown between the doped AlGaAs and the undoped GaAs heterojunction helps confine the 2-D electron gas in the GaAs and, at the same time, further increases the effective mobility of the electrons. The recessed gate allows deposition of the gate electrode closer to the heterojunction. The flow of electrons is controlled by the depletion layer formed by the applied gate voltage. Since the actual physical length of the doped AlGaAs layer is very small (400 Å), the depletion layer can be extended into the undoped GaAs layer with very small gate voltages. This allows the control of the electron gas flow from the source to the drain by small gate voltages. Consequently the transistor can be switched from on-to-off or off-to-on states with very small gate voltages. Since the drain current results through the flow of highly mobile electron gas, the transit time of the electrons in a HEMT is consequently also very short. Switching times in the picosecond range are presently reported.

Since the electron gas is 400 Å away from the gate electrode, a large concentration of charge can thus be controlled with small gate voltages. This also results in increased transconductance of the transistor.

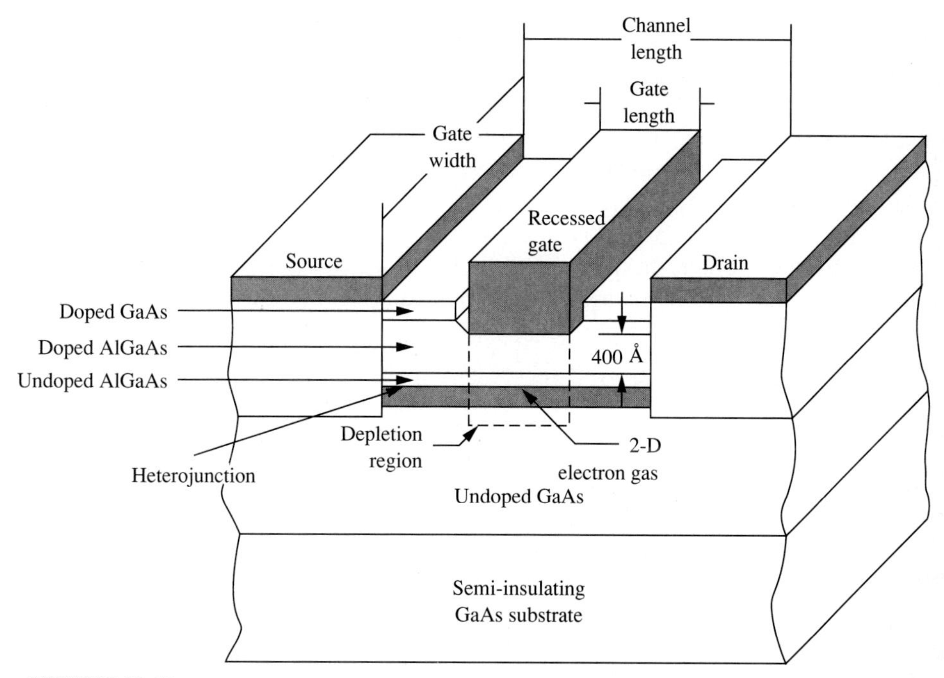

FIGURE 12.12
Schematic drawing of a GaAs High Electron Mobility Transistor (HEMT).

Research in implementing MODFET-type transistors is now being extended to other compound semiconductor materials. One such material is InP-based alloys where the peak electron velocity is higher than GaAs. These MODFETs exhibit unusual current-voltage characteristics at high drain voltages, such as negative differential resistivity.

12.6 BALLISTIC TRANSISTORS

As shown above, in addition to the base width for the bipolar and the channel length for the field-effect transistors, the delay time in transistors depends on the maximum speed with which the electrons can traverse these respective regions. Because of the presence of the impurity atoms and the lattice phonons, the electrons are scattered many times before they reach their final destinations, and in the process significantly change their direction and momenta. This of course leads to a reduction in the drift velocity of the electrons and, consequently, to the increase in the delay time. In GaAs, theoretically predicted maximum velocity in the [100] direction is 10^8 cm/s.

In order to reduce the transit time and thus increase the f_T of the transistor, the best solution would be to reduce further the base width or the channel length of the transistor so that the electrons can traverse these regions with negligible scattering. This will require a reduction of the base width and the channel length to lengths shorter than the mean scattering length of the electrons. In addition, if the electrons are injected into the base or the channel initially, with an excess energy above the thermal energy of the electrons, their transit time will be further shortened. The resulting motion of the electrons inside the semiconductor will be equivalent to the *ballistic* motion of the electrons, that is, as if they were moving in a vacuum. Ballistic motion is common to the electron motion in microwave beam devices where electrons move in a vacuum without any scattering and, thus, the collective motion of the electrons (interaction with other particles in the medium) can be neglected. The transistors whose operations are based on this principle are called *ballistic transistors*.

12.6.1 Metal-Base Transistors

When the base width of the transistor is reduced, the corresponding base resistance r'_b of the transistor increases drastically because of the decreased cross-sectional area of the base. We have seen that $f_{\max}$ of the transistor in Eq. 12.4 depends inversely on the square root of r'_b. Therefore, the resulting large-base resistance will reduce the high-frequency performance of the transistor. The two essential requirements of a small base width and the consequent low resistivity suggest replacement of the base semiconductor by a metal if the metal can form an effective Schottky barrier with the semiconductor and at the same time grow as a single-metal crystal over and in the same orientation with the semiconductor. Fortunately cobalt disilicate ($CoSi_2$) forms single crystal heterostructures with silicon. Small base widths made with metals give base resistivities an order smaller than that of the semiconductors. At the same time the electrons can move in the metal over the required distances with little scattering.

Details of a metal base (n-m-n) transistor are shown in Fig. 12.13. A $CoSi_2$ layer 10 nm thick is sandwiched between an n-type silicon emitter and an n-type silicon collector. The energy-band diagram of the electrons with no bias is shown in Fig. 12.14*a*. Both the emitter-base and the base-collector junctions form Schottky barriers. When a forward bias v_{EB} and a reverse bias v_{CB} are applied across the respective junctions, the energy-band diagram shifts, as shown in Fig. 12.14*b*. The electrons are injected as *hot* electrons into the base, with an excess kinetic energy above the thermal energy of the electrons in the metal. They move across the base ballistically and reach the collector with little scattering. The electrons that reach the base-collector junction do not move into the collector since, even though the energy of the incoming electrons is larger than the base-collector potential barrier, some of these electrons are now reflected back due to the presence of the base-collector potential barrier (see Chapter 1).

Current-voltage characteristics of the metal base transistor are shown in Fig. 12.15. The common-base configuration has i-v characteristics similar to the bipolar junction transistor (Fig. 12.15*a*). The unusual property of the metal-base ballistic transistor in the common-emitter configuration is that even with no collector-emitter voltage there is still current flowing into the collector because of the ballistic nature of the electron motion (Fig. 12.15*b*).

12.6.2 Ballistic GaAs Transistors

A bipolar version of the ballistic transistor can be made from GaAs, which has physical parameters that favorably enhance the high-frequency operation of the transistor.

A bipolar version of the GaAs ballistic transistor is shown in Fig. 12.16*a*. It is made from GaAs-AlGaAs heterostructures. Emitter and base are separated by a very thin layer of AlGaAs barrier. The base is made up of undoped GaAs so that while the electrons are moving toward the collector, their motion is not slowed down by scattering with the impurity atoms, and they can move through the base region with minimal scattering. Also, the total effective base width is made so

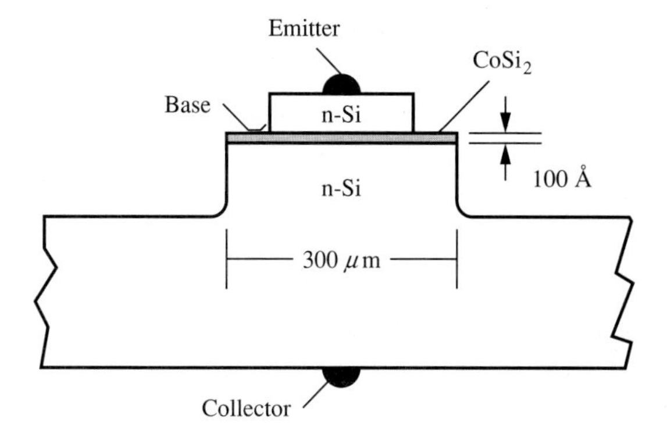

FIGURE 12.13
Schematic cross section of a metal-base transistor (Hensel, *et al.*, 1985).

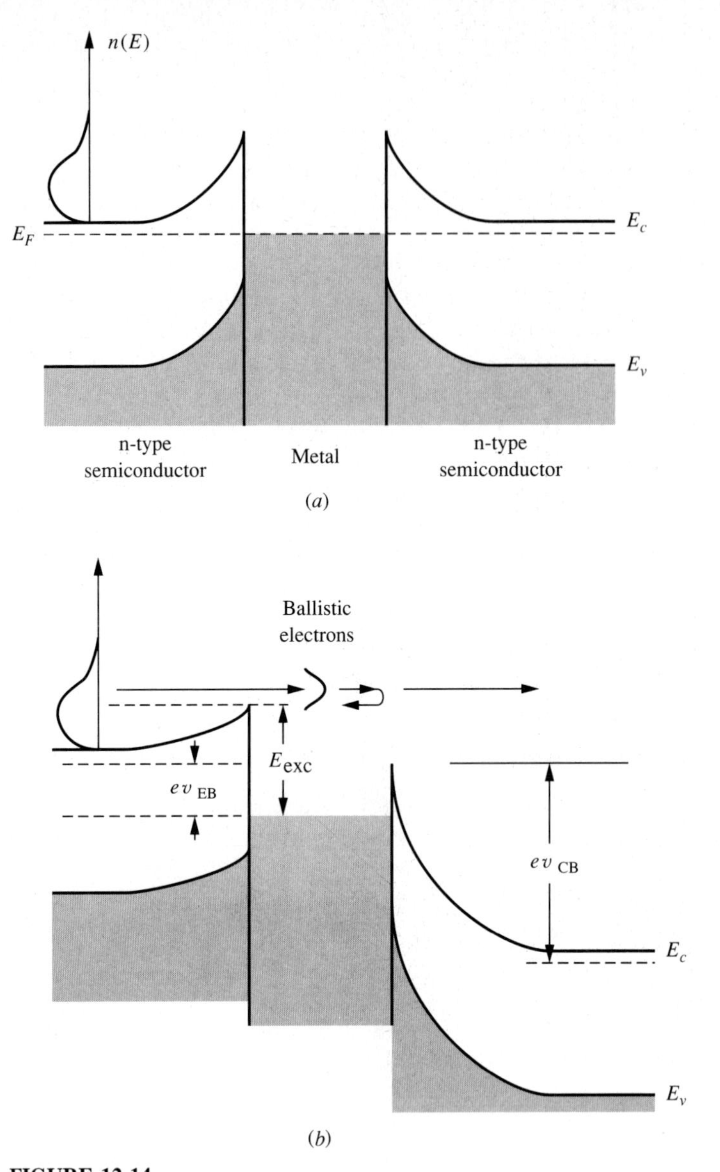

FIGURE 12.14
(*a*) Energy-band diagram of a metal-base (n-m-n) transistor. (*b*) Forward bias on the base-emitter allows "hot" electrons to be injected into the base with an energy of E_{exc} (Hensel, et al., 1985).

small that the electron transit time is comparable to or less than the scattering time of the electrons.

The energy-band diagram of the transistor is shown in Fig. 12.16*b*. Only the conduction bands of the various regions are shown. When the base-emitter junction is forward biased, the electrons from the emitter tunnel through the nar-

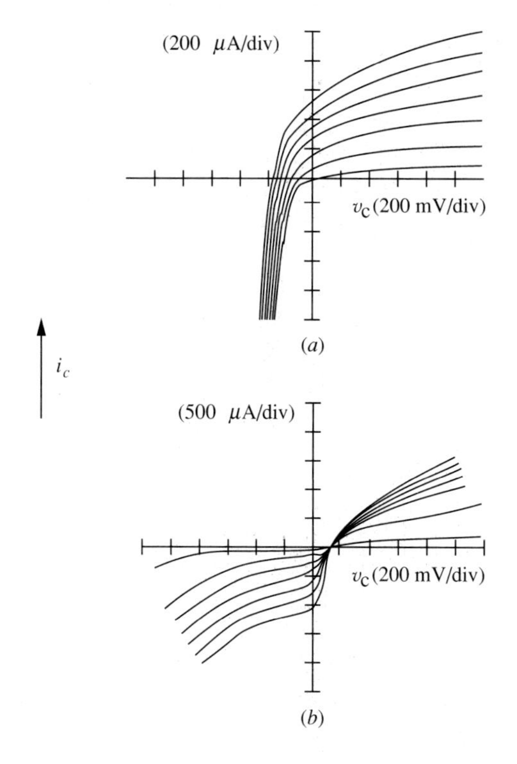

FIGURE 12.15
Current-voltage characteristics of the metal base transistor. (*a*) Common-base and (*b*) common-emitter characteristics (Hensel, *et al.*, 1985).

row AlGaAs barrier into the base. These electrons are injected into the base as hot electrons, with an energy higher than the thermal electrons in the base. They then enter the collector over the lowered base-collector barrier (Fig. 12.16*c*). In addition to controlling the i-v characteristics of the transistor, the energy spectra of the ballistic electrons can be analyzed by raising and lowering the base-collector barrier with the external-bias voltage.

Current-voltage characteristics of a ballistic GaAs transistor in the common-base configuration are shown in Fig. 12.17. Also shown is the distribution of the injected ballistic electrons. As a consequence of the electron tunneling through the AlGaAs barrier, these electrons do not have much of a spread in their energy when they are injected into the base.

A field-effect GaAs ballistic transistor has structural properties similar to its bipolar counterpart, with the exception that the current flow is controlled by the gate rather than by the base. In the vertical structure shown in Fig. 12.18, the electrons are injected into the channel through a heterojunction launcher. These electrons travel through the lightly doped channel of the transistor and reach the drain end of the transistor ballistically.

In general, the i-v characteristics of the ballistic transistor show that the terminal voltages of these transistors are relatively low compared to the conventional transistors. These transistors also have very short sub-picosecond switching times.

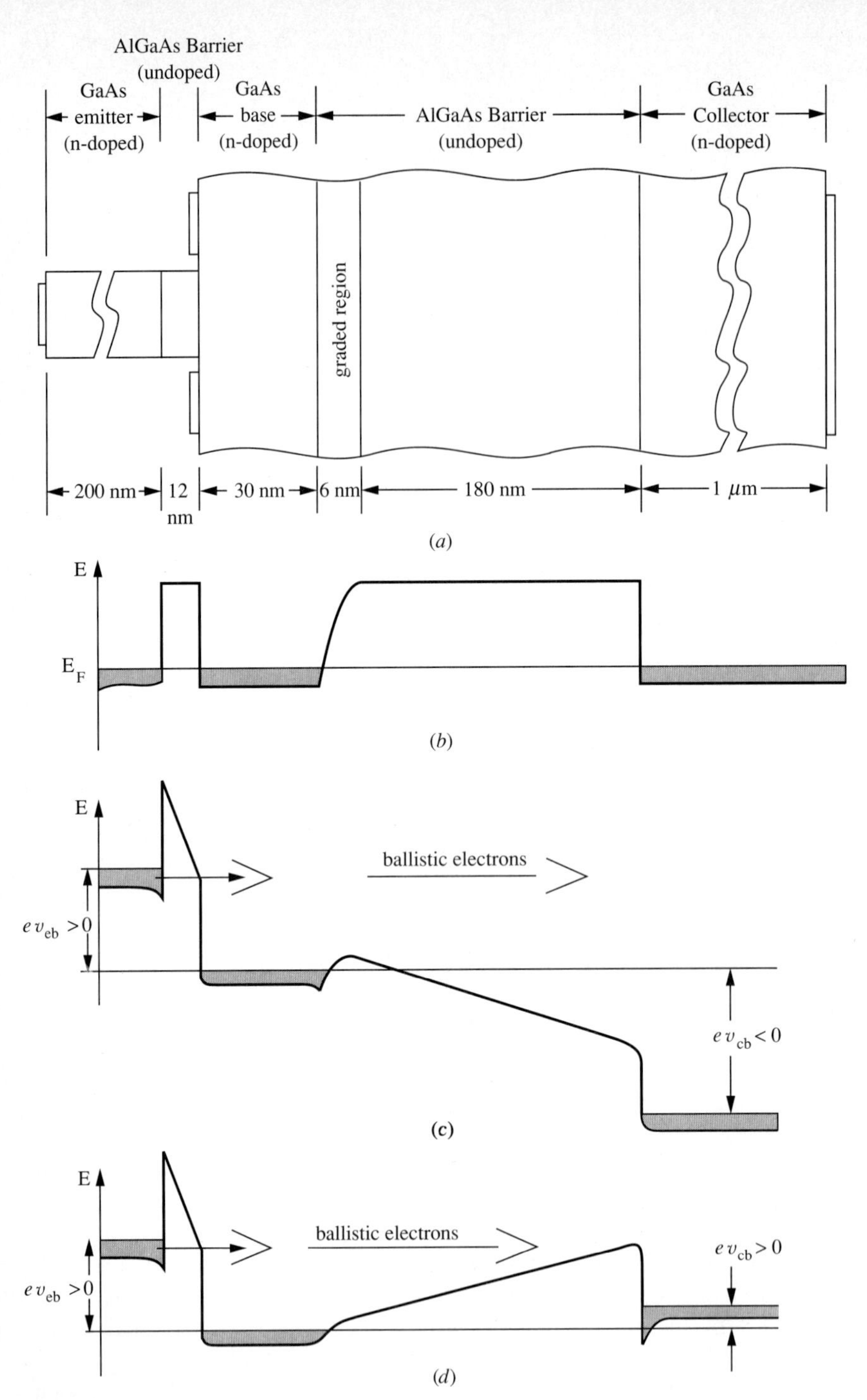

FIGURE 12.16
(*a*) Details of heterojunction GaAs ballistic transistor. (*b*) Energy-band diagram with no bias. (*c*) Forward bias in the base-emitter creates energy levels for the "hot electrons" to tunnel through. (*d*) By applying reverse and forward bias on the base-collector junction, ballistic electron spectrum can also be measured (Heiblum, *et al.*, 1985).

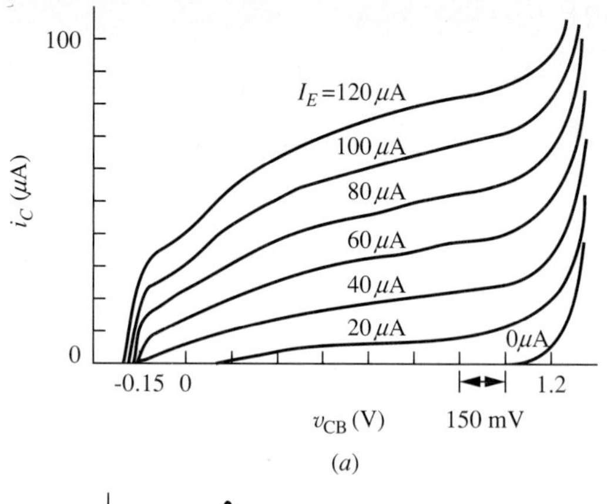

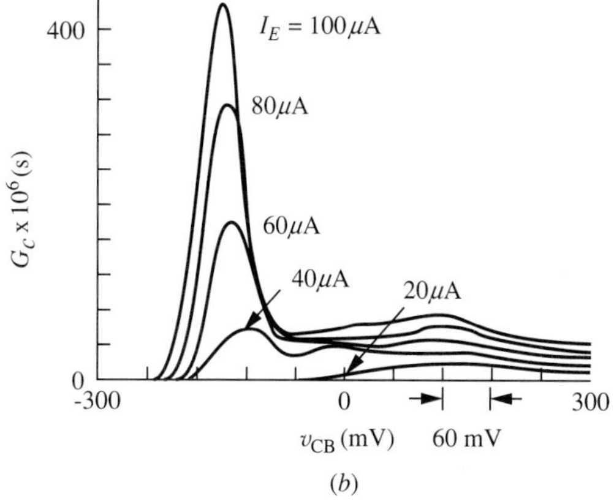

FIGURE 12.17
(*a*) *i-v* characteristics of GaAs ballistic transistor. (*b*) the energy spectra of the ballistic electrons (Heiblum, *et al.*, 1985).

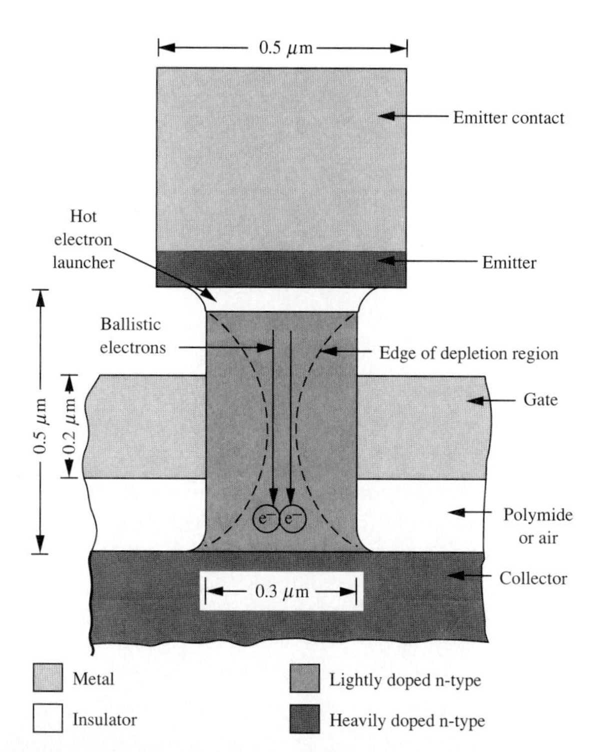

FIGURE 12.18
Vertical structure ballistic GaAs FET (L.F. Eastman, "Ballistic Electrons in Compound Semiconductors," *IEEE Spectrum,* vol. 23, 1986, 42–45, ©1986 IEEE).

Example 12.1. Using the dimensions of the GaAs ballistic transistor given in Fig. 12.16, estimate the transit time and f_T for the transistor if the electrons traversed the effective base distance *(a)* with saturation-limited drift velocity or *(b)* with ballistic motion. $v_{EB} = 0.2\ V$.

Solution. From Fig. 12.16, the effective distance that the electrons traverse before reaching the collector is $W_{beff} = 30\text{ nm} + 6\text{ nm} + 180\text{ nm} = 216\text{ nm}$.

(*a*) Using $v_{sL} = 10^7$ cm/s,

$$\tau_{T1} = \frac{W_{beff}}{V_{sL}} = \frac{(216 \times 10^{-9}\text{ m})}{10^5\text{ m/s}} = 2.26 \times 10^{-12}\text{s} = 2.26\text{ ps}$$

$$f_{T1} = \frac{1}{2\pi\tau_{T1}} = \frac{1}{2\pi(2.26 \times 10^{-12}\text{s})} = 7.37 \times 10^{10}\text{ Hz} = 73.7\text{ GHz}$$

(*b*) Assuming that the excess energy with which the electrons enter into the base region is the emitter-base bias potential and there are no drift fields at the base, the velocity of the electrons entering the base can be calculated from

$$V_{bal} = \sqrt{\frac{2ev_{EB}}{m}} = \sqrt{\frac{2(1.6 \times 10^{-19}\text{J/eV})(0.2\text{ eV})}{(9.1 \times 10^{-31}\text{ kg})}} = 2.65 \times 10^5\text{ m/s}$$

As the electrons move through the base with this constant velocity, the transit time τ_{T2} and the corresponding f_{T2} becomes

$$\tau_{T2} = \frac{W_b}{V_{bal}} = 8.15 \times 10^{-13}\text{ s}$$

$$f_{T2} = \frac{1}{2\pi\tau_{T2}} = 1.95 \times 10^{11}\text{ Hz} = 195\text{ GHz}$$

Thus, there is a marked improvement in the frequency response of a ballistic transistor over a conventional transistor.

12.7 TWO-TERMINAL SOLID-STATE DEVICES

In addition to the three-terminal transistors (usually referred to as *two-port* devices), there are devices with two terminals (*single* or *one-port* devices) that extend the operation of semiconductor devices into the millimeter regions of the microwave spectrum. These can be used as both amplifiers and oscillators, with higher efficiencies and with moderate power output levels up to millimeter wave frequencies. In order to introduce the very important negative differential resistance concept, only three specific examples out of the many forms of these two-terminal devices will be discussed here. These three will be discussed because of their interesting physical bases for operation and their many useful applications.

The drawback of any two-terminal device in implementing an amplifier can be attributed to the use of the same single port for both its input and output terminal. At microwave frequencies, this can be eliminated for certain amplifier configurations by using special microwave elements such as circulators which allow physical separation of the input and output ports.

12.7.1 Gunn Diode

When a bar of GaAs semiconductor is subjected to small electric fields, the drift velocity of the electrons, as expected, increases linearly with the increasing field. But as the field is further increased, the velocity increase deviates from linearity, reaches a maximum velocity, and starts to decrease with further increase of the electric field. It finally reaches a minimum (a valley) and starts to show a gradual increase again with higher fields (see Fig. 12.19). The region between the maximum and the valley is known as the *negative differential mobility* region. This effect is known as the *Gunn Effect*, after the scientist who first discovered and studied it (Gunn, 1963).

In order to explain these experimentally observed results, we use the energy-band diagram of GaAs shown in Fig. 12.20. As we explained in Chapter 5, GaAs is a direct band gap semiconductor but has a second minimum energy at a higher energy near the [100] zone edge. The minimum at the zone center [000] is known as the *central* or *lower valley* and the one near the zone edge at the higher energy is referred to as the *satellite* or *upper valley*. The mobility of the electrons in the lower valley is very large compared to the upper valley, while the effective mass of the electrons in the upper valley is very large compared to the lower valley. The energy difference between the two valleys is $\Delta E = 0.36$ eV.

For small values of the applied electric field, since the thermal energy of the electrons (0.026 eV) is a lot smaller than ΔE, most of the electrons reside in the

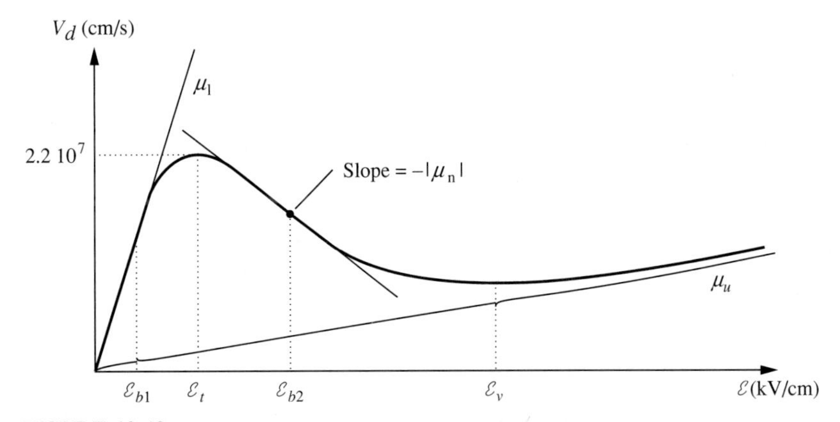

FIGURE 12.19
Drift velocity as a function of electric field applied across a bar of GaAs. Between $\mathcal{E}_t$ and $\mathcal{E}_v$, the bar exhibits negative differential mobility $-|\mu_n| \cdot \mu_u = 180$ cm^{2}V^{-1}s^{-1}, $\mu_l = 8800$ cm^{2}V^{-1}s^{-1}.

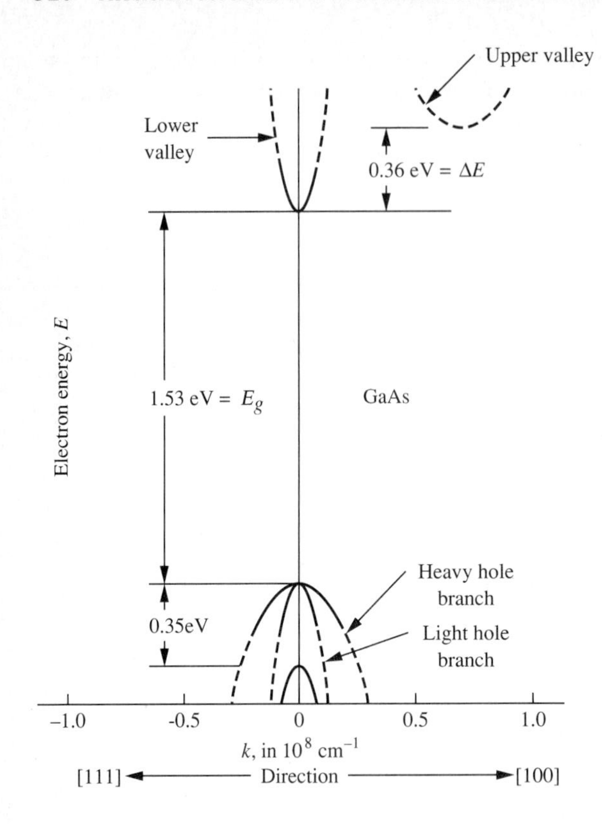

FIGURE 12.20
Energy-band diagram of GaAs showing the parameters that lead to the negative differential mobility. At lower valley, $m^*_{el} = 0.068\ m_{eo}$, $\mu_l = 8800$ cm^2/V s. At upper valley, $m^*_{eu} = 1.2m_{eo}$, $\mu_u = 180$ cm^2/V s (Long, 1964).

lower energy valley where their mobilities are very high. The slope of the linear low-field region is equal to the lower-valley mobility μ_l. As the electric field is increased, the electrons are scattered into the low-mobility upper valley. Since the number of available states is also proportional to the effective mass of the electrons, more electrons are transferred to the upper valley, and thus the overall mobility of the electrons decreases. The changes in mobility are so rapid that the mobility, which is 8800 cm^2/V· s at low fields, reaches zero at the *threshold field* of about $\mathcal{E}_t = 3k$V/cm. With further increase in the field, the negative differential mobility $dv/d\mathcal{E}$ is observed. Finally, at higher fields, the mobility starts to increase with a smaller slope equal to the upper-valley mobility μ_u.

We conclude that to possess negative differential mobility, any other bulk semiconductor should have similar properties to those of GaAs, that is,

1. The energy difference ΔE between the central and satellite valleys should be large compared to kT;
2. The band gap energy should be larger than ΔE, that is, $E_g > \Delta E$; and
3. The mobilities and effective masses of the electrons in the semiconductor at the central and satellite regions should also possess values similar to those of GaAs. InP is another important compound semiconductor that exhibits negative differential mobility.

With the application of large electric fields, the electrons are transferred to the upper valley. Therefore, devices based on this phenomenon are referred to as *transferred electron devices* or TEDs.

12.7.1.1 DOMAIN FORMATION. If the conditions are right, any space-charge fluctuation in a GaAs sample leads to a nonlinear *dipolar-domain* formation as a result of the differential negative mobility. The formation of the domains can be explained with the help of Figs. 12.19 and 12.21. A voltage difference of V_0 is

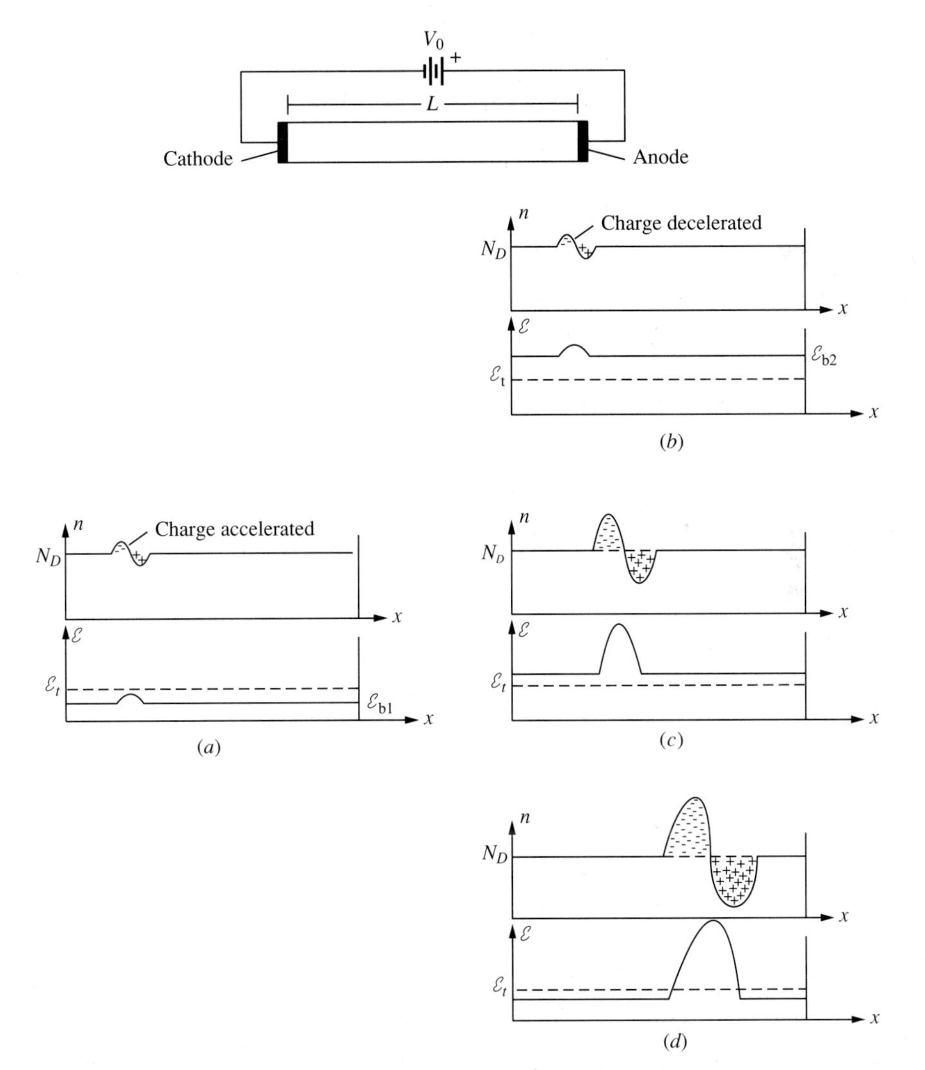

FIGURE 12.21
(*a*) Space-charge distribution and the resulting electric field in a linear device where the disturbance decays down, $\mathcal{E}_{b1} < \mathcal{E}_t$. (*b*) Same disturbance when $\mathcal{E}_{b2} > \mathcal{E}_t$. (*c*) Disturbance and the field at a later time. (*d*) Completed dipole domain and the resulting electric field.

applied across a GaAs bar of length L. The two ends of the bar are labeled as the cathode (− terminal) and anode (+ terminal).

In order to understand the behavior of various charges and the resulting electric fields, we will make use of the one-dimensional Poisson equation in the form

$$\epsilon_s \frac{d\mathscr{E}_x}{dx} = e(N_D - n)$$

$$\mathscr{E}_x = \frac{1}{\epsilon_s}\int_0^x e(N_D - n)dx \tag{12.19}$$

and the definition of potential difference

$$v = -\int_0^L \mathscr{E}_x dx = \text{constant} = V_0 \tag{12.20}$$

where $\mathscr{E}_x$ is the electric field, n is the electron density, N_D is the donor density (assumed to be 100 percent ionized), and ϵ_s is the dielectric permittivity of the semiconductor bar.

At equilibrium and with no disturbance, $n = N_d$ throughout the bar. If, due to any electrical disturbance, the local electron density at the point of the disturbance increases in the sample bar and becomes larger than the ionized donor density (the local relative increase in electron density at that point will be referred to as *excess charge*), to satisfy the charge-neutrality condition, there should be a reduction of electrons (*deficiency of electrons*) in close vicinity of the disturbance (Fig. 12.21*a*). From Eq. 12.19, the local electric field generated by this disturbance is determined by the integration of the charge disturbance over the length of the disturbance. The local field will increase first, then reach a maximum and drop back to zero. The direction of this field is in the $-x$ direction which is also the direction of the externally applied field and thus increases the total electric field in the vicinity of the disturbance. Also, from the integral in Eq. 12.20, if the electric field increases at one point, it should decrease at other points to keep the line integral of the electric field—that is, the total applied-potential difference across the bar—constant.

Depending on the bias conditions, we can analyze the history of the disturbance in the following way:

1. If the bar is biased with $\mathscr{E}_{b1} < \mathscr{E}_t$, and if a charge fluctuation occurs at some point along the bar due to some noise or other disturbance, the electric field will also show a disturbance as a result of Eq. 12.19 (Fig. 12.21*a*). Since the excess charges reside in an increasing electric field region, they will have higher velocities (Fig. 12.19) and will catch up with the deficient electrons, which have relatively slower velocities. Eventually, the disturbance dies out. This is the usual dielectric relaxation in a linear medium.
2. If the bar is now biased so that this time $\mathscr{E}_{b2} > \mathscr{E}_t$, again the charge disturbance will create an electric field disturbance similar to case 1 (see Fig. 12.21*b*). But

since the bar is now biased in the negative differential mobility region, this time the excess electrons are slowed down because of their relatively lower velocity, and the deficient electrons are accelerated, because of their relatively higher velocity. As a result, the magnitude of the disturbance is magnified. The corresponding electric field at the location of the disturbance is proportionately increased (Fig. 12.21*c*). Since the total applied potential across the bar should stay constant, the electric field at other points along the bar should decrease to satisfy Eq. 12.20. As the disturbance propagates, its magnitude grows until the electric field behind and ahead of the disturbance falls below the threshold value $\mathscr{E}_t$ (Fig. 12.21*d*). The growth of the disturbance then stops and, thus, the *dipole domain* that has formed keeps moving toward the positive electrode. Actually the shape of both the dipole domain and the corresponding electric field evolves through a nonlinear process determined by the detailed solution of the Poisson equation.

Due to the manufacturing process, the crystal structure at the cathode end is disturbed more than at any other point along the bar. It is likely that the domain formation will start at the cathode end (negative terminal). In a device of finite length, the domain, once formed, keeps moving toward the positive terminal (anode). When it reaches the anode, the current in the external circuit is accordingly disturbed and an increase in external current is observed. The disturbance in the external current eventually dies out and the process repeats itself by the formation of a new domain. The resulting external current waveform is shown in Fig. 12.22*a*.

The stationary electric field in a TED will not increase and no domain will form if the space-charge injected at the cathode will not have sufficient time to grow before it reaches the anode. In a uniform bar, the temporal dependence of the growth rate of the disturbance is related to the usual dielectric growth rate and is given by

$$Q_a = Q_c \exp\left(-\frac{t}{\tau_d}\right) \tag{12.21}$$

Using the dielectric relaxation time τ_d given by Eq. 12.14 and using the differential mobility $\mu = (dv_d/d\mathscr{E}) = -|_n|$ (see Fig. 12.19), we obtain

$$Q_a = Q_c \exp\left[-\left(\frac{L/v_{\mathrm{sL}}}{\epsilon_s/en_\mu}\right)\right] = \exp\left(\frac{Len|\mu_n|}{\epsilon_s v_{\mathrm{sL}}}\right)$$

$$= Q_c \exp(KnL) \tag{12.22}$$

where Q_a is the charge reaching the anode due to the charge Q_c created at the cathode, $t = (L/v_{\mathrm{sL}})$ (L is the length of the uniform bar and v_{sL} is the velocity with which the domain moves), and K is a constant dependent on the semiconductor material. Note that Q_a grows because of the negative differential mobility.

It has been shown for GaAs that if the nL product in Eq. 12.22 is less than 7.6×10^{11} cm^{-2}, there will not be sufficient time for the domain to form. In an

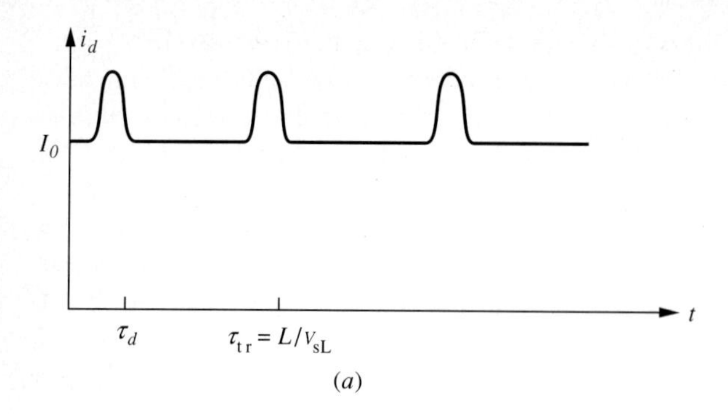

(a)

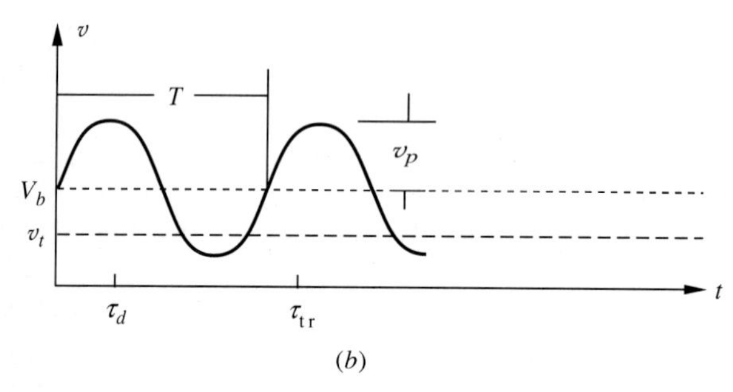

(b)

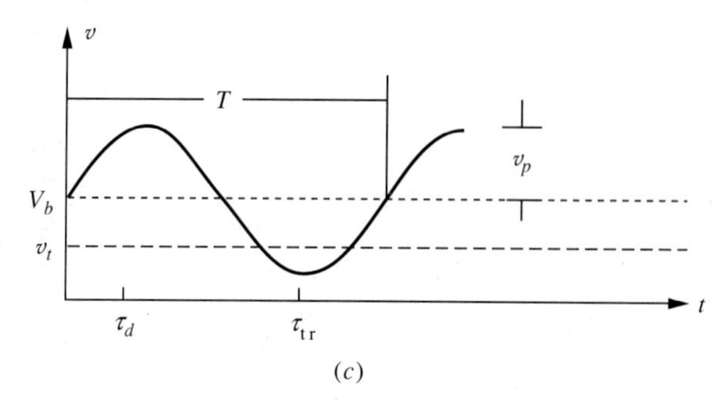

(c)

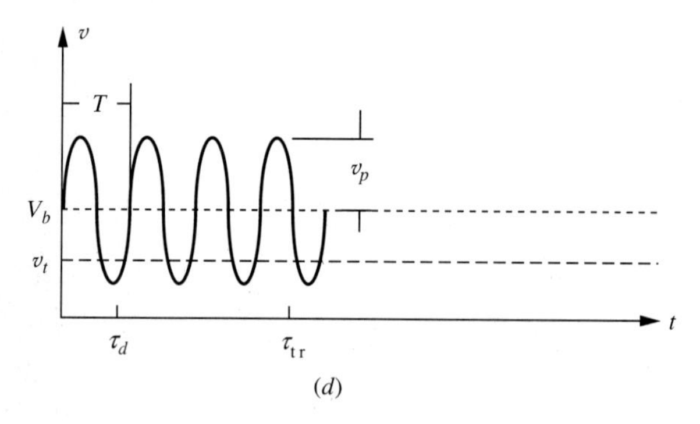

(d)

FIGURE 12.22
Various operating domain modes of a Gunn mode. (*a*) Transit time mode, (*b*) Quenched domain mode, (*c*) Delayed domain mode and (*d*) Limited space charge accumulation mode. τ_{tr} is the transit time of the domain, τ_d is the dielectric relaxation time, and T is the period of RF oscillation.

actual GaAs TED, the required nL product is generally larger than 7.6×10^{11} cm^{-2} because of the simplifying assumptions made here. But still the main parameters that control the domain formation are the length of the bar and the original impurity donor concentration ($n \approx N_D$).

12.7.1.2 MODES OF OPERATION. In order to operate TEDs as microwave devices, it is not always necessary to form a domain in the bar. There are various modes of operation.

Stable negative resistance mode. In this mode, a short length of a bar is biased at the negative differential mobility region. For small amplitude signals, the impedance of the bar can be shown to have a negative real part (negative resistance) over a large frequency range.

Domain modes. If the domain is allowed to fully form before it reaches the anode, the resulting bar will still have a negative real part, but the operation in this mode will depend largely on the properties of the external circuit. Forms of domain modes are

1. *Transit Time Mode*. The TED is operated with a low Q circuit so that the domain moves from the cathode to the anode and generates short current pulses (Fig. 12.22*a*). This mode of operation is not very useful.
2. If the total voltage across the TED swings well below the threshold field during part of the RF cycle, causing the existing domain to discharge or prevent nucleation of new domains, there will result the *Quenched Mode of Operation*, which will occur when a fully formed domain is discharged or *quenched* before it reaches the anode. The period of *RF* oscillation has to be smaller than the domain transit time to quench a domain (Fig. 12.22*b*).
3. *Delayed Domain Mode*. The period of *RF* oscillations is larger than the domain transit time. The domains travel the full length of the bar but the nucleation of the domain is delayed for a fraction of an *RF* cycle while the total voltage across the TED is below the threshold (Fig. 12.22*c*).

These modes need a high Q circuit for their operation. The theoretical efficiency of oscillators operating in the quenched and delayed modes can be as high as 13 percent and 17 percent, respectively.

Limited space charge accumulation (LSA) mode. In this mode, the formation of traveling domain is suppressed by an *RF* voltage whose amplitude is large enough to derive the TED below threshold during every RF cycle, and at the same time the frequency is so high that there is not enough time to form during the part of the RF cycle where the device voltage is above threshold (Fig. 12.22*d*). This mode of operation, independent of the transit time of the carriers, is controlled by external circuitry. The calculated maximum efficiency of the LSA mode is in the range of 18 to 23 percent.

Traveling wave mode. A space-charge wave is excited by means of a radio-frequency coupler at the cathode end of a stable device. It grows exponentially through the TED and excites an amplified RF wave in the output coupler located near the anode end of the bar. In this mode, the TED can provide unidirectional gain without the use of any ferrites and hybrids.

12.7.2 IMPATT (IMPact Avalanche and Transit Time) DIODE

Negative differential resistance can also be obtained if a voltage $V_0 + v$ (ac voltage superimposed on a dc voltage) is applied across a sample of a uniform n-type semiconductor bar (called the *drift region*), as shown in Fig. 12.23. The bar contains a short length of an additional region from which electrons can be injected into the drift region starting at $x = 0$. The applied dc field is so high that the injected electrons move over the length of the bar with the saturation limited velocity v_{sL}.

It will be assumed that due to the applied ac signal v, the total ac current density, which is made up of the ac particle current density and the displacement current density, is independent of the position along the bar. It will also be assumed that the particle current density injected into the bar at $x = 0$ is equal to the total ac current density J but is injected with a relative time phase angle β, that is,

$$J_p(0) = J e^{(-j\beta)} \tag{12.23}$$

The importance of β will be explained shortly. The total current density at any position x can be written as the sum of the particle current density and the displacement current density.

$$J = J_p(x) + j\omega\epsilon_s \mathcal{E}(x) \tag{12.24}$$

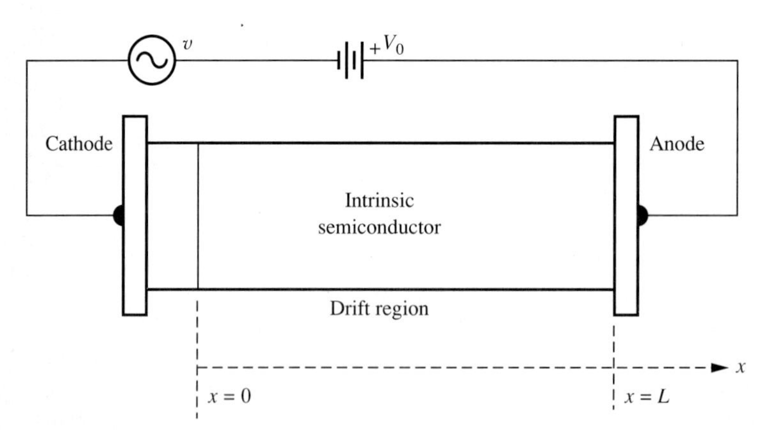

FIGURE 12.23
Charges injected at $x = 0$ drift toward the anode with v_{sL}. Delay in the injected charges and the length of the drift region determines the ac terminal characteristics of the bar.

Solving this for $\mathscr{E}(x)$, we obtain

$$\mathscr{E}(x) = \frac{J - J_p(x)}{j\omega\epsilon_s} \tag{12.25}$$

The spatial dependence of the ac particle current density can be written as

$$J_p\,(x) = J_p\,(0)\,\exp\,(-j\beta_e x)$$
$$J_p\,(x) = J_p\,(0)\,\exp\left[\left(-\frac{j\omega}{v_{sL}}\right)x\right] \tag{12.26}$$

where β_e is the electron phase constant defined by ω/v_{sL}. If both sides of Eq. 12.25 are multiplied by dx and integrated from 0 to L, the left side becomes equal to the applied potential. Using Eq. 12.26 and integrating the right side, we obtain

$$v = \frac{JL}{j\omega\epsilon}\left\{1 - \frac{e^{(-j\beta)}\,[1 - e^{(-j\theta)}]}{j\theta}\right\} \tag{12.27}$$

where θ, the transit angle, is equal to $(\omega L/v_{sL})$. Multiplying and dividing the right side of Eq. 12.27 by the cross-sectional area A of the bar and defining the capacitance of the bar as $C \equiv (\epsilon_s A/L)$, the ac impedance of the bar can be written as

$$Z(\omega) = \frac{v}{i} = \left\{1 - \frac{[e^{(-j\beta)}\,(1 - e^{(-j\theta)}]/j\theta}{j\omega C}\right\} \tag{12.28}$$

Taking the real and imaginary parts of $Z(\omega)$, we obtain

$$R(\omega) = \frac{\cos\beta - \cos\,(\beta + \theta)}{\omega C\theta} \tag{12.29a}$$

and

$$-X(\omega) = \frac{1}{\omega C} - \frac{\sin\,(\beta + \theta) - \sin\beta}{\theta} \tag{12.29b}$$

Due to the causality, the injected electron phase has to be equal to or greater than zero, that is, $\beta \geq 0$. For $\beta = 0$, whatever the value of θ, $R \geq 0$. On the other hand, for $0 < \beta \leq \pi/2$, R will admit negative values. Maximum value of $\cos\beta - \cos(\beta + \theta) = -1$ will occur if $\beta = \pi/2$ and $\theta = 3\pi/2$. Due to the presence of θ in the denominator in Eq. 12.29a, the actual maximum negative resistance is achieved for somewhat lower values of θ, but the optimum value still occurs for $\beta = \pi/2$. The angle θ that maximizes Eq. 12.29b is found from the direct solution of the transcendental equation $\tan\theta = \theta$ which gives $\theta = 257.5$ degrees.

12.7.2.1 MODES OF CREATING INJECTION DELAY

Ohmic contact. The injected-particle current will be in phase with the total current and will be useless in obtaining negative resistance.

Forward-biased Schottky barrier. At any forward bias, the equivalent circuit of such a device is a parallel combination of a conductance and a capacitive admittance as shown in Fig. 12.24. The total current across this junction can be written as

$$i = j\omega C v + G v$$

The phase of the injected current is given by

$$\tan \beta = \frac{\omega C}{G} \tag{12.30}$$

To get $\beta = \pi/2$ implies that $\omega C >> G$. However, this means that the magnitude of the particle current is only a fraction of the total current. This will lead to a very small negative resistance. For this diode, the most favorable condition may be obtained for $\omega C = G$.

Avalanche breakdown. We now consider a reverse-biased p^+-n junction as a possible candidate for creating the right injection phase. As shown in Fig. 12.25, a $p^+ - n$ junction is included next to the intrinsic region. The applied dc field across the junction is high enough that a small ac field perturbation can produce avalanche multiplication across this junction.

The satisfactory generation of the necessary phase relation of $\beta = \pi/2$ as a result of the avalanche breakdown mechanism can easily be shown by using the conservation of charge equation. Rewriting Eq. 4.15 for electrons

$$\frac{\partial n}{\partial t} + \nabla \cdot n\text{v} = <\text{v}> \alpha n \tag{12.31}$$

where $<\text{v}> \alpha n$ is the generation rate of the electrons due to avalanche breakdown, α is the ionization rate which is a function of the electric field $\mathscr{E}$, and $<\text{v}>$ is the constant average electron velocity.

Ac variables will be assumed to be small perturbations over the dc parameters (labeled with subscripts 0), that is,

$$n = n_0 + \tilde{n}$$

$$\alpha = \alpha_0 + \left(\frac{\partial \alpha}{\partial \mathscr{E}}\right)_0 \mathscr{E}$$

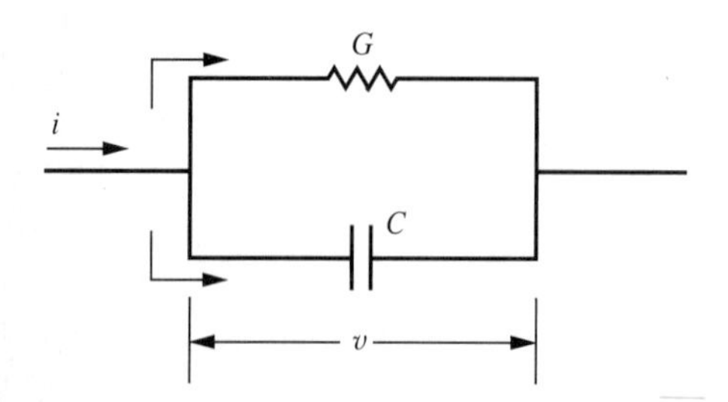

FIGURE 12.24
Equivalent circuit of a Schottky barrier diode. C is the diode capacitance and G is the ac conductance at the operating point.

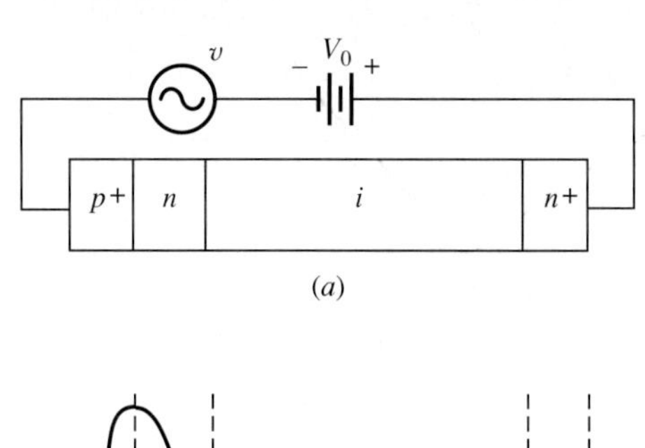

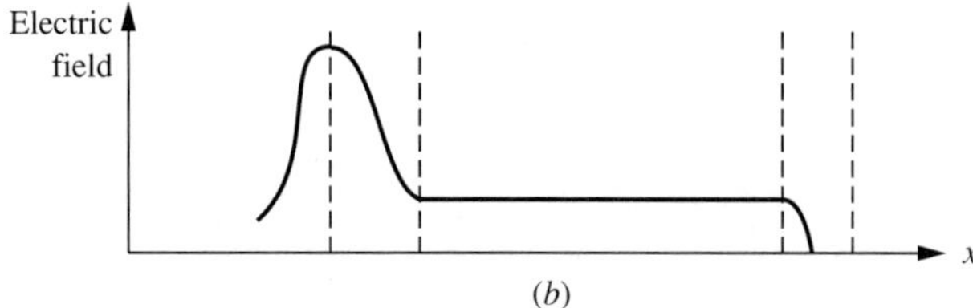

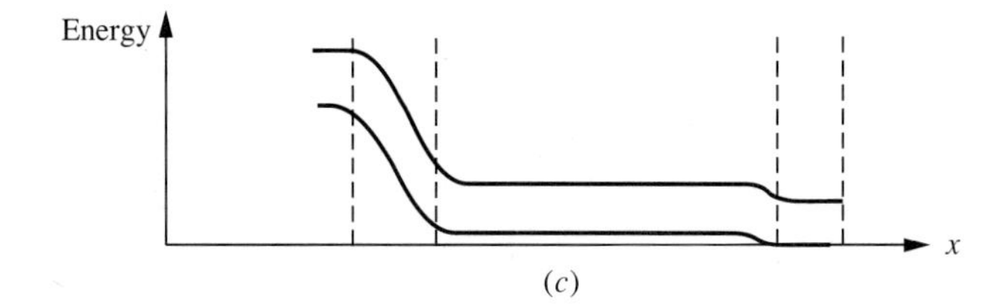

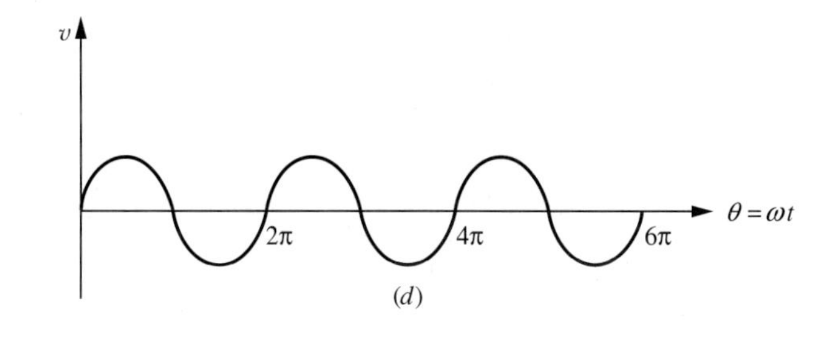

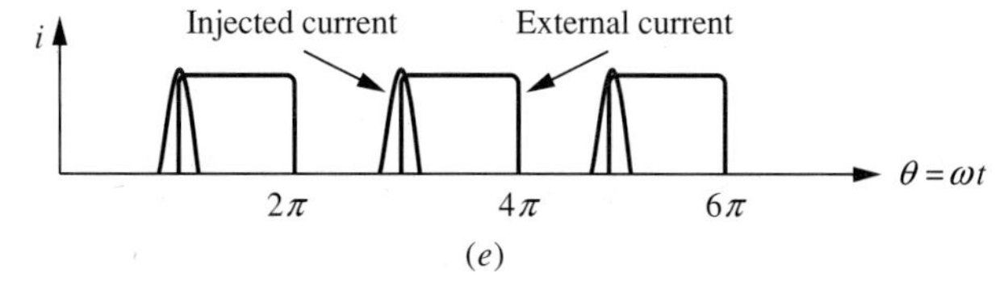

FIGURE 12.25
Parameters related to an IMPATT diode. (*a*) The constructional detail, (*b*) electric field, (*c*) energy, (*d*) ac voltage, and (*e*) the injected and external currents (S. M. Sze and R. M. Ryder, "Microwave Avalanche Diodes," *Proc. IEEE,* vol. 59, 1971, 1140–1154, ©1971 IEEE).

n_0 is the steady-state electron density, $\tilde{n}$ is the ac component of the electron density, and $(d\alpha/d\,\mathcal{E})_0\,\mathcal{E}$ is the change in the ionization coefficient due to the perturbing electric field $\mathcal{E}$.

If we also assume that $\tilde{n}$ is independent of the position in the avalanche region, to a first-order approximation Eq. 12.31 reduces to

$$\frac{\partial \tilde{n}}{\partial t} = j\omega\tilde{n} = \left(\frac{\partial \alpha}{\partial \mathscr{E}}\right)_0 < v > n_0 \mathscr{E} \qquad (12.32)$$

which shows that the phase difference between the $\tilde{n}$ and $\mathscr{E}$ will be exactly equal to 90°. We can, therefore, conclude that the avalanche breakdown should be a very effective injection mechanism.

A widely used avalanche diode is the IMPATT diode. The constructional details, the spatial variation of the electric field, and the electron energy, as well as the time-dependent applied ac voltage and the resulting external current, are shown in Fig. 12.25.

Due to the very high field across the p^+-n junction, electron-hole pairs are generated. The holes quickly enter the p^+-region. The generated electrons are injected into the drift space where they drift toward the n^+ ohmic contact with a constant v_{sL}.

As the ac electric field $\mathscr{E}$, which is directly proportional to the applied ac signal voltage v, varies around an average value, the impact ionization rate per carrier follows field changes almost instantaneously. However, the carrier density does not follow the field change in unison, because the carrier generation also depends on the number of electrons already present at that instant. Even after the field has passed its maximum value, the carrier density keeps increasing because the carrier density is still above the average value. The maximum charge carrier density is reached approximately when the field decreases from the peak value. Although the ionization rate is in phase with the field, the ac variation of the carrier density lags behind ionization by about 90°. This situation is illustrated as the injected current in Fig. 12.25*e*. The peak value of ac field occurs at $\theta = \pi/2$, but the peak of the injected current occurs at $\theta = \pi$. The injected electrons then enter the drift region, where they move with constant v_{sL}. Comparing the ac field with the external current, it is clear that the diode exhibits negative resistance at its terminals since the particle motion through the intrinsic region occurs under the influence of the dc-bias during the negative cycles of the ac signal. That is, the phase difference between the ac voltage and ac current becomes 180°.

12.7.3 Tunnel Diode

The operation of a tunnel diode can only be explained by using quantum mechanical concepts. Although it was one of the earlier two-terminal devices that exhibited negative differential resistance, because of its limited voltage range it did not find wide application areas as did other negative resistance devices. The current-voltage characteristic of a tunnel diode is shown in Fig. 12.26.

The tunnel diode is a p-n junction where both the p- and the n-regions are heavily doped. This is known as degenerate doping, and the impurity concentrations are so high that the respective Fermi levels lie in the valence and the conduction bands. When the junction is first formed, the usual diode potential forms. The resulting energy diagram of the diode is shown in Fig. 12.27*a*. Another important property of the tunnel diode is that the depletion-layer width is very small, that is, on the order of 100 Å or less.

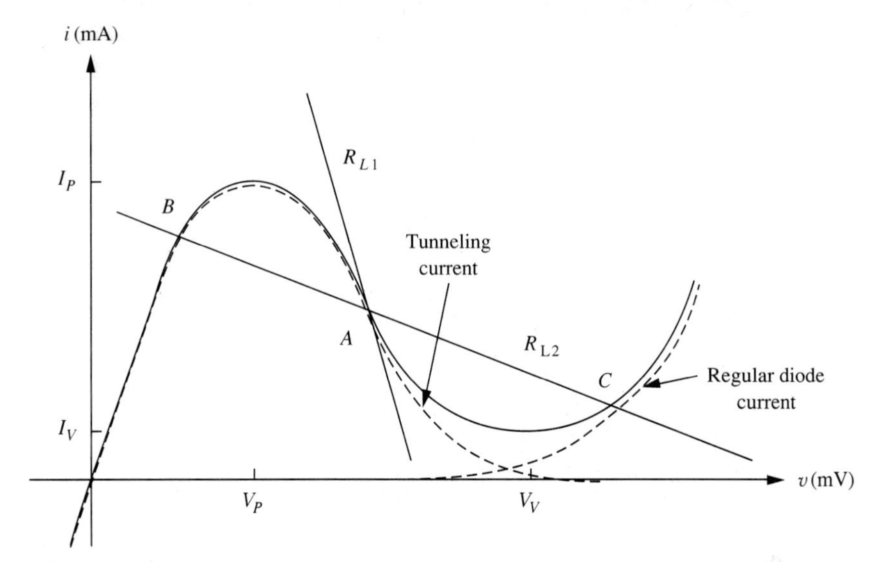

FIGURE 12.26
Current-voltage characteristics of a tunnel diode. The total current is the sum of the tunneling and usual diode currents. Between V_P and V_V the diode exhibits negative differential resistance.

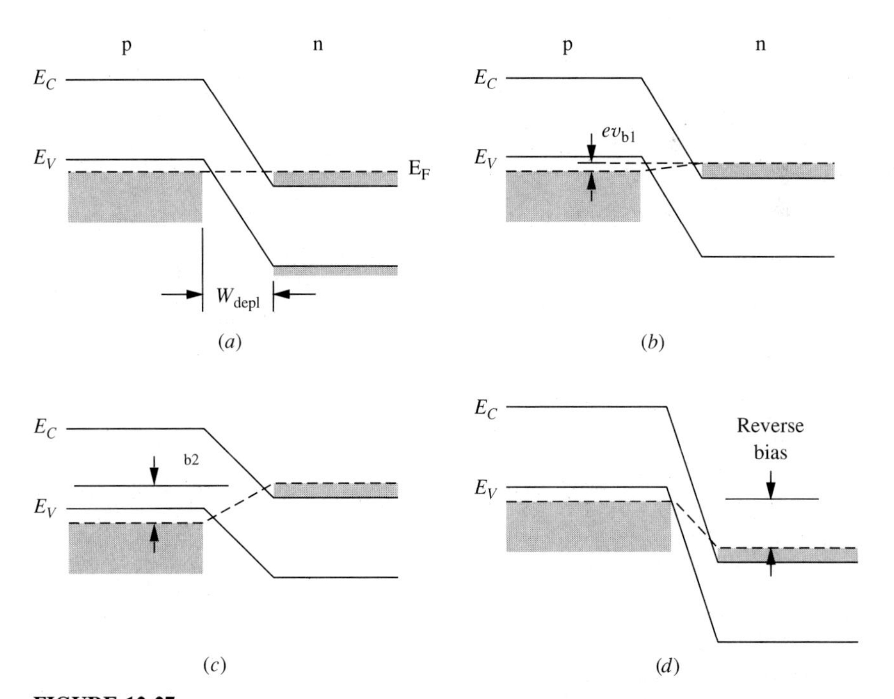

FIGURE 12.27
Energy-band diagram of a tunnel diode. (*a*) No bias, (*b*) with small forward bias, (*c*) with bias where the tunneling current reaches zero, and (*d*) with reverse bias.

When a small forward bias is applied across the diode, the energy levels shift (Fig. 12.27*b*). Some of the electrons in the conduction band of the n-region see empty energy states in the valence band of the p-region. Since the potential barrier width is small, there is a finite probability that these electrons will tunnel through to the p-region. Therefore, we expect an increase in the external current with the applied bias. As the forward bias is increased further, the tunneling current reaches a maximum. Further increase in bias voltage shifts the energy levels further. This leads to a smaller number of available states in the valence band of the p-region for the electrons in the conduction band of the n-region having the same energy level. At the same time, more electrons see the forbidden band gap and thus cannot tunnel through. Eventually, the tunneling electron current stops when the bottom of the conduction band comes across the top of the valence band. In addition to the tunneling current, the usual diode current also starts to flow with increasing bias. The resulting overall current-voltage characteristic is the sum of the tunneling current and the usual diode current (Fig. 12.26). When a reverse bias is applied to the diode (Fig. 12.27*d*), the electrons in the valence band see more available energy states in the conductions band, so there will be a large current flow in the reverse direction.

As can be seen from Fig. 12.26, the diode exhibits negative differential resistance between the peak and the valley voltages. The difference $V_V - V_P$ is usually within a few hundred millivolts. If the diode is biased at point A with a proper load resistance R_{L1}, it can be used as an amplifier or an oscillator. If the diode is biased with R_{L2}, the diode can be used as a switch or a memory element since it will now have two stable states B and C. The switching time of the tunnel diode is very fast, but unfortunately the voltage swing is very limited.

FURTHER READING

Bahl, Inder, and Prakash Bahartia, *Microwave Solid-State Circuit Design*, Wiley, New York, 1988.

Cooke, H. F., "Microwave Transistors: Theory and Design," *Proc. IEEE*, vol. 59, 1971, 1163–1181.

Eastman, L. F., "Ballistic Electrons in Compound Semiconductors," *IEEE Spectrum*, vol. 23, 1986, 42–45.

Eastman, L. F., *Gallium Arsenide Microwave Bulk and Transit Time Devices*, Artech House, Dedham, Mass., 1973.

Gilden, M., and M. E. Hines, "Electronic Tuning Effects in Read Microwave Avalanche Diodes," *IEEE Trans*, ED-13, 1966, 168–175.

Gossard, A. C., "Quantum Effects at GaAs/Al_xG_{1-x}As Junctions," *Thin Solid Films*, vol. 104, 1982, 279–284.

Gunn, J. B., "Microwave Oscillations of Current in III-V Semiconductors," *Solid-State Communications*, vol. 1, 1963, 88-91.

Heiblum, M., M. I. Nathan, D. C. Thomas, and C.M. Knoedler, "Direct Observation of Ballistic Transport in GaAs," *Phys. Rev. Lett.*, vol. 55, 1985, 2200–2203.

Hensel, J. C., A. F. Levi, R. T. Tung, and J. M. Gibson, "Transistor Action in Si/$CoSi_2$ /Si Heterostructures," *Appl. Phys. Lett.*, vol. 47, 1985, 151–153.

Laskar, L., et al., "Gate-Controlled Negative Differential Resistance in Drain Current Characteristics of AlGaAs/InGaAs/GaAs Pseudomorphic MODFETs," *IEEE Elect. Dev. Let.*, vol. 10, 1989, 528–530.

Liechti, Charles A., "Microwave Field Effect Transistors-1976," *IEEE Trans. Micr. Theory and Tech.*, vol. MTT-24, 1976, 279–290.

Long, D., "Energy-Band Structure of Mixed Crystals of III–V Compounds," in *Semiconductors and Semimetals*, vol. 1, R. K. Willardson and A. C. Beer, eds., Academic Press, New York, 1964, 143–158.

McCumber, D. M., and A. G. Chenoweth, "Theory of Negative Conductance Amplification and of Gunn Instabilities in 'Two Valley' Semiconductors," *IEEE Trans.*, ED-13, 1966, 4–21.

Morkoc, K., and P. M. Solomon, "The HEMT: a Superfast Transistor," *IEEE Spectrum*, vol. 21, 1984, 28–35.

Pucel, R. A., H. A. Haus and H. Statz, "Signal and Noise Properties of GaAs Field Effect Transistors," in *Advances in Electronics and Electron Physics,* vol. 38, L. Marton, ed., 1975, 195–265.

Sze, S. M., and R. M. Ryder, "Microwave Avalanche Diodes," *Proc. IEEE,* vol. 59, 1971, 1140–1154.

Sze, S. M., *Physics of Semiconductor Devices,* 2d ed., Wiley, New York, 1981.

Teszner, S., and J. L. Teszner, "Microwave Power Semiconductor Devices I. Critical Review," in *Advances in Electronics and Electron Physics*, vol. 39, L. Marton, ed., Academic Press, New York, 1975, 291–378.

Teszner, S., and J. L. Teszner, "Microwave Power Semiconductor Devices II. Critical Review," in *Advances in Electronics and Electron Physics*, vol. 41, L. Marton, ed., Academic Press, New York, 1977, 141–219.

Tsang, W. T., "MBE for III-V Semiconductors," in *Semiconductors and Semimetals*, vol. 22, Part A, R. K. Willardson and A. C. Beer, eds., Academic Press, New York, 1985.

Watkins, E. T., "GaAs FET Amp uses one-quarter Micron Gate Heralding MIC Opportunities at up to 60 GHz," *Microwave System News,* vol. 13, 1983, 52–62.

White, M. H., and M. O. Thorston, "Characterization of Microwave Transistors," *Solid-State Electronics*, vol. 31, 1970, 523–542.

PROBLEMS

12.1. For a Si n-p-n transistor, the following parameters are given:

$$C_{\text{TE}} = 2 \text{ pF}, \quad r_e = 12.5\,\Omega, \quad \mathcal{E}_b = 0, \quad C_{\text{C}} = 0.01 \text{ pF},$$
$$r_c = 1\,\text{k}\Omega, \quad W_{\text{dc}} = 1\,\mu\text{m}, \quad \text{and} \quad W_b = 0.7\,\mu\text{m}$$

Find f_T for the transistor. What would f_T be if a p-n-p transistor was used under similar conditions?

12.2. In an n-channel InP FET, what is the dielectric relaxation time of electrons? How does it compare with the channel transit time if the channel length is 0.5 μm and electron concentration is 5×10^{17} cm^{-3}? ($\epsilon_s/\epsilon_o = 12.4$ $\mu = 0.46$ m^2/V · s)

12.3. Assume that the domain formation time is very short compared to the time it takes for the domain to travel in a bar of GaAs. Assume also that as one domain reaches the anode, another domain is launched at the cathode. Find the length of the bar so that the pulse repetition rate can be made equal to 1 GHz. If the bias field is 10^4 V/m, what should be the applied potential?

12.4. The negative resistance of a tunnel diode is -35Ω. This is shunted by a capacitor of 10 pF. If there is also a series lead inductance and series lead resistance of 1.3 nH and 2Ω, find the complex input impedance of the diode. Does it always show a negative resistance throughout the frequency range? Find the frequencies at which the real and the imaginary parts of the impedance become zero.

12.5. If a Gunn diode is made from GaAs, find the minimum length of the intrinsic region for proper operation at (*a*) 10 GHz and (*b*) 100 GHz. If $n = 10^{16}$ cm^{-3}, will domains form for each case?

12.6. Assume that the base width of a GaAs ballistic transistor is 30 nm.

(*a*) If the electrons enter the base region with thermal velocity, find the time it will take for these electrons to traverse the base region.

(*b*) If the electrons are ballistically injected into the base with a bias voltage of 1.5 V, what is the transit time of the electrons? Assume that no field exists at the base region.

(*c*) If a uniform field of 1000 V/cm exists at the base region directed toward the emitter, find the resulting transit times for (*a*) and (*b*). In order for the ballistic approximation to hold, what should be the scattering time of electrons at the base?

12.7. For Problem 12.6, estimate the scattering length using the mobilities given in Fig. 12.10 at (*a*) room temperature, (*b*) liquid nitrogen temperature (77 K), and (*c*) liquid helium temperature (4 K). Check to see if any of these numbers come close to that required in Problem 12.6.

12.8. Suppose you were to design a JFET by using (*a*) GaAs or (*b*) Si as the semiconductor. If the dimensions of the two transistors are the same, explain which transistor will have a higher-frequency response.

12.9. Using the conservation of charge, Ohm's law, and the Poisson equation, derive Eq. 12.14 and show that the charge relaxation occurs with a time constant of τ_d.

12.10. Design a GaAs Gunn diode that will operate at 100 GHz. Are there limiting parameters in implementing such a diode?

12.11. Derive Eq. 12.27.

12.12. Holding β constant, find θ that will give the maximum negative resistance given by Eq. 12.29*a*.

CHAPTER

13

ELECTRO-OPTIC DEVICES

The interaction of photons with semiconductors is of fundamental importance in the study of the physical properties of semiconductors, as well as in implementing many useful optical devices. These devices range from photon detectors to coherent light generators and are used in many areas of technology and communication systems. Different semiconductors interact with photons in ways that reflect the properties of each semiconductor. For example, although the detection of an optical signal may be achieved by a certain semiconductor, that semiconductor may not be useful as a source of a light-emitting diode (LED) or a laser. In order to appreciate these finer details, a brief but extensive introduction to the optical properties of semiconductors will be given first. Although there are a large number of optical semiconductor devices, only the important and illustrative ones will be explained in this chapter.

13.1 PHOTON ABSORPTION AND EMISSION IN SEMICONDUCTORS

When photons of frequency ν are incident on a semiconductor, the intensity of light inside the semiconductor decreases exponentially as a function of distance according to

$$I_\nu(z) = I_\nu(0)e^{-\kappa(\nu)z} \tag{13.1}$$

where $\kappa(\nu)$ is the absorption coefficient (nepers m^{-1}), and $I_\nu(0)$ is the intensity of radiation at $z = 0$ inside the semiconductor. Note that under normal conditions and normal incidence, the light that is transmitted into the semiconductor is only a fraction of the incident intensity as a result of the high index of refraction of the semiconductors. Knowledge of $\kappa(\nu)$ as a function of frequency provides the necessary information of how a semiconductor can be incorporated into a useful optical device.

Absorption spectra of a typical semiconductor are shown in Fig. 13.1. At far infrared frequencies, the *restrahlen absorption* is due to the ionic binding of the crystal and becomes stronger as the ionic binding of the crystal becomes stronger. As the frequency is increased, multiphonon absorptions occur which show themselves as small peaks. Very strong absorption results when the photon energy reaches the band gap energy E_g. This sharp increase in absorption is known as the *absorption edge*. The onset as well as the steepness of this edge depends largely on the physical properties of the semiconductor. As the energy of the incident photon is further increased, interband transitions as well as transitions from the bound layers take place.

Some of the important sources of the absorption processes can be explained by the following.

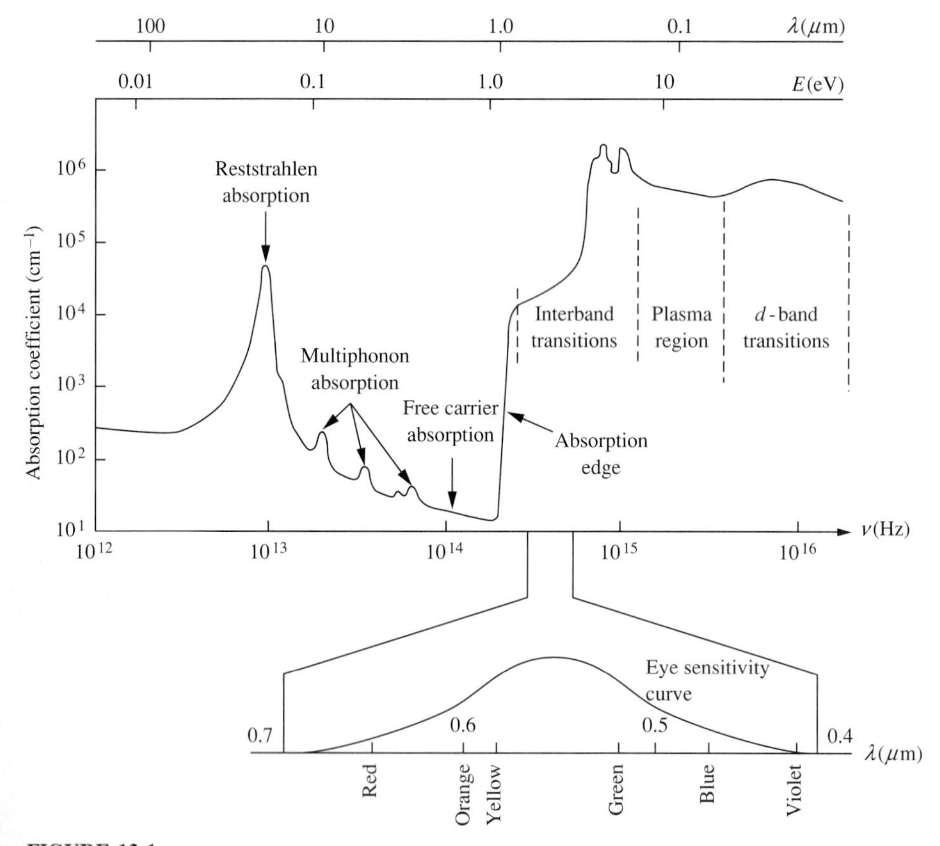

FIGURE 13.1
Absorption coefficient of a typical intrinsic semiconductor. The expanded view shows the sensitivity of the eye relative to different colors. The maximum intensity of the eye response is around 0.556μm.

1. An electron from the valence band is excited into the conduction band, leaving a hole in the valence band, thus producing an electron-hole pair (Fig. 13.2*a*). The condition necessary for this to occur is that the photon energy $h\nu$ should be equal to or greater than the band-gap energy E_g. This is one of the most important mechanisms of photon absorption and determines the absorption edge of the semiconductor.
2. An electron from the valence band is captured by an impurity acceptor atom, producing a hole in the valence band (Fig. 13.2*b*). Similarly, an electron from a donor atom is excited into the conduction band, producing an additional electron in the conduction band (Fig. 13.2*c*). Similar transition processes occur between the defect centers and between the defect centers and the nearest bands in the crystal. Because of the closeness of the impurity energy levels to the band edges, these processes require low photon energies and are useful in detecting far infrared signals at low temperatures so that most of the impurity atoms are not ionized.

   By means of thermal energy, electrons and holes may be generated by multiple excitations through the impurity and the defect levels and produce a current in the external circuit even without the presence of any optical signal. This is one of the major contributors to the dark noise in a semiconductor.
3. A bound electron from the inner orbits is excited into the conduction band. This requires high photon energies and takes place at the far ultraviolet or X-ray frequencies. This frequency range is not crucial for the optical devices that will be discussed in this chapter.
4. Photons are absorbed by the free charges, which are also effective at the higher energies.
5. At low photon energies, photons may be absorbed by phonons and restrahlen.

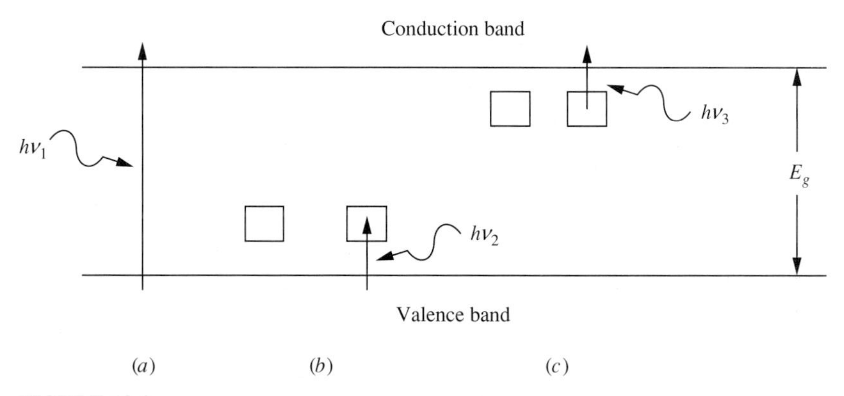

FIGURE 13.2
Absorption of photons by different mechanisms. (*a*) The transition of an electron from the valence band to the conduction band. (*b*) The transition of an electron from a valence to an impurity or trap level. (*c*) The transition of an electron from an impurity level into the conduction band.

A very important consequence of the absorption coefficient is that the low-frequency refractive index of a material can be written as an integral of the absorption coefficient, that is,

$$n_0 - 1 = \frac{1}{2\pi^2} \int_0^\infty \kappa(\lambda) d\lambda \tag{13.2}$$

where $\kappa(\lambda)$ is the wavelength-dependent absorption coefficient, and n_0 is the low-frequency index of refraction of the material. Thus n_0 depends on the area under the absorption coefficient curve and is independent of the details of the absorption. As can be seen from Fig. 13.1, in a semiconductor, this is largely determined by the absorption at higher frequencies beyond the absorption edge. It can also be shown that the contribution to the above integral is largest when the slope, that is, $d\kappa/d\omega$, of the absorption edge is highest.

Figure 13.3 shows the energy-band diagrams in direct and indirect band gap semiconductors. When an interaction occurs, both energy and momentum should be conserved for the overall system. The photon carries negligible momentum. For the direct band gap semiconductor, both the minimum of the conduction band and the maximum of the valence band occur at $\mathbf{k} = 0$. Therefore both momentum and energy are easily conserved. On the other hand, for an indirect semiconductor, in order for momentum to be conserved, the phonons of the lattice should take part in the interaction process. This occurs only if in the process a phonon is either absorbed or emitted. Depending on the phonon energy, these two cases require an energy $h\nu > E_g - E_p$ (phonon emission) or $h\nu > E_g + E_p$ (phonon absorption). Here, E_p is the corresponding phonon energy.

Quantum mechanically, it can be shown that the direct transition is a first-order process and the indirect transition is a second-order process. Therefore, the probability of absorption for the direct transition will be very much higher than the indirect case. Since emission and absorption are related by a constant, emission of a photon is also dependent on these probabilities. An electron making a transi-

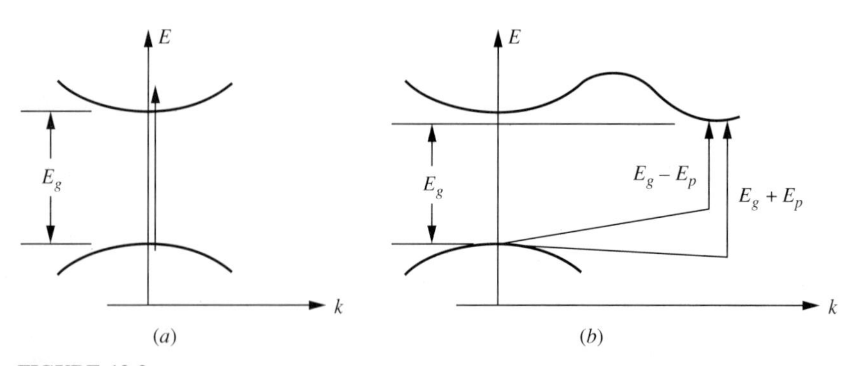

FIGURE 13.3
Absorption of a photon in a direct (a) and an indirect (b) band gap semiconductor. E_p is the phonon energy. When $h\nu > E_g + E_p$, phonon is absorbed and $h\nu > E_g - E_p$, phonon is emitted.

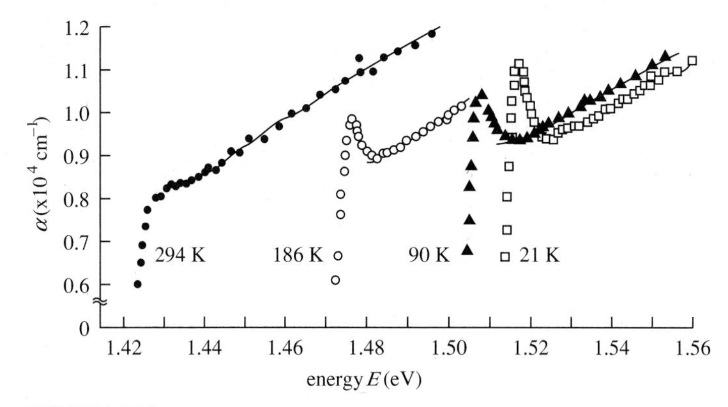

FIGURE 13.4
Exciton states modify the absorption edge of GaAs (Sturge, 1962).

tion from the conduction band into the valence band is more likely to emit a photon of frequency $\nu = E_g/h$ in a direct band gap semiconductor than in an indirect band semiconductor. Thus direct band gap semiconductors are most often used as materials for light-emitting diodes and lasers. In an indirect band semiconductor, inclusion of certain impurities may increase the probability of the emission of a photon and can, in certain cases, be used in making light-emitting diodes (LEDs).

When a photon with sufficient energy is incident on a semiconductor, an electron from the valence band may be excited into a higher energy level. This electron leaves behind a hole in the valence band. Instead of the electron being excited into the conduction band, as in case 1, this electron and the corresponding hole may form a loosely bound system, as in a hydrogen atom. The energy required to form this type of state will be slightly less than the band gap energy E_g. This is not in violation of the definition of E_g, which is the minimum energy necessary to excite an electron into the conduction band so that the electron can move freely and contribute to the electrical current of the material. The bound electron-hole pair is known as an *exciton* and it requires very little energy to break this bond between the electron and hole. Excitons play an important role in determining the absorption edge of a semiconductor. Figure 13.4 shows the effects of excitons on the absorption edge of GaAs.

13.2 LIGHT SOURCES

In order to appreciate the differences between various light sources such as a gaseous discharge or an LED or a laser, it is helpful to discuss the general properties of the radiation emitted by an atomic system as a result of electrons making optical transitions from upper to lower energy levels.

One of the identifying properties of a light-emitting source is the *coherence* of the generated optical radiation. In order to understand this concept, consider a plane wave of single frequency ν_0 given by

$$\psi(z, t) = \psi_0 \exp\left\{ j \left[2\pi\nu_0 \left(t - \frac{z}{v} \right) - \delta \right] \right\} \tag{13.3}$$

where v is the speed of propagation, and δ is an arbitrary phase constant. The Fourier transform of this wave, which is related to the power spectrum of the wave, is a delta function in the ν space (Fig. 13.5*a*). If the radiation is emitted within a finite time interval τ (Fig. 13.5*b*), or decays in amplitude with a time constant τ (Fig. 13.5*c*), the corresponding power spectra have a frequency distribution whose half power points are given by $\Delta\nu \approx 1/\tau$. The larger the τ, the smaller $\Delta\nu$. τ is known as the *coherence time* of the radiation. With the help of the coherence time, we define the *coherence length* as

$$l = c\tau = \frac{c}{\Delta\nu} = \frac{\lambda_0^2}{\Delta\lambda} \tag{13.4}$$

where λ_0 is the wavelength corresponding to ν_0.

Consider two sources of radiation, M_1 and M_2, having the same frequency ν (Fig. 13.6*a*). The time it takes for these radiations to reach the point P are $\theta_1 = M_1P/v$ and $\theta_2 = M_2P/v$. The intensity of radiation reaching P can be written as

$$I = I_1 + I_2 + 2\sqrt{I_1}\sqrt{I_2} Re[\gamma_{12}(\theta)] \tag{13.5a}$$

The parameter γ_{12} is given by

$$\gamma_{12}(\theta) = \frac{\psi_1(t + \theta)\psi_2^*}{\sqrt{I_1}\sqrt{I_2}} \tag{13.5b}$$

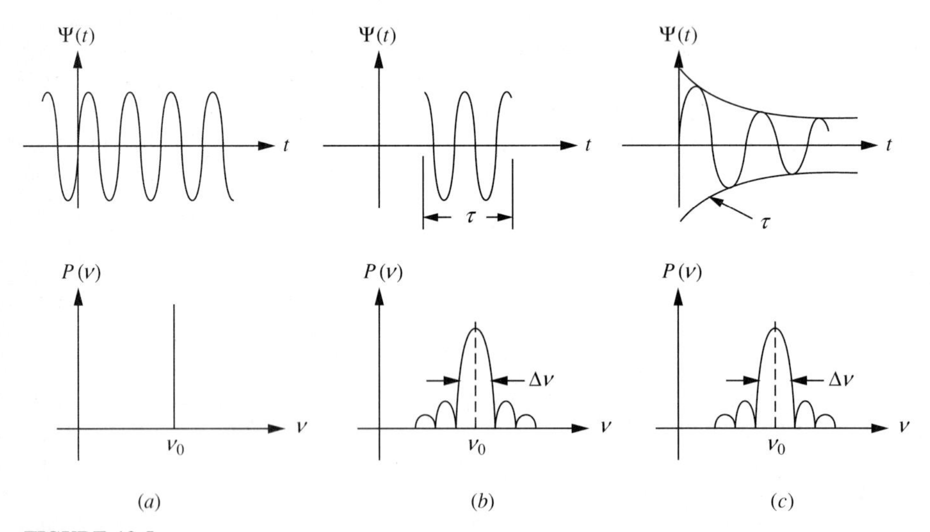

FIGURE 13.5
The temporal and power spectra of various emitted radiation. (*a*) A single frequency infinitely long plane wave. (*b*) Same wave with a time duration of τ, and (*c*) with a decay time constant τ.

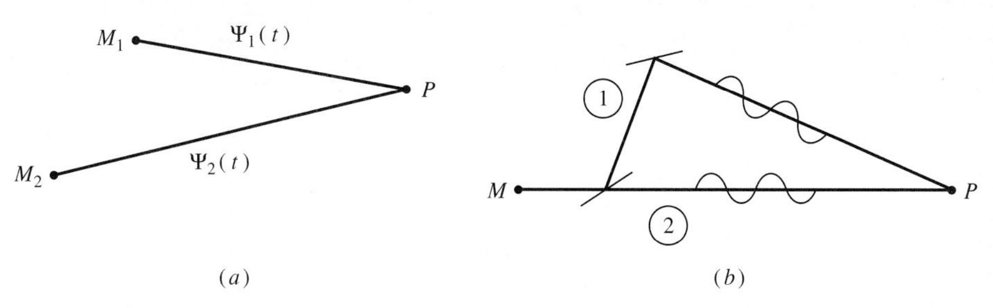

FIGURE 13.6
(*a*) Two sources are completely coherent if the degree of coherence is 1.0. (*b*) Radiation through the two paths will produce interference if the coherence length of the source is longer than the largest of the path lengths.

and is known as the *complex degree of coherence*, where $\theta = \theta_1 - \theta_2$. When $\lambda_{12} = 0$, the intensity of light at P will be the linear sum of the two vibrations. If $\lambda_{12} = 1.0$, the intensity of light at P will be controlled by both the intensity and the phase of the two vibrations. Depending on the value of the complex degree of coherence, the following definitions are made:

$\gamma_{12} = 0$	incoherent
$\gamma_{12} = 1.0$	complete coherence
$0 < \gamma_{12} < 1.0$	degree of coherence

The radiations from two different atoms are incoherent. The radiation emitted at two closely separated frequencies by the same atom is incoherent if the time of observation is longer than the coherence time.

If the radiation emitted by a source having a coherence time τ reaches the point P through two different path lengths, as shown in Fig. 13.6*b*, the two radiations will interfere at P if the coherence time of the source is long. If τ is short, the radiation reaching point P from Path 1 will have no relation to the one from Path 2. For interference experiments, the coherence length of the source should be longer than the total path length of the experimental system so that the two signals remain in phase with one another up to and beyond the point P. For example, light emitted by an atom at $\lambda_0 = 0.5\mu$m has $\Delta\lambda = 0.5\times 10^{-2}$ A, the coherence time $\tau \approx 10^{-9}$ s, and the coherence length $l \approx 15$ cm. For lasers, $\tau \approx 10^{-2}$ s, and $l \approx 3 \times 10^6$ m. Among the optical spectra emitted by various atomic systems, only those with the narrowest line widths are generally used for optical interference experiments. Because of the large coherence lengths, lasers are replacing other sources for optical communication, holography, and many other similar applications.

13.3 PHOTON EMISSION

Consider two distinct energy levels of an atomic system. Let these energies be identified as E_2 and E_1 with $E_2 > E_1$. When a photon of frequency $\nu_{21} =$

$(E_2 - E_1)/h$ is incident on the atom, an electron from the level E_1 will be excited to the level E_2 by absorbing the photon (Fig. 13.7*a*). Once the electron is in level E_2, after a time τ_{21} called the *life time* of the state E_2 relative to E_1, it will make a transition back to E_1 by emitting a photon of frequency ν_{21}* (Fig. 13.7*b*). This process is called *spontaneous emission*. If τ_{21} is very long, the electron will occupy that state for a very long time. Such a state is called a *metastable state*, which means that the probability of excitation to that state is also very low. Such a state can be excited by other nonradiative excitation mechanisms: collision with another atom, free electrons, and so on.

While an electron occupies the state E_2, if a photon of frequency ν_{21}, is incident on the atom, the electron in state E_2 will be forced to make a transition to state E_1 by emitting a photon of frequency ν_{21}, as shown in Fig. 13.7*c*. The emitted photon will have the same direction, phase, and polarization as the incident photon. This mechanism is known as the *stimulated emission*. In the system, there are now two photons with the same properties.

Let us consider a large number of the same atoms and again concentrate on the two arbitrary energy levels of the atom E_2 and E_1 with $E_2 > E_1$. If photons of frequency ν_{21} are incident on the system of atoms, and if stimulated emission is taken into account, the absorption coefficient of the system is given by

$$\kappa(\nu) = \left(N_1 \frac{g_2}{g_1} - N_2\right) \frac{c^2}{8\pi\nu^2\tau_{21}} g(\nu - \nu_0) \tag{13.6}$$

where $g(\nu - \nu_0)$ is the normalized line shape function of the radiation around ν_0, τ_{21} is the lifetime of state 2 relative to the lower state 1, and g_1 and g_2 are the multiplicities of the two levels. N_2 and N_1 are the number of atoms possessing energies E_2 and E_1, respectively. For the present discussion and practical purposes g_2/g_1 can be taken to be equal to unity.

The line shape function, which is dependent on a given atomic system, and the corresponding probability of transition between energy levels E_2 and E_1 also

* This excitation can take place only if the transition between the two levels is allowed. Quantum mechanical selection rules determine these transition probabilities, which in general depend on the quantum numbers n, l, m, and m_s (see Chapter 2).

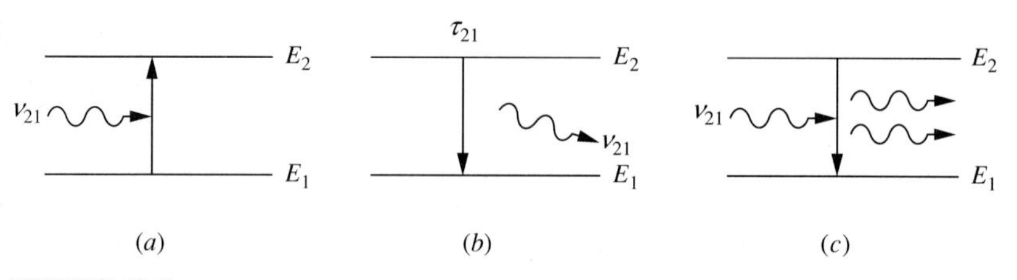

FIGURE 13.7
(*a*) Absorption of a photon. (*b*) Spontaneous emission of a photon. (*c*) Stimulated emission of a photon when a photon of ν_{21} is incident on the atom. The emitted photon has the same phase, direction, polarization, and frequency as of the incident photon.

determine the width of the emitted line spectra. The wider this width, the shorter is the equivalent coherence time of the radiation. The $\Delta\nu$ is in general inversely proportional to the lifetime of the state E_2. In semiconductors, this is determined by the interaction between the atoms and the phonons in the crystal.

As can be seen from Eq. 13.6, if $N_1(g_2/g_1)$ is greater than N_2, κ will be positive; thus, the radiation will be absorbed as it moves through the system of atoms. On the other hand, if N_2 is made greater than $N_1(g_2/g_1)$, then due to the large influence of stimulated emission the radiation intensity will increase along the chosen direction (z axis in Eq. 13.1). The negative of κ given by Eq. 13.6 is also known as the *gain coefficient*.

At equilibrium, the excitation of energy levels of a very large number of the same atom follows the Boltzmann distribution function, that is, as the energy level increases, the number of electrons excited into the higher energy levels decreases exponentially (Fig. 13.8*a*). Therefore, if by some means this trend is reversed (Fig. 13.8*b*), there will be an amplification of the radiation as it goes through the system of atoms. The reversal of the population-level density is known as *population-inversion,* and the corresponding process as the *pumping* of the upper state.

In a gaseous discharge (see Chapter 2), identifying electrons are thermally or collisionally excited into the higher energy levels and make transitions to lower allowed states by emitting characteristic frequencies of the atoms. In order to obtain a specific frequency of the radiated signal, filters or spectrometers are used to select the desired wavelength of radiation. The frequency bandwidth of the resulting light is determined by the radiative transition between the two levels,

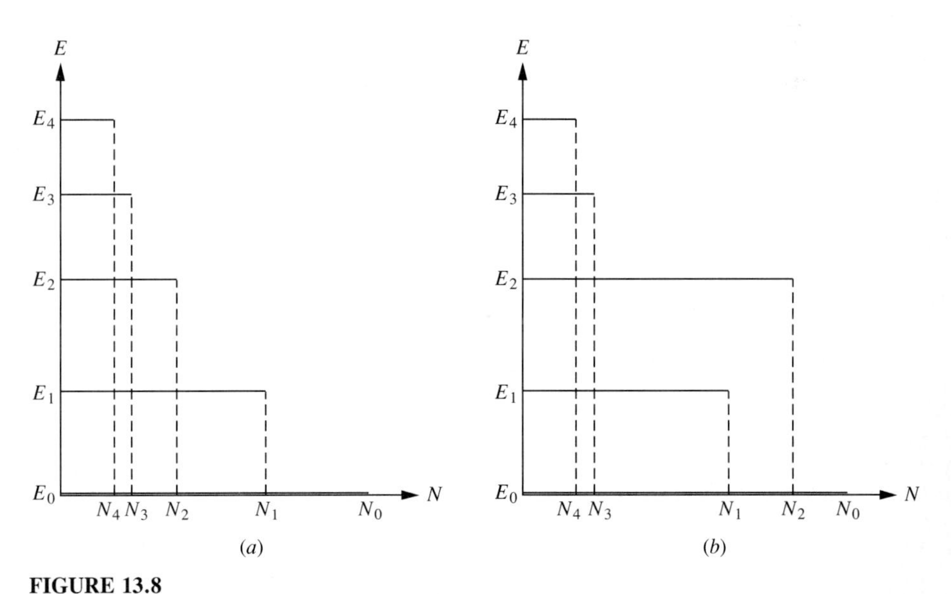

FIGURE 13.8
(*a*) Energy-level distribution for a gaseous system at equilibrium. (*b*) Population inversion is created under nonequilibrium conditions.

the resolving power of a spectrometer, or the band-pass characteristic of a filter. In any case, there is always a finite $\Delta\nu$ resulting in a relatively short coherence length.

13.4 LIGHT-EMITTING DIODES

In a semiconductor, the electrons can only make radiative transitions between the conduction and the valence bands. The probability of the emission of a photon is largest when there are a large number of both electrons and holes present at the same **k** value. As we know from the discussions in Chapter 6, these conditions cannot be created easily in an extrinsic semiconductor alone. Either flash-light optical pumping should be used or we should resort back to a p-n junction to create these favorable conditions.

Before explaining the operation of the light emitting p-n junction diode, we should look closely to the band structure of a degenerately doped semiconductor whose band structure is modified by the large concentrations of impurity atoms. When impurity atom concentration is low, we can represent the corresponding energy levels as discrete shallow levels appropriate to those impurity atoms in that semiconductor (Fig. 13.9*a*). When the impurity concentration is increased, because of the large number of the impurity atoms and their statistical distribution within the host semiconductor, the impurity levels now show a spread in energy.

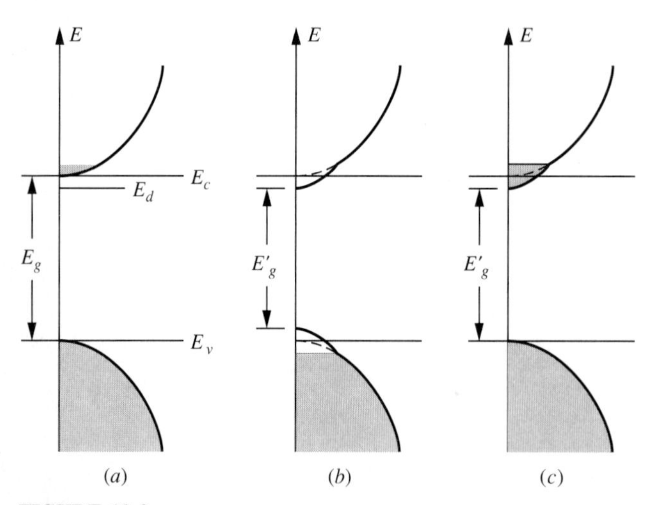

FIGURE 13.9
Formation of tail states. (*a*) Impurity concentration is represented by a delta function at a discrete energy level inside the forbidden gap in a lightly doped semiconductor. (*b*) Impurity energy levels spread into the band edge and decrease the effective band gap in a degenerately doped semiconductor. Tail states form in both the conduction and valence bands in a p-type semiconductor, if p-type is formed by diffusing acceptor atoms into a degenerately doped n-type semiconductor. (*c*) Tail states form only in the conduction band if only impurity donor atoms are diffused into an intrinsic semiconductor.

As the impurity density is increased further, the spread in energy extends into the band edges and the two bands overlap. The overall bands then extend into the forbidden gap and form what are known as *tail states*. This results in a reduction of the effective band gap energy of the semiconductor. If acceptor atoms are diffused into an n-type semiconductor that already contains a large number of donor atoms, the random distribution of donor atoms and acceptor atoms produces tail states both in the conduction and the valence bands (Fig. 13.9*b*). On the other hand, if in an n-type semiconductor there are only donor atoms, there are thus only conduction-band tail states (Fig. 13.9*c*).

A light-emitting diode is simply made by growing an epitaxial p-layer on an n-type semiconductor (GaAsP or GaAs) (Fig. 13.10*a*). Both sides of the junction are degenerately doped so that the equilibrium Fermi levels lie in the respective bands (Fig. 13.10*b*). Charge-density distributions for the unbiased diode are shown in Fig. 13.10*d*.

When the diode is highly forward biased, electrons are injected from the n-side into the p-side. These electrons make radiative recombination with the holes in the valence band of the p-side and emit photons around the frequency $\nu = E'_g/h$ (Fig. 13.10*c*). Experiments show that actually the electrons from the conduction band tunnel through the junction barrier to the tail states of the conduction band on the p-side and make radiative recombinations with the holes in the tail states of the valence band (Fig. 13.10*e*). It is also possible that some of the electrons in the conduction band recombine with the holes diffusing into the n-region (Fig. 13.10*c*). Since the available states for the holes in the n-region are very small, the possibility of this process occurring is very small. The experiments also show that the radiation is emitted close to the p-side of the junction.

The internal efficiency η_{int} of the diode is defined as the number of photons generated per each injected electron. η_{int} can be written as

$$\eta_{\text{int}} = \frac{1}{1 + (\tau_{rr}/\tau_{nr})} \tag{13.7}$$

where τ_{rr} and τ_{nr} are the radiative and nonradiative lifetimes of the minority carriers in the semiconductor.

Since the probability of radiative recombination is largest for direct gap semiconductors (τ_{rr} is very small), compound semiconductors are generally used in making light-emitting diodes (LED). To generate visible light, ternary or quaternary compounds are used to obtain the required bandgap for the semiconductor (see Chapter 5).

The external efficiency of an LED is determined by the amount of light that can escape from the semiconductor. If we imagine the semiconductor to be a box inside which radiation is generated, we know from the fundamental laws of electromagnetic theory that not all the light will be transmitted to the outside world. Beyond the critical incident angle $\theta = \sin^{-1}(1/n_s)$, where n_s is the index of refraction of the semiconductor, the emitted radiation will be completely reflected back into the semiconductor. Because of the high index of refraction, even at normal incidence, most of the light reaching the surface will

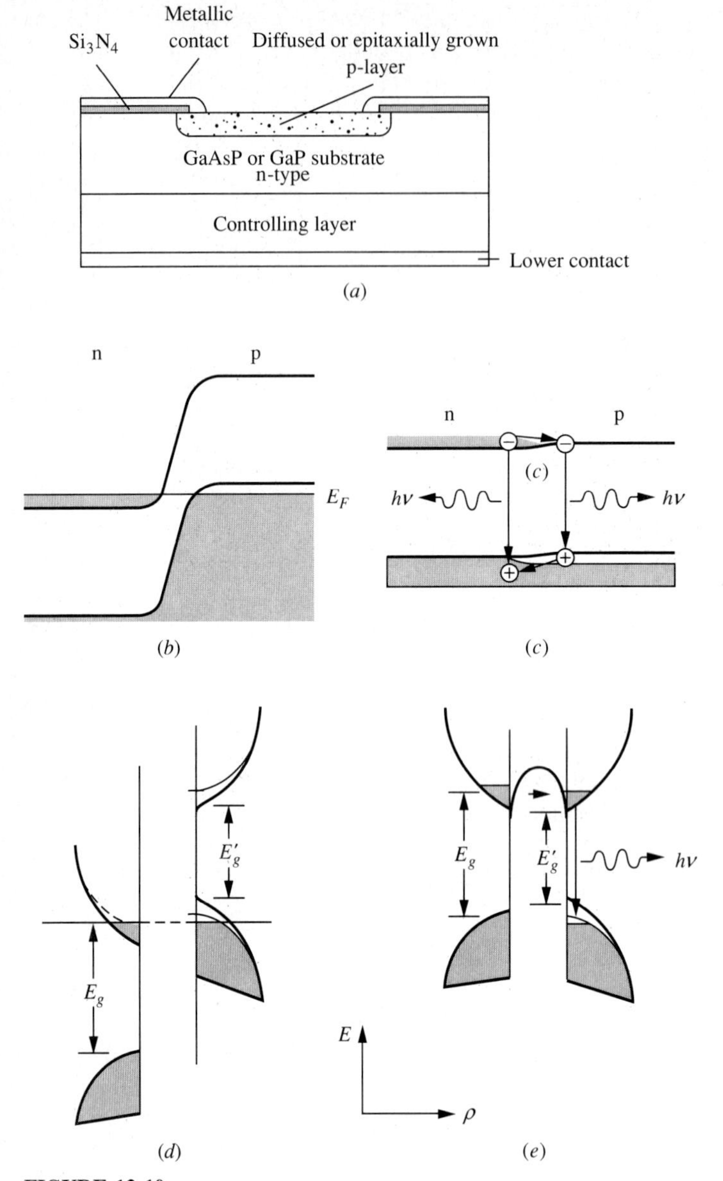

FIGURE 13.10
(*a*) Light-emitting diode. The resulting energy-band diagrams with no bias (*b*) and with forward bias (*c*). (*d*) and (*e*) are the corresponding charge distributions for a diffused p-layer. When the diode is highly forward biased, the electrons tunnel through to the conduction band of the p-side and make radiative transitions to the valence band.

also be reflected back. Overall, the light reaching the outside will be a small portion of that generated. The external efficiency of the diode is very small and is approximately given by

$$\eta_{\text{ext}} \approx \frac{1}{n_s(n_s + 1)^2} \tag{13.8}$$

The layer labeled as the controlling layer in Fig. 13.10*a* can be either (*a*) a transparent medium made up from GaInP where the back-emitted radiation travels toward the back side without absorption and is reflected back toward the front or (*b*) an absorbing layer where the back-radiated layer is completely absorbed. The amount of light transmitted to the outside is a very small fraction of the generated light. Thus, the external efficiency of an LED is very low. For example, in a GaAs LED, η_{ext} is about 1.5 percent. The emitted light also has a very short coherence length. By modifying the surrounding medium of the LED and/or shaping the semiconductor into a spherical dome, the light transmitted to the outside world can in some cases be dramatically increased.

Before we can explain the difference between a laser and any other light source, such as an LED, we need to understand the propagating characteristic of optical signals in optical waveguides.

13.5 OPTICAL WAVEGUIDES AND CAVITIES

The optical waveguides are made up of layers of different dielectric materials or, as in the case of optical fibers, from various forms of glasses, especially quartz glass in circular form. Optical waveguiding is based on the total internal reflection of the electromagnetic signal and requires for the propagating medium a higher dielectric constant relative to the *cladding* or surrounding material.

The optical fiber is the most commonly used light-guiding structure in long-haul optical communications. As shown in Fig. 13.11*a*, it is made up of a circular glass fiber encased with a plastic cladding material. The fiber can have a uniform index of refraction or it can have what is known as a graded index whose index decreases radially with radius.

Figure 13.11*b* shows a typical semiconductor optical waveguide. It is made by diffusing or by ion implanting a strip of finite width of a layer of atoms into a semiconductor substrate.

Another low-loss version of a semiconductor waveguide is shown in Fig. 13.11*c*. It is made up of a substrate, generally a semiconductor, over which thin layers of semiconductors with different dielectric constants are grown. The second layer is the guiding layer and has the highest dielectric constant. The third layer can be omitted but it is usually another grown layer with a lower dielectric constant than the guiding layer. A ridge is etched on this layer. The reduced thicknesses on each side of the ridge are instrumental in reducing the effective indices of those regions so that the wave is confined to propagate in the high effective index region, that is, under the ridge.

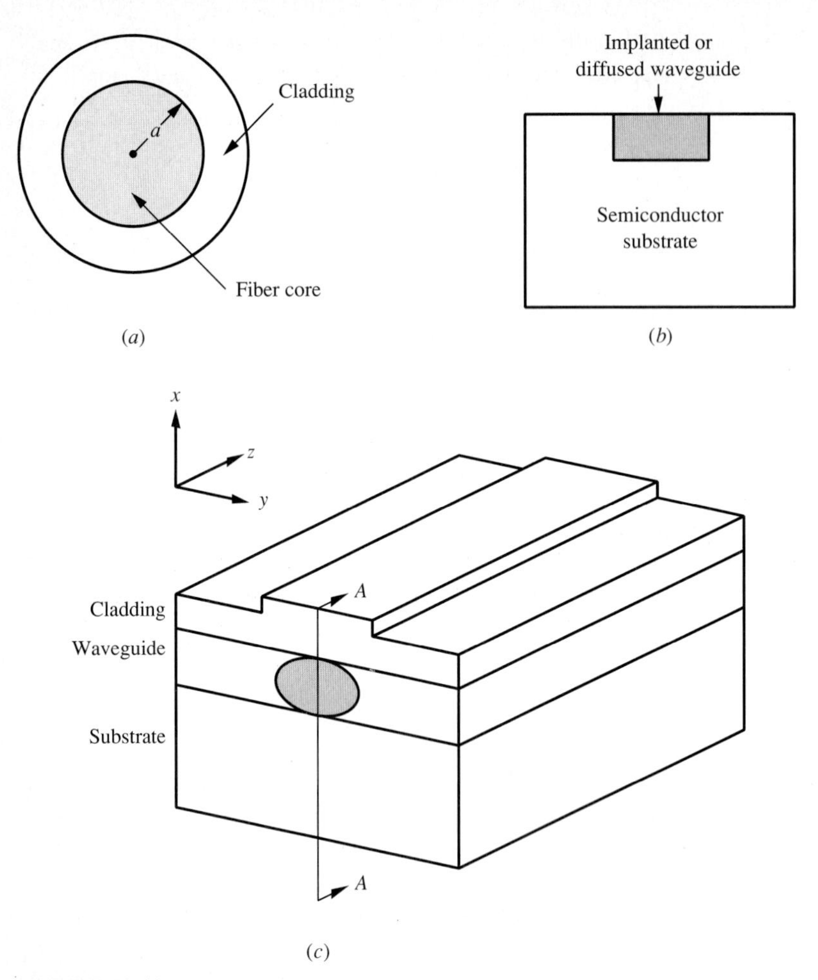

FIGURE 13.11
Various forms of optical waveguides. (*a*) Fiber optic. (*b*) Semiconductor waveguide with a diffused or ion implanted layer. (*c*) A three-layer low-loss waveguide using a ridged cladding layer.

The formal solution for finding the possible modes of propagation and the resulting propagation constants in optical waveguides or in fibers requires the solution of Maxwell's equations, subjected to proper boundary conditions. These are very complicated and are beyond the scope of this book. Instead of looking for a rigorous solution of Maxwell's equations, a simpler physical interpretation of the wave propagation in optical waveguides will be given, using an analogy with the potential well problem in quantum mechanics (see Chapter 2). Although a one-dimensional solution will be considered here, the results can easily be extended to multiple dimensions and other geometries.

The electric (and the magnetic) fields in a waveguide satisfy the homogeneous wave equation

$$\nabla^2 \mathscr{E} = \mu_0 \epsilon \frac{\partial^2 \mathscr{E}}{\partial t^2} \tag{13.9}$$

Consider a wave propagation in the z direction in the waveguide shown in Fig. 13.11*c*. For simplicity, we take the cross-sectional plane *AA* and assume that the y dimension is infinitely long. We also assume a solution for the fields in the form given by

$$\mathscr{E}(x, z, t) = F(x) \exp[j(\omega t - k_z z)]\mathbf{a}_y \tag{13.10}$$

where k_z is the propagation constant in the z direction. Substituting Eq. 13.10 into 13.9 and arranging terms, we can write

$$\frac{d^2 F(x)}{dx^2} + k_0^2 \kappa_e(x) F(x) = k_z^2 F(x) \tag{13.11}$$

where $k_0 = \omega\sqrt{\mu_0 \epsilon_0}$ is the free space propagation constant, $\omega = 2\pi\nu$ is the radian frequency, μ_0 is the free space permeability, and $\epsilon = \kappa_e \epsilon_0$ is the dielectric permittivity of material with a dielectric constant κ_e, which is assumed to vary in the x direction.

In order to get a physical insight into the solution of this equation, we recall the one-dimensional time-independent Schrodinger equation that was introduced in Chapter 1.

$$-\frac{\hbar^2}{2m^*}\frac{d^2\Psi}{dx^2} + \Phi(x)\Psi = E\Psi \tag{13.12}$$

This can be written as

$$\frac{d^2\Psi}{dx^2} + \left[-\frac{2m^*}{\hbar^2}\Phi(x)\right]\Psi = -\frac{2m^*}{\hbar^2}E\Psi \tag{13.13}$$

If we compare Eq. 13.11 with Eq. 13.13, we observe that the two equations are identical if we identify respectively the propagation and dielectric constants, other than some multiplying constants, in terms of energy and potential functions of a one-dimensional quantum mechanical well problem, that is,

$$\begin{aligned} -\kappa_e(x) &\Rightarrow \Phi(x) \\ -k_z^2 &\Rightarrow E \end{aligned} \tag{13.14}$$

We can look for physical solutions of Eq. 13.11 in terms of the parameters of the potential well problem using Fig. 13.12. The dielectric constant varies as a function of distance x in steps with the highest dielectric constant of the waveguiding layer determining the depth of the well. If the upper layer, referred to as the cladding layer, has a different dielectric constant than the substrate, the well is nonsymmetric. The conditions we impose on *F(x)* and its first derivative are that they should be continuous across the dielectric discontinuity.*

* For TE modes, the continuity of *F(x)* and its first derivative implies the continuity of the electric and magnetic fields. For TM modes, it requires the continuity of $(1/\epsilon)\ dF/dx$ in addition to the continuity of $F(x)$.

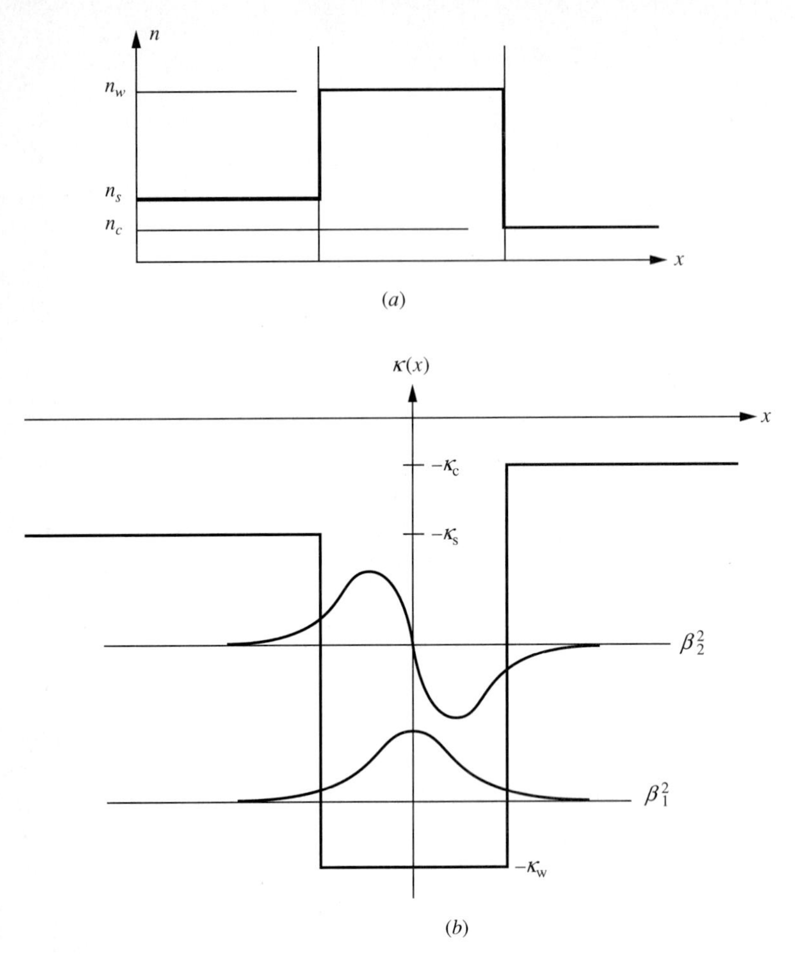

FIGURE 13.12
(*a*) The index of refraction variation along the x direction at the cross section *AA* of the waveguide of Fig. 13.11*c*. (*b*) The corresponding potential well model. The dielectric constant variation and the resulting possible propagation constants are equivalent to the potential energy variations and discrete energy values in a quantum well. w, s and c stand for waveguide, substrate and cladding.

The electric-field solutions will be similar to the solutions of the wave function. In the well problem, there were solutions only for discrete energies. Here, the propagation constant takes the place of the energy, and the discrete propagation constants will correspond to the modes of propagation in the waveguide. The number of modes will be determined by the depth of the waveguiding layer, as well as the magnitudes of the various dielectric constants. The fields also extend into the cladding and the substrate layers and decay exponentially away from the well. These also put restrictions on the thicknesses of the cladding and substrate layers. If the fields extend outside these thicknesses, they will represent losses from the waveguide. The deeper and wider the well is, the more will be the number of possible modes of propagation that the waveguide will support. For many communication applications, it is necessary to have a single mode of propagation, especially in the optical fiber. This requires a small diameter (on the order of $5\mu m$) for the fiber and imposes technological refinements into the manufacture of fiber optics. It is also possible to produce graded index fibers that will support single modes. The solutions that give the spatial variation of the electric field over the transverse plane (plane perpendicular to the axis of the waveguide) are known as the *transverse modes*.

In the quantum well analogy, it was assumed that the materials that make up the waveguide are lossless. This is not true in reality. The losses depend on the material and are also largely frequency dependent. Figure 13.13 shows the attenuation constant of quartz optical fiber as a function of wavelength. The minimum attenuation occurs around 1.5 μm. This also imposes restrictions on the type of laser to be used with optical fibers over long hauls.

In order to produce laser action and select the operating frequency of the laser, it is necessary to place the lasing medium inside an optical cavity. Optical cavities are made up of two reflecting mirrors forming a Fabry–Perot type resonator. The two mirrors have radii of curvatures of R_1 and R_2 respectively. The stable configuration and the Q of an optical cavity depend on the choice of reflectivity of the two mirrors, as well as the radii of curvatures R_1 and R_2. Two flat mirrors, although they do not have the minimum cavity loss, are used for many laser systems, especially in semiconductor lasers. The frequency selectivity

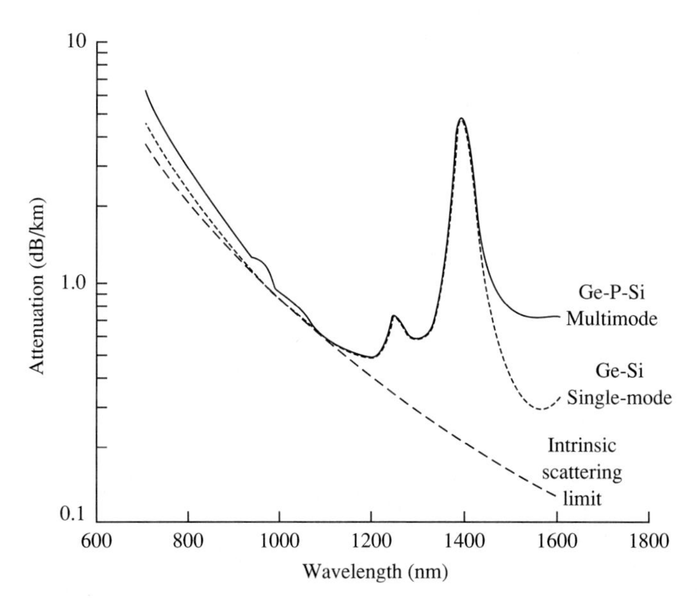

FIGURE 13.13
Attenuation spectrum for a high silica fiber optic waveguide between 700 nm and 1600 nm. The dotted curve indicates the single-mode total light scattering. Clearly visible are the OH absorption bands (Keck, 1981).

is obtained by multiple coating of the mirrors with alternating layers of differing dielectric materials. With this technique, reflectivities close to 100 percent can be obtained. For semiconductor lasers, since the frequency of oscillation is dependent on E_g and the internal gain is very high, a cleaved polished surface is adequate and no coating is necessary.

The cavity resonance frequency is determined by the condition that the axial length of the cavity should be integer multiples of half guide wavelength λ_g of the waveguide. This can be written as

$$m\frac{\lambda_g}{2} = D \tag{13.15}$$

where m is an integer and D is the separation distance between the two mirrors. Since D is very large, the integer m is also very large (for gas lasers, m $\approx 10^6$). The approximate difference in frequency between two axial modes is given by

$$\nu_{m+1} - \nu_m = \frac{c}{2\sqrt{\kappa_e}D} \tag{13.16}$$

where κ_e is the dielectric constant of the lasing medium between the mirrors. This assumes that the whole space between the mirrors is completely filled with the same lasing material.

13.6 LASERS

After the transistor, the laser is the second most important invention of the latter part of the twentieth century. Lasers have had an impact on many areas of our lives from the supermarket check-out counters, where they are used as a holographic means of reading the codes on the groceries, to hospital operating rooms, where they are used as precision knives in laser surgery. The technological applications of lasers are extensive, but in this section, we will restrict ourselves to the basic operating principles of lasers.

To generate laser action in a material, the fundamental aim is to produce population inversion by some external means. This is obtained for different materials by different means. The concepts of producing laser action in gases and solids are similar, and these will be discussed next.

Figure 13.14 shows the four energy levels of a typical laser system. By an external excitation agent, such as optical pumping or collisional excitation, the electrons from level E_0 are excited into level E_3. The electrons at level E_3 can make a nonradiative transition to level E_2. If level E_2 is a metastable state, that is, τ_{21} is very large ($\tau_{21} >> \tau_{10}$ is sufficient), that means that an electron excited to level E_2 will not immediately make a transition to level E_1. Therefore, if the excitation from level E_3 to level E_2 continues, there will be an accumulation of atoms in level E_2. Thus, *population inversion* in level E_2 takes place. Level E_3 usually belongs to a different atom such as an additive to a gas or an impurity dopant in a solid.

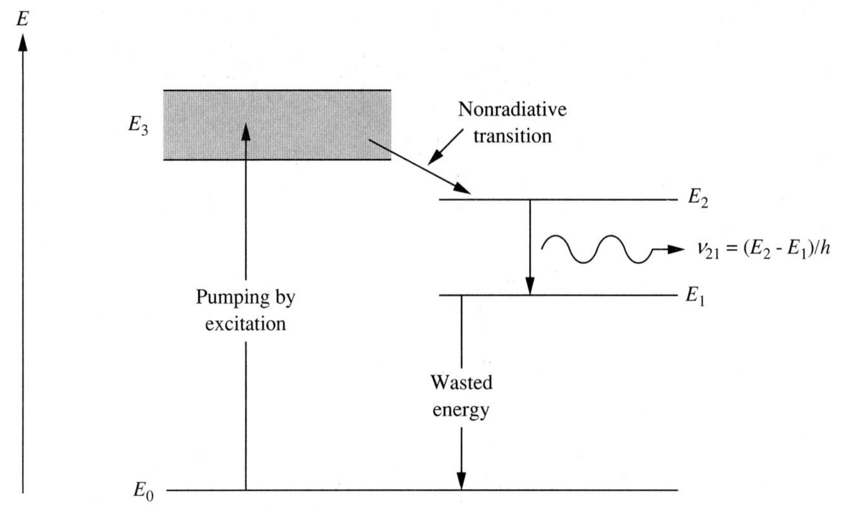

FIGURE 13.14
(*a*) Four-level laser when $(E_1 - E_0) >> kT$. (*b*) Three-level laser when $(E_1 - E_0) \leq kT$.

Figure 13.15 shows the typical components of a laser. The lasing medium is located between two highly reflective mirrors. It will be assumed that the population inversion is already created in the medium by an external means, usually by a gas discharge in a gas laser, by a flash light in a solid laser, or by injection in a semiconductor laser. If a photon of frequency ν_{21} is generated by noise or by spontaneous emission, it will initiate stimulated emission in one of the atoms, and a second photon with the same characteristic as the initial photon will be generated. If the direction of the initial photon is along the axis of the mirrors, the two photons will continue to stimulate additional atoms and produce additional photons in the same direction with the same phase and polarization. The photons reaching the mirrors will be reflected back by the highly reflecting mirrors and continue to generate additional stimulated photons while moving in the reverse direction. Any other photon initially generated along any other direction except along the axis, even if it makes stimulated emissions, will be lost from the system. If the overall gain of the system, including the optical cavity, is greater than the total losses in the system, laser action will take place and some of the photons will be transmitted through the partially reflecting mirror M_2 to the outside, as shown in Fig. 13.15.

13.6.1 Gaseous and Solid Lasers

The first laser that was invented was the solid laser, which is the flashlight-pumped Cr-doped Al_2O_3 (Ruby) laser. The other widely used solid laser is the Nd^{3+}-doped $Y_3Al_5O_{13}$ laser, which is called the Nd:YAG laser. These usually generate pulsed high power laser lights.

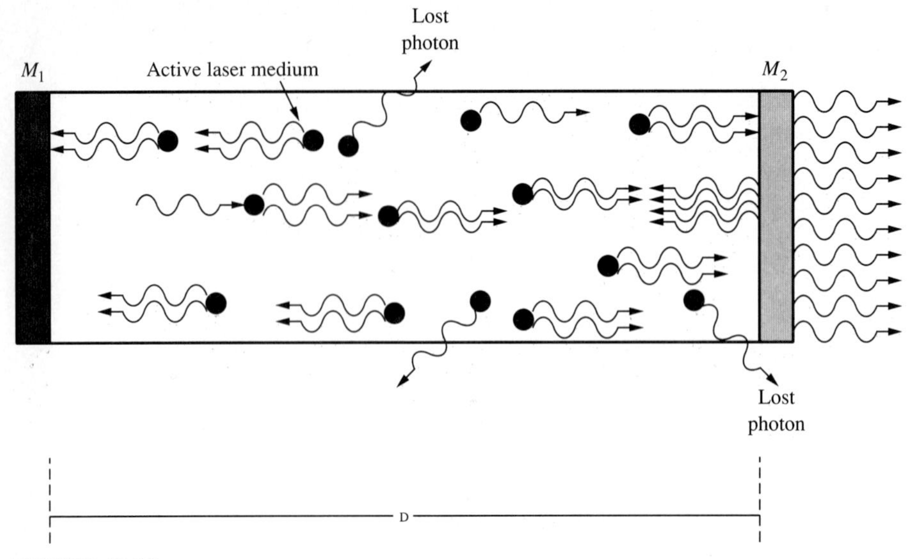

FIGURE 13.15
Photons emitted by stimulated emission increase in number along the axis of the mirrors. Other photons generated in arbitrary directions are lost through the side walls.

The helium neon gas laser was the first gas laser that was demonstrated to have a laser action in a gaseous medium, and it is still one of the most widely used gas lasers. It can generate laser light at three different wavelengths: 3.39 μm (far infrared), 1.15 μm (infrared), and 0.6328 μm (red, visible). Helium atoms are excited through electron collisions to the metastable 2^3S and 2^1S states. There are $2S$ and $3S$ energy states of Ne that are very close in value to these He metastable states. Through nonradiative transitions, the electrons are transferred to the $2S$ and $3S$ states of neon. Electrons making transitions from the $3S$ state to the $3p$ state of neon generate 3.39 μm, from the $3S$ to the $2p$ generate 6328 A, and from the $2S$ to the $2p$ generate 1.15 μm laser lights. The frequency of operation is determined by the proper choice of mirror coatings at each end of the optical cavity.

Other important gas lasers are the powerful CO_2 laser which works with the transitions among vibrational states (around 10.6 μm), the argon laser which emits at different wavelengths in the visible, and N_2 laser which emits near ultraviolet.

13.6.2 Semiconductor Lasers

Unlike the common gaseous and solid laser systems, the transitions in a semiconductor laser occur between energy bands rather than between discrete levels because of the closeness of the atoms constituting the semiconductor. Although in a semiconductor population inversion was originally achieved by optical pumping, almost all of the present day semiconductor lasers use minority carrier injection in a p-n junction as a means of population inversion.

Figure 13.16*a* shows the constructional details of a typical semiconductor laser. The major difference between a laser and an LED is the way the optical signal is manipulated in each case. Figure 13.16*a* shows the way the light is emitted from a laser. The emitted photons initially may travel in arbitrary directions, but the ones that are along the optical axis, that is, parallel to the p-n junction, continue to stimulate more radiative transitions. Once they reach the cleaved edges, they are reflected back into the semiconductor and continue generating photons. When the gain exceeds the losses, oscillation sets in and the laser action reaches a steady value. Since the gain coefficient is very high due to the very high charge concentrations, one does not need the highly reflective mirrors that are necessary for gas lasers. Here, the reflection at normal incidence from the cleaved end surfaces is sufficient to generate the necessary gain from the system.

As shown in Fig. 13.16*b*, light is emitted in all directions from an LED, and there are no provisions to guide the light in a specific direction to enhance stimulated emission. Whatever fraction of light can escape from the front surface of the semiconductor appears as useful light. The coherence length of LEDs is relatively short but can be used as pulsed-light sources in some digital communication systems.

Similar to an LED, a p-layer is generally epitaxially grown over an n-type semiconductor to form a p-n junction laser diode. Both sides of the junction are degenerate and, at equilibrium, the Fermi levels of each side lie in their respective bands similar to that shown in Fig. 13.10*b*. Again, by forward bias, electrons are injected into the p-side and make radiative transitions to the valence band, thus emitting photons with a frequency $\nu = E_g'/h$.

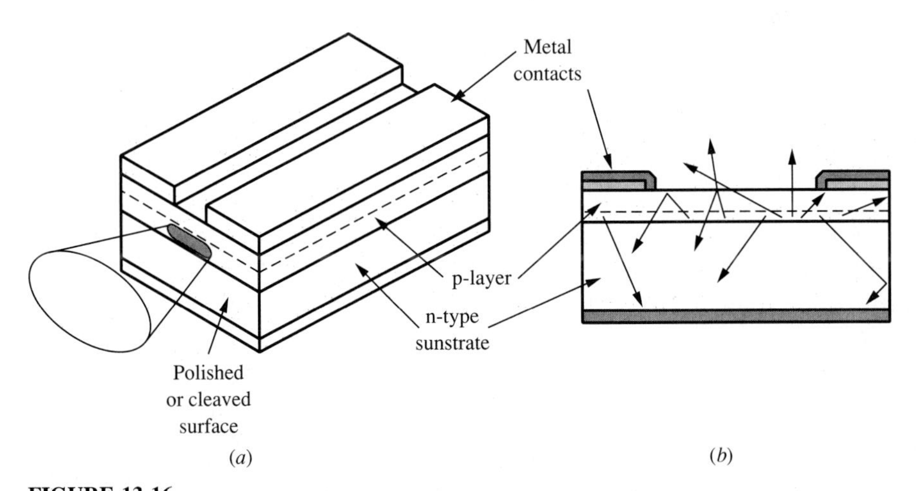

FIGURE 13.16
Constructional details of a typical p-n junction diode used as a light source. The light paths in a semiconductor laser (*a*) and an LED (*b*) are different.

As shown in Fig. 13.17, the variation of the index of refraction across the homojunction is very small. Even with the additional absorption introduced by the adjacent n- and p-regions, the mode containment across the junction is relatively poor. This is reflected in the high start-oscillation currents* necessary for these lasers. The emitted radiation also has a large half-angle, and the beam spread is very large compared to that of a gas laser.

Figure 13.18 shows a heterojunction laser made up of a p-type GaAs layer sandwiched between two layers of n- and p-types of GaAlAs. The band structure of the unbiased junction is shown in Fig. 13.18*a*. When forward bias is applied, electrons are injected by tunneling through the junction barrier into the p-type GaAs. The electrons make radiative transitions into the valence band and emit photons through stimulated emission (Fig. 13.18*b*). The addition of Al on the

* Start-oscillation current is the current at which the gain exceeds the losses, so that the laser is at the threshold of emitting light.

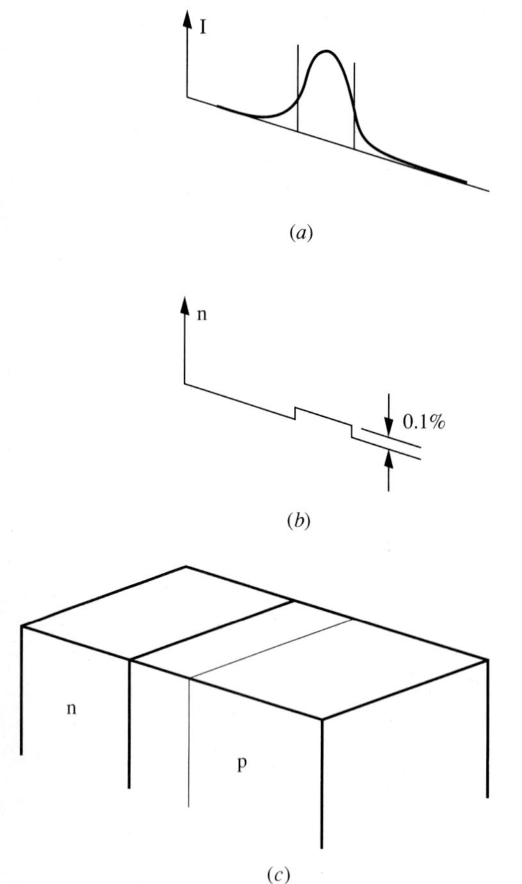

FIGURE 13.17
In a homojunction semiconductor laser, (*a*) the intensity of emitted radiation and (*b*) the change in index of refraction in (*c*) the corresponding regions of the laser.

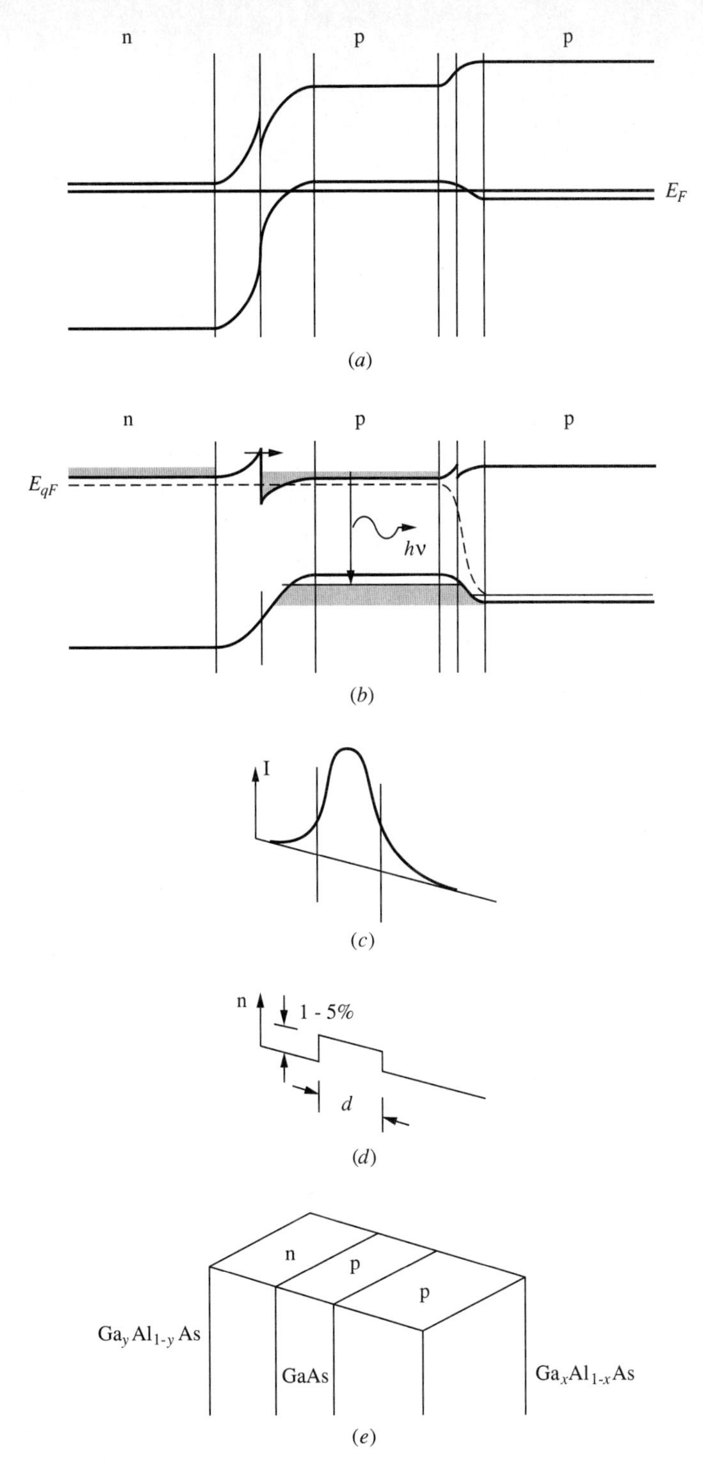

FIGURE 13.18
Energy-band diagram of a heterojunction laser with no bias (*a*), and with forward bias (*b*). Electrons are injected into the p-type GaAs and make radiative transitions with the holes. (*c*) Radiated intensity profile. (*d*) Spatial change in the index of refraction. (*e*) Corresponding regions of the diode.

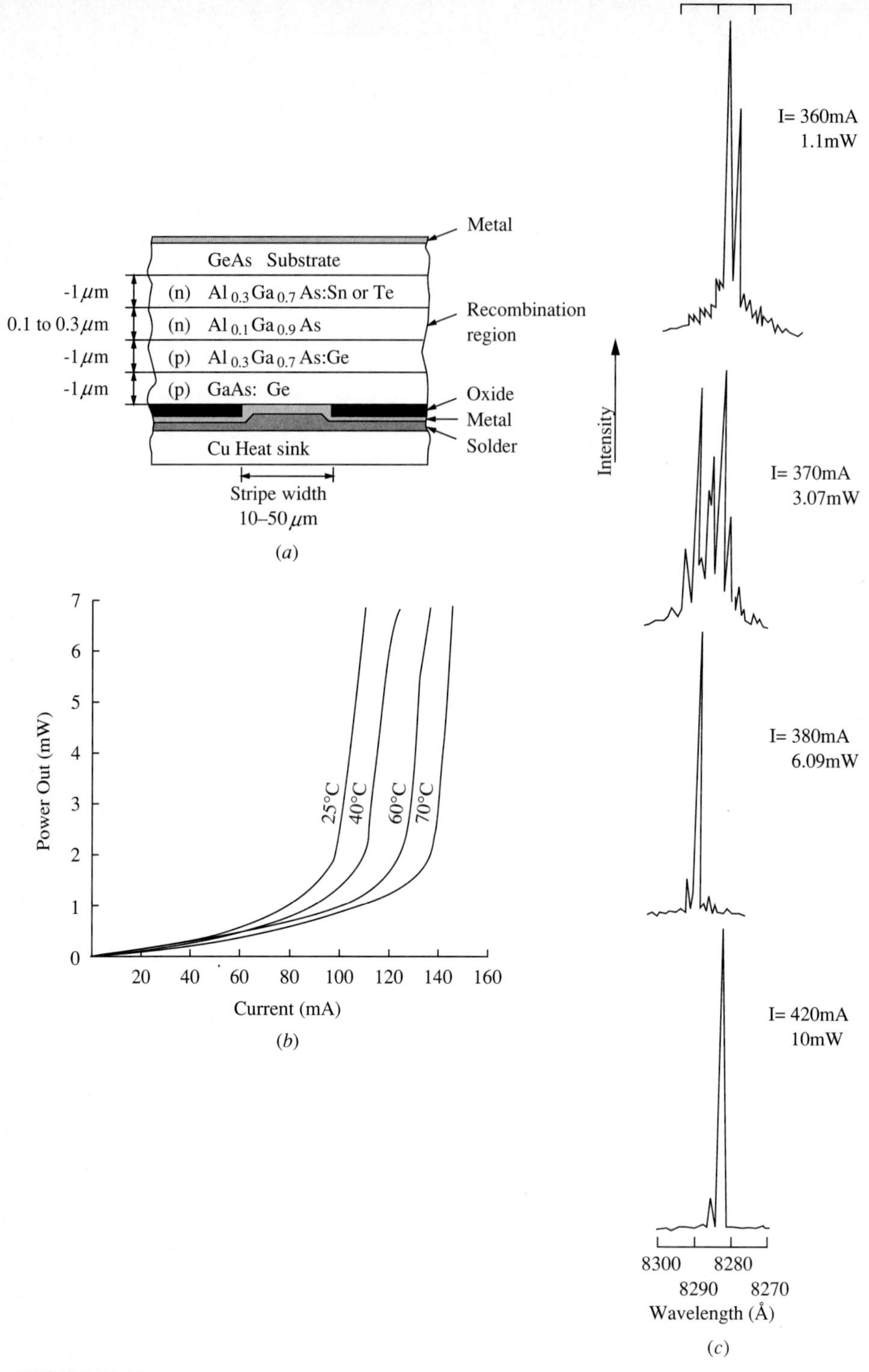

FIGURE 13.19
(*a*) Typical cross section (not to scale) of oxide-isolated stripe contact cw AlGaAs laser diode (0.8–0.83 μm). (*b*) Power output at various heat-sink temperatures as a function of current and (*c*) spectral output as a function of current. The lowest trace shows a relatively high drive with one dominant longitudinal mode (Kressel and Butler, 1979).

outside layers reduces the index of refraction of these layers and results in a noticeable index change (Fig. 13.18*d*). The Al addition also shifts the absorption edge of the two outside regions to higher energies, thus reducing considerably the absorption losses in these layers. This results in better wave containment and overall low cavity losses. The start-oscillation currents are generally low and these lasers can operate at room temperature without too much increase in junction temperature.

Various combinations of III–V ternary and quaternary compounds are used for heterojunction lasers. Since the wavelength of the radiated signal is a direct function of band gap energy, it is possible to obtain the desired band gap by proper combination of these compounds. As can be seen from Fig. 13.13, the minimum of attenuation for a quartz fiber optic occurs at 1.5 μm wavelength. There is also a slightly higher but low attenuation region near 1.3 μm. InGaAsP quaternary alloys represent a wide range of band gap energy. Using InP as a substrate allows the generation of laser wavelengths near 1.5 and 1.3 μm. Using Ga(AsP) substrates, the lasers can also emit visible radiation. Other common laser sources are produced by using GaAs and AlAs alloys. These emit radiation within the 0.8–0.9 μm range (Fig. 13.19).

13.7 OPTICAL DETECTION

There are various forms of optical detection using semiconductors. These can be classified as

1. Extrinsic photoemission
2. Photoconductivity
3. p-n junction
4. Avalanche photodiodes
5. Phototransistors
6. Superlattices or quantum well detectors

Extrinsic photoemission occurs when the incident photon energy is larger than the work function of the semiconductor. Photoemission was discussed in detail in Chapter 7 and will not be discussed here further.

In addition to the response of the photodetector to the wavelength of the incident photon, they are compared with each other for their

1. Quantum efficiency η—number of charge carriers generated per incident photon;
2. Frequency response—time response to a pulsed source or to rapid changes in optical light intensity; and
3. Noise output without (dark current) and with (shot noise generated internally) optical illumination.

These properties are incorporated into the following figures of merits:

1. Responsivity: the ratio of the photocurrent to the optical power; and
2. Noise equivalent power; minimum rms optical power with a signal-to-noise ratio of 1.0 at a bandwidth of 1 Hz.

Responsivity is directly proportional to the quantum efficiency, and the wavelength of the incident radiation. High responsivity is obtained with detectors of high quantum efficiency. Noise equivalent power is inversely proportional to the quantum efficiency and decreases with increasing η.

13.7.1 Photoconductivity

A photoconductor is a uniform piece of material, usually an intrinsic, an n-type, or a p-type semiconductor, whose conductivity is changed by the optical generation of charges. Figure 13.20 shows a photoconductor using a semiconductor bar having two ohmic contacts at each end that are connected to a battery. Electrons can be excited from the valence band into the conduction band to generate electron-hole pairs. At low temperatures, if the frequency of the incident photons is low, electrons from the valence band can be excited into the acceptor impurity levels in a p-type semiconductor, thus creating additional holes in the semiconductor. In an n-type material, the electrons from the donor levels can also be excited into the conduction band, thus creating additional electrons. In any case, the additional charges created by the photons change the conductivity of the sample bar. This appears as a change in electrical current in the external circuit. The output current is linearly dependent on the intensity of the incident radiation at low power levels.

Since the impurity levels are generally shallow, that is, close to the respective band edges, the photon energies needed for these interactions are low. This

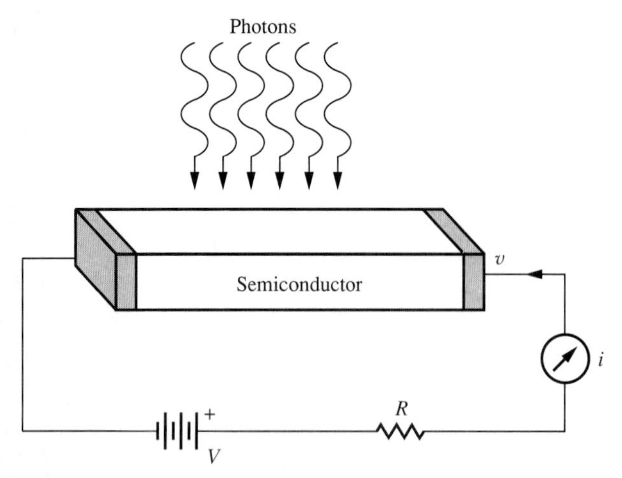

FIGURE 13.20
A simple photoconductor. Photons incident on the semiconductor produce charge carriers which move in the semiconductor under the application of an electric field.

means that longer wavelengths, that is, far infrared wavelengths, can be detected with these detectors. At room temperature, the impurity levels may be fully ionized. In order to prevent this and to increase the sensitivity of the system, the semiconductor is cooled to liquid nitrogen or liquid helium temperatures. These cooled detectors can be made to be very sensitive and theoretically, using heterodyne detection techniques, it is possible to approach the quantum limits of optical sensitivity.

13.7.2 p-n Junction Diodes

When a p-n junction is illuminated by photons, three things can occur, assuming that the photons penetrate the three regions of the junction shown in Fig. 13.21. A photon can excite an electron from the valence to the conduction band on the p-side. Since the holes are already majority carriers in the p-side, the generated hole does not affect the hole concentration and has little effect on the electrical characteristic of the junction. On the other hand, the electron excited into the conduction band increases the minority carrier concentration on the p-side. If the electron is generated within an electron diffusion length, it may appear at the junction where the internally generated diode electric field sweeps the electron across the junction, thus producing a current in the external circuit. Similarly, if a photon generates an electron-hole pair on the n-side, this time the hole that is generated can be swept across the junction, again provided that it is generated within a hole-diffusion length. The diffusion lengths imply that the holes or the electrons are not lost from the system by recombination with the corresponding majority carriers in their respective regions before they reach the junction.

The largest contribution to the current occurs when the electron-hole pair is generated in the junction itself. In this case, both the electron and the hole contribute to the current since they are immediately swept in opposite directions by the built-in electric field of the junction.

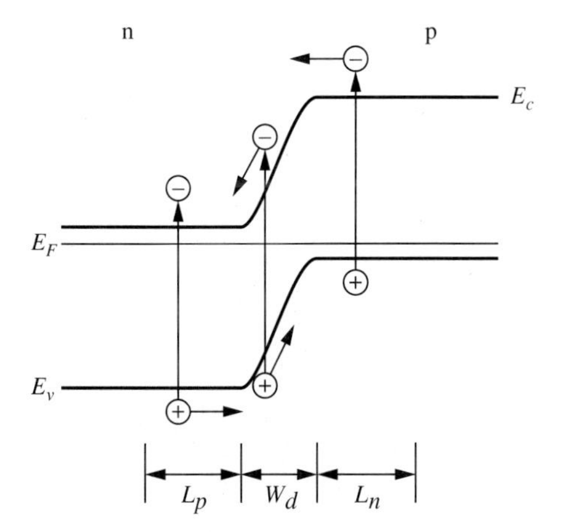

FIGURE 13.21
p-n junction under photon illumination with no external bias voltage.

A p-n junction can be used both with and without any bias. When no bias is applied, it is referred to as a *photovoltaic detector*. In this case, the internally generated diode voltage is modified by the presence of the photoelectrically generated charges. In other words, the charges reduce the space-charge potential. This is reflected as a shift in the equilibrium Fermi levels of both sides of the junction. This is then equivalent to a voltage difference between the two sides of the junction and appears as a voltage difference at the external terminals. The maximum voltage that can be generated under this case can approach the original built-in diode potential. For a given illumination, the maximum voltage V_{oc} appears when the terminals are open, and maximum current I_{sc} results when the terminals are short-circuited (Fig. 13.22). When there is an external load in series with the diode, neither V_{oc} nor I_{sc} can be obtained. When the diode is illuminated with solar energy, the aim is to maximize the resulting output from the diode. This is the basis for *solar cells*. The power-conversion efficiency—defined as the electrical power output generated relative to the incident light intensity—of solar cells is relatively low. At present, considerable research is directed toward increasing the efficiency of solar cells.

13.7.3 PiN Diodes

The diode shown in Fig. 13.21 can also be used as a photodetector by reverse-biasing the diode. The most common p-n junction photodetector is usually made by introducing an intrinsic region between the n- and p-regions. Because of the high resistivity of this region, most of the applied voltage appears across the intrinsic region. The region is made long enough so that most of the photons are absorbed within this region. If the intrinsic region is made too long, the transit time effects reduce the frequency response of the diode, and if it is made too short, the capacitance of the diode reduces this response. These diodes are known as PiN diodes.

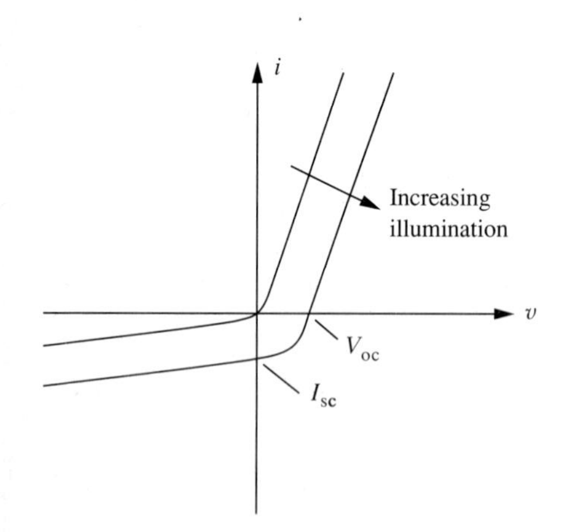

FIGURE 13.22
Current-voltage characteristics of a solar cell. V_{oc} and I_{sc} are the open-circuit voltage and short-circuit currents.

13.7.4 Schottky Barrier Diodes

Schottky barrier (metal-semiconductor) diodes were discussed in Chapter 8. The band diagram of the diode is now redrawn in Fig. 13.23. In a Schottky diode, charge can be generated in two ways. If the photon energy is greater than the barrier height, electrons from the metal can gain enough energy to overcome the barrier height Φ_b and be swept across the junction, thus producing current. If the photon energy is greater than the band gap energy of the semiconductor, electron-hole pairs can be generated on the semiconductor side of the junction, and current flow can be created. Schottky diodes are easier to manufacture, and by making the equivalent cross-sectional area small, very high-frequency response can be obtained.

The metal of the Schottky barrier diode, which is generally evaporated onto the semiconductor, can be replaced by a metallic whisker. Although the concept is very old, the resulting diode still has one of the highest-frequency responses of all the photodiodes. This is due in part to the very small capacitance of the junction. In reality, there may be a very thin layer of an oxide present between the metal and the semiconductor, which changes the characteristic and response of the diode noticeably.

13.7.5 Avalanche Photodiodes

When a p-n junction is negatively biased, most of the applied voltage appears across the depletion layer provided the resistivity of the two sides are relatively low. This is also reflected as a large electric field through the junction. If photons are incident on the junction, the hole-electron pairs are swept across the junction and produce the usual photocurrent. On the other hand, if the reverse bias and, thus, the electric field, is large enough, the photoelectrically generated charges can gain enough energy from the field to generate additional electron hole-pairs in the depletion layer collisionally before reaching the junction boundary. These newly

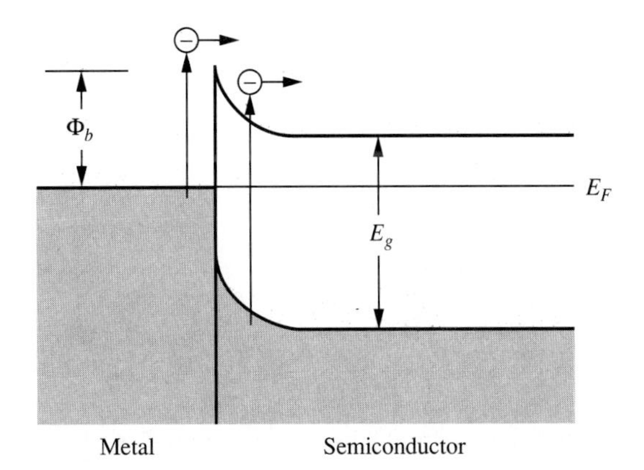

FIGURE 13.23
Energy-band diagram of a Schottky barrier photodiode.

created charges will also contribute to the external current. If the field is so large that the newer charges can generate additional electron-hole pairs within the depletion layer, there will be a dramatic increase in the total charge crossing the junction. This type of avalanche charge generation produces charge multiplication, and the response of the diode increases dramatically to the optical illumination. The diode is said to have gain, that is, more than one electron or hole is generated and collected for each incident photon. Figure 13.24 shows the typical voltage dependence of the avalanche multiplication in Si for a given light input of two different crystal orientations.

The time or frequency response of an avalanche photodiode depends on the ratio of the ionization coefficient of the electrons to that of the holes in a given semiconductor. If the two coefficients are the same, as in Ge or GaAs, the response time is long. Assume that a photon initially creates a hole-electron pair near the p-side of the depletion layer. The hole is immediately swept to the p-side. The electron moves toward the n-side and, in the process, can create a second electron-hole pair. Both of these charges moving in opposite directions can create additional electron-hole pairs. Even though the initially generated electrons have

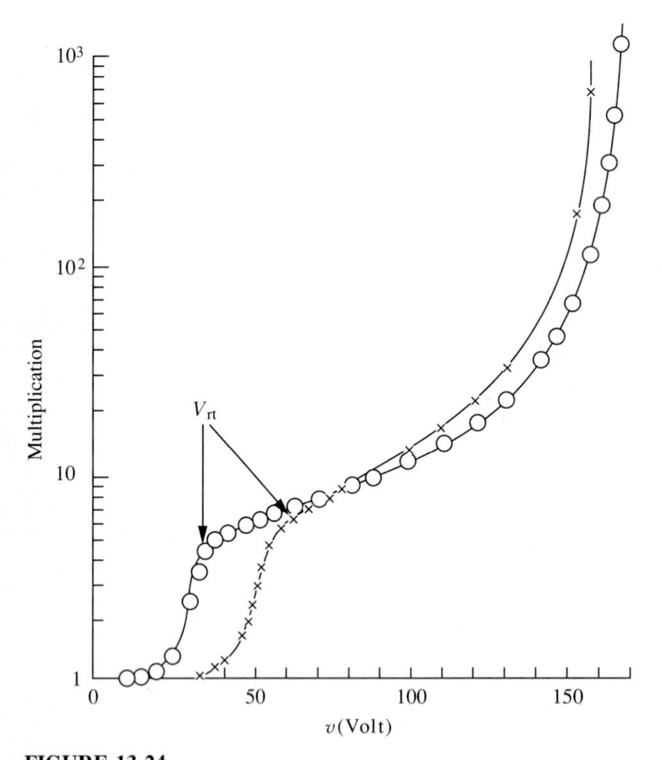

FIGURE 13.24
Avalanche multiplication of charges in silicon with increasing voltage at 1 kHz at different crystal orientations (circles; random, cross; < 110 >) (Kaneda, 1985).

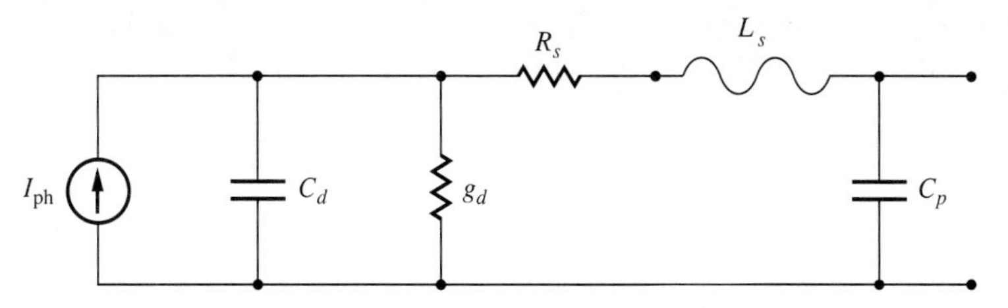

FIGURE 13.25
Small signal equivalent circuit of a photodiode.

already been swept across the junction, the hole that moved near to the p-side can create another electron-hole pair and start the avalanche process over again. This can be thought of as a positive feedback process. Although the number of charge carriers generated per photon is very large, the time duration of the charge multiplication can be very long. On the other hand, if the electron ionization coefficient is very large relative to the holes as in silicon, electrons moving toward the n-region can initiate successive avalanche breakdowns. When the electrons are all swept across the junction, the holes created in the avalanche process will also be swept to the p-side. Since probability of their creating additional electron-hole pairs is low, no noticeable electron-hole pairs will be generated by these holes. The current will be stopped. Thus, the response time of the detector improves considerably. There is still gain in the avalanching process. The noise in this process is also reduced.

The equivalent circuit of a photodiode can be approximated by the circuit shown in Fig. 13.25. The diode is represented by a current source, whose magnitude is proportional to the photon intensity, in parallel with the diode capacitance C_d and the differential conductance g_d. R_s is the series resistance presented by the bulk material, and L_s is the series inductance of the leads. C_p is the parasitic capacitance associated with the packaging. In general, diode conductance g_d, and the diode capacitance C_d are effective in determining the frequency response of the diode.

Example 13.1. Find the maximum frequency response of a silicon PiN diode. The parameters associated with the diode are

$$W_{\text{intrinsic}} = 2.5\ \mu\text{m}$$
$$v_{\text{sL}} = 10^{5}\ \text{m/s}$$
$$\text{Junction area} = 20\mu\text{m} \times 20\mu\text{m square}$$
$$r_d = 100\ \Omega$$

The frequency response can be related to the transit time by

$$\omega_t = 2\pi f_t = \frac{1}{\tau_t}$$

The time response of the diode, in addition to the $r_d\, c_d$ time constant, depends on the transit time of charge carriers through the junction depletion layer.

where

$$\tau_t = \frac{W}{v_{sL}}$$

Solution. Using the values given, the frequency associated with the transit time is

$$f_t = \frac{v_{sL}}{2\pi W_{int}} = \frac{10^5 \text{m/s}}{2\pi 2.5 \times 10^{-6} m} = 6.37 \times 10^9 \text{Hz} = 6.37\text{GHz}$$

The frequency associated with the capacitance charging time is calculated from

$$\tau_c \approx r_d C_d$$

The capacitance is given by

$$C_d = \epsilon \frac{A}{W_{int}}$$

$$= \frac{11.8 \times (8.854 \times 10^{-12} \text{F/m}) \times (20 \times 20 \times 10^{-12} \text{m}^2)}{2.5 \times 10^{-6} \text{m}} = 1.67x10^{-14}\text{F}$$

The corresponding frequency is

$$f_c = \frac{1}{2\pi\tau_c} = 9.52x10^{10}\text{HZ} = 95.2GHz$$

Since the frequency associated with the transit time is lower than that of the charging time, the frequency response is determined by the lowest value, that is, by the transit time.

13.7.6 Phototransistors

The phototransistor is a bipolar junction transistor with the base lead omitted and the base-emitter junction exposed to light. The photons generate photoelectrons, which are injected into the base region, and the regular transistor action follows. By using a cascading scheme known as the Darlington connection, the sensitivity of the phototransistor is enhanced at the expense of time response.

13.8 QUANTUM WELL AND SUPERLATTICE DEVICES

Very interesting and unusual electronic devices can be created by growing alternating layers of high and low band gap energy materials over an intrinsic substrate. A typical multilayer system consisting of GaAs and $Ga_xAl_{1-x}As$ layers grown over a GaAs substrate is shown in Fig. 13.26. Because of the band gap differences between the subsequent layers, alternating layers of high- and low-energy gaps

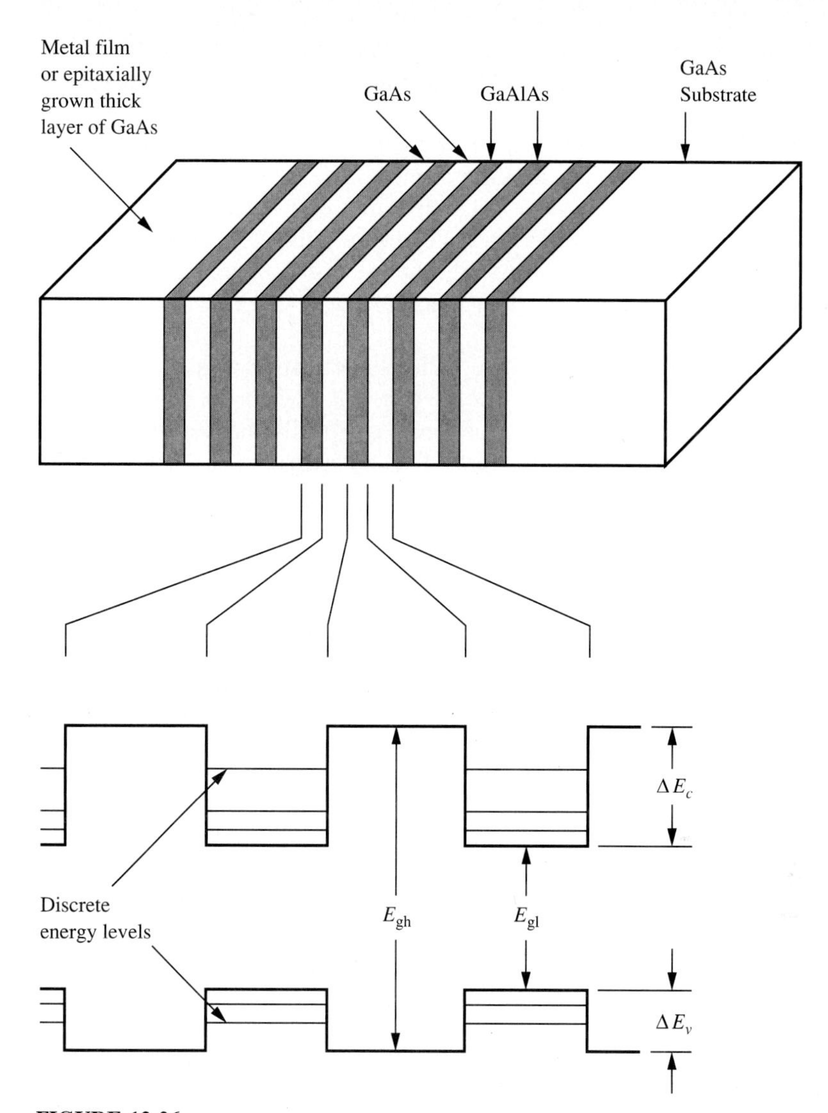

FIGURE 13.26
Multilayer heterojunction systems. Layers of GaAs and AlGaAs are grown one over the other on a GaAs substrate. Resulting energy diagram shows discrete energy levels in the plane perpendicular to the layers.

occur. The result is the generation of *quantum wells* where electrons can now reside.

If the high band gap material is doped with donor atoms, electrons diffuse over to the quantum wells of the low band gap GaAs layers. The motion of the electrons in the quantum wells has very interesting features. In the plane of the lattice, since the GaAs is free of impurity atoms the electrons do not encounter any impurity scatterers. This leads to a reduction in the effective mass of the electrons, thus to a substantial increase in electron mobility. The motion of the electrons in this plane is relatively free, and the electrons are referred to as a *two-dimensional electron gas*. In the plane perpendicular to the lattice, because of the presence of potential wells the electrons have discrete energies and these appear as peaks in the absorption measurements. If the thicknesses of the layers are relatively large, so that the wave function of the electrons is confined to the well (Fig. 13.27*a*), the configuration is generally referred to as a *quantum well device*. If the layer thicknesses are very thin, the electronic wave functions extend into the neighboring wells and result in unusual features. This configuration is known as a *superlattice* (Fig. 13.27*b*).

In a quantum well or a superlattice, since the effective mass is very small and, at the same time, the effective dielectric constant of the system is increased, the effective ionization energy of the electron is drastically reduced. Contrary to this, even when an energy many times greater than the ionization energy is applied, an exciton in a superlattice is not ionized because of the very closeness of the quantum well walls. The exciton peaks are now observable at room temperature.

When an electric field is applied perpendicular to the lattice, the electrons and hole wave functions move apart. There is also a tilting of the band energy of the overall lattice and a shift in the energy levels as shown in Fig. 13.27*c*. Note that the effective energy gap is not the actual gap of the low band gap semiconductor, but the energy difference between the available discrete energy levels in the conduction and valence wells. The electric field is, therefore, able to control the effective absorption coefficient of the multilayer system, as shown

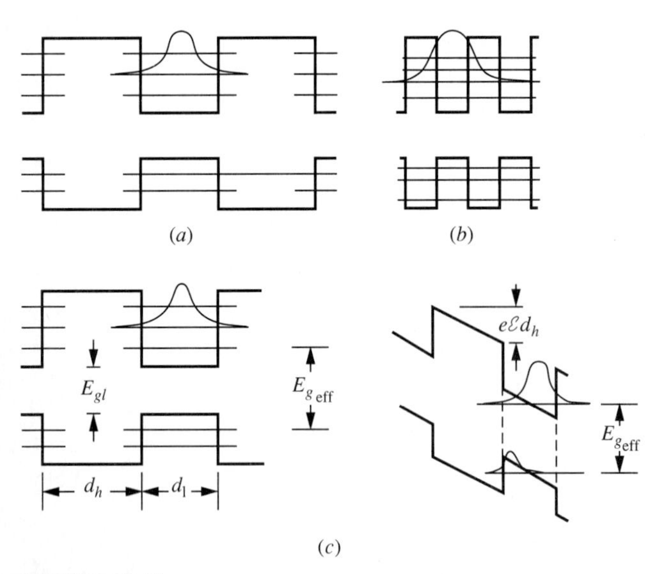

FIGURE 13.27
(*a*) Quantum well device and (*b*) superlattice. $E_{g_{\mathrm{eff}}}$ is different than the lower gap meterial band gap E_{gl}. (*c*) When an electric field is applied perpendicular to the lattice, the wave functions for the holes and electrons are displaced, the energies are shifted, and the band gap energy changes.

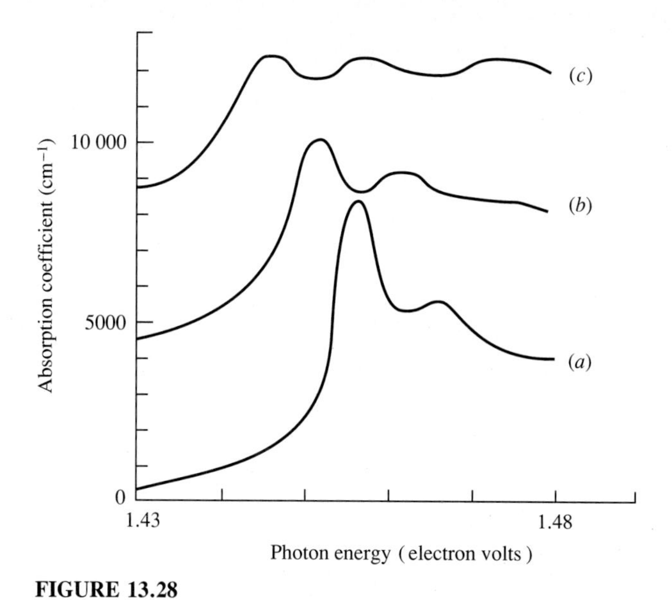

FIGURE 13.28
Absorption coefficient of a quantum well device as a function of photon energy. Curves (*a*) 10^4 V/m, (*b*) 5×10^4 V/cm, and (*c*) 7.5×10^4 V/cm (Chemla, 1985).

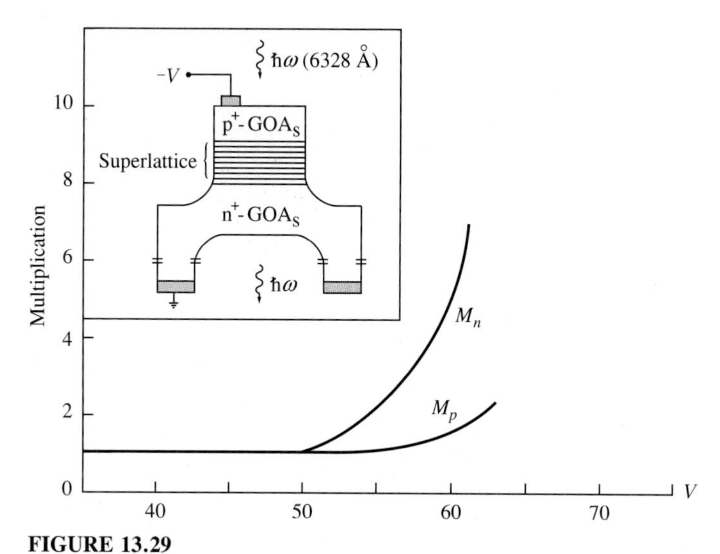

FIGURE 13.29
Electron (M_n) and hole (M_p) multiplications initiated versus voltage. Inset shows a schematic diagram of the experimental configuration (Capasso, 1985).

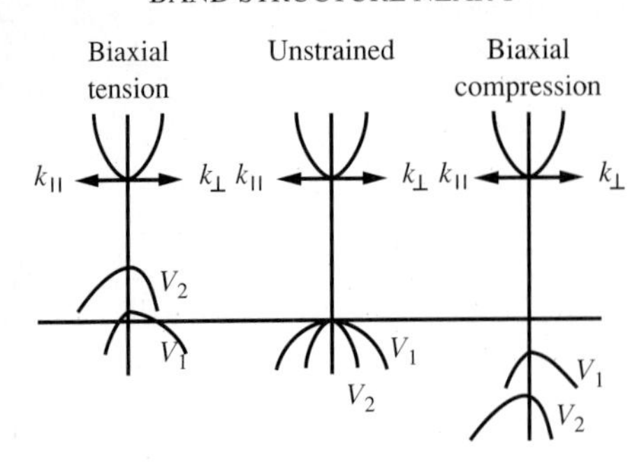

FIGURE 13.30
Band structure of typical zincblend semiconductor near $k = 0$ for the case of no strain (center), biaxial tension (left), and biaxial compression (right). V_1 and V_2 correspond to heavy and light holes (R.E. Kolbas, *et al.*, "Strained-Layer InGaAs-GaAs-AlGaAs Photopumped and Current Injection Lasers," *IEEE J. Quant. Electr.*, vol. 24, 1988, ©1988 IEEE).

in Fig. 13.28. This effect is used to modulate laser light. The lattice can be made to be absorbing or transparent by the application of a time-varying electric field. The time response is limited by the capacitive circuit effects and can be made to be very short.

By profiling the quantum wells, that is, by making a gradual transition from GaAlAs to GaAs for each of the wells, it is possible to preferentially increase the impact ionization coefficient of electrons relative to the holes. Therefore, one can make low-noise fast-responding avalanche photodiodes using multilayer quantum wells, as shown in Fig. 13.29.

It is also possible to introduce strain into the crystal structure by growing layers with different lattice constants. As a result of this internally generated strain, it is possible to shift the potential energies of the heavy and light holes in the crystal (Fig. 13.30). One can then utilize a p-type semiconductor and use the light holes as charge carriers. This allows fabrication of CMOS devices where both the n-channel and the p-channel have the comparable time responses.

These devices are already important in semiconductor lasers. They are expected to be even more important in future electronic devices.

FURTHER READING

Alferness, Rod C., "Waveguide Electrooptic Modulators," *IEEE Trans. Mic. Theory and Tech.*, MMT-30, 1982, 1121–1137.

Boyd, J. T., *Integrated Optics: Devices and Applications*, IEEE Reprint Volume, LEOS Series, IEEE Press (in preparation).

Campbell, J. C., "Phototransistors for Lightwave Communications," in *Semiconductors and Semimetals*, vol. 22.D, W. T. Tsang, ed., Academic Press, New York, 1985, 389–447.

Capasso, F., "Physics of Avalanche Photodiodes," in *Semiconductors and Semimetals*, vol. 22.D, W. T. Tsang, ed., Academic Press, New York, 1985, 2–172.

Chemla, Daniel, "Quantum Wells for Photonics," *Physics Today*, vol. 38, 1985, 56–64.

Conwell, Esther M., "Integrated Optics," *Physics Today*, May 1976, 48–59.

Forrest, Stephen R. "Optical Detectors: Three Contenders," *IEEE Spectrum*, vol. 18, May 1986, 76–84.

Gowar, John, *Optical Communication Systems*, Prentice-Hall, Englewood Cliffs, N.J., 1984.

Johnson, Ernest J., "Absorption Near the Fundamental Band Edge," *in Semiconductors and Semimetals*, vol. 3, R. K. Willardson and Albert C. Beer, eds., Academic Press, New York, 1967, 153–258.

Kaneda, T., "Silicon and Germanium Avalanche Photodiodes," in *Semiconductors and Semimetals*, vol. 22. D, W. T. Tsang, ed., Academic Press, New York, 1985, 247–328.

Keck, Donald B., "Optical Fiber Waveguides," in *Fundamentals of Optical Fiber Communication*, M. K. Barnoski, ed., Academic Press, New York, 1981, 1–108.

Kolbas, R. E., N. G. Anderson, W. D. Laidig, Y. Sin, Y. C. Lo, K. Y. Hsieh, and Y. J. Yang, "Strained-Layer InGaAs-GaAs-AlGaAs Photopumped and Current Injection Lasers," *IEEE J. Quant. Electr.*, vol. 24, 1988, 1605–1613.

Kressel, Henry, and Jerome K. Butler, "Heterojunction Laser Diodes," in *Semiconductors and Semimetals*, vol. 14, R. K. Willardson and Albert C. Beer, eds., Academic Press, New York, 1979, 66–194.

Madan, Arun, "Amorphous Silicon: From Promise to Practice," *IEEE Spectrum*, vol. 18, Sept. 1986, 38–43.

Moss, T. S., G. J. Burrell, and B. Ellis, *Semiconductor Opto-Electronics*, Wiley, New York, 1973.

Nishihara, H., M. Haruna, and T. Suhara, *Optical Integrated Circuits*, McGraw-Hill, New York, 1989.

Panish, M. B., and I. Hayashi, "Heterostructure Junction Lasers," in *Applied Solid-State Science, Advances in Materials and Device Research*, vol. 4, E. Wolfe, ed., 1974, 235-328.

Pearsall T. P., and M. A. Pollack, "Compound Semiconductor Photodiodes," in *Semiconductors and Semimetals*, vol. 22.D, W. T. Tsang, ed., Academic Press, New York, 1985, 174–246.

Physics Today, Special Issue on Optoelectronics, May 1985.

Sturge, M. D., "Optical absoprtion of Gallium Arsenide between 0.6 and 2.75 eV," *Phys. Rev.*, vol. 127, 1962, 768–773.

Yariv, Ammon, *Optical Electronics*, 3d ed., Holt, Rinehart & Winston, New York, 1985.

Yariv, A. *Quantum Electronics*, Wiley, New York, 1967.

PROBLEMS

13.1. The line-shape function (the broadened frequency output of spontaneous emission) has a band width of $\Delta\lambda = 0.1\text{Å}$ at $\lambda_o = 6328$ Å of a He-Ne laser (see Fig. P.13.1). Over the band width, it is possible that successive modes separated in frequency $c/2D$ (c = speed of light, D = mirror separation) can oscillate. This is referred to as hole burning. Find

(*a*) the number of oscillation modes if the length of the laser is 50 cm

(*b*) the shortest length of the cavity so that a single mode will oscillate. Is this practical from dc discharge point of view?

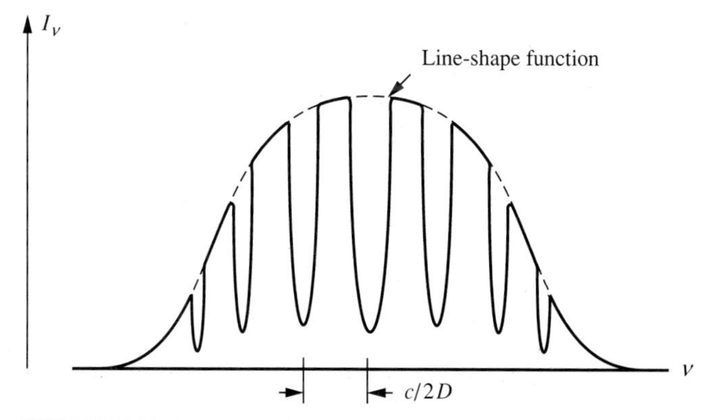

FIGURE P.13.1

13.2. In a semiconductor laser, the perpendicular distance between the two cleaved surfaces is 10 μm. What is the frequency separation between adjacent TE modes for a GaAs laser operating at 0.82 μm? What is the closest integer m for the cavity oscillation?

13.3. If GaAs is used as a LED, find the external efficiency of the diode.

13.4. If you are to design a laser to emit photons at 1.3 μm, what combination of materials can you use by referring to Fig. 5.26? Can you find roughly the molar concentration of the elements from the same graph?

13.5. Light from a pulsed source, which lasts for 5 ns, travels through two different paths and reaches a point P, as shown in Fig. 13.5b. The corresponding path lengths are 50 cm and 100 cm.

(*a*) Find whether the two paths will interfere at P.
(*b*) Find the fraction of time during which they will be interference.
(*c*) What should be the maximum path difference so that no interference will take place?
(*d*) If the lengths of the two paths were the same, would there still be a limitation on the path lengths for constructive interference?

13.6. A mirror is made up of multilayers of GaAlAs and GaAs. The requirement is that the path lengths through each of the layers be a quarter of a wavelength in that material. The respective index of refraction of GaAs and $Al_{0.2}Ga_{0.8}As$ is 3.66 and 3.49. Find the necessary thicknesses for making mirror. Note: if the high index layer is first, the result is a mirror. If the low index material is the first layer, the result is band-pass filter.

13.7. You are to design a photodetector to work in the vicinity of 3.5 μm. Find at least one semiconductor that will detect this radiation.

13.8. In a 20 layer quantum well device the total thickness is 3000 Å. If the low band gap material is twice as thick as the high band gap material, and if 6.0 V appears between the two end layers, find the electric field that appears across each layer and the change in the band-edge energy between the beginning and end of each layer.

13.9. In Problem 13.8, what is the maximum voltage that can be applied across the layers, and the corresponding maximum potential energy shift between the ends of each layer? The applied field should be less than the breakdown field $\mathscr{E}_B$. (Assume GaAs and GaAlAs for the layers and that both have the same $\mathscr{E}_B$ of GaAs.)

CHAPTER

14

SEMICONDUCTOR AND INTEGRATED CIRCUIT PROCESSING TECHNOLOGY

In the previous chapters, the physical bases for the operation of some of the major semiconductor devices were given. Although it is a challenging task to design a component, implementing that component in real life is more of a challenge in itself. The successful operation of a component depends on how well the processing technology produces the desired geometry and the physical characteristics of the device. We can easily talk about an n-region with a uniform donor concentration of 10^{17} cm^{-3} or an ohmic contact on a semiconductor, but in reality how closely can we achieve these desired results? We also want our device to last for a long time and maintain its terminal characteristics during its lifetime. All these considerations show the importance of the process technology.

Semiconductor processing is a highly sophisticated and mature technology. It has been improved over the years to such a level that on a single chip within 10×10 mm^2 we are able to put tens of thousands of transistors and other related components. Producing such a large number of components in such a small area also reduces the throughput (number of devices produced per hour) of the resulting process. These are still being perfected and new processing technologies are being developed. Mono-crystal-related technology, such as silicon, is a highly sophisticated and very well-established technology. Compound semiconductor technologies such as the manufacture of GaAs or InP based devices are also being vigorously developed. It is expected that these will reach such sophisticated stages in the coming years that it will be possible to commercially mass produce reliable, low-cost devices.

A typical insulated gate InGaAs/InP FET (Fig. 14.1), picked arbitrarily from a recent journal article, reflects many of the pertinent manufacturing requirements of modern-day semiconductor devices. Although it seems to be a relatively simple device, it shows the range of various layer growths, doped regions, and insulator and electrode requirements, as well as the relative dimensions necessary for implementing desired electrical characteristics. Excerpts from the article state the manufacturing steps in the following way.

> InP/InGaAs heterojunctions were grown undoped and lattice matched by atmospheric OMVPE. . . . The growth sequence began with deposition of 2000 Å of $In_{0.53}Ga_{0.47}As$ followed by 200 Å of InP. . . . After growth, the devices were mesa isolated using a bromine-based etchant. The source and drain were patterned lithographically and implanted with Si^{29}. The implanted samples were capped with PECVD-deposited SiO_2 and annealed. . . . The SiO_2 capping was removed and ohmic metallization was deposited by e-beam evaporation of 200-Å Ni/1200-Å AuGe/1500-Å Au. After metallization the sample was again capped with PECVD SiO_2 and the contacts were sintered at 375°C using thermal annealing. . . . The deposited capping layer was removed, and after a native oxide etch, the sample was returned to the PECVD system where a 300-Å SiO_2 was deposited at 250°C to form the gate insulator. Device fabrication was completed by deposition of the gate metal, consisting of 300-Å Ti/3000-Å Au and post metallization in H_2. . . .

Because of the scope of this book, in this chapter only the important processing techniques will be briefly explained to give the reader a tutorial background. There are excellent textbooks and review articles that describe the various semiconductor processing technologies. Interested readers can refer to these sources (see Further Reading).

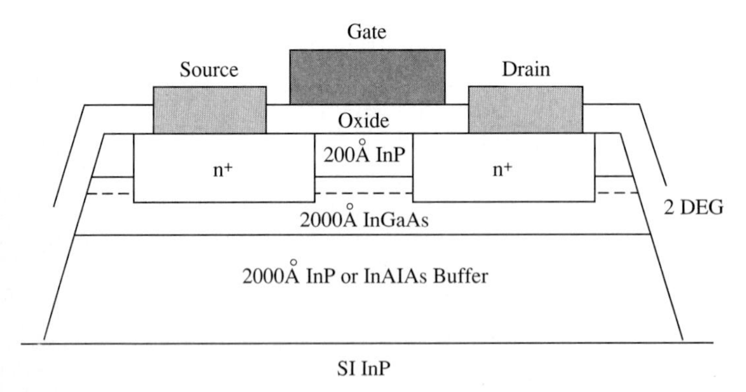

FIGURE 14.1
Typical state-of-the-art semiconductor insulated-gate In GaAs/InP FET. Gate width is 1.5 μm (E.A. Martin, *et al.*, "Undoped InP/InGaAs Heterostructure Insulated Gate FETs Grown by OMVPE with PECVD Deposited SiO_2 as Gate Insulator," *Electr. Devices Lett.*, vol. 9, 1988, © 1988 IEEE).

14.1 CRYSTAL GROWTH AND DOPING

In order to start the process of a semiconductor device, one needs a *wafer*, which is a slice of a semiconductor material, and is either in intrinsic form or has one kind of impurity atom already introduced uniformly in the form of donor (n-type) or acceptor (p-type) atoms. The wafer resistivity is determined by the amount of the impurity atom concentration introduced during the growth of the crystal. The intrinsic types are usually referred to as semi-insulating substrates with the possible highest purity that can be obtained during the growth process.

Silicon, which exists in nature in the form of SiO_2, is first converted in large arc furnaces to metallurgical grade Si, which is about 98 percent pure. At high temperatures, it is further purified by a second process to a higher purity polycrystalline form in the shape of thin rods. In the Czochralski method, these are then melted in a quartz crucible with thermal or induction heating. A sample seed-crystal, inserted into the melt, is gradually lifted out while the seed is rotated slowly, in a highly controlled manner. The crystal grows in the same orientation as the seed sample. The diameter and the size of the resulting crystal depend on various parameters of the growing environment and very large crystals can be grown in this manner. If desired, the crystal can be purified further by additional processing. It is also possible to introduce into the crucible the desired type of impurity atoms during the crystal growth. Once the desired type of specifications of the crystal are obtained, the crystals are sliced into wafers and polished; they are then ready for shipment. These processes are highly specialized processes, and one can obtain a large number of wafers from a single batch. The wafers are usually 100 μm thick, and their diameters vary from two to about eight inches.

14.1.1 Diffusion and Ion Implantation

We saw in Chapter 4 that the diffusion process and ion implantation were two fundamental ways of introducing impurity atoms into a crystal to change the electrical characteristics of a semiconductor.

The *diffusion* process is one of the oldest techniques, but a highly developed one, with which impurity atoms can be introduced into a semiconductor. As shown in Fig. 14.2, the wafers are placed on a holder inside a quartz oven tube and heated by induction or thermal heating, and a carrier gas containing the impurity atoms flows through the chamber. The diffusion process is usually of the constant

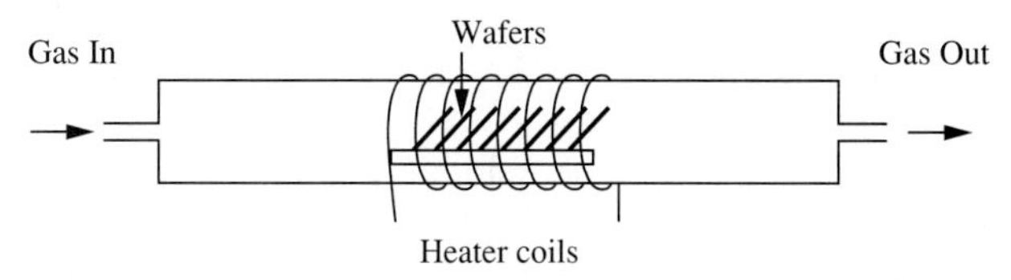

FIGURE 14.2
Diffusion furnace for impurity doping of semiconductor wafers.

replenishment type (see Section 4.7). Even though the electrical characteristics of the entire surface of the wafer can be changed by the diffusion process, through the use of masking techniques, only selected portions of the wafer are intentionally doped by a given impurity atom. The diffusion process produces density profiles that vary with distance as a function of complementary error function. Therefore, across the semiconductor, the change in impurity density from one type to the other is not as abrupt as assumed in many of the idealized models previously discussed, but the resulting impurity profiles yield highly successful device characteristics not too different than depicted by the idealized models.

Ion implantation is obtained by using a high-energy ion source. A discharge creates a plasma from which ions are extracted and accelerated by electrostatic fields. Using a magnetic field-mass separator, the desired ion species are selected. With additional accelerating and focusing potentials, ions reach their final energies and are directed onto the crystals as a beam of positively charged particles. Ion implantation is a low temperature process, and thus will not alter the doping profiles of previously doped regions in the semiconductor. Impurity profiles can easily be controlled in the ion implantation process and precise amounts of material can be introduced into the semiconductor. With the advent of microprocessors and microcomputers, the profile shaping can be programmed into the process and precise process control achieved. With proper care, semiconductor damage by the incident high-energy ions can be reduced to a minimum. After implantation, the implanted atoms have to be annealed to be electrically activated. This requires temperatures in the range of 600 to 800°C for Si. This is generally achieved by a process known as rapid thermal heating, obtained by precisely controlling the heating and cooling of the substrate within seconds by subjecting the wafer to radiant heat from either a discharge lamp or a laser light. Also, if different energy ions are used to produce the desired depth of the impurity profile, thermal annealing will smooth out the variations in the density profile of the subsequent implantations, as shown in Fig. 4.11 (see Chapter 4).

14.1.2 Liquid and Vapor Phase Epitaxy

Epitaxy is the growth of a crystal layer (n-type, p-type, or intrinsic) over another crystal usually of the same kind and in the same orientation as the base crystal. The process has two forms: liquid phase epitaxy (LPE) and vapor phase epitaxy (VPE). In the liquid phase, the base crystal is placed in a quartz or graphite boat and covered by the liquid from which the layer will be deposited. The boat, the sample, and the liquid—which contains the dopant impurity atoms—are heated to the proper temperature and a layer of crystal grows in the same crystal direction as the base crystal. The thickness of the growth layer depends on the temperature as well as to the duration of growth.

There are many variations of the VPE technique, but they have essentially the same principle shown in Fig. 14.3. The reactor is made up of a quartz tube where various heated regions produce the necessary temperature profiles. HCl gas flows over the metals to be deposited so that metal chlorides will form. Other materials are brought in through separate tubes. All input reactant flows are controlled

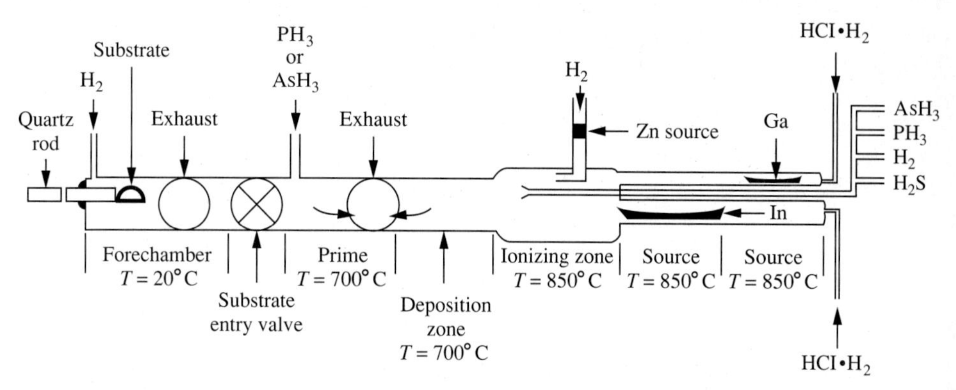

FIGURE 14.3
Typical components of VPE system to grow III–V compound semiconductors (Reprinted with permission from G.H. Olsen, "Vapor Phase Epitaxy of Group III–V Compound Optoelectronic Devices," Chapter 12 in *Integrated Circuits: Chemical and Physical Processing*, P. Stroeve, ed., ACS Symposium series 290, ©1985 American Chemical Society).

by electronic mass flow controllers. The substrate temperature is held to within $\pm 0.1°C$. The reaction takes place at the heated surface of the wafer. Once the growth is completed, the boat is pulled toward the fore-chamber without stopping the gas flows. The wafer is cooled to room temperature and removed from the chamber. With VPE processing, very high purity layers can be grown.

14.1.3 Molecular Beam Epitaxy (MBE)

Molecular beam epitaxy (MBE) is one of the most recently developed techniques of producing high quality, uniformly thin layers (within atomic layers) of single or compound crystals.

A typical MBE machine is shown in Fig. 14.4. In a highly evacuated chamber ($p < 10^{-11}$ torr), the mean free path of atoms is very large compared to the chamber dimensions. Thus, any energetic atomic or molecular particles introduced into the chamber will move in straight lines without making any collisions with the background gas atoms. The beam of particles emitted from ovens located at different regions of the chamber are deposited on a substrate that is heated to a high temperature for proper crystal formation. The sources of the beams are generally effusion ovens, which are crucibles surrounded by heaters supported in alumina rods. Additional heat shielding and thermocouples allow precise control of the temperature at which beam materials evaporate. These ovens can be made to be very compact, and four or more ovens can be distributed within a vacuum chamber to include as many components of compound semiconductors necessary for crystal growth.

To obtain significant evaporation from Si, temperatures on the order of 1600°C are required. This would introduce out-gassing problems from the crucible and the heating wires if an effusion oven were used. Therefore, an electron beam evaporation technique is used to produce Si beams in MBE machines. This

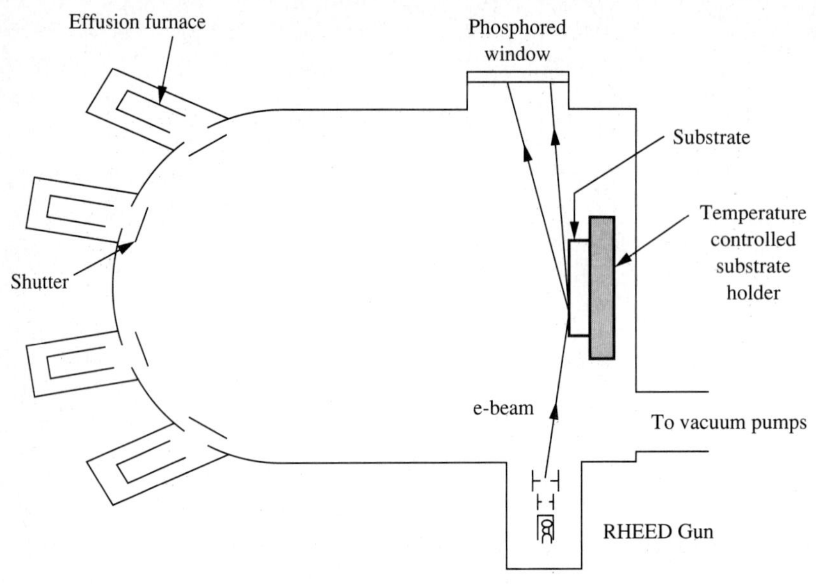

FIGURE 14.4
Molecular beam epitaxy (MBE) machine for deposition of thin layers of semiconductor materials.

is shown in Fig. 14.5. A heated filament biased at a high negative voltage emits electrons that are accelerated toward the grounded anode. But a steady magnetic field bends the electron beam nearly 270°. These electrons then impinge on a silicon slug located inside a copper crucible. Silicon is melted at the location of the electron impact. Since the rest of the silicon is cooled by the flowing water in the crucible, the molten silicon is isolated from the other contaminants by the unmolten silicon itself.

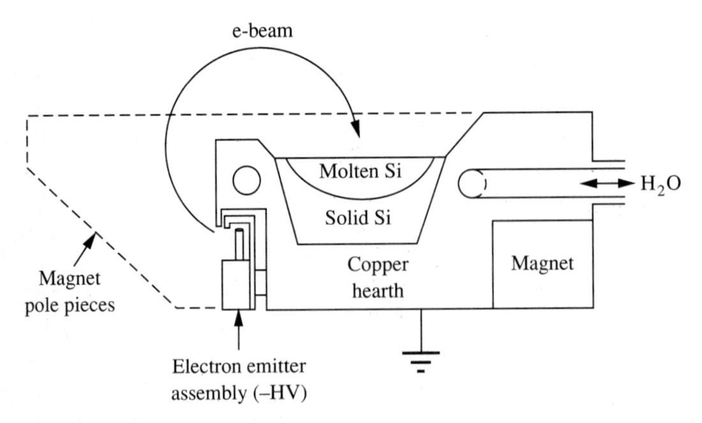

FIGURE 14.5
Schematic cross section of e-beam silicon evaporation source. Magnetic field is perpendicular to the page. Typical slug dimensions 2 inches diameter by 1 inch deep (40 cc) (Bean, 1981).

Precise control of the temperatures of the ovens and the use of mechanical shutters allow precise deposition of intrinsic as well as doped layers. Mechanical shutters allow fast introduction or removal of a beam from the system thus providing for abrupt growth changes from one semiconductor type to another. It is also possible to grow monolayers of atoms using this process. In compound semiconductors, the growth of the base crystal depends on how well the ratio of the component atoms are to be maintained to produce the required crystal. The MBE process incorporates extremely precise control of the temperatures, since the amount of evaporation of a given material depends on the temperature and thus regulates the amount to be deposited.

There are also diagnostic systems built as an integral part of the MBE system. One system commonly incorporated into MBE machines is the reflection high-energy electron diffraction (RHEED) apparatus. In RHEED, a high-energy electron beam impinges on the substrate near grazing angle between 0 and 10°. Electrons that are diffracted impinge on a phosphore-treated chamber window that emits light in proportion to the electrons impinging on it. The observed diffraction pattern is an indication of the crystalline structure of the surface of the semiconductor. RHEED is used to monitor the semiconductor while growth is actually taking place.

14.1.4 Chemical Vapor Deposition (CVD)

Chemical vapor deposition (CVD) is a process in which high-quality thin layers of intrinsic or doped layers of semiconductors can be grown. The substrate is heated to high temperatures where chemical decomposition, called pyrolysis of a gas, generally takes place directly on the surface of the heated substrate. CVD can be used to grow Si, SiO_2, Si_3N_4, and various forms of compound semiconductors.

In growing Si, the following typical decomposition of the gases takes place:

$$SiCl_4(gas) + 2H_2(gas) \rightleftharpoons Si(solid) + 4HCl(gas) \tag{14.1}$$

In growing SiO_2 and Si_3N_4, typical reactions are:

$$SiH_4(gas) + O_2(gas) \rightleftharpoons SiO_2(solid) + 2H_2(gas) \tag{14.2}$$

$$3SiH_4(gas) + 4NH_3(gas) \rightleftharpoons Si_3N_4(solid) + 12H_2(gas) \tag{14.3}$$

In growing compound semiconductors, hydrides (containing H) and chlorides (containing Cl) are used as the chemical gases. A typical example is

$$Ga(CH_3)_3(gas) + AsH_3(gas) \rightarrow GaAs(solid) + 3CH_4(gas) \tag{14.4}$$

There are various forms of CVD equipment. These are:

1. **Atmospheric** pressure chemical vapor deposition (APCVD). The reaction takes place at atmospheric pressure and requires large volumes of carrier gases.
2. Low-pressure chemical vapor deposition (LPCVD) allows easy removal of carrier gas and uses small amounts of gas at reduced pressures at the expense of slower growth rate.

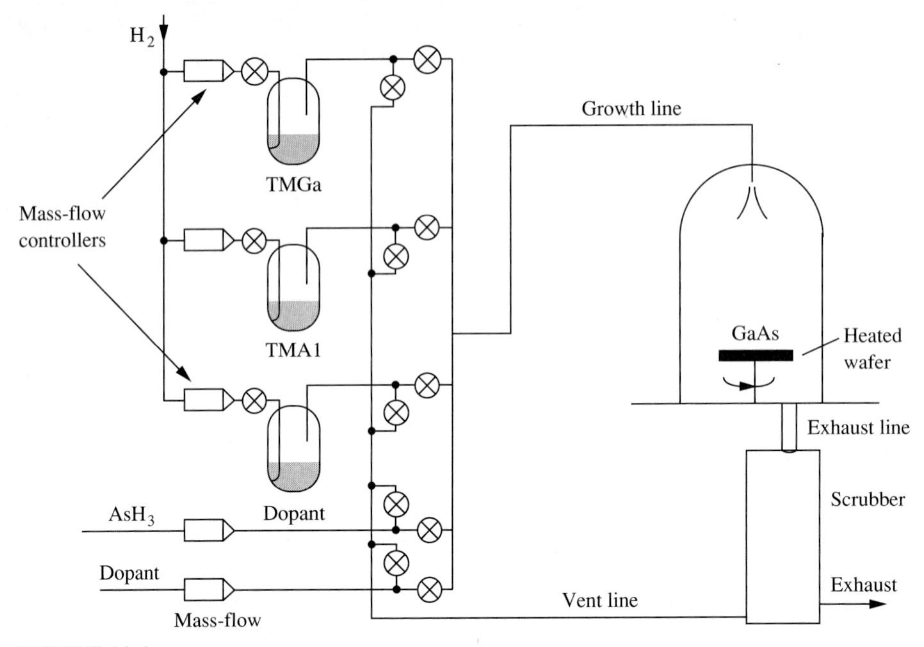

FIGURE 14.6
Schematic diagram of an MOCVD system employing alkyds [trimethyl gallium (TMGa) and trimethyl aluminum (TMAI)] and metal hydride (arsine) material sources, with hydrogen as a carrier gas (Schumacher, 1988).

3. Plasma-enhanced chemical vapor deposition (PECVD) initiates reactions easily, allows lower substrate temperatures to be used, and allows other gases to be used where otherwise the reaction would not have been possible.
4. Metallo-organic chemical vapor deposition (MOCVD) uses alkyds and hydrides of metals and is generally used in compound semiconductor manufacturing. A typical arrangement for an MOCVD system for growing III–V compounds is shown in Fig. 14.6. Flow rates of necessary gases are regulated by precision mass flow controllers. These gases and other dopant materials are mixed together and enter the growth chamber. Extreme precautions have to be taken to remove the unused dangerous chemicals from the system before they are pumped out of the system.

The CVD systems are also further classified with regards to the cooling of the walls of their containers, that is, hot or cold wall. The physics and chemistry of some of the CVD processes are well understood. Others require additional research and development to improve the growth uniformity and quality.

14.2 LITHOGRAPHY AND MASK MAKING

One of the fundamental processes of integrated circuit manufacturing, which determines the degree of complexity, density of component distribution, and frequency of operation of a given device, is lithography, that is, the success with which patterns of precise dimensions can be transferred onto the semiconductors for processing. Once the final geometrical configuration of a device is made, the purpose of lithography is to transfer this configuration onto a mask within micron or submicron resolution. The resulting mask should have a high contrast ratio between the opaque and the transparent regions, should have, in principle, no imperfections, and should have the same precision throughout its complete surface area. Since more than one mask is usually required for the manufacture of a given device, the masks should have the same tolerances, same resolution, and contain markings so that precise placement of one layer can be superimposed over the next.

Once the design of a given device is completed, a large scale (100× or more) replica of the mask, called the artwork, is laid out on paper. Once the final modifications are made to the artwork, the patterns are transferred to a rubylith, which is a transparent sheet of plastic base coated with a second very thin layer of removable red-colored plastic. The pattern is generally cut on the rubylith by a computer-controlled cutting machine, and the unwanted upper red layers are removed. Using a precision reduction machine, the resulting image is reduced photographically at least once or, in many cases, twice to its final size. It is necessary that the final mask should contain the required dimensional tolerances of the device to be manufactured. There are now computer software programs that allow the design of masks in faster and more efficient ways. These programs are known as Computer Aided Design (CAD) software and allow the mask exposure directly through a computer.

Once the masks are ready, the manufacturing process starts. In general, a thin oxide or other dielectric layer ($\approx 1\ \mu$m thick) is grown over the wafer. A chemical polymeric compound, called a *photoresist*, is spread uniformly over the semiconductor wafer by a process called spin coating. First, a drop of photoresist is placed over the wafer, which is held by a vacuum chuck that is spun between 2500–8000 rpm. The centrifugal force spreads the resist over the wafer and the excess resist spills out from the edges of the wafer. A very uniform thin layer (on the order of 1 μm thick) of photoresist coats the wafer. The resist is softbaked (90°C) to increase the adhesion of the resist to the oxide surface. The mask is then placed over and in contact with the resist and, using a monochromatic light source, the image of the mask is transferred to the photoresist by contact printing. Sometimes projection printing can be used for this process.

As shown in Fig. 14.7, there are two types of photoresists: negative and positive. When exposed to light, the photons interact with the exposed parts of the resist. Depending on the type, the exposed or the unexposed portions of the resist are removed by the development process. In the positive (negative) photoresist, the interaction is such that the exposed (unexposed) portions are soluble in the developer. Thus, depending on the type of resist, either the image of the mask (negative resist) or the opposite of the mask (positive resist) can be transferred to the photoresist.

The photoresist is then hardbaked at a higher temperature ($\approx$ 150°C). Finally, using a wet or dry process, the parts of the oxide unprotected by the resist

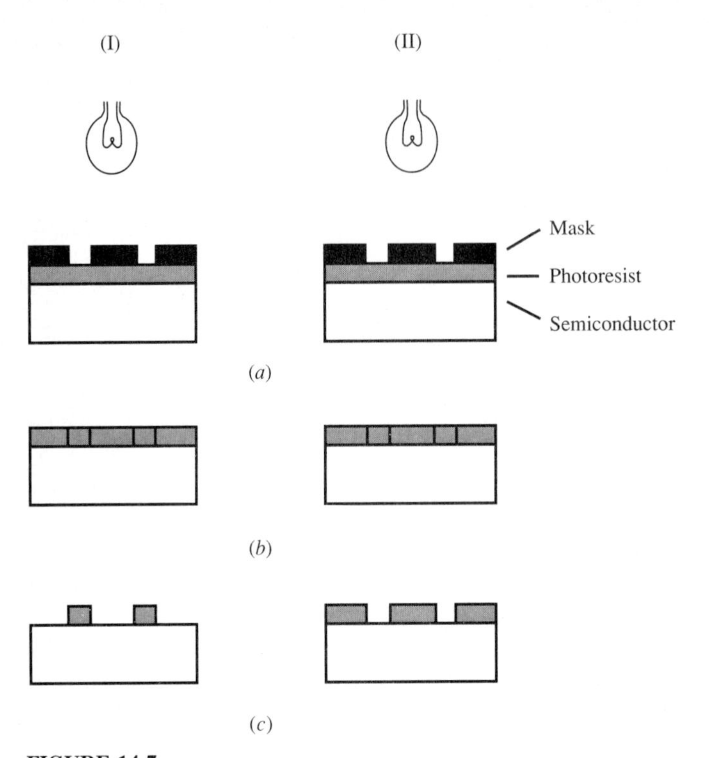

FIGURE 14.7
(I) Negative-acting and (II) positive-acting photoresists: (*a*) exposure to light, (*b*) before development, and (*c*) after development.

are removed (etched away) so that the exposed portions of the semiconductor can be processed further.

Since there is usually plenty of space to put more than one device (in many cases, many devices) on a given size wafer, there are two methods with which the image on the mask can be repeated at specified intervals so that as many devices as possible can be manufactured on the same chip. In the most common method, using a step-and-repeat camera, the image is repeatedly copied photographically as many times as necessary onto chosen locations on the final master mask. In a less used method, using a projection system, the wafer is stepped and the image is repeatedly projected directly on the substrate.

Mask preparation and the transfer of the image onto the substrate, especially one that contains submicron dimensions, push the necessary optical components to their limits and necessitate newer and novel lithographic techniques. In order to appreciate these, one has to review some of the fundamental principles of optics.

1. Resolving power of an optical system. The Rayleigh criterion states the diffraction-limited resolution of an instrument (related to the resolution of two Airy discs) is given by the angular separation

$$\Delta\theta = \frac{\Delta x}{D} = \frac{0.61\lambda}{r_0} \tag{14.5}$$

where $2r_0$ is the diameter of the circular aperture and D is the distance away from the opening to the location of the image. $\Delta\theta$ is the angular and Δx is the spatial minimum resolution between two lines that can be optically resolved for a given wavelength of light. Equation 14.5 gives the ultimate resolution that can be obtained from any optical system. It also applies to the images produced by lenses that are used in mask-making equipment. The larger the lens opening or the shorter the wavelength of light, the better will be the ability to separate the two lines within the distance Δx given above.

2. Diffraction from a knife edge. When plane waves are incident on a knife edge (an abrupt discontinuity in going from the opaque to the transparent regions of a mask), light is also diffracted and appears under the geometric projection of the knife edge. A typical line on a mask is shown in Fig. 14.8. Because of the diffraction of light by the sharp edges, one does not see the true projection of the line on the screen. This becomes more pronounced as the width of the line becomes comparable to the wavelength of the light source, thus, lines may actually disappear from the image projected onto the photoresist. As can be seen from these optical limitations, one needs shorter wavelengths of illumination as the line widths get narrower.

There are three lithographic techniques that are used in the manufacture of integrated circuitry. These are (a) optical, (b) e-beam, and (c) X-rays.

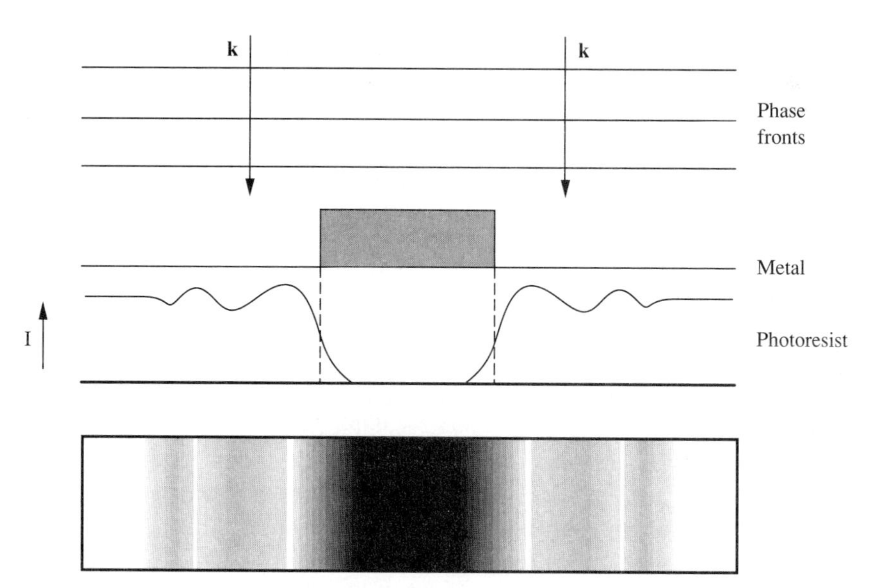

FIGURE 14.8
When light of wavelength λ falls on an opaque material (metal), light diffracted by the edges of the metal appear at the regions under the projected width of the metal mask.

14.2.1 Optical Lithography

Optical lithography is the most widely used technique at present in the manufacture of integrated circuits. It is being pushed to its limits as the device dimensions get smaller and smaller. It is highly developed, yet relatively simpler than other lithographic techniques, and line widths close to less than one micrometer are possible with optical lithography. The mask is an optically flat glass plate coated with a high-contrast photographic emulsion. For exposure, a monochromatic light source (usually an ultraviolet line radiation of a mercury lamp), highly precision-ground optics, and very stable mechanical components are used. As the wavelength gets smaller, requirements on optical components also get more stringent. The optical aperture, the plane of focus of the optical lenses, the planarity of the mask, as well as the smoothness of the semiconductor wafer surface limit the smallest line that can be resolved by the optical system. Going to very short ultraviolet sources require specialized optical components with high uniformity and high transparency.

14.2.2 Electron-Beam Lithography

Electron beams (e-beams) can be used as an exposure source of making $1x$-size photomasks or as a means of direct exposure of the photoresist.

The base glass plate of an e-beam mask is coated with chromium metal. A positive photoresist is applied over the metal film. By moving the e-beam through computer control, the pattern is exposed electronically by the e-beam on the resist. Instead of an optical repeat-and-step system, the e-beam itself can be used to repeat the pattern over on the mask by mechanically displacing the plate and repeating the electronic exposure. Next, the resist is developed, then the exposed portions of the chromium metal are etched away. Finally, the remaining photoresist is stripped away and the mask is ready to be used. With this system, line resolution down to less than 0.1 μm is possible.

The mask can be eliminated altogether completely by directing the e-beam onto the photoresist that is applied directly over the semiconductor wafer. Similar to the mask-making process, the substrate holder can be moved by stepping motors for large displacements. For smaller displacements, the e-beam can be scanned by applying electrostatic potentials to the deflecting plates of the electron gun. Figure 14.9 shows the major components of an e-beam lithographic system. Since the corresponding DeBroglie wavelength for energetic electrons is very short, there are practically no diffraction problems associated with e-beam lithography. The electron beam can easily be focused to very small spot sizes and moved within very small distances with high precision.

The e-beam system can easily resolve submicron size lines. When the e-beam impinges on the resist, it is scattered and limits the line resolution to a certain extent. Although it eliminates the mask-making process, there are certain disadvantages in using direct-writing e-beam lithographic systems. Since the beam has to scan individually the geometry of the device on the photoresist,

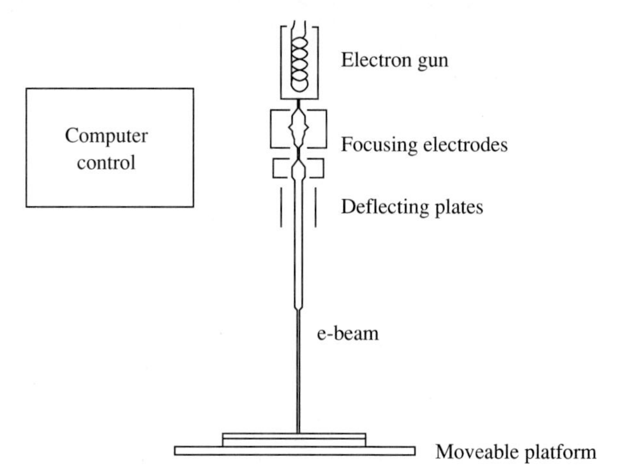

FIGURE 14.9
Electron beam lithography. Electron beam is precisely moved by computer control.

the process is a serial process and is very slow. If the pattern has to be repeated over and over on similar substrates, the increased time to transfer the images decreases the throughput of the manufacturing process. With further research on faster-responding photoresists, e-beam lithography will complement the optical lithography in submicron device fabrication. At present, it is not economical to use e-beam lithography for mass production of complicated device geometries. It is used mainly as a mask-making facility and in the manufacturing of specialized discrete components such as microwave devices.

14.2.3 X-ray Lithography

Because of the wavelength dependence of diffraction, it is necessary to use shorter wavelengths of radiation for higher resolution. X-rays, whose wavelength can be as low as one Angstrom, can be used for lithographic purposes provided proper conditions are satisfied for the components of X-ray lithography.

In principle, although X-ray lithography is an extension of optical lithography to shorter wavelengths, there are severe technological and material limitations in its present form. Some of these can be summarized as follows.

SOURCES. There are three types of X-ray sources that are used at present for X-ray lithography. These are

1. Continuous wave (CW) low-power X-ray sources, generated by high-energy electrons impinging on metal targets and emitting line radiation or continuous spectra;
2. Pulsed sources, generated either from a dense plasma discharge or by pulsed irradiation of a metal by a high-power laser; and

3. X-rays generated by a synchrotron, a high-energy accelerator accelerating electrons in circular orbits to very high energies which emit radiation at X-ray wavelengths in the process.

For a given resist, it is necessary to have a minimum amount of dosage (power-time product) so that the resist can be properly exposed with the X-rays. For CW sources, the power levels are low, and it therefore takes a longer time to expose a given resist. This introduces problems with vibration, thermal dimensional changes, alignment of the mask, and so on. On the other hand, for pulsed sources, the spectral purity and the extent (size) of the source become important. If the source is not a point source, there is the problem associated with what is known as the penumbra-effect, shown in Fig. 14.10*a*. Here, the extended source projects a wider shadow of the line, thus reducing the line resolution of the system. On the other hand, a point source projects a nearly true image of a line. In order to increase the power level at the surface, the source is usually brought closer to the substrate. But this also produces line tilting effects as shown in a highly exaggerated manner in Fig. 14.10*b*, so that different portions of the mask are exposed to different equivalent line widths. Plasma-generated or laser-generated pulsed sources come close to being point sources.

The synchrotron radiation emits X-ray radiation whose wavelength spectra can be controlled by the energy of the accelerated electrons. But the physical size as well as the cost of these accelerators are very large and make them at present impractical for industrial production.

Because of the interaction of X-rays with matter, it is almost impossible to make conventional imaging components, such as lenses, that will operate at Angstrom wavelengths. Thus, there is a need for novel components that will operate at these wavelengths. In the future, with further research and development on sources, resists, and components, X-ray lithography will be a dominant processing technique at submicron-size device manufacturing.

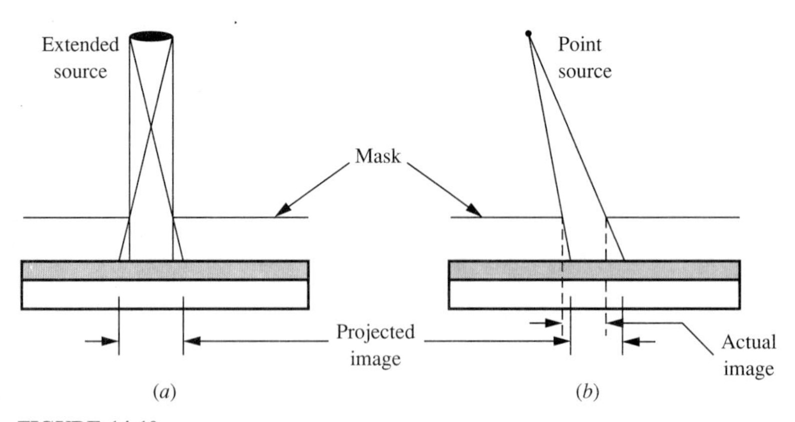

FIGURE 14.10
(*a*) Penumbra effect due to an extended X-ray source. (*b*) Point source close to the substrate projects a tilted shadow near the edges of the mask.

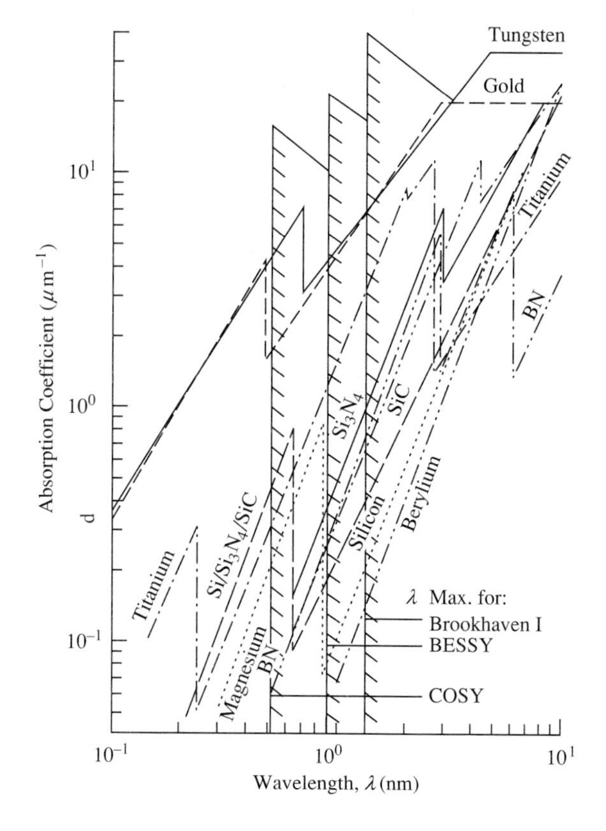

FIGURE 14.11
Absorption length for some materials as a function of X-ray energy (Heuberger, 1986. Reprinted with the permission of *Solid State Technology*).

MASKS. Masks for X-ray lithography should be properly chosen. A metal that may be opaque to optical radiation can be transparent to X-ray radiation, and if used as a mask, it will be useless. Also, the extent of the transparency of a material will be determined by the thickness of the material as well as the wavelength of the X-ray radiation. The shorter the wavelength, the more radiation will be transmitted through the material, as shown in Fig. 14.11. Thus, one has to use the right material at the right wavelength for proper exposure of the resist.

PHOTORESISTS. Although some of the resists developed for optical lithography are also used for X-ray lithography, new and improved resists are needed to be used specifically for X-ray lithography. These resists should be highly sensitive to X-ray radiation to reduce exposure time. They should also be sturdy enough to stand X-ray radiation damage for the duration of exposure and easily strippable once the etching of the oxide is completed.

14.3 ETCHING

In many instances, controlled amounts of material should be removed from the wafer so that other layers or connecting leads can be added to the device during the manufacturing process. A simple example will be the removal of an exposed oxide layer from a specific region on the semiconductor. This can also be an unprotected region of the semiconductor by the photoresist so that impurity doping by diffusion or ion implantation, or connecting electrodes can be deposited over the region in question.

The etching process can be divided into two general categories: wet etching and dry etching. In either of these cases, it is possible to etch a layer isotropically, as shown in Fig. 14.12*a*, or uniformly (anisotropically), as shown in Fig. 14.12*b*. In isotropic etching, material under the projected area of the photoresist is also removed due to the penetration of the etching material into these regions. In the ideal uniform case, the etch produces a true projected image of the opening throughout the depth of the dielectric. As the dimensional tolerances get critical, uniform etching becomes desirable.

14.3.1 Wet Process

In the wet etching process, the wafer is exposed, by immersion or spraying, to a chemical that reacts with the material to be removed from the wafer. A variety of chemicals are used to wet etch wafers, and these depend on the types of semiconductor, the dielectric, and the metal to be etched away. The resulting etch is generally isotropic since there is no control of the chemical to directionally etch the material.

14.3.2 Dry Process

In the dry etching process, the etching gas is chemically neutral under normal conditions but becomes chemically active when it is in an ionized state. The discharge that ionizes the gas is produced by a dc or RF source applied between two plates. The substrate is placed on one of the plates, the discharge is started, and the chemically active ions react with and remove those unwanted exposed portions of the substrate that are not protected by the previously deposited dielectric layer.

There are various forms of dry etching processes with regards to the arrangement of the electrodes, background pressure, and the energy of the reacting ions. The physics of ionization, chemistry of reaction of the gas with the substrate, and

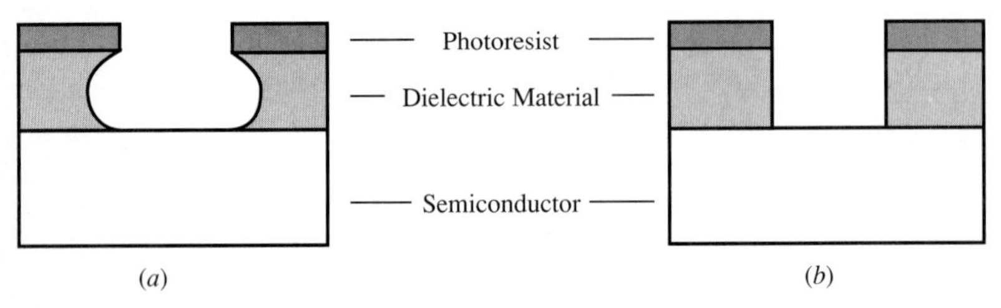

FIGURE 14.12
Results of etching (*a*) isotropic and (*b*) uniform (anisotropic).

the removal of material is a very complicated process, but the resulting processes can be classified into three basic categories.

1. *Plasma etching*. In this configuration, the plate that holds the substrate is at ground potential and the background pressure is between 100 to 200 mtorr. The etch rate is high and uniform and can be isotropic, but there is the danger of radiation damage due to plasma being in contact with the substrate, and the process sometimes leaves behind some unwanted residues. (See Fig. 14.13*a*.)
2. *Reactive ion beam etching (RIBE)*. Because of the relatively low background pressure of the chamber, in the order of 10^{-2} to 10^{-3} torr, the mean free path of the reacting ions is longer. This allows ions to hit the substrate at nearly normal incidence. Although the etch rate is lower, the resulting etch is more anisotropic. RIBE is a widely used and developed process in the manufacturing of very large scale integrated (VLSI) circuitry. (See Fig. 14.13*b*.)
3. *Ion beam milling*. In this process, the ions are extracted from the plasma and directed toward the substrate with relatively high energies (500–1000 eV) as a beam of ions. The etch process is a physical sputtering process where impinging ions remove material from the substrate by impact. The resulting etch is highly anisotropic, with high resolution and free of radiation damage because of the isolation of the substrate from the plasma. There is the danger of redeposition of the removed material back onto the wafer. (See Fig. 14.13*c*.)

14.4 DIELECTRIC LAYERS

Dielectric layers are one of the most essential components of integrated circuit manufacturing, as well as an integral part of many discrete solid-state devices. Their applications range from (*a*) a means of isolating different layers of an integrated circuit to (*b*) being the interelectrode dielectric in the capacitor of an MOS device to (*c*) an agent in reducing the surface states at the surface of a semiconductor.

Most of the device components are built within a few micrometers of the surface of the semiconductor wafer. Since there is a discontinuity of the crystalline structure at the surface, this leads to a phenomenon known as *dangling bonds*. This results in generation of surface states that highly modify the electrical characteristics of the semiconductor near the surface. If a passive dielectric layer is deposited over the wafer, it is possible to reduce these surface states to negligible levels. This procedure is known as *passivation*.

The ideal dielectric is the native oxide of the semiconductor itself such as SiO_2 over Si. But unfortunately many semiconductors, especially the compound semiconductors, do not have native oxides. Even for these, silicon dioxide, SiO_2, and silicon nitride, Si_3N_4, are the most widely used dielectrics in the manufacture of semiconductor components.

There are various ways of obtaining SiO_2, native oxide of Si. In the wet and dry processes, the respective reactions are

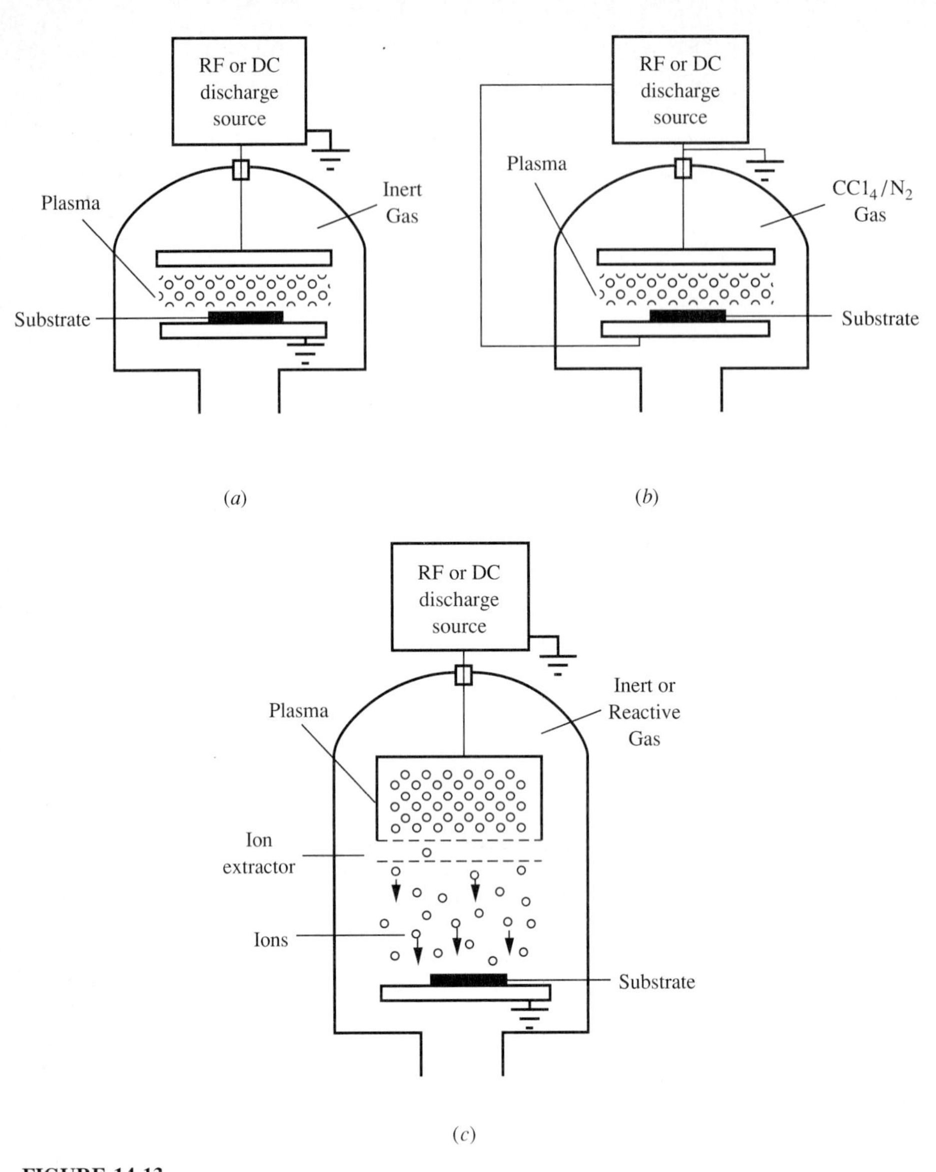

FIGURE 14.13
(*a*) Plasma etching, (*b*) reactive ion beam etching, (*c*) ion beam milling.

$$\mathrm{Si} + 2\mathrm{H_2O} \rightarrow \mathrm{SiO_2} + 2\mathrm{H_2} \tag{14.6}$$

$$\mathrm{Si} + \mathrm{O_2} \rightarrow \mathrm{SiO_2} \tag{14.7}$$

In the wet process, water vapor, and in the dry process, oxygen, reacts with the silicon wafer. The rate of oxide formation depends on the temperature of the Si wafer. Since the oxide forms by a reaction with the silicon, there is a continuous consumption of Si near the surface of the wafer. This leads to lowering of the physical Si level where the oxide forms. Also, once the initial layers of the oxide start to form, in order to continue the growth of the oxide oxygen has to diffuse toward Si to continue the oxidation process. This leads to relatively slow oxide growth rates.

Chemical vapor deposition is another way of growing dielectric layers with improved efficiency and faster growth times. Typical sample reactions for growing SiO_2 and Si_3N_4 are

$$SiH_4(\text{gas}) + O_2(\text{gas}) \rightleftharpoons SiO_2(\text{solid}) + 2H_2(\text{gas}) \tag{14.8}$$

$$3SiH_4(\text{gas}) + 4NH_3(\text{gas}) \rightleftharpoons Si_3N_4(\text{solid}) + 12H_2(\text{gas}) \tag{14.9}$$

Different CVD processes using different gas combinations lead to different dielectric growth rates. The advantage of these processes is that these dielectric layers can be deposited over any semiconductor. This is especially useful for those that do not have native oxides, such as the compound semiconductors.

14.5 INTERCONNECTS, LEADS

Once various parts of a device are formed, it is necessary to connect these regions to the outside world so the device can be utilized in a circuit. These connections should form ohmic contacts so the device characteristics will not be modified by additional nonlinearities introduced at the connections.

The connecting links are generally made by highly conducting elements. These are pure metals such as Al, Au, and so on, their mixtures with other materials, such as Ge, highly doped polysilicon, and silicides of metals such as tungsten, molybdenum, and tantalum. With recent advances in high critical temperature superconductors, these connecting leads may be replaced by superconducting leads to reduce power consumption due to ohmic losses.

There are three methods by which conducting materials can be deposited over semiconductors: evaporation, sputtering, and chemical vapor deposition (CVD). In each case, the adhesion of the film onto the material over which it is deposited is very important.

14.5.1 Vapor Deposition

Figure 14.14 shows typical vapor deposition equipment. A vacuum chamber is evacuated to below 10^{-5} torr where the collision mean free path of atoms with the gas molecules is negligible. A source produces the metallic vapor that moves in straight lines and is deposited on the surface of the substrate. The evaporation material sources are

1. A heated filament (Fig. 14.14*a*), wetted by the molten metal which then evaporates and coats the substrate. There is the danger of contamination from the filament itself.
2. A heated crucible in which the metal to be deposited melts and evaporates (Fig. 14.14*b*).

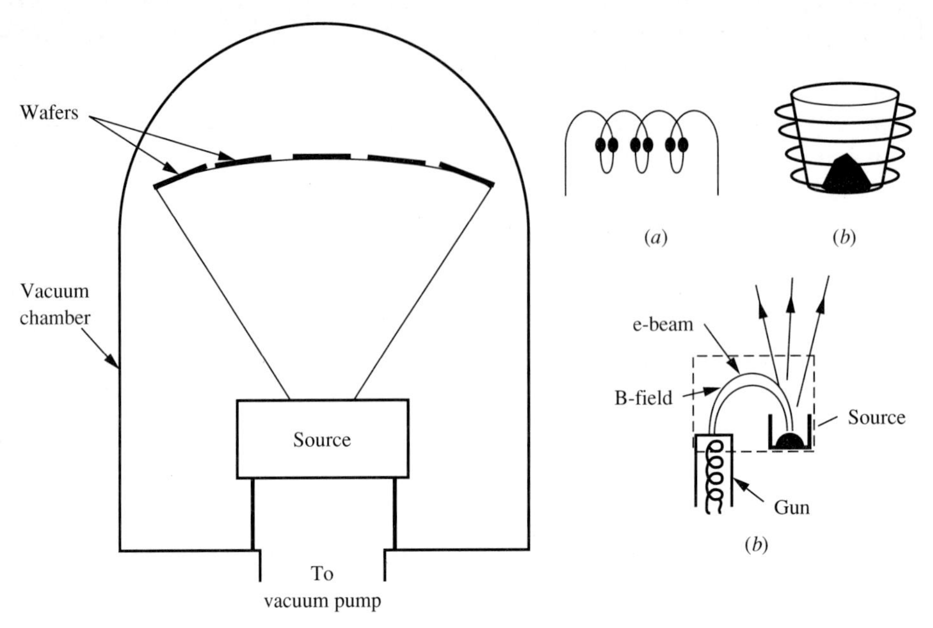

FIGURE 14.14
Vapor deposition equipment. Sources are (*a*) filament, (*b*) crucible, and (*c*) e-beam evaporation.

3. An e-beam heated source where an electron beam impinges on a metal in a crucible and melts the metal inside the crucible. By controlling the location where the e-beam hits the metal, melted metal can be surrounded by the unmolten metal, thus allowing relatively impurity-free coatings (Fig. 14.14*c*).

The growth rate of the evaporated metal depends on the temperature of the source, pressure of the chamber, type of metal to be evaporated, and the distance from the source to the location of the wafer. By placing many wafers on a spherical holder, it is possible to coat many wafers simultaneously. The thickness of the deposited material can be monitored during the evaporation process by using a quartz crystal whose resonant frequency changes in proportion to the material that is deposited onto it. Precautions have to be taken to protect the other portions of the vacuum chamber since the coating takes place wherever the evaporated particles travel.

14.5.2 Sputtering

If ions with high enough energy impinge on a metal, they may sputter (remove) material from the metal which can then be deposited on a different material. Figure 14.15 shows typical components of a sputtering system.

In Fig. 14.15*a*, a dc discharge is formed using an inert gas such as Ar. The ions of the Ar atom move toward the cathode which is also the target material.

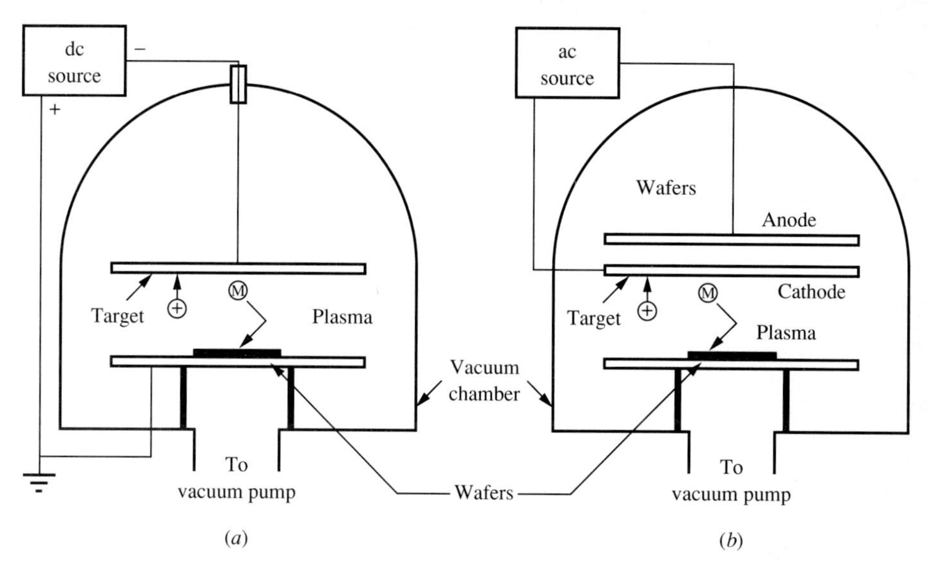

FIGURE 14.15
(*a*) dc and (*b*) ac sputtering systems.

When they impinge on the target metal, they dislodge metal atoms from the target. These diffuse toward and coat the anode plate where the wafer to be metallized is located. In the ac sputtering system (Fig. 14.15*b*), a discharge is formed between the anode and the cathode, but the pressure of the chamber as well as the separation between the anode and the cathode can be so adjusted that the plasma discharge spills over to the other side of the cathode into the space between the cathode and the region where the wafer to be metallized is located. In a process similar to the dc case, sputtered metal particles from the target then diffuse toward the wafer and metallize the wafer.

Sputtering is highly recommended for metals having very high boiling points where evaporation will not be practical. It is also very important that the gas used in the process should be very pure, otherwise contamination of the film takes place by imbedding these impurities within the metal while the deposition takes place.

14.5.3 Chemical Vapor Deposition

Chemical vapor deposition can be used to deposit doped polysilicon, tungsten, molybdenum, tantalum, titanium, and other metals. Typical reactions for tungsten silicide and tungsten films are

$$\mathrm{WF_6(gas) + 2SiH_4(gas) \rightleftharpoons WSi_2(solid) + 6HF(gas) + H_2(gas)} \tag{14.10}$$

$$\mathrm{WF_6(gas) + 3H_2(gas) \rightleftharpoons W(solid) + 6HF(gas)} \tag{14.11}$$

Similar reactions take place for the other types of metals.

FURTHER READING

Bean, J. C., "Growth of Impurity Doped Silicon Layers," Chapter 4 in *Impurity Doping Processes in Silicon*, F.F.Y. Wang. ed., Vol. 2, North Holland Publ. Comp. New York, 1981.

DeForest, W. D., *Photoresists, Materials and Processes*, McGraw-Hill, New York, 1975.

Heuberger, A., "X-Ray Lithography," *Solid-State Technology*, vol. 29, 1986, 93–101.

Holland, L., ed. *Thin Film Microelectronics*, Wiley, New York, 1965.

Martin, E. A., O. A. Aina, A. A. Illadis, M. R. Mattingly, and L. H. Stecker, "Undoped InP/InGaAs Heterostructure Insulated Gate FETs Grown by OMVPE with PECVD Deposited SiO_2 as Gate Insulator," *Electr. Devices Lett.*, vol. 9, 1988, 500–501.

Olsen, G. H., "Vapor Phase Epitaxy of Group III–V Compound Optoelectronic Devices," Chapter 12 in *Integrated Circuits: Chemical and Physical Processing*, P. Stroeve, ed., ACS Symposium series 290, American Chemical Society: Washington, D.C., 1985.

Powell, C., J. H. Oxley, and J. M. Blocher, Jr., *Vapor Deposition*, Wiley, New York, 1966.

Powell, R. A., ed., *Dry Etching for Microelectronics*, North Holland Physics Publ., Amsterdam, 1984.

Reinhard, D. K., *Introduction to Integrated Circuit Engineering*, Houghton Mifflin, Boston, 1987.

Ruska, W.S., *Microelectronic Processing: An Introduction to the Manufacture of Integrated Circuits*, McGraw-Hill, New York, 1987.

Schumacher, N. E., "Epitaxy Technology, State of the Art," *Microelectronic Manufacturing and Testing*, vol. 11, Nov. 1988, 26–27.

Sherman, A., and W. M. Coney, "Modern Developments in Chemical Vapor Deposition," *Microelectronic Manufacturing and Testing*, vol. 11, May 1988, 6–9.

PROBLEMS

14.1. The vacuum chamber of an MBE machine is spherical in shape, has a diameter of 60 cm, and has N_2 as a background gas. What should be the minimum pressure inside the chamber so that the mean free path of the atoms is at least 1000 times larger than the size of the vacuum chamber?

14.2. A lens has a diameter of 5.0 cm and the lens-to-image distance is 7.0 cm. If the wavelength of infrared light is 3500 Å, what is the minimum line resolution that can be obtained with this optical system?

14.3. A resistor to be used in an integrated circuit can be implemented in two ways: (*a*) by depositing a resistive material whose resistance depends on the thickness of the deposited material which produces so many ohms/square, and (*b*) by diffusing an impurity atom into a semiconductor (Fig. P.14.3). Design a 100 Ω resistor using the diffusion technique. The total length of the resistor should be 10 μm. Estimate

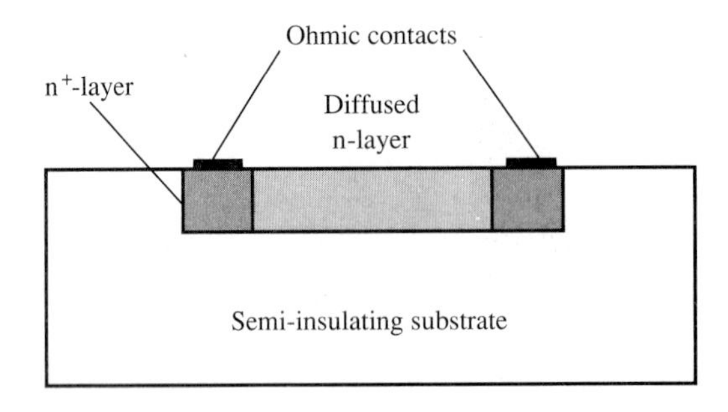

FIGURE P.14.3
Integrated circuit resistor.

the physical size (width and depth) and the average doping concentration of arsenic impurity atoms in silicon in order to produce 100 Ω of resistance. What kind of metal will you use to make contacts to the n^+ contact elements? ($T = 300$ K). What steps will you use in implementing the resistor?

14.4. Design an MOS capacitor of 10 pF that uses a silicon nitride as an insulator. The cross-sectional area is not to exceed 10 μm × 10 μm square. Assume that there is no band banding at the interface. Describe the steps in implementing the capacitor.

14.5. Describe the steps in manufacturing the FET transistor shown in Fig. 12.7.

14.6. A GaAs wafer is to be deposited with silicon by using an ion implanter. The Si source contains isotopes of Si(28) (most abundant), Si(29), and Si(30) (the numbers in parentheses are the atomic weights of different isotopes). If the incoming energy of the ions is 100 keV, and they enter a uniform magnetic field of 0.1 T (1000 G), whose extent is 10 cm × 10 cm, find the minimum distance that a 5 inch wafer should be placed away from the separator so that only the Si(29) can be used for doping purposes.

Note: the cyclotron frequency for a charged particle q in a magnetic field **B** is given by $\omega_c = (q|\mathbf{B}|/m)$. m is the mass of the particle.

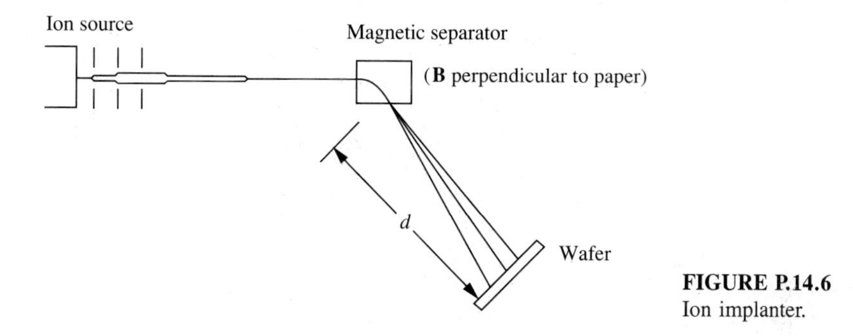

FIGURE P.14.6
Ion implanter.

14.7. Find the minimum distance between two lines that can be resolved if the source of light is (*a*) 10.6 μm (far infrared, CO_2 laser light), (*b*) 0.63 μm (red, He-Ne laser light), (*c*) 0.35 μm (ultraviolet), and (*d*) 5 A (X-rays). Assume in all cases the aperture is 10 mm in diameter and the distance from source to the screen is also 10 mm.

14.8. The requirements on X-ray lithography is that X-rays masks should be absorbing the X-rays (not to transmit radiation) at the opaque regions of the mask. From Fig. 14.11 find the best material and its corresponding thickness if the opaque part is to be 10 percent transparent.

CHAPTER 15

GAS DISCHARGES, ELECTRON BEAMS, AND RELATED DEVICES

Most of the topics covered in the previous chapters were related to devices where charged particles move within a solid material under the application of an external potential difference between two points within or at the ends of the solid. In Chapter 12, we saw that in order to raise the frequency response of a bipolar transistor or an FET, it is necessary to reduce the base width or the gate length of the respective transistors to achieve higher frequency of operation. These shorter dimensions introduce power dissipation limitations on the transistors. Since the charge motion occurs within a very short distance, the i^2R drop occurs within a small region of the transistor as well. As the current is increased, the power dissipation increases and heat should be transferred away from the active region to maintain the operational characteristic of the transistor. For these very short dimensions, the active region looks like a point heat source, and the amount of heat transferred from that region is limited by the heat transfer coefficient of the substrate material as well as by how effectively the heat removal is incorporated into the device geometry. One can thus easily expect a proportionate decrease in power output as the dimensions get shorter.

While covering the topics of ballistic bipolar transistors, we stated that if the injected charge carriers were given an initial energy to start with before entering the base region, they would then move faster through the base region, thus increasing the frequency response of the transistor. There are many high-frequency, high-power electron beam devices that operate as ballistic devices. Unfortunately, they have the disadvantage of needing filament supplies that consume additional power, and require vacuum envelopes and bulky permanent magnets.

15.1 GASEOUS DISCHARGES

Gaseous discharges appear around us in many forms, from uncontrolled lightning in stormy weather to neon signs used for advertisements. The range of parameters, as well as the physical mechanisms responsible for the production and maintenance of gaseous discharges can be understood by considering the current-voltage characteristics of a laboratory discharge tube, as shown in Fig. 15.1. An evacuated glass tube is filled with a low-pressure inert gas. External current is measured by an ammeter, and a load resistor R_L is added to limit the current flow. The voltage v_A that appears across the tube is controlled by the battery voltage V_A. The observed current-voltage characteristic of a typical discharge tube is shown in Fig. 15.2.

As the voltage is increased, the current increases gradually until point A is reached and stays constant beyond A. Beyond B, the current starts to increase, but at a faster rate than before. At point C, the tube reaches a sparking condition at $v_A = V_s$ and, at the same time, a voltage drop across the tube is also observed. Beyond point D, the current in the tube keeps increasing with very little increase in the voltage drop across the tube. As point E is reached, the voltage drop again starts to increase with increasing current. As point F is reached, an arc across the tube occurs and there is a dramatic increase in current accompanied by a drastic reduction of voltage drop across the tube. The actual current that flows in the external circuit is limited by the load resistor R_L connected in series with the discharge tube. The effect of the load resistor on the overall tube i-v characteristic is shown as the load line in Fig. 15.2. The slope of the load line is equal to -1/R. Two different load lines for the same applied potential $V_A = V_1$ are shown in

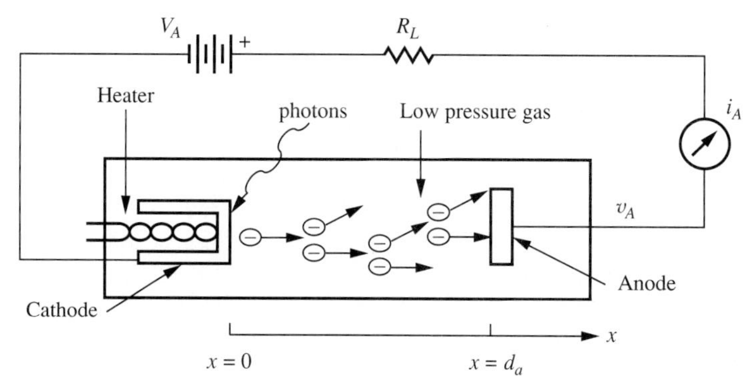

FIGURE 15.1
Gaseous discharge in a laboratory glass envelope.

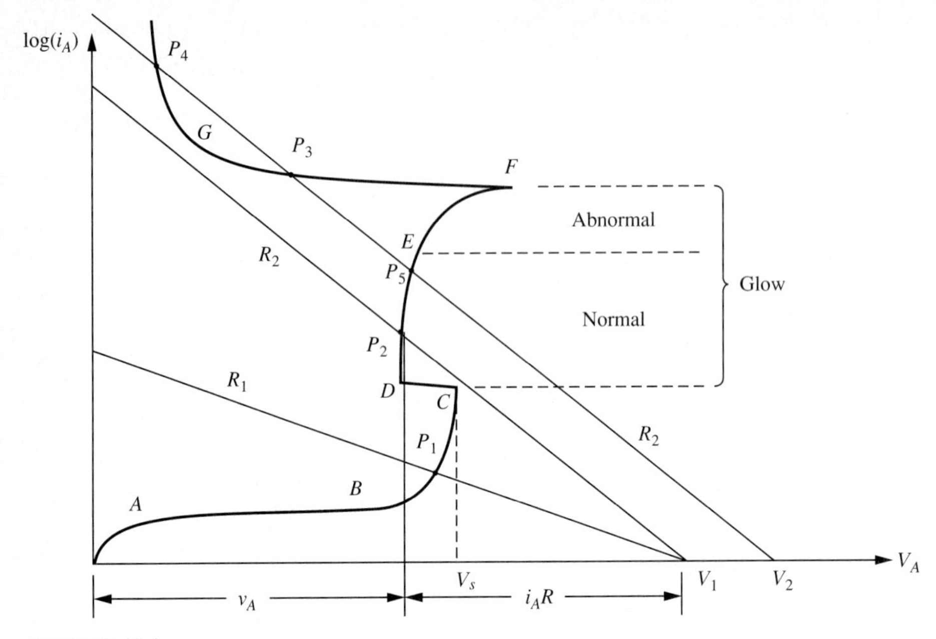

FIGURE 15.2
Current-voltage characteristics of a typical gaseous discharge tube. $-1/R$ is the slope of the load line.

Fig. 15.2. If the load R_2 is kept constant and the external source voltage V_A is increased from V_1 to V_2, the load line moves to the right with a constant slope. From Fig. 15.1, the relation between various voltage drops is given by

$$V_A = v_A + i_A R_L \tag{15.1}$$

If the load line intersects the current-voltage ($i_A - v_A$) curve at more than one point, the stability of the operating point is determined by considering the resulting changes in v_A as a function of i_A. The operating point is stable if

$$\frac{dv_A}{di_A} + R_L > 0 \tag{15.2}$$

This means that if dv_A/di_A is considered to be the equivalent resistance of the operating point, the total resistance of the circuit, including the load resistor, should be greater than zero. This condition applies to the points P_5, P_3, and P_4 where there are three possible operating points for the given load. P_5 is stable since the slope of the curve is positive to start with. For a single point of interaction, such as P_1, the load line will be stable and the condition (Eq. 15.2) does not apply. For the current-voltage characteristics of a typical discharge tube given in Fig. 15.2, points P_1, P_2, P_4 are also stable and P_3 is an unstable point of operation.

15.2 TOWNSEND IONIZATION COEFFICIENTS

In the $i_A - v_A$ characteristics of a discharge tube (Fig. 15.2), the region up to the point C is referred to as the Townsend discharge, named after J. S. Townsend, who studied gas discharges extensively.

Suppose that the cathode of the discharge tube (Fig. 15.1) is illuminated by photons so that the cathode emits electrons. Under the application of an electric field, the electrons move through the gas and make collisions with the gas atoms. As the voltage across the tube is increased, almost all of these electrons reach the anode. Further increase in the potential across the tube has very little effect on the external current. During this interval, the electrons move under the resulting electric field $\mathcal{E}$ produced by the applied potential v_A. The electrons gain energy between each collision, which can be approximately written as the electric field multiplied by the mean free path of the electrons. The energy gain between collisions thus increases with the increasing electric field. Once point B is reached, it is then possible that some electrons have gained enough energy from the field that they can make ionizing collisions with the gas atoms and generate additional electrons which then contribute to the current. Thus, with increasing voltage, the current again starts to increase beyond the point B.

At a point x in the discharge tube, let the number density of electrons be $n(x)$. The increase in the number of electrons dn due to field-assisted ionization per unit area occurs within a distance dx. This increase is proportional to the number of electrons present at x, that is,

$$dn = \alpha n \, dx \tag{15.3}$$

Here, the proportionality constant α is called Townsend's first ionization coefficient. It depends on the average energy gain of an electron between each collision, $f(\mathcal{E}\lambda_m)$, and the number of new electrons produced per unit length, which is inversely proportional to the electron mean free path λ_m of the accelerated electrons.

$$\alpha = \frac{f(\mathcal{E}\lambda_m)}{\lambda_m} \tag{15.4}$$

Neglecting recombination and diffusion, Eq. 15.3 can be written as

$$\frac{dn}{n} = \alpha \, dx$$

or

$$n(x) = Ae^{\alpha x} \tag{15.5}$$

If at $x = 0$, n_0 is the electrons emitted by the cathode per unit area per second, the electrons reaching the anode per unit area are

$$n = n_0 e^{\alpha d} \tag{15.6}$$

Multiplying both sides by the charge of the electrons, we obtain

$$j_x = j_0 e^{\alpha d} \tag{15.7}$$

where j_0 is the saturation photoelectric current density at the cathode.

As the field is further increased, there are other factors that contribute to the ionization of the gas molecules and therefore to the increase in the external current. These are

1. The volume photoelectric effect. It is possible that the ionizing radiation incident on the cathode can also produce volume photoelectric effect and produce additional electrons from the gas atoms. Equation 15.7 is modified for the resulting current density as

$$j_x = \frac{j_0(e^{\alpha x} - 1)}{\alpha d} \tag{15.8}$$

2. As the electric field is further increased, the ions produced by ionizing electron collisions gain enough energy to make ionizing collisions with the gas atoms themselves. This leads to Townsend's second ionizing coefficient, β. The resulting steady-state current is then given by

$$j_x = \frac{j_0(\alpha - \beta)e^{(\alpha-\beta)x}}{\alpha - \beta e^{(\alpha-\beta)x}} \tag{15.9}$$

where β, in addition to depending on the electric field, is also a function of the number of particles per unit length, or

$$\beta = \frac{g(\mathcal{E})}{\lambda_m} \tag{15.10}$$

Note that if the denominator in Eq. 15.9 is zero at $x = d$, then

$$\alpha - \beta e^{(\alpha-\beta)d} = 0 \tag{15.11}$$

determines the breakdown condition (point C in Fig. 15.2) for the gaseous discharge. From the functional dependence of α and β on the applied electric field, the sparking (or spark breakdown) potential V_s can be determined.

3. The emission of electrons at the cathode by positive ion bombardment is also a probable process in the intensification of the discharge current. From an experimental point of view, the volume ionization and electron emission by the cathode due to ion bombardment are not distinguishable and β takes both processes into account.

For many gaseous media $\alpha >> \beta$, therefore the condition for breakdown reduces to

$$\frac{\alpha}{\beta} = e^{(\alpha-\beta)d} \approx e^{\alpha d} \tag{15.12}$$

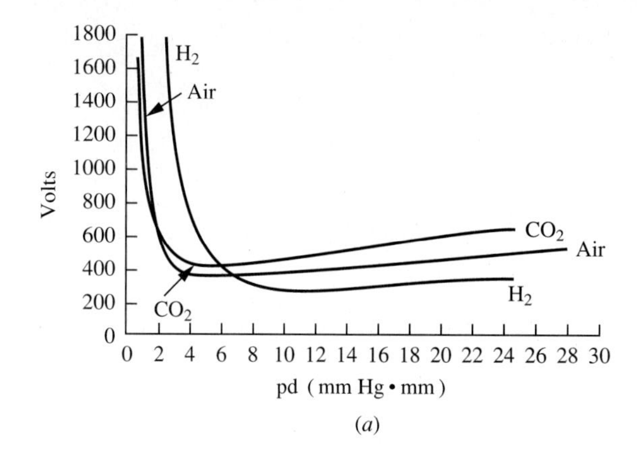

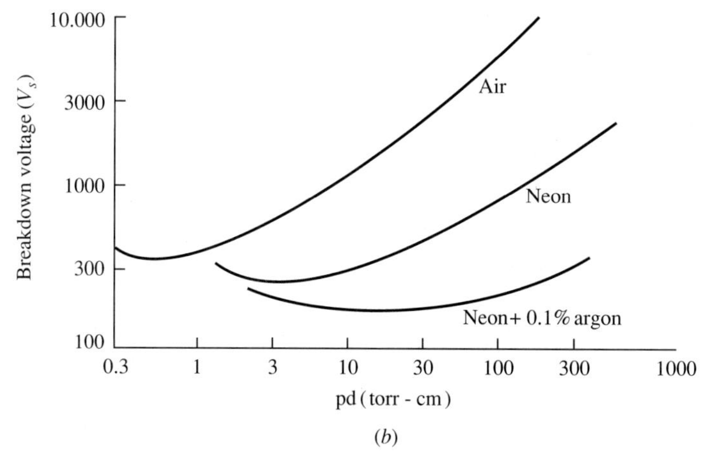

FIGURE 15.3
Spark breakdown voltages for various gases and gas mixtures for plane parallel electrodes. (*a*) From (Cobine, 1958), (*b*) From (Sherr, 1979).

The voltage breakdown may be initiated either by keeping the voltage constant and varying the gap between the electrodes or by keeping the gap constant and varying the voltage applied across the tube. The sparking voltage can be shown to be only a function of the pressure-gap (pd) product. For various gases, the resulting V_s for parallel plane electrodes is shown in Fig. 15.3. Note that for a particular gas there is a distinct minimum spark-breakdown potential as (pd) is varied.

15.3 POSITIVE COLUMN

In a gaseous discharge, the region from zero current to the point C in Fig. 15.2 is known as a non-self-sustaining discharge. When the source of electrons, that is, photons, is removed, the discharge is extinguished. On the other hand, once the region beyond point D is reached, the discharge is self-sustained even when the initial electron source is removed. If there are no externally applied ionization sources, self-sustained discharges can still be produced because the stray radiation, such as X-rays, cosmic rays, and so on, that is present everywhere, can generate sufficient photoelectrons to start a discharge. It has also been shown experimentally that the presence of external sources of electrons greatly helps in producing self-sustained discharges with lower sparking, as well as lower operating potentials.

When the discharge is operated between the regions D and F, the resulting discharge is known as a *glow discharge*. In general, the glow discharges are low-

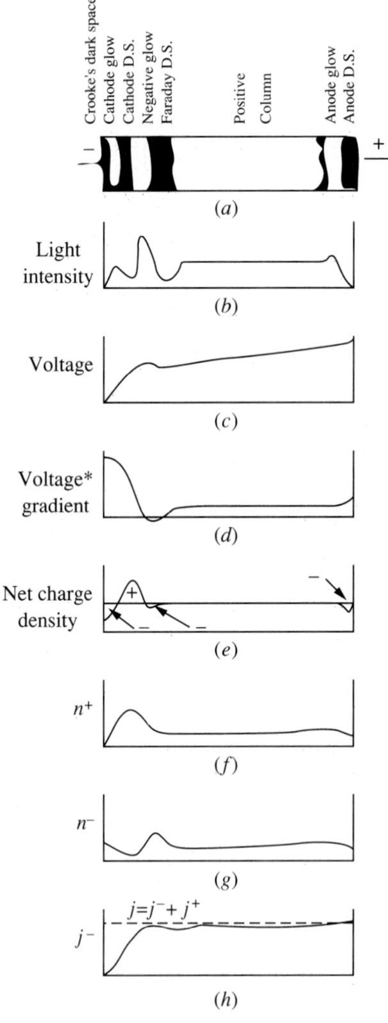

FIGURE 15.4
Approximate characteristics of a glow discharge (Cobine, 1958).

pressure discharges and some of their properties can be summarized by referring to Fig. 15.4. Depending on the gas and the pressure, the discharge consists of dark and bright regions with different identifying names. If the background gas pressure is changed, the extent of the light and dark regions changes considerably, and in some cases one or more of the regions may shorten or lengthen or even disappear from the discharge. The colors of various glowing regions depend on the background gas and may also vary from one glow region to the other. These are determined by the electronic excitation of the bound states of the electrons in the gas and may differ due to different excitation mechanisms of electronic energy levels in each of the glow regions.

The voltage drop, the electric field variation, electron and ion concentrations are shown in Fig. 15.4 as a function of distance. One of the most important characteristics of the discharge is the region referred to as the *positive column*. In this region, the axial voltage variation and the resulting voltage gradients are very small, and the electron and ion concentrations are equal to each other. The resulting discharge current is mostly due to the electrons. Because of the conditions in the positive column of the discharge, the electrons, ions, and gas atoms are practically in thermodynamic equilibrium and the emitted spectra of light are used to study the electron excitation mechanisms in the gaseous media. The positive column is also used as a light source for spectroscopic analysis.

In general, gas discharges are called plasma, the fourth state of matter.

15.4 VACUUM DIODE AND SPACE-CHARGE LIMITED CURRENT

A vacuum tube diode consists of an evacuated envelope containing an electron-emitting cathode (generally a thermionic cathode) and an anode that collects the electrons, as shown in Fig. 15.5. When the cathode is cold and does not emit electrons, the electric field between the cathode and the anode is constant and equal to $-(v_A/d_a)\mathbf{a}_x$ and the potential between the two electrodes varies linearly from zero at the cathode to v_A at the anode.

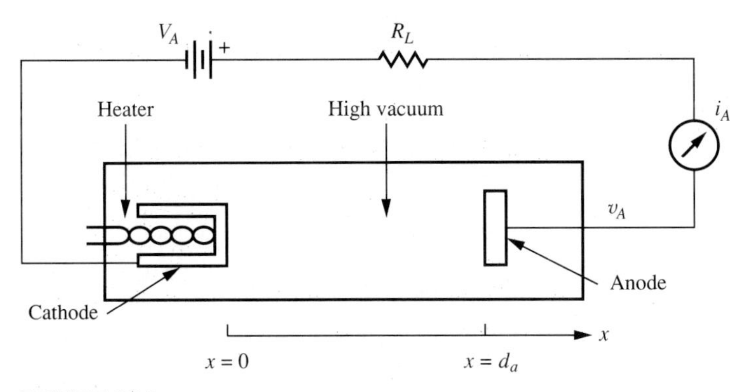

FIGURE 15.5
Elements of a vacuum diode.

We know from the Poisson equation that the presence of charges at a point in space modifies the total effective field that exists in the vicinity of those charges. By definition, we also know that the electric field lines are directed from positive toward negative charges.

When the temperature of the cathode is raised, emitted electrons are subjected to the electric field produced by the externally applied potential. As long as the number density of the emitted electrons is very low, the electric field forcing the electrons to move toward the anode can be assumed to be not too much different than the no electron value.

As the number of emitted electrons is increased further, the density of electrons near the cathode also increases due to the almost zero initial velocity of the electrons. Some of the electric field lines that used to end up at the cathode now end up at some of these electrons. This results in a reduction of the electric field near the cathode. As the emission current density is further increased, more and more field lines end up at the negatively charged electrons. This, in effect, further modifies the electric field near the cathode. If the density of the electrons is high enough, it is likely that the direction of the electric field near the cathode can be reversed in the space between the cathode and a short distance away from the cathode. This phenomenon is known as *space-charge* buildup near the cathode. The modification of the electric field and the potential distribution between the cathode and the anode due to the space-charge buildup is shown in Fig. 15.6. This implies that the potential near the cathode becomes negative and extends to a distance of $x_{\min}$ in front of the cathode. The potential of Fig. 15.6*c* is now the space-charge potential variation that exists between the cathode and the anode. A potential barrier is formed near the cathode for the electrons.

Any electron that is emitted by the cathode with an initial velocity can only move toward the anode provided it has enough initial kinetic energy to overcome this potential barrier. Otherwise, it will be repelled back toward the cathode. As the cathode temperature is increased further, although more electrons are emitted, the space-charge potential in front of the cathode gets larger in that proportion and only those electrons that overcome the potential barrier reach the anode. The result is a balancing effect and the current flow to the anode is independent of the cathode temperature provided the cathode can supply an unlimited number of electrons. The flow of this type of current is referred to as the *space-charge limited current*.

The average kinetic energy of a normally directed electron is $< mv^2/2 > = kT$. The potential buildup in front of the cathode is expected to be within this average value. For practical cathode temperatures, this is less than one volt. Also, the distance $x_{\min}$ from the cathode, where the space-charge potential is a minimum and the electric field is zero, is very small compared to the actual cathode-anode spacing d_a. Location of $x_{\min}$ is called the *virtual cathode*. The virtual cathode-anode separation is not too much different from the actual cathode-anode spacing. Thus, these two distances will be taken to be equal in subsequent calculations. Although only the electric field is zero, the electric potential is also assumed to be zero at the virtual cathode.

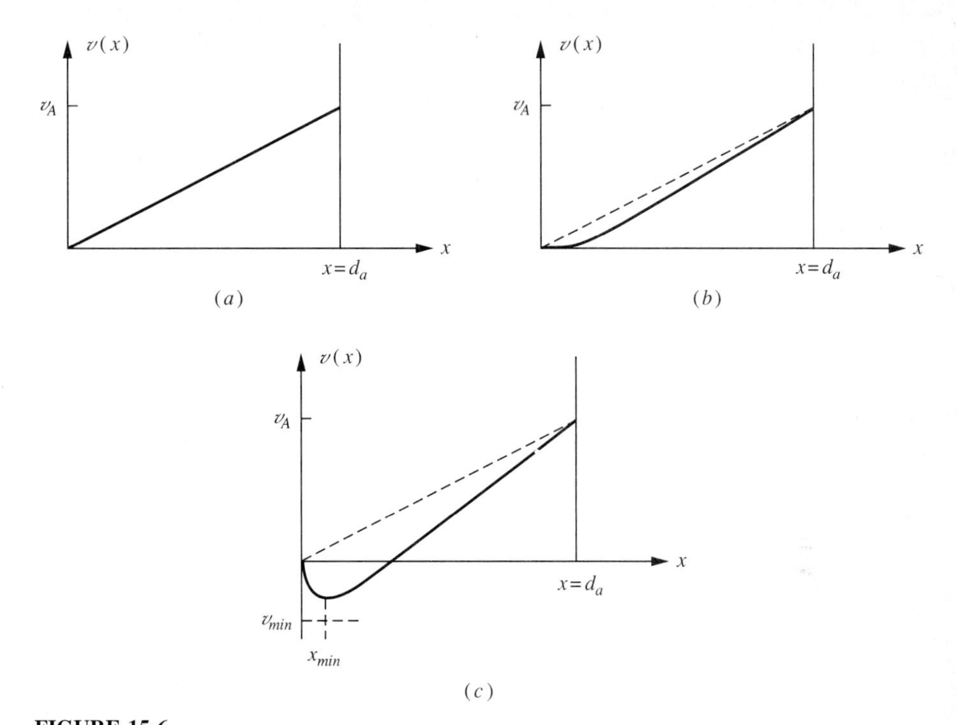

FIGURE 15.6
Potential variation through a vacuum diode with no electron emission (*a*), with a few emitted electrons (*b*), and with large number of electrons (*c*).

The space-charge current flow in a diode is derived by assuming that the electron energy and the diode current are conserved. Solving a one-dimensional problem, the current density at any plane at x away from the cathode can be written as

$$J_x = \frac{4}{9}\epsilon_0 \left(\frac{2e}{m}\right)^{1/2} \frac{v_A^{3/2}}{x^2} \tag{15.13}$$

Since the current density is constant everywhere, Eq. 15.13 should also hold at $x = d_a$, that is,

$$J_a = \frac{4}{9}\epsilon_0 \left(\frac{2e}{m}\right)^{1/2} \frac{v_A^{3/2}}{d_a^2} \tag{15.14}$$

This is known as the *Child–Langmuir* law, sometimes referred to as the three-halves power law. The resultant current density is independent of the temperature of the emitting cathode.

The diode current i_A (Eq. 15.14 multiplied by the area of the cathode) is plotted in Fig. 15.7 as a function of v_A, the terminal voltage between the anode and the cathode. As long as the cathode can supply an unlimited number of electrons, the current-voltage relationship follows the three-halves power law. But at a given

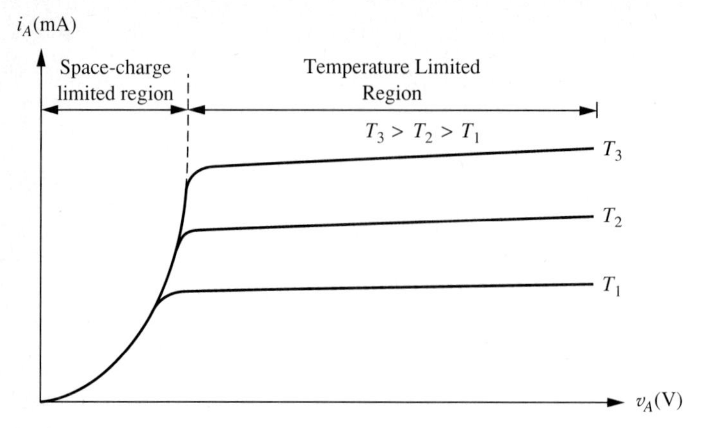

FIGURE 15.7
Diode current as a function of anode voltage at various temperatures. $T_3 > T_2 > T_1$.

temperature of the cathode, the supply of electrons is limited. Once these are collected, there are no additional electrons that will contribute to the current, and the external current saturates and stays nearly constant with increasing v_A. As the temperature is raised, the cathode supplies more electrons and the three-halves power law extends to higher currents. Eventually, i_A reaches saturation but at a higher current value. Figure. 15.7 shows saturated-current regions that are referred to as the *temperature-limited* region of operation of the diode.

When Eq. 15.14 is multiplied by the area of the anode A, it can be written as

$$\frac{i_A}{v_A^{3/2}} = K \tag{15.15}$$

The ratio $i_A/v_A^{3/2}$ is called the *perveance* of the diode and relates to the maximum space-charge current that can be drawn from the cathode for a given applied potential v_A and is an indication of the power-handling capability of electron beam devices. Here, i_A is the maximum current before saturation. The constant K depends only on the geometry of the diode configuration.

A gridded electrode, when placed close to the cathode and biased at a negative potential relative to the cathode, can be used to control the current flow through the device. This is the basis of the vacuum *triode*, which was the workhorse of many electronic systems before transistors replaced them. Triodes may still be found in electronic systems where high power and high frequency of operation are required.

15.5 MICROELECTRONIC TRIODE

Recent advances in semiconductor manufacturing technology allow solid-state versions of the vacuum triodes to be built using semiconductor materials. A typical semiconductor triode is shown in Fig. 15.8. A triangular metal emitter with a sharp

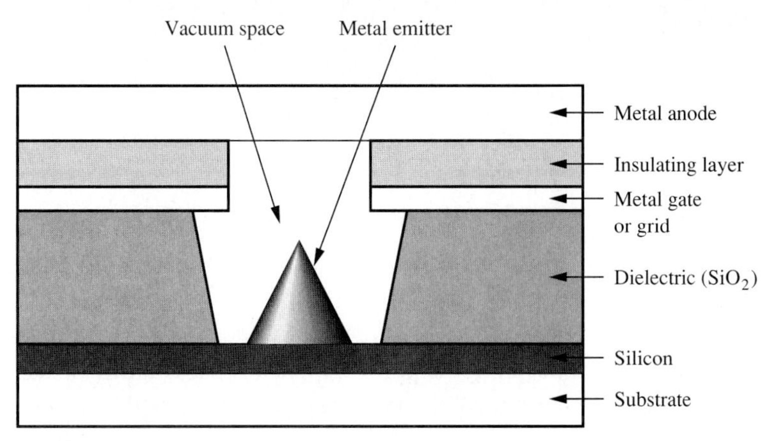

FIGURE 15.8
Solid-state version of a vacuum triode (Skidmore, 1988).

tip is grown over a silicon wafer. A dielectric insulator such as SiO_2 is grown over the silicon. Next, a metal gate electrode is deposited over the insulator. A second insulator is grown over the gate electrode. A final metallic coating acts as the electron collector (or anode). The sharp tip of the cathode emits electrons by field emission. The potential on the gate electrode allows the control of the electron flow to the anode. Since the distances involved are in the order of micrometers, the electron transit times are very short. Since also the anode cathode spacing is very short, a high vacuum is not necessary, since the mean free path of electrons in the region may be longer than the transit time of the electrons. The main disadvantage of the solid-state triodes is the very small currents that are emitted by the cathodes. Present day research activity is directed at paralleling these devices or even stacking them in three-dimensional arrays to increase the current-carrying capabilities of these devices.

15.6 ELECTRON BEAM DEVICES

Similar to the BJTs and FETs, as the signal frequency is increased the behavior of triodes also deteriorates and, in addition to the tube and circuit effects, electron transit time also reduces the effectiveness of these devices. In space-charge-controlled electron tubes, the electrons leave the grid region with almost no initial velocity, but gain kinetic energy under the applied anode voltage. The power delivered to the electrons by the external battery is converted into amplified signal energy.

Fundamental limitations of the electron tubes can be avoided if, instead of at the grid location where the initial electron velocities are practically zero, electromagnetic signals are applied to an electron stream at a location where the electrons have already acquired very high initial velocities. Figure 15.9 shows a device based on this concept, called an *electron beam device*. It has three basic components: an electron gun, an interaction region, and a collector. The

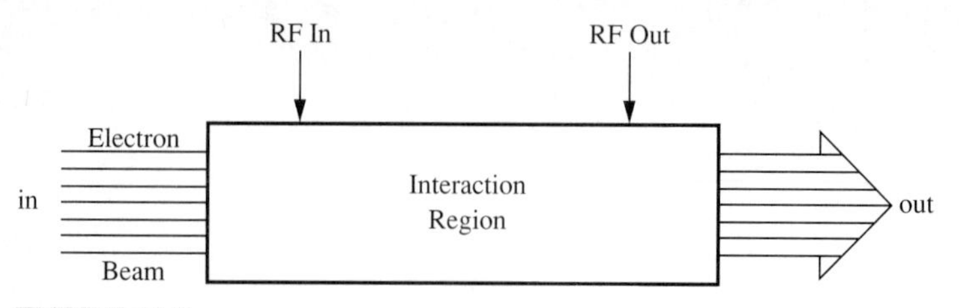

FIGURE 15.9
Essential components of an electron beam device.

energetic electron beam is produced by a thermionic cathode and an associated accelerating structure. The electrons enter the interaction region with an initial velocity $u_0 = (2eV_0/m)^{1/2}$ where V_0 is the final accelerating voltage.* Once the electrons enter the interaction region, they travel in almost a field-free region (the only fields that may be present are the fields associated with the electromagnetic signal). While traversing the interaction region, kinetic energy of the electrons are converted into electromagnetic energy. Electrons finally leave the interaction region and end up in a collector.

In a classical electron tube, once the electrons leave the grid, they are constantly under the influence of a very high intensity electric field. The distance between the grid and the anode is usually very short. On the other hand, in a beam device, once the electrons enter the interaction region, they move in a region where there are no more accelerating fields. As shown in Fig. 15.10*a*, due to the repulsive Coulomb forces between like charges, electrons repel each other and spread out radially as they move throughout the interaction region toward the collector if no provisions are made to neutralize these space-charge forces.

One of the main requirements of a beam device is to maintain the constant cross section of the electron beam throughout the interaction region. There are various techniques in balancing the space-charge forces. These are referred to as *beam focusing*. Although both magnetostatic and electrostatic fields are used for focusing purposes, most practical devices incorporate a magnetic field structure for this purpose. Although the magnets that produce the necessary focusing fields are bulky and heavy, they are superior to electrostatic focusing systems. They can be designed to be external to the tube envelope so that changing or aligning the tube can be done easily. Although the focusing mechanisms incorporated into electron beams are not ideal, there exist a few different field configurations for effective beam focusing.

---

* This is for nonrelativistic velocity. If the electron is accelerated through a high-potential V_0, the correct velocity is given by

$$u_0 = c\left\{1 - [1 + (V_0/V_n)]^{-2}\right\}^{1/2}$$

where $V_n = m_0c^2/e = 0.512$ MeV is the equivalent rest mass potential and c is the speed of light in vacuum.

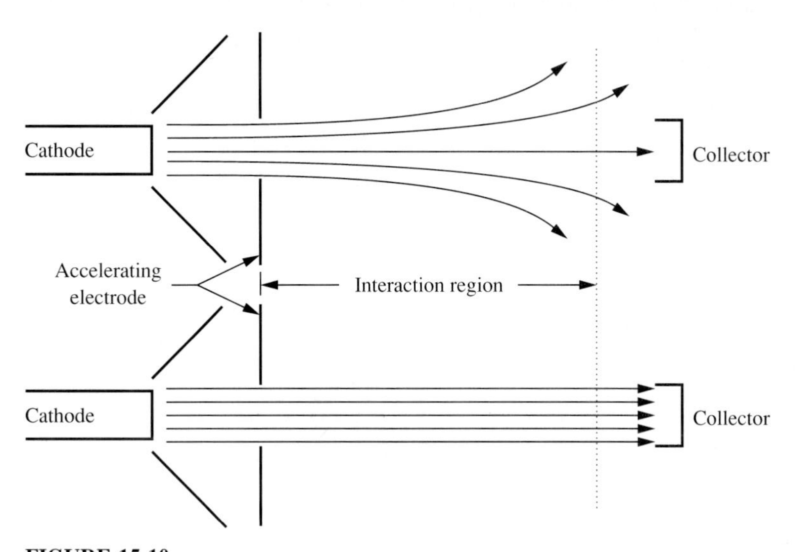

FIGURE 15.10
(*a*) Spreading of an electron beam due to space-charge forces. (*b*) Idealized (focused) flow of the electron beam in the interaction region.

We now assume that a specific field configuration is chosen so that an ideal focusing of the beam is accomplished and the beam moves with a uniform cross section through the interaction region. The next major goal is to provide an efficient way for the electromagnetic signal to interact with the beam. The signal should be optimally coupled to the beam, and through exchange of energy, it should be amplified and finally extracted out of the interaction region. If a proper feedback path is introduced into the interaction region or a portion of the output signal is fed into the input, the device can be used as a very efficient oscillator. In general, we are dealing with electromagnetic fields whose frequencies fall in the microwave spectrum, and the practical dimensions of the interaction region are comparable to the wavelength of the electromagnetic signal which is on the order of a few cm's and below. Thus, microwave principles have to be incorporated into the field-guiding structures as well as into the analysis of the beam-field interaction.

In many beam devices, the optimum interaction between an electromagnetic signal and an electron beam requires that the phase velocity v_p of the electromagnetic signal be nearly equal to the electron velocity u_0. In most of the beam devices, practical electron beams acquire velocities that are smaller than the speed of light, that is, $u_0 << c$. Therefore, the electromagnetic field-guiding structures should be chosen among those that have phase velocities much smaller than c. Many periodic waveguiding structures, known as slow-wave structures, satisfy this criterion.

Actual vacuum tube devices contain residual gases at their operating pressures. A typical device is usually evacuated to a pressure of 10^{-7} torr. Even at these pressures, the residual gas density in the vacuum envelope is still very high. When an electron beam goes through such a medium, a great number of gas atoms are ionized by the electron impact of the beam electrons. Therefore, we may have a partial neutralization of the space-charge forces in the beam, but this neutralization is never complete and external focusing structures are still needed. One has to keep in mind that although there are ions in the interaction region, they have masses that are very large compared to the electron mass. They are not influenced by the rapidly varying high-frequency fields, and the interactions that take place in the beam devices are only between the electromagnetic fields and the beam electrons. Thus, electron beam device concepts are effective up to millimeter wavelengths.

Efficient beam-device designs require that the electron beams maintain their constant cross section throughout the interaction region and be guided extremely accurately over large distances. Such stringent goals are difficult to achieve in practice. The difficulties associated with focusing fields are

1. The production of perfectly uniform focusing magnetic fields or fields with a given configuration is very difficult.
2. The electron gun has to be perfectly aligned with the focusing field. In many instances, the gun design and focusing field configuration are considered as a whole for efficient focusing.
3. Practical cathodes never produce the desired uniform current density for ideal focusing even if the electrode system is so well designed that the electric field at the cathode is perfectly uniform.
4. With magnetic field focusing, any radial velocity purposely or accidentally imparted into the beam will result in electrons spiraling around the magnetic fields.

15.7 ELECTRON GUNS

Electron guns in beam devices should be capable of producing beams with a uniform current density and cross section and low noise. Efficient signal amplification requires beam parameters with uniform beam cross section and uniform current density. Since the origin of the beam is the cathode, any nonuniformities in the emitted electrons are thus reflected in the beam itself. Electrons thermally emitted by the cathode have a velocity distribution, and this is also reflected on the accelerated electron beam distribution. This velocity spread is reflected as an increase in the noise level of the device.

The electron beam guns can be divided into two groups: nonconvergent (parallel flow) and convergent. Depending on the focusing magnetic field, the gun configuration may be immersed or nonimmersed in the magnetic field.

In a nonconvergent gun, the electrons emitted by the cathode move in straight lines, and the accelerating structure is designed in such a way that the electrons moving between the cathode and the accelerating anode do not have any radial velocity.

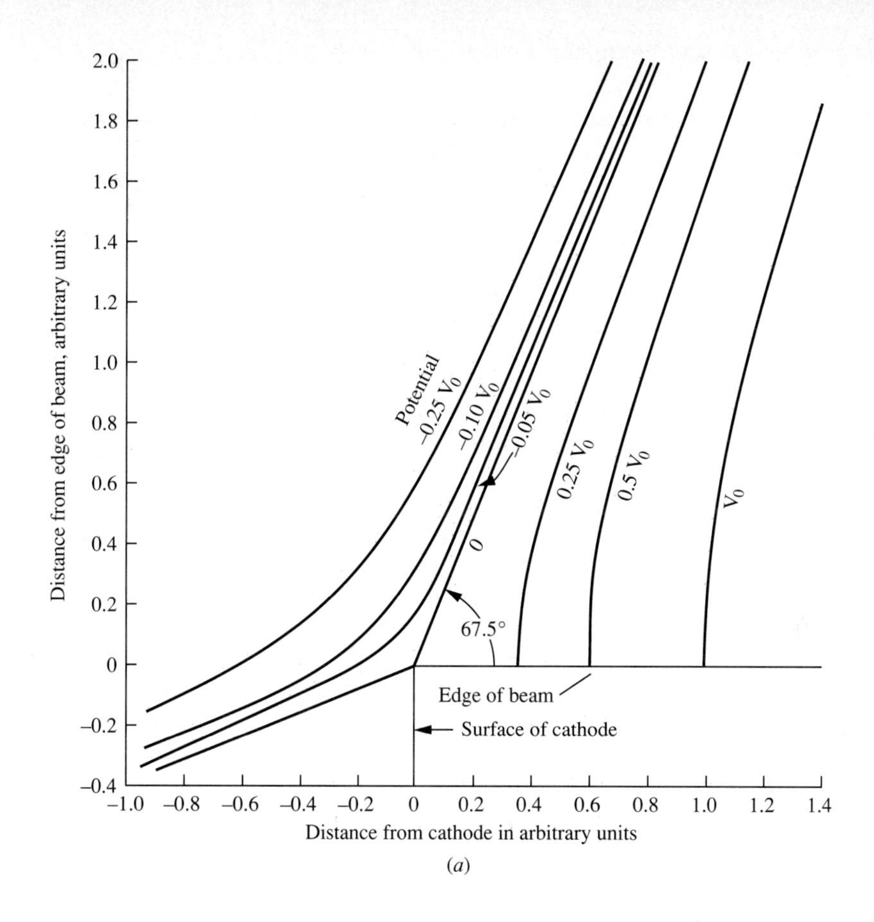

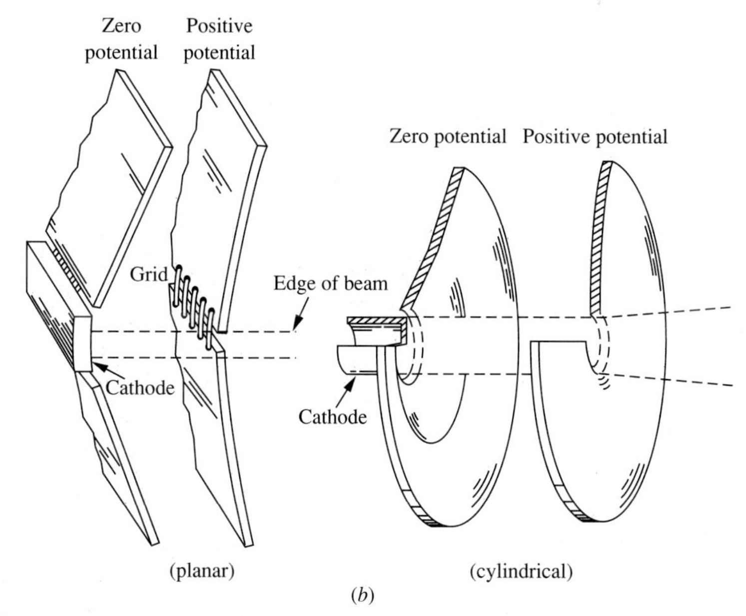

FIGURE 15.11
(*a*) Equipotential lines in the cathode anode region of an electron gun to produce non-divergent electron beam. (*b*) Pierce gun (planar and cylindrical) (Pierce, 1940).

If there are no magnetic fields (nonimmersed), electron trajectories between the cathode and the accelerating electrodes are controlled by the electric fields and the space-charge forces between the electrons. Motion of electrons in such a system is studied extensively. If a simple plate at the cathode potential is placed at an angle of 67.5° with respect to the cathode, electrons moving from the cathode toward the anode will have only axial motion, provided the anode is shaped to the special equipotential profile V_0, as shown in Fig. 15.11*a*. This type of electron gun is known as the Pierce gun and is used extensively in beam devices (Fig. 15.11*b*).

If the outer one of two concentric spheres is used as a cathode and a hole is drilled in the inner one, the electrons emitted by the cathode will move in radial directions and converge toward the axis. As they bunch together, the space-charge force will prevent further convergence. As they move out further, they will now start to diverge since there are no fields that will compensate the space-charge forces. The convergent gun is shown in Fig. 15.12. If proper focusing fields are applied at the beginning of plane A, where electron flow is entirely parallel, a dense electron beam will form and move without divergence. Convergent guns also increase the beam current density necessary for high-power applications.

15.8 FIELD-ELECTRON INTERACTION

To understand the interaction of an electron beam with electromagnetic fields and their resultant motion, a simplified physical model will be used. If more than one electron is used, this model may not be a good representation of the collective interaction mechanisms. But the basic concepts that will be explained here using a single electron are still applicable to the actual beam of electrons.

Figure 15.13*a* shows two parallel plates separated by a distance d. A potential difference v whose polarity can be reversed is connected across two plates. An electron with a longitudinal velocity u_0 is injected into the field region between the plates through a hole at $z = 0$. The equation of motion for the electron is

$$m_e \frac{d v_z}{dt} = \pm e |\mathcal{E}_z| \tag{15.16}$$

The sign $\pm$ depends on the polarity of the plates and $|\mathcal{E}_z| = |v/d|$.

Solution of Eq. 15.16, with the initial condition $v_z = u_0$ at $t = 0$, yields

$$v_z = u_0 \pm \eta |\mathcal{E}_z| t \tag{15.17a}$$

$$= u_0 \pm \Delta v \tag{15.17b}$$

The second term $\Delta v = \eta |\mathcal{E}_z| t$ is the change in velocity due to the influence of the $\mathcal{E}_z$ field. Here $\eta \equiv e/m$. If we now assume that $\Delta v << u_0$, we can find the time it takes for the electron to reach the second plate at $z = d$. Using

$$t \approx \frac{d}{u_0} \tag{15.18}$$

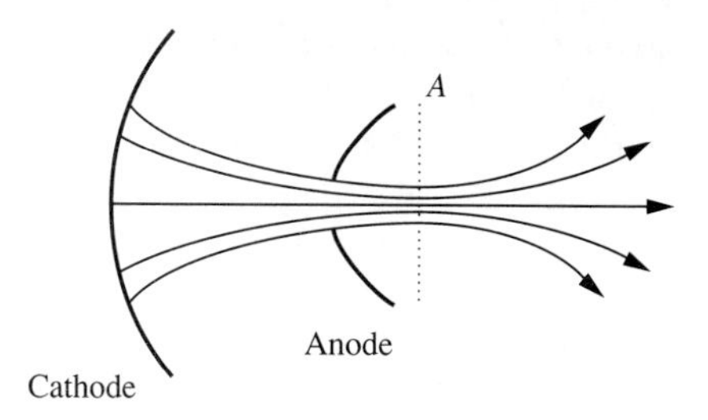

FIGURE 15.12
Convergent electron gun. At plane A the electron flow is entirely parallel.

Equation 15.17a can be written as

$$
\begin{aligned}
V_z &= u_0\left[1 \pm \frac{\eta|\mathcal{E}_z|}{u_0}\left(\frac{d}{u_0}\right)\right] \\
&= u_0\left[1 \pm \frac{(1/2)e|\mathcal{E}_z|d}{(1/2)mu_0^2}\right] \\
&= u_0\left[1 \pm \frac{1}{2}\frac{e(|\mathcal{E}_z|d)}{KE_0}\right] \qquad (15.19a) \\
&= u_0\left[1 \pm \frac{1}{2}\frac{(ev)}{KE_0}\right] \qquad (15.19b)
\end{aligned}
$$

where we have substituted $KE_0 = m_e u_0^2/2$ for the initial kinetic energy of the electrons and written $|\mathcal{E}_z|d = (v/d)d = v$. Note that at $z = d$, the velocity of

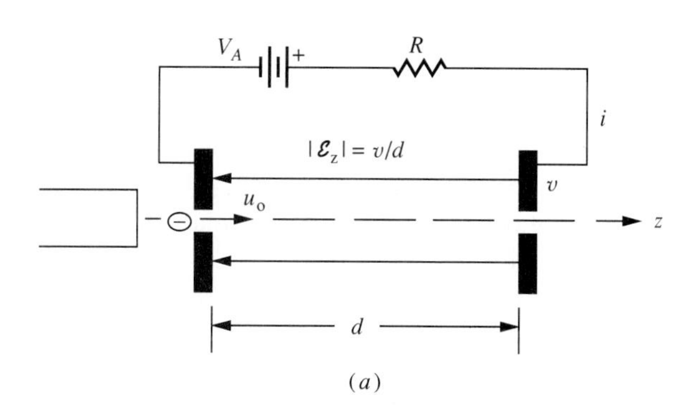

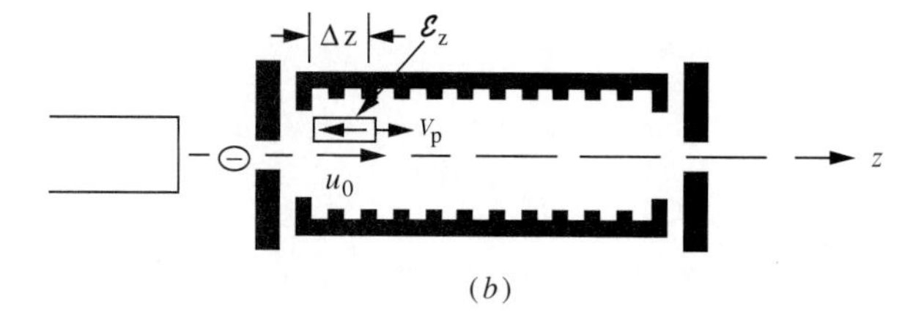

FIGURE 15.13
Interaction of an electron with a field. (a) Uniform $\mathcal{E}$ field, constant between $0 < z < d$. (b) Moving field $\mathcal{E}$ of extent Δz but acting continuously through $0 < z < d$.

the electron is modified by the ratio of the potential energy the electron loses or gains between the plates compared to the initial kinetic energy of the electron. The potential energy between the plates is independent of plate separation d. The larger the separation d, the lower is the electric field $|\mathcal{E}_z| = v/d$ that is acting on the electron, but since the electron traverses a longer longitudinal distance d, the potential energy is still ev.

Consider a different situation, shown in Fig. 15.13*b*. Again, an electron moving in a longitudinal direction (z axis) is injected into an interaction region with an initial velocity u_0. This time, instead of the constant electric field $|\mathcal{E}_z| = v/d$ due to the externally applied potential difference between the plates of Fig. 15.13*a*, we assume that a $-z$-directed externally applied electric field $\mathcal{E}_z$ of the same magnitude, v/d, exists but the extent of this field is within a short distance Δz, that is, the field is zero everywhere except within Δz. We now assume that the finite electric field $\mathcal{E}_z$ of extent Δz moves with a velocity $v_p = u_0$, that is, in synchronism with the electron. This means that the electron is constantly under the action of this synchronously moving electric field from $z = 0$ to $z = d$. At $z = d$, the velocity of the electron is the same as Eq. 15.19*b*

$$v_z = u_0 \left[1 \pm \frac{1}{2} \frac{(ev)}{KE_0} \right]$$

but we now write it as

$$v_z = u_0 \left[1 \pm \frac{1}{2} \left(\frac{W}{KE_0} \right) \right] \tag{15.20}$$

where

$$W = \pm e \int_0^d \mathcal{E} \cdot dl \tag{15.21}$$

is the work done by the field on the electron. The negative sign signifies that the electron loses energy. This implies that the energy loss of the electron should be converted to a gain of energy by the field. On the other hand, when the work done is positive, it means that the field does work on the electron and the energy of the electron increases at the expense of the field energy. Even if $|\mathcal{E}_z|$ is small, since the moving field acts constantly on the electron, the accumulated gain or loss of the electron energy is the same as if the electron were moving through a large distance d under an applied potential v.

Note that in this model the electron velocity will exceed or recede the phase velocity v_p of the field, and the electron will lose synchronism if the interaction length is long enough to produce appreciable velocity changes in the electron motion.

15.9 ELECTRON BUNCHING

In an actual device, we are interested in rapidly varying high-frequency fields. We should, therefore, investigate the motion of an electron by replacing the constant potential v in Fig. 15.13 by a sinusoidally varying signal. In this case, the motion of a single electron traversing the distance d is not interesting. Instead, we now consider a stream of electrons, and assume that all these electrons move in the axial direction. All electrons in the stream enter the interaction region with the same speed u_0 and initially are equally spaced from one another. For clarity, only a linear stream of electrons will be considered.

In Fig. 15.14, the constant electric field intensity $\mathscr{E}_z$ of Fig. 15.13*b* is replaced by a time-varying electric field $\mathscr{E}_z$ whose only single cycle is shown. Δz is the extent of the electric field in a small portion of the cycle where within Δz the field can be considered to be constant. The electrons, which are equally spaced initially, enter the interaction region labeled (I) and are subjected to the spatially varying field intensity. It will be assumed that the stream of electrons under the electric field and every point of the electric field will move synchronously together throughout their existence in the interaction region. Referring to the upper expanded figure of the cycle, the electric field magnitude is zero at A, C, and E, and the electrons that are synchronized with these points will move through the whole interaction region d with velocity u_0, that is, with no change in their velocities. But the electrons that are synchronized with the field within the region $A < \xi < C$ will be subjected to a z-directed decelerating force because of $\mathbf{F}_z = -e\mathscr{E}_z$. Depending on the magnitude of $\mathscr{E}_z$, the decelerating force on the electrons in this region will vary according to the magnitude of the field at a specific point within $A < \xi < C$. The maximum decelerating force occurs in the vicinity of point B, the maximum of the sinusoid. On the other hand, the electrons that are synchronized with the electric field within the region $C < \xi < E$

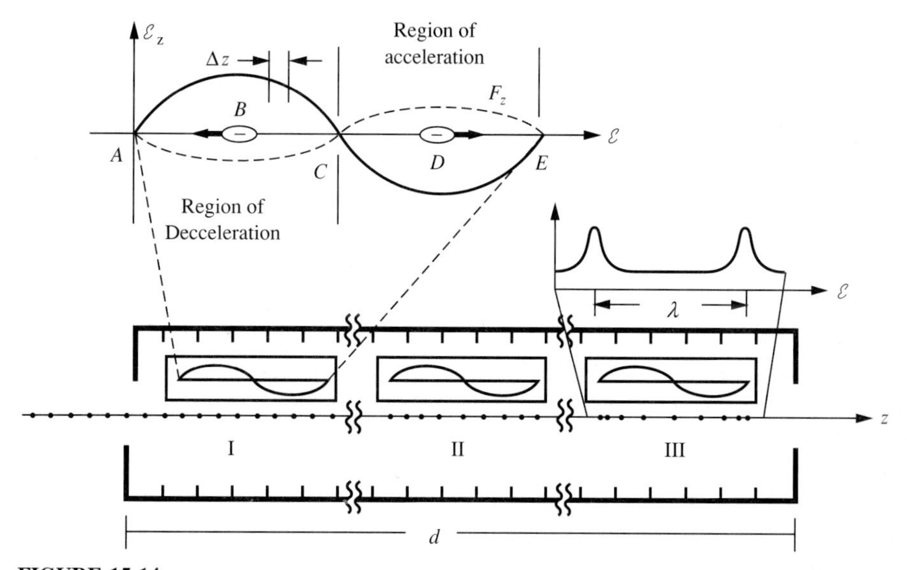

FIGURE 15.14
Electron bunching in a slow-wave structure. The expanded view shows the acceleration and deceleration regions of the electrons over a cycle.

are subjected to an accelerating force $\mathbf{F}_z = e\mathcal{E}_z$ and the maximum acceleration occurs around point D.

We now assume that a certain number of equally spaced electrons are initially synchronized with the electric field $\mathcal{E}_z$ and that these electrons follow the electric field as they move axially in the interaction region. While the electric field moves from location I to location II, electrons within $A < \xi < C$ are continuously slowed down, their rate depending on the magnitude of the decelerating force at their respective locations. Similarly, the electrons synchronized within $C < \xi < E$ are constantly accelerated. The accelerated electrons catch up with the decelerated electrons and start bunching around points A and E. As the electron stream reaches location III, this bunching becomes more pronounced. The electron density close to $x = d$ is shown in the upper expanded portion of Region III in Fig. 15.14. The electrons cannot all bunch to a point because as they get closer, repulsive space-charge forces become important and further bunching is prevented. As we shall see shortly, electron bunching is essential for the operation of the beam devices as signal amplifiers or as oscillators.

There is a second important class of electron beam devices whose working principles are based on the velocity modulation of the electron stream. A sinusoidally varying signal ($v \sin \omega t$) is applied between two plates that are separated by a small distance d (Fig. 15.15). An electron stream, with equal spacing and initial velocity u_0, enters through a gridded hole into the region between the plates. Let us assume that the transit time of the electrons traversing the plate separation d is small compared to the period of the sinusoidal signal. As the electron stream leaves the second plate, their velocities will be given by

$$v_z = u_0 + (\Delta v)_{\max} \sin(\omega t) \tag{15.22}$$

Thus, the velocities of the electrons are modulated by the sinusoidally varying input signal. Once the electrons enter the region $z > 0$, which is called the *drift space* where no fields exist, each electron moves throughout this region with the velocity that it acquired just before leaving the second plate.

The history of each electron can be followed from the space-time plot for each electron. In Fig. 15.15, the second plate is taken as the reference point, and on the vertical axis (downward) the change in the velocity of the electrons leaving $z > 0$ is plotted as a function of time. From Eq. 15.19, the maximum velocity change the electrons can have is $(\Delta v)_{\max} = u_0(ev)/2(KE_0)$. Any line drawn from the time axis is a straight line with a slope equal to $v_z^{-1} = [u_0 + (\Delta v)_{\max} \sin(\omega t)]^{-1}$. Point A represents an electron that has left the gap at an earlier time $t_A = T$ (T = period of the input signal) and has $v_z = u_0$ since $(\Delta v)_{\max} \sin(\omega t) = 0$. Similarly, electrons that left the gap at $t_B = 2T$ and $t_C = 3T$ also have velocities equal to u_0.

The slopes corresponding to the other electrons leaving the gap at any arbitrary times are equal to the reciprocal of their respective v_z velocities given by Eq. 15.22. If we follow these electrons along the z axis in the drift space, we can find the location of any electron on the z axis at time t where t is the time that has elapsed between that particular electron leaving the gap at $z = 0$ and reaching

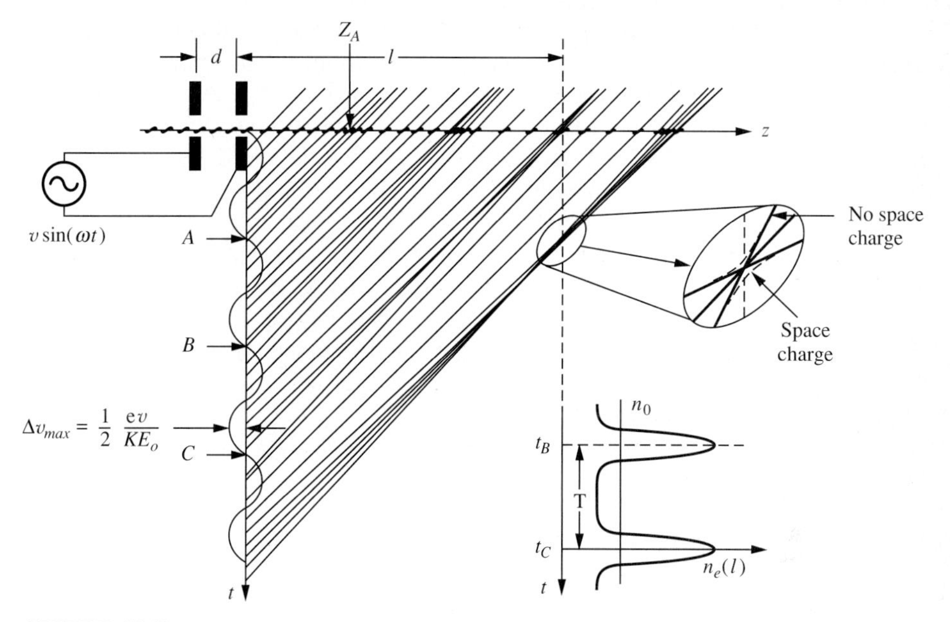

FIGURE 15.15
Electron bunching in velocity modulated devices (t versus z plot is known as the Applegate plot). Slopes of the lines are inversely proportional to the velocity of the electrons. Insert shows the trajectories of the electrons with (dashed) and without (solid) space charge.

the point z. For example, z_A is the location of an electron that has left the gap at $t_A = A$. In this way, we can follow the stream of electrons in the drift space. We see immediately that, due to the velocity modulation of electrons by the input signal, the electrons start bunching together as they move along the drift space. This occurs at $z = l$ (Fig. 15.15). The electrons cannot bunch to a point nor can their trajectories cross one another. The space-charge forces counteract these. Instead of bunching further, the electrons start repelling each other and, as they move further down the drift space, they spread apart.

From the two examples given, we can see that electron bunching is an important property of the beam devices. Since bunching implies the presence of space-charge forces, it is reasonable to assume that space-charge forces also play a very important role in the operating principles of all beam devices. The existence of electromagnetic waves in electron beams subject to space-charge forces will now be discussed.

15.10 SPACE-CHARGE WAVES

We have shown that in a beam device the bunching of electrons is a consequence of the electrons being accelerated or decelerated. The action of the time-varying electromagnetic signals favors bunching, and the Coulomb forces between electrons oppose this bunching. In an actual device, the beam will be confined in the axial direction by an externally applied focusing magnetic field. As the charge density increases due to bunching, it becomes exceedingly difficult or even impossible for an electron to overtake the electron just ahead of it. Also, the electrons may not have gained sufficient velocity to overcome the Coulomb forces, and the large confining magnetic fields will prevent the electrons from scattering in the radial direction. Thus, the beam will only be bunched in the longitudinal direction.

The consequence of bunching is, then, a one-dimensional problem, as can be seen from Figs. 15.14 and 15.15. Bunching occurs around unperturbed electrons at A and E of Fig. 15.14 and at A, B, and C in Fig. 15.15. In either case, each bunching of electrons takes place with a separation equal to the wavelength of the electromagnetic signal. The bunching and debunching in the beam can also be seen as oscillations taking place in an elastic medium subjected to axial compression and rarefaction forces.

In studying the properties of space-charge waves, we make the following simplifying assumptions:

1. Since bunching occurs only axially, a one-dimensional problem is considered.
2. The beam is infinite in cross section.
3. There are only axial electric fields, and all parameters vary with z only.
4. The ions produced as a result of the ionized background atoms completely neutralize the electron beam.
5. All variables have $\exp[j(\omega t - \beta z)]$ variation.
6. All variables can be written as the sum of zeroth-order (time and space independent) and first-order (time and space dependent) quantities.
7. Only the first-order terms are retained in the linearized equations.

When Maxwell's equations are solved in conjunction with the equations of motion, the linearized theory, which uses the approximations listed above, gives the following dispersion equation for the axial electric field $\mathscr{E}_z$

$$\left\{(k_0^2 - \beta^2)\left[1 - \frac{\omega_p^2}{(\omega - \beta u_0)^2}\right]\right\} \mathscr{E}_z = 0 \tag{15.23}$$

where $k_0 = \omega\sqrt{u_0\epsilon_0}$, ω_p is the electron plasma frequency, $\omega = 2\pi f$ is the electromagnetic signal frequency, and β is the propagation constant of the electromagnetic wave.

For a finite $\mathscr{E}_z$, Eq. 15.23 is satisfied only if the term inside the braces is equal to zero. There are four characteristic roots (propagation constants) of this equation for β; that is, the medium can support four waves. The corresponding propagation constants are

$$\beta_1 = k_0 \tag{15.24a}$$

$$\beta_2 = -k_0 \tag{15.24b}$$

$$\beta_3 = \frac{\omega - \omega_p}{u_0} \tag{15.24c}$$

$$\beta_4 = \frac{\omega + \omega_p}{u_0} \tag{15.24d}$$

The first two solutions are waves traveling with the speed of light, that is, $v_{p1,2} = \omega/\beta_{1,2} = \pm c$. The phase velocities of the last two waves are

$$v_{p3} = \frac{\omega}{\beta_3} = \frac{\omega}{(\omega - \omega_p)/u_0} = \frac{u_0}{1 - (\omega_p/\omega)} \tag{15.25}$$

and

$$v_{p4} = \frac{\omega}{\beta_4} = \frac{\omega}{(\omega + \omega_p)/u_0} = \frac{u_0}{1 + (\omega_p/\omega)} \tag{15.26}$$

The plasma frequency of the electrons can be written as

$$\omega_p^2 = \frac{\eta J_0}{\epsilon_0 u_0} \tag{15.27}$$

$\eta = e/m$ and J_0 is the current density of the electron beam (see Problem 15.13).

We see that out of the four roots of Eq. 15.23, two give wave solutions that are traveling in opposite directions with the speed of light. The phase velocity of these waves is a lot faster than the velocity u_0 of the electron beam. On the other hand, the other two waves both travel in the same direction as the beam particles with speeds which are close but faster (v_{p3}) or slower (v_{p4}) than the speed of the electrons u_0. Since for practical beams and signal frequencies $(\omega_p/\omega) << 1$, these differences in wave phase velocities are not too much different than u_0. These two waves are called fast and slow waves respectively.

15.11 KLYSTRONS

Devices based on bunching due to velocity modulation are called klystrons. A typical klystron amplifier is shown in Fig. 15.16. It consists of a thermionic cathode, an accelerating gun structure, two microwave cavities, and a collector. The electrons that are accelerated to a velocity u_0 by the gun enter the input cavity called the *buncher cavity*. A sinusoidally varying input signal applied to this first cavity velocity modulates the electron beam. Bunching of the electrons occurs while the electrons traverse the drift space between the two cavities. The second cavity is placed at the location of the maximum bunching.

As the electrons pass through the second cavity, which is called the *catcher cavity*, they induce currents in the second cavity whose output signal is coupled to the load R_L. The catcher cavity is placed at $z = l$ away from the buncher cavity to optimally couple the maximum power to the load. The electrons fulfilling their mission end up at the collector. The focusing magnetic field, which prevents

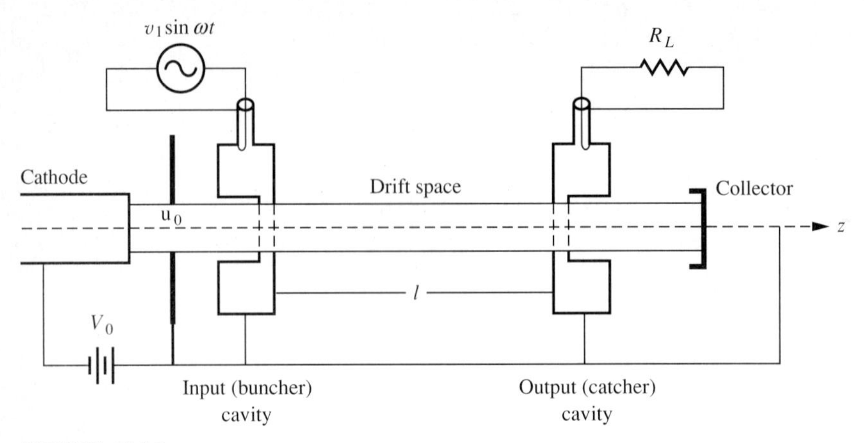

FIGURE 15.16
Essential components of a two-cavity klystron amplifier.

space-charge spreading and thus maintains the uniform cross section of the beam, is assumed to exist but is not shown in the figure.

Klystrons can be analyzed in two different ways. For low beam-current densities, the individual electron motion approach, known as the *ballistic theory*, is utilized to explain the velocity modulation and location of the resulting maximum electron bunching. As the beam density is increased, space-charge forces become essential to the operation of the device and certain major differences become apparent in the design of useful klystron amplifiers and oscillators.

15.11.1 Electromagnetic Resonant Cavities

Although many different geometrical structures are used as resonant cavities in the microwave frequency range, cylindrical cavities producing maximum electric fields along their axis are preferred over the others when used in klystrons. The cross section of a typical microwave cavity, known as the re-entrant cavity, is shown in Fig. 15.17*a*. Holes or grids located on the cavity axis allow the electron beam to pass through the cavity. Although grids intercept some of the electrons, they are generally used in low-power tubes since they produce more uniform axial electric fields along the axis of the cavity compared to the cavities with holes. When high-density electron beams are to be transmitted through the cavity, the holes become essential to avoid burning of the grid wires in high-power tubes.

The cylindrical cavity may be looked upon as an equivalent shorted transmission line, as shown in Fig. 15.17*b*. The cavity gap is the open end of a transmission line and the outer circumference is the location of the short. The cavity is made up of a large number of these shorted lines connected in parallel forming the final cavity. Depending on the length of the transmission line, the shorted line can be equivalent to a series or parallel resonant circuit. If the equivalent length is equal to the multiples of half wavelength, the line is a series

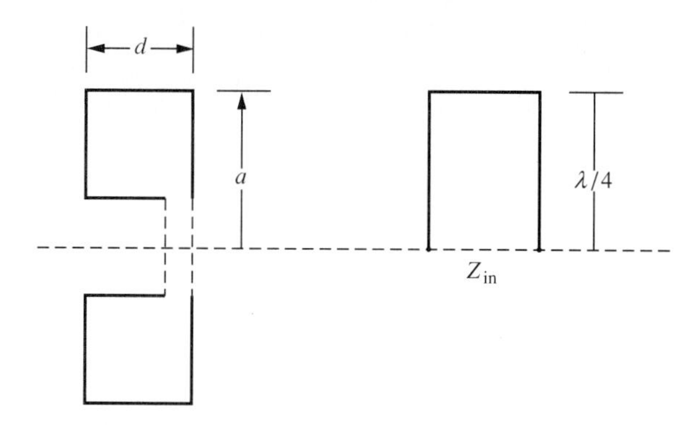

FIGURE 15.17
(*a*) Typical cylindrical resonant cavity used in a klystron. (*b*) Equivalent short-circuited transmission line.

resonant circuit and the current is a maximum at the input. For a line that is a quarter wavelength long, it is equivalent to an antiresonant circuit (parallel resonance) and the voltage is a maximum at the input of the line or the gap of the cavity.

As the electrons cross the catcher cavity they induce currents in this cavity as shown in Fig. 15.18. When an electron enters the cavity, it induces positive charges q' and q'' on the two opposite plates of the cavity. Initially, since the electron is near the first plate, the amount of positive charge q' induced in the first plate is larger than the positive charge q'' induced on the second plate—that is, the first plate is more positively charged than the second. (Positive charge implies that the electrons are pushed away from those regions by the repulsive Coulomb force on the electrons by the passing beam electron.) As the electron moves through the distance d between the plates, the induced charge q' decreases and q'' increases. Beyond $z > d/2$, q'' becomes more positive than q'. When the electron is between $0 < z < d/2$, the electric field due to the charges q' and q'' is in the $+z$ direction and when $d/2 < z < d$, it changes direction and points in the negative direction. If a velocity modulated beam of electrons now passes through the cavity, it will induce alternating electric fields in the cavity. In other words,

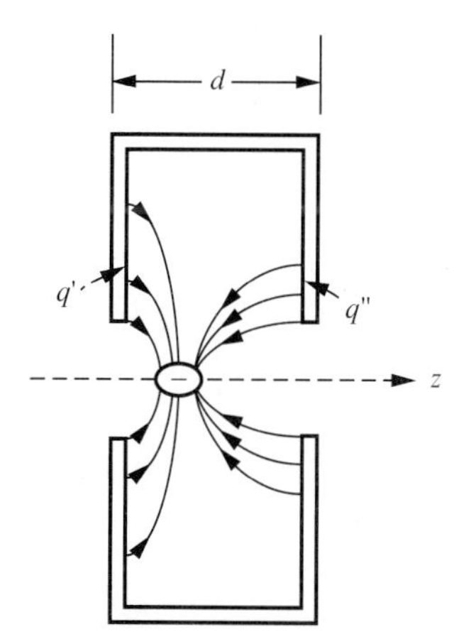

FIGURE 15.18
Charge induction by a passage of an electron between the two walls of a cavity.

passage of the electron beam will induce fields in the cavity in proportion to the charge passing between its plates. The higher the charge density, the higher will be this excitation. Thus, if weak signals are applied to the buncher cavity of Fig. 15.16, the induced fields in the catcher cavity will be proportionately larger than the input cavity, due to the bunched electrons crossing the catcher cavity. Thus, amplification of the electromagnetic signal takes place. When a uniform e-beam without bunching traverses through the cavity, no fields are induced between the plates since q' and q'' will be equal and remain equal with time.

15.11.2 Ballistic Theory of a Klystron

At low beam currents, the theory of operation of a klystron can be explained using the ballistic theory, based on the individual trajectory of an electron. For this, we refer to Fig. 15.19, where relevant time scales and distances are shown. Two identical cavities, with axial plate separation distances of d, are used as the buncher and catcher cavities. The separation between the cavities is l. The electrons enter the cavity at $t = t_1$ and leave the first cavity at $t = t_2$. The electrons reach the center of the first cavity at $t = t_0$, and reach the center of the catcher cavity at $t = t_c$. Nonrelativistic electron motion is considered in the following analysis.

When an electron enters the buncher cavity, it is subjected to an alternating electric field $\mathscr{E}_z(t) = -(v_1/d)\sin(\omega t)$. The resulting force is

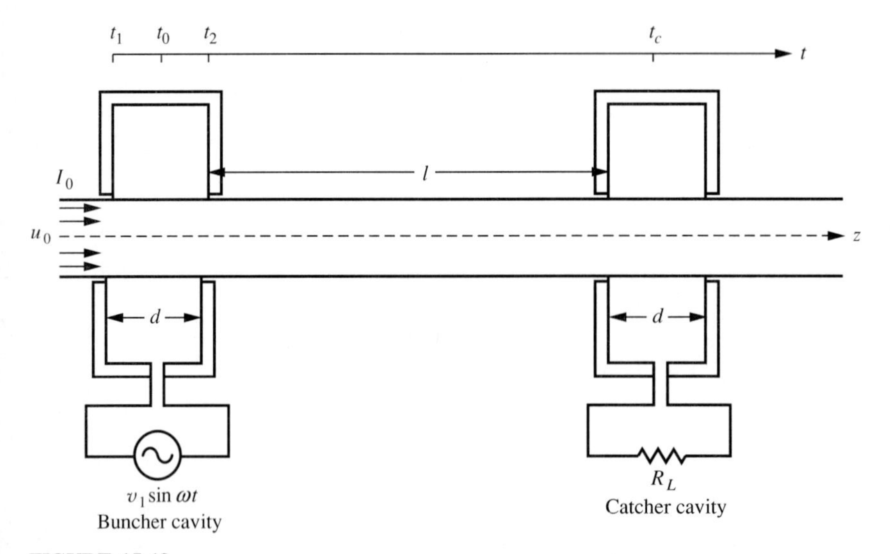

FIGURE 15.19
Typical two-cavity klystron amplifier parameters.

$$F_z = m_e \ddot{z} = -e \mathcal{E}_z \tag{15.28}$$

or it can be written as

$$\ddot{z} = \eta \frac{v_1}{d} \sin \omega t \tag{15.29}$$

Integrating with respect to time and using the initial condition that at $t = t_1, v_z = u_0$, we get

$$v_z = u_0 - \frac{\eta v_1}{\omega d}(\cos \omega t - \cos \omega t_1) \tag{15.30}$$

If the electrons leave the cavity at $t = t_2$ with velocity v_{z0}, we get

$$v_{z0} = u_0 - \frac{\eta v_1}{\omega d}(\cos \omega t_2 - \cos \omega t_1) \tag{15.31}$$

We assume that the time for the electron to cross the input cavity gap is mostly determined by the initial electron beam velocity u_0. This implies that $v_1 << V_0$. We can thus write

$$t_1 \approx t_0 - \frac{d}{2u_0} \tag{15.32a}$$

$$t_2 \approx t_0 + \frac{d}{2u_0} \tag{15.32b}$$

Substituting Eq. 15.32 into 15.31, we obtain

$$v_{z0} = u_0 + \frac{\eta v_1}{u_0} \left[\frac{\sin (\omega d/2u_0)}{(\omega d/2u_0)} \right] \sin \omega t_0 \tag{15.33}$$

The quantity

$$M \equiv \frac{\sin (\omega d/2u_0)}{\omega d/2u_0} = \frac{\sin (\theta/2)}{\theta/2} \tag{15.34}$$

is called the *cavity-coupling coefficient* and $\theta \equiv \omega d/u_0$ is the *transit angle* across the cavity. As d goes to zero, M approaches unity.

The time for an electron to reach the center of the catcher cavity can be written as

$$t_c = t_0 + \frac{d + l}{v_{z0}} \tag{15.35}$$

Substituting Eq. 15.35 into 15.33

$$t_c = t_0 + \frac{d + l}{u_0 + (\eta v_1/u_0)M \sin (\omega t_0)} \tag{15.36}$$

If $(v_1/2V_0) << 1$, we can expand the denominator and write

$$t_c = t_0 + \frac{d + l}{\mu_0 \, [1 + (\eta v_1/u_0^2) M \sin(\omega t_0)]}$$

$$t_c \approx t_0 + \frac{d + l}{u_0} \left[1 - \frac{v_1}{2V_0} M \sin(\omega t_0) \right] \tag{15.37}$$

To find the current at the center of the catcher cavity, we use the conservation of charge and assume that the number of electrons crossing the buncher cavity within a time Δt_0 should be equal to the electrons crossing the catcher cavity within a time Δt_c, that is,

$$I_0 \Delta t_0 = i_z \Delta t_c \tag{15.38}$$

where i_z is the value of current at the center of catcher cavity. We can write Eq. 15.38 as

$$i_z = I_0 \frac{dt_0}{dt_c} \tag{15.39}$$

Differentiating Eq. 15.37 with respect to t_0

$$\frac{dt_c}{dt_0} = 1 - \left[\frac{(d + l)\,\omega}{u_0} \frac{v_1}{2V_0} M \right] \cos(\omega t_0) \tag{15.40}$$

substituting into Eq. 15.39, we find

$$i_z = I_0 [1 - k \cos(\omega t_0)]^{-1} \tag{15.41}$$

where

$$k \equiv \frac{(d + l)\omega}{u_0} \frac{v_1}{2V_0} M \tag{15.42}$$

is called the *bunching parameter* for the cavity. In deriving Eq. 15.41, it was assumed that the modulating voltage v_1 is a lot smaller than the accelerating voltage V_0.

Using Eq. 15.41, we can plot the current at the catcher cavity location. Note that i_z depends on the bunching parameter k. If $k = 0$, there is no bunching or the initial dc beam current does not change. i_z for various values of k is plotted in Fig. 15.20. For $k = 1$, the current becomes infinite at $\omega t_0 = (2\pi)n$ $(n = 0, 1, 2, \ldots)$. For $k > 1$, the current becomes infinite at two points over one cycle. In an actual device, these are prevented by the space-charge forces.

The induced voltage v_c at the catcher cavity can be written as

$$v_c = M i_z R_{sh} \tag{15.43}$$

where R_{sh} is the shunt resistance and M is the beam-coupling coefficient of the equivalent catcher cavity.

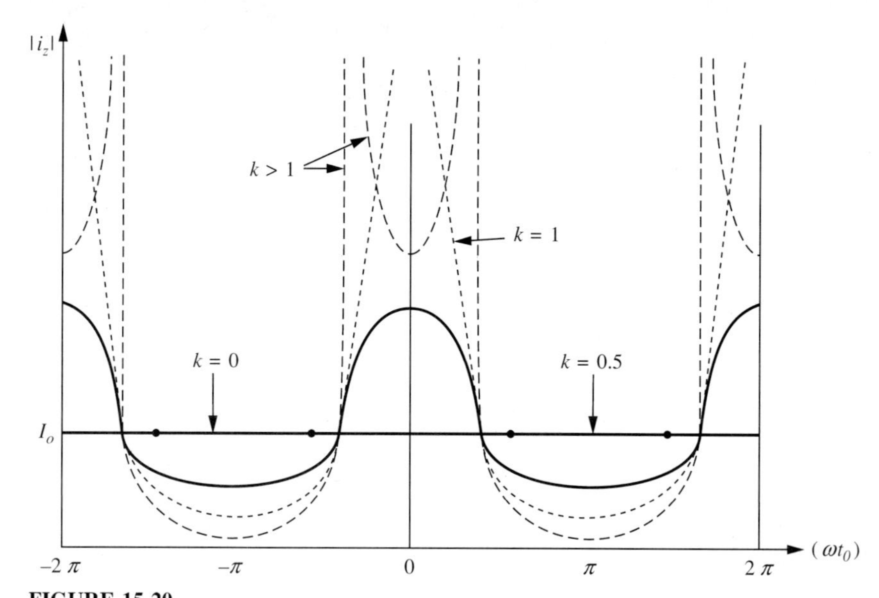

FIGURE 15.20
Output current in the catcher cavity load as function of time for various values of k.

For $k \neq 1$, we can expand Eq. 15.41 in terms of Bessel functions in the form

$$i_z = I_0 + 2I_0 \sum_{n=1}^{\infty} J_n(nk) \sin\left[n\left(\omega t - \frac{\omega(d+l)}{u_0} + \frac{\pi}{2}\right)\right] \tag{15.44}$$

where $J_n(kr)$ is the $n'th$-order Bessel function of the argument (kr).

We see that the current distribution at the catcher cavity contains the fundamental as well as the harmonics of the original signal frequency ω. The catcher cavity can be tuned to any of these harmonics, and the klystron then becomes a frequency multiplier. If the catcher cavity is designed so that it resonates at the fundamental frequency $\omega(n = 1)$, the ac current at the catcher cavity becomes equal to

$$i_z = 2I_0 J_1(k) \sin(\omega t - \beta) \tag{15.45}$$

where $\beta = [\omega(d+l)/u_0] - \pi/2$ is a phase factor. We observe that the maximum current occurs when $J_1(k)$ is a maximum. This occurs when $k = 1.84$ so that $J_1(1.84) = 0.584$. Therefore

$$i_{z\max} = 1.168 I_0 \tag{15.46}$$

The optimum location of the catcher cavity is found from Eq. 15.42,

$$k = \frac{(d + l_{\text{opt}})\omega}{u_0} \frac{v_1}{2V_0} m = 1.84$$

or

$$l_{\text{opt}} = \left(\frac{1.84 \times 2 \times V_0 \times u_0}{M \omega v_1} \right) - d \tag{15.47}$$

We see that the optimum distance depends inversely on the input signal amplitude and frequency. If one of these changes, the location of the second cavity should also be changed.

When the tube is operating under ballistic conditions, maximum first harmonic ac current is equal to 1.168 I_0. The voltage appearing at the output gap is dependent on the shunt resistance of the output cavity, but it is usually assumed that the peak ac output cannot be greater than the dc beam voltage V_0. When both voltage and current are the peak values, the ac power out is given by

$$P_{\text{out}} = \frac{1.168\, I_0 V_0}{2} = 0.584\, I_0 V_0 \tag{15.48}$$

The dc power input to the beam is $P_{\text{in}} = I_0 V_0$ so that the maximum electronic efficiency of the klystron is 58.4 percent. Although this is a highly simplified model, it gives the ultimate efficiency that can be achieved in klystron devices.

15.12 TRAVELING WAVE TUBES

In order to understand the basic principles of traveling wave tubes, some of the parameters related to waveguiding structures will be reviewed.

15.12.1 Phase and Group Velocity

As was shown in Chapter 1, the group and phase velocities for a wave packet were given by

$$v_p = \frac{\omega}{\beta} \tag{15.49a}$$

$$v_g = \frac{\partial \omega}{\partial \beta} \tag{15.49b}$$

The phase velocity is the phase with which a fictitious point of the wave front moves. The group velocity is the velocity with which energy is transported. Although in ordinary transmission lines the group velocity is important for energy transport, we need the phase velocity of the electromagnetic signal for the analysis of the beam devices.

In Section 15.9, the bunching of electrons due to an electric field moving in synchronism with the electron was discussed. It was assumed that the electron is acted upon continuously by $\mathcal{E}_z$ of extent Δz. For a particular electron, the field magnitude may be that corresponding to point A in Fig. 15.21. For continuous

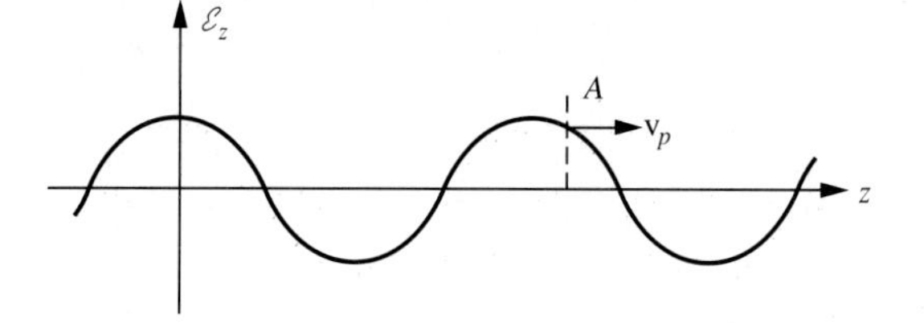

FIGURE 15.21
An infinitely long traveling wave at an instant of time. The velocity with which phase point A travels is v_p.

interaction, the electric field magnitude at A should move in synchronism with the electron. Since the speed with which the point A moves through the guiding medium is the phase velocity of the wave, for efficient and continuous interaction between the beam and the signal, v_p should be close to the initial electron speed u_0. In beam devices, we use those waveguiding structures whose phase velocities are close to the beam velocity. These structures are called *slow-wave guides*.

Solution of Maxwell's electromagnetic equations for a given medium leads to a *dispersion relation* that gives the dependence of the propagation constant on the applied signal frequency and the geometrical parameters of the guiding structure. In general, the solutions leading to the dispersion equation and the resulting relation are lengthy. Without resorting to a specific solution, we can illustrate the general features of waveguiding structures by referring to Fig. 15.22, which shows the dispersion relation for various transmission lines.

Curve A is the dispersion relation for a typical waveguide. The phase velocity is always greater than the speed of light c but the group velocity is always less than c. ω_c is the cutoff frequency for that particular mode of propagation in the waveguide. For a plane wave (Curve B), the group and phase velocities are both equal to c. Curve C is the dispersion relation for a slow-wave structure. Note that at point P_1, both the phase and the group velocities are positive and smaller than c. It is also possible to have group and phase velocities in opposite directions,

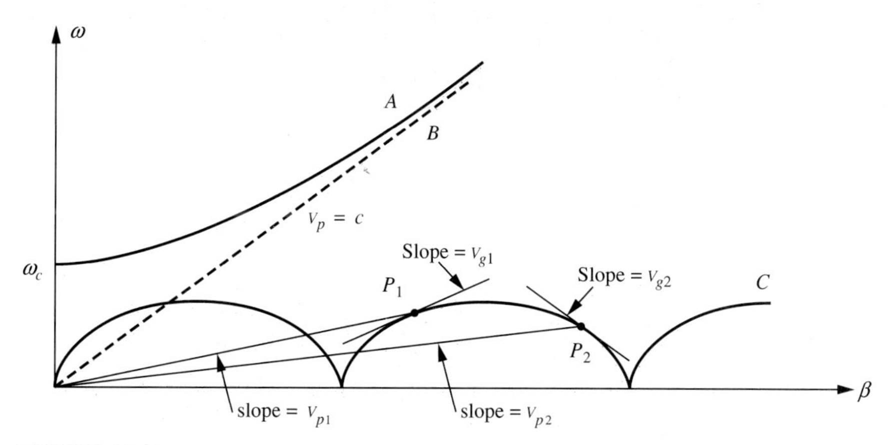

FIGURE 15.22
Dispersion relations for (*a*) waveguide, (*b*) plane wave, and (*c*) slow-wave structure.

as shown for point P_2. These are called *backward waves*. Periodic structures generally support slow waves.

One of the common slow-wave guides used in traveling wave tubes is a helix (Fig. 15.23). An approximate model for the helix is that the wave follows the wire at about the speed of light, so that the effective propagation constant on the axis corresponds to a phase velocity

$$v_p = c \sin \psi \qquad \text{where} \qquad \psi = \tan^{-1}\left(\frac{p}{2\pi a}\right) \tag{15.50}$$

$$v_p \approx c\frac{p}{2\pi a} \qquad \text{for small} \quad \psi$$

where p is the pitch and ψ is the pitch angle. It is rather surprising, but Eq. 15.50 represents a good approximation to the helix over a wide range of parameters. The length of the helix determines the high-frequency cutoff properties of the helix. The helix also has adequate power-handling capacity.

15.12.2 Convection Current

When an electron moves in space between two plates, as shown in Fig. 15.24*a*, it induces current in the external circuit. As the electron moves under the influence of $\mathscr{E}_z$, the incremental work done on the electron is

$$\Delta W = \mathbf{F} \cdot \Delta \mathbf{z} = -e\,\mathscr{E}_z \Delta z = e\left(\frac{v}{d}\right)\Delta z \tag{15.51}$$

This work must be balanced by the battery in moving a charge Δq between the plates. Thus

$$\Delta W = v \Delta q = e\left(\frac{v}{d}\right)\Delta z \tag{15.52}$$

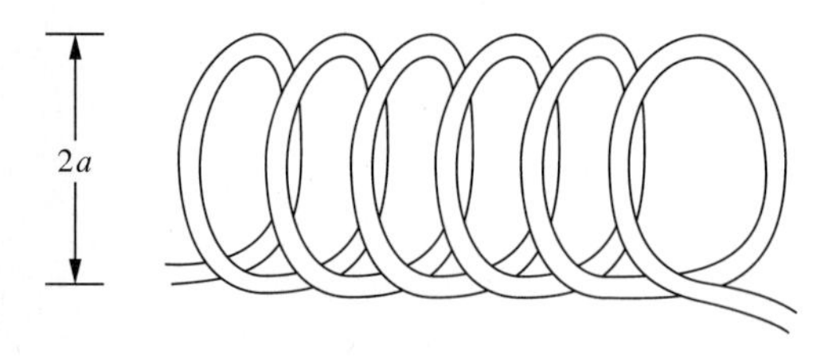

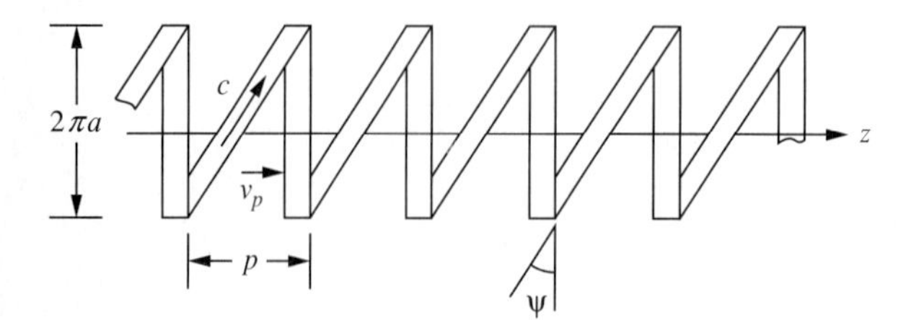

FIGURE 15.23
Typical helix and dimensions in determining the propagation parameters. Phase velocity is the projection of c on the z axis.

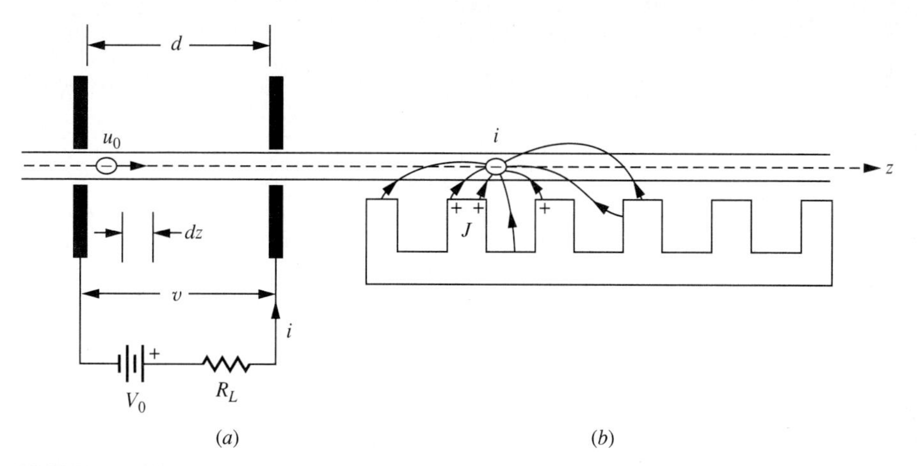

FIGURE 15.24
Induced currents (*a*) between parallel plates, and (*b*) in a slow-wave structure.

Dividing both sides by Δt, we obtain

$$i = \lim_{\Delta t \to 0} \frac{\Delta q}{\Delta t} = \frac{dq}{dt} = \frac{e}{d}\frac{dz}{dt} = \frac{e}{d} v_z \tag{15.53}$$

where v_z is the velocity of the electron at z.

The motion of the electron between the plates induces a conduction current in the external circuit made up of two plates, a load resistor, and a battery. If the continuity of the total current in the overall circuit is considered, the electron motion between the plates constitutes a current. This is called the *convection current*. Even if the battery is removed from the circuit, the equivalent amount of current given by Eq. 15.53 still flows in the external circuit. The current lasts for the duration of the electron traversal between the plates. If the motion of the electrons is uniform or time varying, the corresponding induced current will also follow similar time dependence.

Similar induced currents will flow in the arrangement of Fig. 15.24*b*, where an electron beam flows very close to the walls of a slow-wave structure. Thus, the electron motion, which leads to convection current i in the beam, induces current density J in the slow-wave structure at a rate

$$J = -\frac{\partial i}{\partial z} \tag{15.54}$$

If the velocity of the electron moving in the axial direction, and thus i, is a function of position, the induced current will also be a function of position in the slow-wave structure. This follows from Eq. 15.54. This concept is essential to understanding the coupling between the electron beam and the fields of the slow-wave guide structure.

15.12.3 Traveling Wave Tubes (TWT)

A typical traveling wave tube, shown in Fig. 15.25, consists of an electron gun, a slow-wave interaction region in the form of a helix with corresponding input-output waveguide couplers, and a collector. Depending on the electron gun and the externally applied dc magnetic field, these tubes are classified as O (ordinary) and M (magnetron) type traveling tubes. In the O-type, the electron beam is guided through the slow-wave structure by the externally applied focusing fields. In the M-type, crossed electric and magnetic fields guide the beam while the electrons traverse the interaction region.

A schematic cross section and the corresponding equivalent circuit representation of a traveling wave tube are shown in Fig. 15.26. The equivalent circuit representation using distributed inductances and capacitors has the same slow-phase properties as that of a periodic slow-wave guide.

There are two separate but coupled systems in Fig. 15.26*b*. One of these is the electron beam and the other is the periodic structure. In the slow-wave structure, a voltage wave v is assumed to travel in the positive z direction. This is related to the electric field through $\mathscr{E}_z = -dv/dz$. As the circuit wave moves along the guide, it induces a convection current i in the electron beam. If the electrons are moving close to the slow-wave structure, this convection current, in return, will induce back a field in the slow-wave structure. If the induced field is in such a phase that it enhances (adds to) the field of the slow-wave structure, the enhanced field will induce a higher convection current in the beam, which will

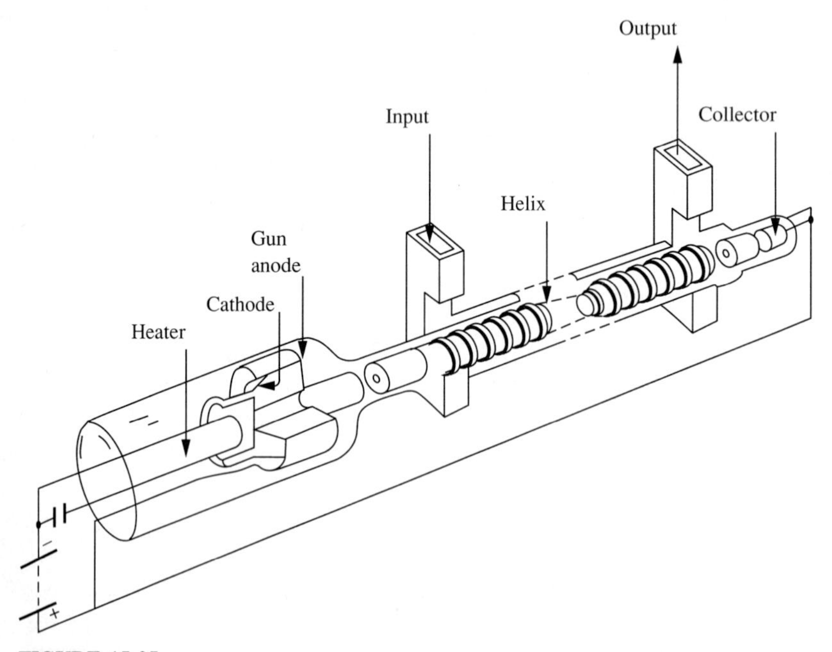

FIGURE 15.25
Constructional details of a typical traveling wave-tube amplifier.

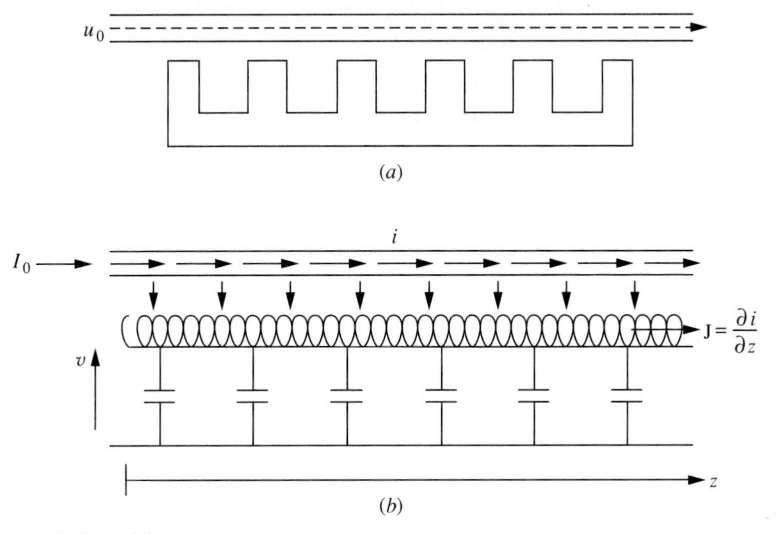

FIGURE 15.26
(*a*) Schematic axial cross section of the traveling wave tube and (*b*) its equivalent circuit.

further enhance back the field. As the electron beam and the field move along the z axis, this interaction will transfer energy from the electrons to the field or vice versa. Although energy is exchanged between the circuit and the beam, the total energy of the system will remain constant. Whether the field gains or loses energy depends on the detailed relation between the electron beam and the space-charge fields.

Assuming the electromagnetic fields vary as $e^{-\Gamma z}$, the analysis of the interaction between the beam electrons and the slow-wave circuit leads to a fourth-order dispersion relation containing the frequency, the propagation constant, and the beam and circuit parameters. The resulting first three propagation constants are

$$\Gamma_1 = j\beta_e\left(1 + \frac{C}{\sqrt{2}}\right) - \frac{\sqrt{3}}{2}\beta_e C \tag{15.55a}$$

$$\Gamma_2 = j\beta_e\left(1 + \frac{C}{\sqrt{2}}\right) + \frac{\sqrt{3}}{2}\beta_e C \tag{15.55b}$$

$$\Gamma_3 = j\beta_e(1 - C) \tag{15.55c}$$

where $\beta_e = \omega/u_0$ is the phase constant for the electrons. C is known as Pierce's gain parameter and is given by

$$C \equiv \left(\frac{ZI_0}{4V_0}\right)^{1/3} \tag{15.56}$$

Z is the interaction-wave impedance, I_0 is the dc beam current, and V_0 is the beam-accelerating voltage. The fourth root of the dispersion relation, a backward wave without amplitude change, is not given here.

All of the propagation constants given by Eq. 15.55 are traveling waves moving in the positive z direction. The phase velocities of the first two are less than the speed u_0 of the electrons, and the third wave moves faster than the electrons. On the other hand, the amplitude of the first wave is increasing and the second wave is decreasing. The third wave moves without any attenuation.

The increase in the amplitude of the wave as a function of distance (in decibels) is given by

$$20 \log \left[\exp \left(\frac{\sqrt{3}}{2} \beta_e C \right) z \right] \quad \text{dB} \tag{15.57}$$

Input signal has no distinction in launching the four possible waves. Thus, all four modes of propagating waves are launched at the input of the helix with the same probability, and with the exception of the one that is used for the amplification of the electromagnetic signal, the others are ignored in the helix. Thus, excitation of nonamplifying waves are referred to as the *launching losses*. The gain of a simple helix type TWT becomes

$$G(\text{dB}) = 20 \log \left(\frac{1}{3} \right) + 20 \log \left[\exp \left(\frac{\sqrt{3}}{2} \beta_e C \right) L \right] \tag{15.58}$$

$$G(\text{dB}) = -9.54 + 8.68 \left[\left(\frac{\sqrt{3}}{2} \beta_e C \right) L \right]$$

The loss factor of -9.54 dB comes from the equal excitation of the input signal among the three nonamplifying waves at the input.

15.12.4 Energy Exchange

To understand why the slow waves give rise to amplification, energy exchange between the beam and the wave should be considered. The kinetic input power of the beam is

$$P_{\text{in}} = I_0 V_0 \tag{15.59}$$

Assuming $u_1 << u_0$, the kinetic energy of a particle can be written as

$$\frac{1}{2} m_e (u_0 + u_1)^2 \approx \frac{1}{2} m_e u_0^2 + m_e \mu_0 \mu_1 \left(\frac{-e}{-e} \right) = E_0 + (-e) \left[-\frac{u_0 u_1}{(e/m_e)} \right]$$

The term in brackets has the units of potential. We define the ac kinetic potential of the beam as

$$v_{\text{ac}} = -\frac{u_1 u_0}{\eta} \tag{15.60}$$

where u_1 is the ac part of the total velocity of the beam. The average power at the output can be written as

$$\tilde{P} = I_0 V_0 + \frac{1}{2}\text{Re}(i_{ac} v_{ac}^*) \tag{15.61}$$

The ac kinetic power is given by

$$\tilde{P}_{\text{kac}} = \frac{1}{2}\text{Re}(i_{ac} v_{ac}^*)$$

$$= -\frac{u_0}{2}\left(\frac{m_e}{e}\right)\text{Re}(i_{ac} u_1^*) \tag{15.62}$$

For energy to be conserved within the framework of the linear theory, the average kinetic power at the output must be less than $I_0 V_0$, the kinetic power at the input, by an amount equal to the gain in ac power of the slow-wave circuit.

When the corresponding ac velocity of the slow and fast space-charge waves are substituted into Eq. 15.62, for the kinetic ac power of the corresponding waves, we find

$$\tilde{P}^s_{\text{kac}} = -\frac{u_0}{2}\omega/\omega_p|\mathcal{E}_z|^2 \tag{15.63a}$$

$$\tilde{P}^f_{\text{kac}} = \frac{u_0}{2}\frac{\omega}{\omega_p}|\mathcal{E}_z|^2 \tag{15.63b}$$

The ac kinetic power of the beam is negative for the slow space-charge wave, and this energy appears as a positive ac kinetic power in the slow-wave circuit. Phase velocity of the wave effective in amplifying the RF signal travels more slowly than the electrons. This satisfies the energy conditions because RF energy is obtained only by slowing down the electrons to extract their kinetic energy. Thus, the electrons must be traveling faster than the wave to give energy to the electromagnetic wave.

15.13 M-TYPE TRAVELING WAVE TUBES (MAGNETRONS)

If detailed analyses of space-charge forces are considered, amplification for the case of high space-charge fields is obtained for only a very restricted range of beam velocities. If, through the interaction process, the electron velocity is reduced to a value below a lower limit of this range by giving its energy to the field, further amplification is not obtained, and the tube saturates whatever the length of the tube is. In any case, the electron beam that emerges out of the interaction region still possesses considerable kinetic energy which is, in general, dissipated in the collector electrode. The tube is thus inefficient in its present form. There are two different ways in which the efficiency of the TWT can be increased. These are obtained by

1. Reducing the beam velocity after the interaction has taken place by applying retarding potentials through specially shaped electrodes before the electrons

reach the collector. This technique is known as *depressed collector* operation; or

2. Enabling the electrons to give up a higher fraction of their kinetic energy in the interaction process. This is known as an M-type interaction process.

The traveling wave tube shown in Fig. 15.25 was assumed to have an axial magnetic focusing field. If this field is replaced by a magnetic field perpendicular to the paper, as shown in Fig. 15.27, and a dc potential applied between the lower electrode (called a sole) and the upper slow-wave structure to produce an electric field in the direction shown, with no RF field, an electron beam entering the interaction region will drift in this space with a drift velocity given by $v_d = (\mathscr{E}_{dc} \times \mathbf{B})/(\mathbf{B} \cdot \mathbf{B}) = (\mathscr{E}_{dc}/\mathrm{B})\mathbf{a}_z$. The overall trajectory of electrons will depend on the initial velocity with which the electrons are injected into the interaction region. In general, this leads to a cycloidal trajectory for the electrons. The drift velocity of the electrons will be perpendicular to both the $\mathscr{E}$ and $\mathbf{B}$ fields and independent of the injection velocity. Thus, any traveling wave whose phase velocity is equal to the drift velocity and is launched at the input will travel in synchronism with the electron beam until the end of the interaction region. Thus, no matter how much energy is extracted from the beam by the wave, the electrons will draw sufficient replenishment energy from the transverse dc electric field to maintain their longitudinal drift velocity. If the initial electron injection velocity is made equal to the drift velocity, the electron trajectories become straight lines, that is, the force due to the electric field cancels the force due to the magnetic field. Under dc conditions, the magnetic field (provided it is strong enough) prevents an electron from moving toward the anode under the action of the transverse electric field. An electron in the crossed field region at any point between the electrodes thus possesses considerable potential energy. Without the magnetic field, an electron would move immediately to the anode, arriving with a kinetic energy exactly equal to the potential energy that it would have possessed when the magnetic field was present. The action of the slow wave is to draw energy from the electron stream

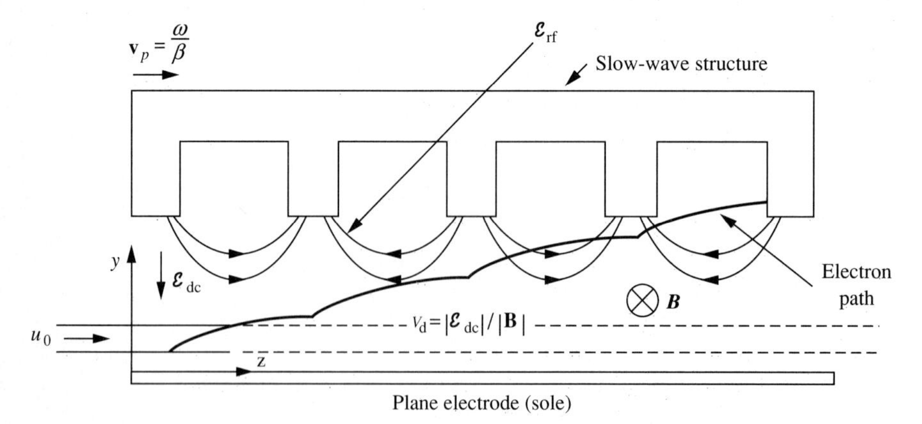

FIGURE 15.27
M-type traveling wave tube. Uniform magnetic field **B** and the crossed electric field produce a drift velocity $\mathscr{E}_{dc}/B$ equal to v_p.

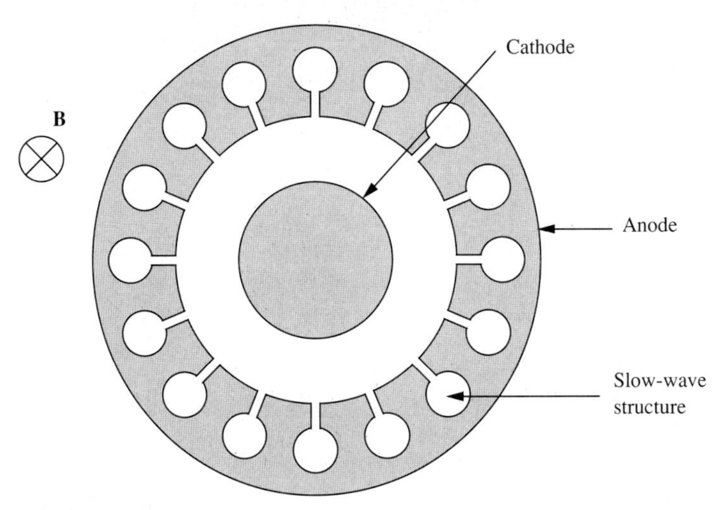

FIGURE 15.28
Magnetron oscillator.

by trying to reduce its longitudinal velocity. Due to the axial RF electric field, the electrons slowly drift toward the anode while moving in the interaction region. The drift velocity is compensated by the energy drawn from the transverse field. While moving toward the anode, the electron can only give up potential energy to compensate for the losses in the longitudinal drift energy which is caused by the slow wave. The process ends when an electron reaches the anode. Whereas O-type devices function as a result of electron beam giving up kinetic energy, M-type devices function as a result of the electrons yielding potential energy.

By far, the most widely used M-type device is the circular, continuous cathode form of the forward-wave oscillator. This is known as the *magnetron*, shown schematically in Fig. 15.28. The slow-wave structure of Fig. 15.27 is closed on itself to form a circular guide and the sole is replaced by a continuous circular cathode. The slow-wave field in the region between the cathode and the anode increases in magnitude because of the continuous feedback and oscillations build up. Magnetrons are highly efficient oscillators.

15.14 BACKWARD-WAVE TUBES

We have seen that a slow-wave structure can also support a backward wave. In a helix, this could be achieved by choosing a large enough radius for the helix winding.

A schematic of backward-wave tubes is shown in Fig. 15.29. An electron beam moves inside the helix, whose velocity is determined by the final accelerating voltage of the electron gun. Since the helix is a broad-band structure, the backward-wave phase velocity in the circuit will be determined by the beam velocity. Thus as the electron moves through the helix, it will synchronize with the backward wave whose velocity will be equal to its own. The frequency of operation of the tube will be determined by the final accelerating voltage of the

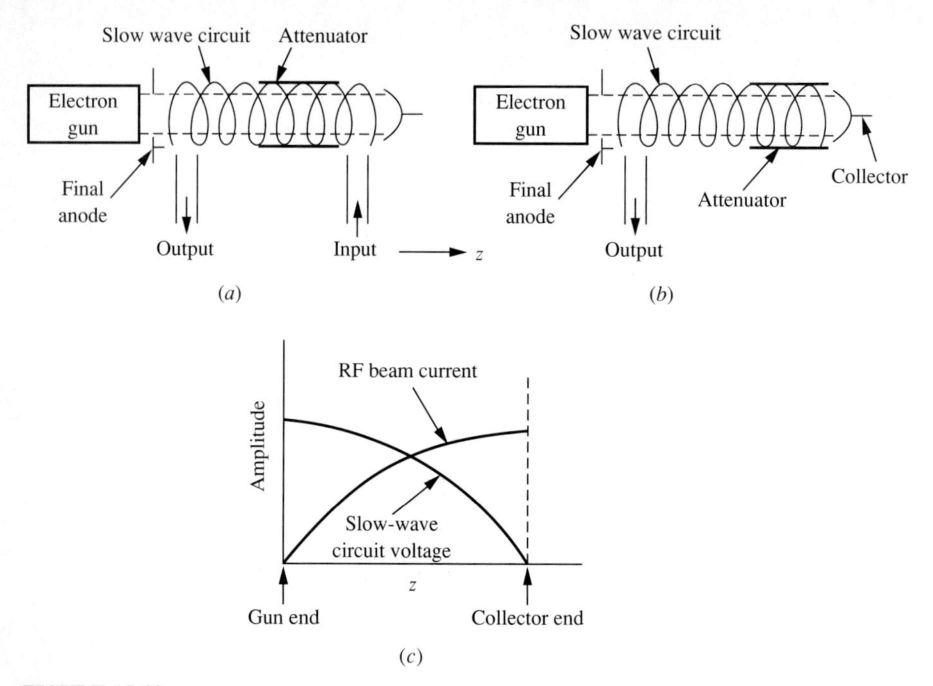

FIGURE 15.29
Backward-wave tubes (O-type). (*a*) amplifier, (*b*) oscillator, and (*c*) variation of slow-wave circuit voltage and RF beam current.

electron gun. The phase velocity of the wave will be in the forward direction, but the slow-wave circuit voltage will increase in the opposite direction because of the negative group velocity of the backward wave. Thus, in a backward-wave amplifier, the signal input will be from the end of the helix near the collector, and the output will be from the end near the gun (Fig. 12.29*a*). Although all the other space harmonics of the slow-wave structure will influence the gain of the tube, other traveling space harmonic waves may lead to unwanted couplings, which could lead to oscillations. Therefore, an attenuator placed in an optimum location prevents oscillation of the tube when working as an amplifier. On the other hand, if the input is removed and proper feedback is maintained so that the tube is forced to oscillate, the device becomes a backward-wave oscillator (Fig. 12.29*b*). Backward-wave amplifiers have a narrow frequency band but wide tuning range. Backward-wave oscillators also have a wide tuning range.

15.15 GYROTRONS

Gyrotrons are the only devices at present that can generate hundreds of kilowatts of RF power at millimeter wavelengths. The most general configuration of a gyrotron is cylindrical in shape, as shown in Fig. 15.30. The electron gun, referred to as a Magnetron Injection Gun (MIG), consists of a cathode that is slanted at an angle with respect to the axis, and emits electrons in the radial direction. The

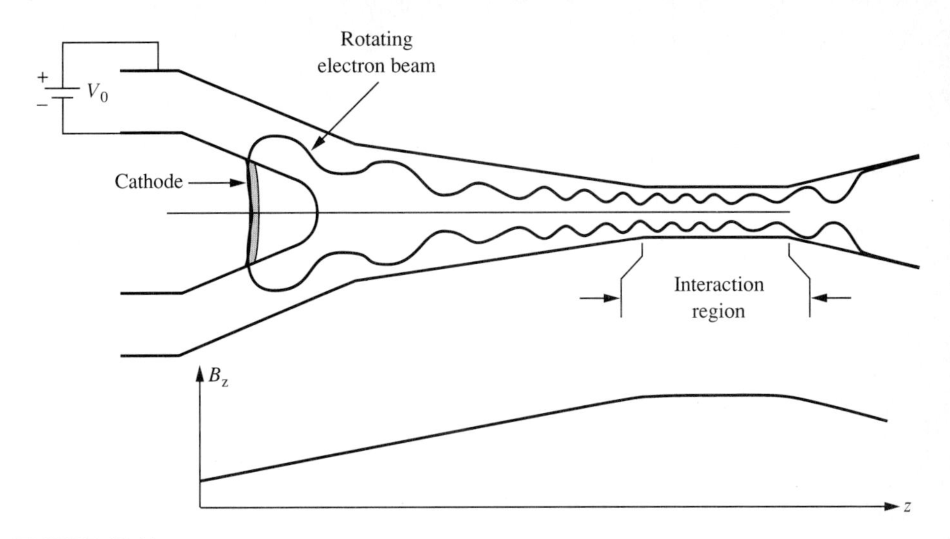

FIGURE 15.30
A conventional gyrotron oscillator uses a MIG gun and a cylindrical cavity for the interaction region. The spatial dependence of the axial **B** field is shown below.

combined electric and axial magnetic fields give the electrons a rotating motion. As the electrons move in the axial direction, their rotational motion increases with increasing magnetic field, and they enter the interaction region with a rotational speed greater than their longitudinal speeds. Since the voltages involved are very high, the electron motion is quasirelativistic and the electron velocities are close to the speed of light, that is, $v_{\perp} = 0.4c$. Therefore, gyrotrons are fast-wave devices contrary to the slow-wave devices discussed in the previous sections.

The phase velocity of the wave is nearly equal to the Doppler-shifted cyclotron frequency of the electrons. As the electrons move through the interaction region, they give up their rotational kinetic energy to the electromagnetic field. The interaction region is an overmoded cylindrical cavity operated close to its cutoff frequency. The power is coupled out from the right side of the tube in Fig. 15.30. The efficiency of conversion of kinetic energy to the field energy is very high. The operating frequency is determined by the cyclotron frequency of the electrons

$$\omega_c = \frac{eB}{m_e} \tag{15.64}$$

By choosing the right magnetic field and the proper dimensions of the cavity, the operating frequency of gyrotrons can be extended up to millimeter wavelengths. By increasing the current, the pulsed-power output can reach hundreds of kilowatts. For operation at millimeter wavelengths, the gyrotrons require superconducting magnets. Since the motion of the electrons are quasirelativistic, the gyrotrons can also operate efficiently at the higher harmonics of the cyclotron frequency.

FURTHER READING

Cobine, J.D., *Gaseous Conductors, Theory and Engineering Applications*, Dover, 1958, Chapters 7 and 8.
Collin, R. E., *Foundations of Microwave Engineering*, McGraw-Hill, New York, 1966.
Flyagin, V. A., A. V. Gaponov, M. I. Petelin, and V. K. Yulpatov, "The Gyrotron," *IEEE Transactions on Microwave Theory and Techniques*, vol. MTT-25, June 1977.
Hutter, R. G. E., *Beam and Wave Electronics in Microwave Tubes*, Van Nostrand, New York, 1960.
Pierce, J. R., "Rectilinear Electron Flow in Beams," *J. Appl. Phys.*, vol. 11, 1940, 548–554.
Pierce, J. R., *Theory and Design of Electron Beams*, Van Nostrand, New York, 1949.
Sherr, Sol, *Electronic Displays*, Wiley, New York, 1979.
Sims, G. D., and I. M. Stephenson, *Microwave Tubes and Semiconductor Devices*, Blackie and Sons, New York, 1963.
Skidmore, K., "The Comeback of Vacuum Tube: Will Semiconductors Versions Supplement Transistors?," *Semiconductor International*, Aug. 1988, 15.
Watson, G. N., *Theory of Bessel Functions*, Cambridge University Press, New York, 1922.

PROBLEMS

15.1. If the recombination coefficient for air is 2×10^{-7} cm^3/s at 100 torr of pressure, find the time t it takes for the ion density to drop to 10 percent of its original value of $n_0 = 10^{10}$ cm^{-3} after the ionizing source is turned off.

15.2. The first Townsend coefficient is given by

$$\alpha = Ap\,\exp\left(-\frac{Bpd}{v_A}\right)$$

where A and B are constants and p is the gas pressure. If we define $\gamma = \alpha/\beta$, show that using the sparking condition Eq. 2.45, the sparking potential is given by

$$V_s = \frac{B(pd)}{\ln\{[A(pd)]/[\ln(1/\gamma)]\}}$$

where V_s is the potential v_A when sparking condition is obtained.

15.3. Find the charge density, electron number density, and velocity of electrons at distance x away from the cathode for the case of space-charge flow. Note: the potential variation as a function of distance is given by $\phi(x) = Kx^{3/4}$. Here K is a constant. Use conservation of energy and charge.

15.4. Find the transit time for an electron going from the cathode to the anode if
(*a*) there is no space-charge field (very few electrons)
(*b*) there is space-charge flow between the electrons

15.5. For a microelectronic vacuum triode, find the longest cathode-anode spacing if the device is to operate at atmospheric pressure. Assume N_2 gas in this space.

15.6. Consider a klystron with an electron beam current of 100 mA and a beam diameter of 5 mm. The accelerating voltage is 2 kV. The two cavities are identical and the cavity gaps are 1 mm. If the signal applied to the buncher cavity is

$$v_1 = 10 \sin(2 \times 10^{10} t)\,(V)$$

(*a*) Find the location of the optimum catcher cavity location.
(*b*) The output voltage at the catcher cavity can be calculated from $v_0 = Mi_zR_{sh}$, where R_{sh} is the equivalent shunt resistance of the cavity ($R_{sh} = 1000\ \Omega$). Find the efficiency of the klystron.

15.7. Under the same conditions as in Problem 15.6, what is the optimum location of the catcher cavity if the applied frequency is changed to (*a*) 60 Hz or (*b*) 100 MHz? Explain why it is not practical to use velocity modulation at low frequencies.

15.8. A helix has 100 turns per inch and a diameter of 0.06 in. Assume that the velocity of the wave along the helix axis is given by

$$v_p \approx c \frac{p}{2\pi a}$$

where a is the radius and p is the pitch of the helix. Calculate the V_0 at which the helix should be operated above the cathode potential to achieve traveling wave tube gain (synchronism).

15.9. The helix in Problem 15.8 is to be used in a TWT with a beam of current density 600 mA/cm^2. Calculate the beam voltage that will give a maximum gain at a frequency of 3 GHz. Hint: in the space-charge theory, the velocity of the slow space-charge and the velocity of the electromagnetic wave should be approximately equal.

15.10. A traveling wave tube uses a helix for its slow-wave structure. The following parameters are given: $I_0 = 30$ mA, diameter of the beam 2mm, number of turns of the helix per millimeter = 1.0, diameter of helix = 2.2 mm, and length of the helix = 25 cm.

(*a*) What is the phase velocity of the helix?

(*b*) What is the necessary beam velocity of the electrons for proper operation (i) including space charge and (ii) excluding space charge.

15.11. Figure P.15.11 shows the approximate dispersion curve of a typical helix employed in a backward-wave amplifier. This graph is analogous to the E versus k diagrams of quantum mechanics. Points A and B are possible points of operation of the backward-wave amplifier. Suppose that the backward-wave amplifier operates at 6×10^9 rads/s at point A with a beam voltage of 1000 V. What should be the beam voltage in order to operate the tube at point B at a frequency of 2×10^9 rads/s?

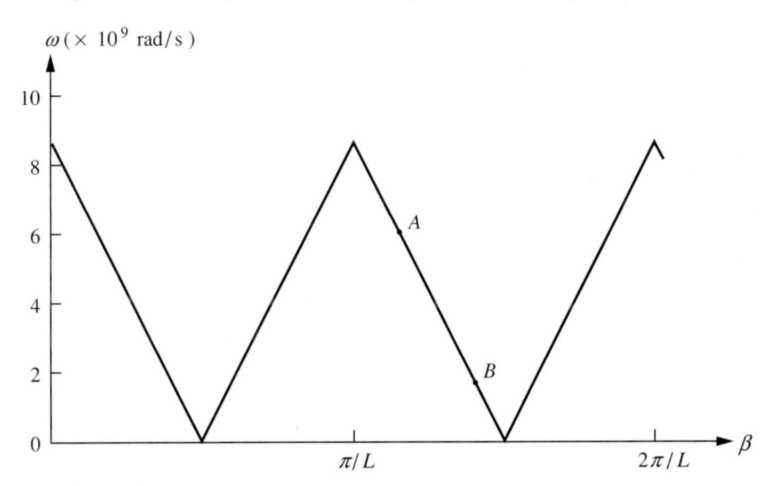

FIGURE P.15.11

15.12. Explain why it is not possible to have velocity modulation or traveling wave interaction in a semiconductor, whatever the resistivity of the sample is.

15.13. Derive Eq. 15.27 for an electron beam using the definition of ω_p given in Chapter 2.

APPENDIX A

NOISE

The goal of the information sciences is to transport information from one point to another. The information may be in many forms, and signals in the form of acoustic vibrations and electromagnetic waves constitute the main means of transporting information. When two people talk, information is transferred from one person to the other through acoustic vibrations. When a radio station transmits sound or music, the sound is carried by electromagnetic signals from the transmitting antenna through space to the receiving antenna and, through electronic processing equipment, to a loudspeaker which converts the signals back into acoustic vibrations so that we can hear the sound. The aim in information transmission is to transport the information with minimal interference from other sources.

When two people are talking in a quiet room, they don't have to shout in order to hear one another. If there is a television on in the room with the volume turned up high, the two people will have to raise their voices to hear one another. If you attend a crowded party, sometimes you have to raise your voice almost to a shouting level in order to be heard even by a person sitting next to you. The reason is that there are other people in the room talking to each other. The speech of one person thus becomes noise to another person. Whenever additional information interferes with the desired information and they reach the final destination together, the unnecessary information is considered noise for the party receiving that information.

Assuming that with some magic we were able to remove all manmade noise, consider a voice-modulated electromagnetic signal entering a receiver through an antenna. Even if the signal does not contain any noise itself, it is possible that the individual components of the receiver may add noise to the signal while the signal is being amplified and processed by the receiving electronics. If the received signal amplitude is low, it is possible that the noise generated within the electronic components of the receiver may be comparable to or greater than the information signal amplitude. Eventually, although the signal amplitude is amplified along the way, so is the noise, and the signal may not be discernable among the larger noise amplitude.

The noise generated within the amplifying system that generally affects the overall signal performance is due to noise contributions from the first few stages of an amplifier. By considering the noise properties of an electronic device, the noise contributions by the device and its low-level signal-amplifying properties can be determined. By optimizing the parameters of the device, the noise contribution can be minimized within the limitations of that device.

A.1 NOISE POWER

Let us assume that a signal is transmitted across space and reaches a receiving antenna. The signal then enters the first stage of the receiving amplifier. We know that the receiving antenna has an equivalent input resistance, and we can assume that this resistor is connected across the input terminals of the amplifying stage. As a matter of fact, we can assume that any other resistor may be connected across the amplifier input terminals.

A resistor is made up of a material that contains atoms of positive charges and negative electrons. Many of the electrons are bound to the atoms, but a few are free to move within the solid. Even if there are no external fields applied to the resistor, we know that at a given temperature, due to their internal thermal energy, these electrons are in thermal motion within the solid. Generally, the surrounding medium is at room temperature, but if power is dissipated within the resistor, the temperature of the resistor increases. This is reflected as a higher internal energy and more vibrant motion for the electrons. We also know that if charged particles are accelerated, they radiate electromagnetic energy. If the energy of the radiated signal is $h\nu$, the power radiated per unit time will be given by the probability that the electrons will have the energy $h\nu$ multiplied by the energy $h\nu$ within the frequency interval $d\nu$. The probability of having an energy $h\nu$ is given by the Bose–Einstein probability function (Eq. 3.9, Chapter 3). The power radiated is

$$P(\nu) = \frac{h\nu d\nu}{e^{h\nu/kT} - 1} \tag{A.1}$$

Electromagnetic signals that cover present day communication channels range from low frequencies to millimeter wavelengths. If we take $kT = 0.026$ eV (room temperature), we can practically assume that $h\nu/kT \ll 1$. Therefore, expanding the denominator of Eq. A.1, we obtain

$$P(\nu) = kT\, d\nu \tag{A.2}$$

Therefore, the power radiated at the frequency ν within the frequency range $d\nu$ is independent of frequency and depends on temperature T only. At very high frequencies, Eq. A.1 should be used.

Whenever the noise power over a very wide frequency range is constant, the resulting noise is called *white noise*. Note that any signal that has an amplitude below the power level given by Eq. A.2 will not be detected and will be lost in the background noise.

A.2 JOHNSON NOISE

We can apply the concepts given to individual device components. Since one of the major dissipating components in an electronic device is a resistive element, we can represent the noise generated by the thermal electrons by an equivalent noise source.

Since noise is a result of the random motion of charged particles, the resulting noise signal has no dc component. Therefore, we express the currents and voltages associated with noise by their root mean squares values. We can assume that the internally generated noise voltage is connected in series with a resistor R as shown in Fig. A.1*a*. Let the rms voltage be v_{rms}. For maximum power transfer, the current fed into a second resistor of the same value R is $v_{\text{rms}}/2R$, and the power dissipated in this resistor is $v_{\text{rms}}^2/4R$. If this is equated to the thermal power given by Eq. 11.2, then

$$\tilde{v}_{\text{rms}}^2 = 4RkTB \tag{A.3}$$

where we have used B instead of $d\nu$ where B is the bandwidth.

Similarly, using Thevenin's equivalent theorem, a current noise source given by

$$\tilde{i}_{\text{rms}}^2 = 4GkTB \tag{A.4}$$

can be used in parallel with a conductance G as shown in Fig. A.1*b*. The noise voltage or the current represented by Eqs. A.3 and A.4 are known as Johnson noise.

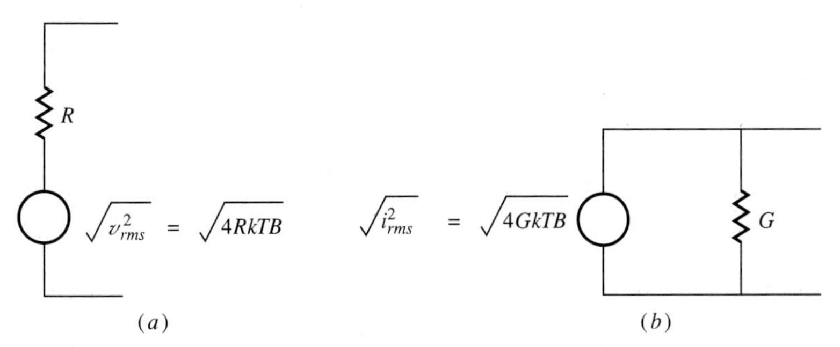

FIGURE A.1
Equivalent voltage and current noise sources in a resistor, $R = 1/G$.

A.3 SHOT NOISE

The random generation and directed motion of charge carriers leads to shot noise. Here the random fluctuation of moving charge carriers constitutes the shot noise and as expected depends on the amount of charge flowing, that is, to current. It is given by an equivalent noise current

$$\tilde{i}_s^2(\nu) = 2eIB \tag{A.5}$$

where I is the average current produced by the moving charges.

A.4 1/*f* NOISE

There is another source of noise that becomes important at low frequencies. This is referred to as $1/f$ noise. The origin of this noise can be attributed to random fluctuations of gain, capacitance, voltage source, and so on. Although one can question the validity of these arguments, by modeling these drifts properly one can show an increase in noise as the frequency decreases. Experimental measurements also prove the existence of $1/f$ noise.

A.5 NOISE FIGURE

Whenever signals are applied to amplifiers, the noise generated by an amplifier far exceeds the thermal noise discussed earlier. The electrons within the device that determine the amplifying characteristics of the device contribute to the internal noise to the device. Depending on the device configuration and mode of operation, the noise sources and their effective contributions vary from one device to the other. The noise contribution by an amplifier is identified by a noise figure F defined by

$$F = \frac{\text{signal to noise ratio at input}}{\text{signal to noise ratio at output}} \tag{A.6}$$

In order to simplify the analysis, the noise generated by an amplifier is assumed to be applied to an idealized amplifier with no internal noise, and the resulting noise figure is calculated.

In Fig. A.2, an input signal S and an input noise N_o are connected across the terminals of an amplifier. Also, the noise N_1 generated by the amplifier is connected as a source at the input of the amplifier. The gain of the amplifier is now G_1 and the amplified noise is $(N_o + N_1)G_1$. The noise figure of the amplifier is given by

$$F_1 = \frac{S/N_0}{SG_1/[(N_0 + N_1)G_1]}$$

$$F_1 = 1 + \frac{N_1}{N_0} \tag{A.7}$$

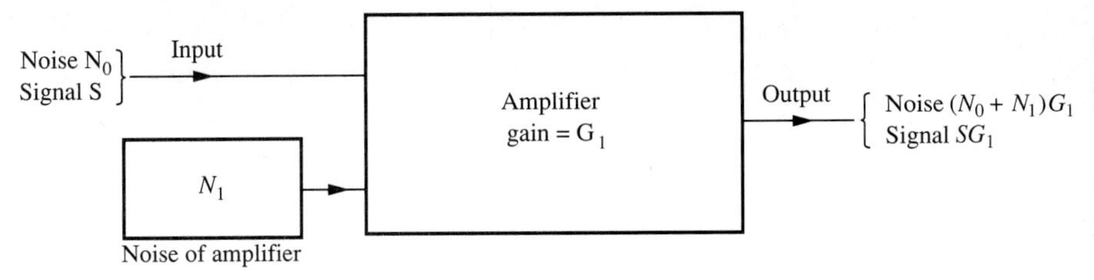

FIGURE A.2
Ideal representation of an amplifier with noise N_1.

If we measure every noise source with respect to kTB, we can write $N_i = N_i' kTB$ and Eq.A.6 becomes

$$F_1 = 1 + N_1' \tag{A.8}$$

Therefore, the contribution to the overall noise by the amplifier is given by N_1'.

Whenever two amplifiers are connected in cascade, the overall noise figure can be calculated by referring to Fig. A.3. Using the indicated signal and noise at the output, the overall noise figure of a two-stage amplifier can be written as

$$F_{\text{overall}} = \frac{S/N_0}{SG_1G_2/[(N_0 + N_1)G_1 + N_2]G_2}$$

$$F_{\text{overall}} = \frac{N_0 + N_1 + (N_2/G_1)}{N_0}$$

$$F_{\text{overall}} = 1 + N_1' + \frac{N_2'}{G_1} \tag{A.9}$$

Using the noise figure for each stage

$$F_{\text{overall}} = F_1 + \frac{F_2 - 1}{G_1} \tag{A.10}$$

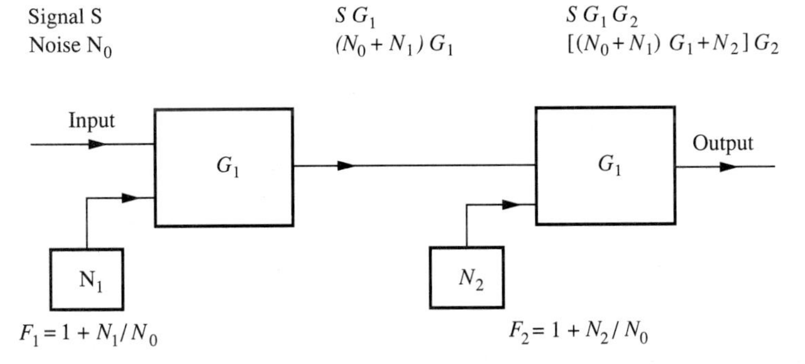

FIGURE A.3
Overall noise figure of a two-stage cascade amplifier.

Therefore, the overall noise figure is determined by the gain of the first amplifier stage in addition to the noise figures of both stages. But if the gain of the first stage is high, the contribution of the second stage on the overall noise figure of the system can be neglected.

If we let $T_o = N_o'T = 1$ in $N_o = N_o'kTB$ be our reference temperature, the temperature added by the two stages is

$$T_1 + \frac{T_2}{G_1} \tag{A.11}$$

and is known as the excess temperature added by the amplifiers. Sometimes the reference temperature is taken to be 190 K or 300 K and the excess noise temperature is measured relative to this. If the source is connected to a source T_s other than the reference temperature, the overall noise temperature of the receiver is

$$T_s + T_1 + \frac{T_2}{G} \tag{A.12}$$

The noise figures of various electronic devices and their origins are covered in many textbooks. Detailed descriptions of these are beyond the scope of this book; interested readers are referred to the Further Reading at the end of this Appendix.

FURTHER READING

Johnson, J. B., "Thermal Agitation of Electricity in Conductors," *Phys. Rev.*, vol. 32, 1928, 97–109.

Nyquist, H., "Thermal Agitation of Electrical Charge in Conductors," *Phys. Rev.*, vol. 32, 1928, 110–113.

Senturia, S. D., and B. D. Wedlock, *Electronic Circuits and Applications,* Wiley, New York, 1975.

van der Ziel, A., *Noise: Sources, Characterization, Measurement,* Prentice Hall, Englewood Cliffs, N.J., 1970.

PROBLEMS

A.1. Calculate the power level generated per Hz of white noise at room temperature.

A.2. If a resistor $R = 1\ k\Omega$ and at room temperature is connected across the terminals of a preamplifier, find the minimum signal amplitude that can be detected by the receiver. Note: a signal- to-noise ratio of greater than one can be considered a detectable signal.

A.3. A microwave antenna is listening to outer space. The equivalent noise temperature of outer space can be considered to be $T = 15$K. Even if a signal amplitude above the noise of outer space is incident on the antenna, will the receiver be able to detect such low-level signals? Explain.

APPENDIX B

PHYSICAL CONSTANTS*

Constant	Symbol	Value	Standard deviation (ppm)
Permeability of vacuum	μ_o	$4\pi \times 10^{-7}\text{Hm}^{-1}$	
		$1.2566370614 \times 10^{-6}\text{Hm}^{-1}$	
Speed of light	c	$299\ 792\ 458\ (1.2)\text{ms}^{-1}$	0.004
Permittivity of vacuum	$\epsilon_o = 1/\mu_o c^2$	$8.85418782(7) \times 10^{-12}\text{C}^2\text{J}^{-1}\text{m}^{-1}$	0.008
Gravitational constant	G	$6.6720(41) \times 10^{-11}\text{N} \cdot \text{m}^2\text{kg}^{-2}$	615
Elementary charge	e	$1.6021892(46) \times 10^{-19}\text{C}$	2.9
Planck constant	h	$6.626176(36) \times 10^{-34}\text{JHz}^{-1}$	5.4
	$\hbar = h/2\pi$	$1.0545887\ (57) \times 10^{-34}\text{J} \cdot \text{s}$	5.4
Ratio, proton mass to electron mass	$m_{\text{po}}/m_{\text{eo}}$	$1836.15152\ (70)$	0.4
Boltzmann constant	k	$1.380662\ (44) \times 10^{-23}\text{JK}^{-1}$	32
		$0.861735\ (28) \times 10^{-4}\text{eVK}^{-1}$	32
Stefan-Boltzmann constant	σ	$5.6032\ (71) \times 10^{-8}\text{W} \cdot \text{m}^2\text{K}^{-2}$	125
Avogadro constant	N_A	$6.022045\ (31) \times 10^{23}\text{mol}^{-1}$	5.1
Electron rest mass	m_{eo}	$9.109534(47) \times 10^{-31}\text{kg}$	5.1
Proton rest mass	m_{po}	$1.6726485\ (86) \times 10^{-27}\text{kg}$	5.1
Standard volume of a perfect gas (1 atm, $T_o = 213.15$ K)	V_m	$22.41383\ (70) \times 10^{-3}\text{m}^3\text{mol}^{-1}$	31
Gas constant	R	$8.31441\ (26)\ \text{Jmol}^{-1}\text{K}^{-1}$	31
Electron charge to mass ratio	e/m_{eo}	$1.7588047\ (49) \times 10^{11}\text{Ckg}^{-1}$	2.8
Bohr radius	$\alpha_o = a/4\pi R_\infty$	$0.52917706\ (44) \times 10^{-10}\text{m}$	0.8

UNITS USED IN THE ABOVE CONSTANTS

A = ampere; C = coulomb; G = gauss; H = henry; Hz = hertz = cycles/s;

J = joule; T = tesla; K = kelvin (degrees kelvin); W = watt; Wb = weber

* From *Physics Today*, September 1974

INDEX

FIAT LVX